Textile Engineering

Textile Engineering

Edited by **John Tegmeyer**

Clanrye International
New Jersey

Published by Clanrye International,
55 Van Reypen Street,
Jersey City, NJ 07306, USA
www.clanryeinternational.com

Textile Engineering
Edited by John Tegmeyer

International Standard Book Number: 978-1-63240-544-9 (Hardback)

Printed in China.

Contents

Preface

I am honored to present to you this unique book which encompasses the most up-to-date data in the field. I was extremely pleased to get this opportunity of editing the work of experts from across the globe. I have also written papers in this field and researched the various aspects revolving around the progress of the discipline. I have tried to unify my knowledge along with that of stalwarts from every corner of the world, to produce a text which not only benefits the readers but also facilitates the growth of the field.

Discovery and existence of a large number of artificial and natural fibers led to the birth of a discipline known as textile engineering. Textile is commonly referred to as a cloth which is formed using a yarn. Different cultures across the world are responsible for the formation of a diverse variety of fibers and their varied usage. Demand for new and improved fabrics like wrinkle free and heat resistant, has revolutionized this field. This book brings forth some of the most innovative concepts and elucidates some unexplored aspects of this area. It is a vital tool for all researching or studying textile engineering as it gives incredible insights into emerging trends and concepts.

Finally, I would like to thank all the contributing authors for their valuable time and contributions. This book would not have been possible without their efforts. I would also like to thank my friends and family for their constant support.

Editor

Effect of Some Fabric and Sewing Conditions on Apparel Seam Characteristics

A. K. Choudhary[1] and Amit Goel[1,2]

[1] *Department of Textile Technology, National Institute of Technology, Jalandhar, Punjab 144011, India*
[2] *Punjab Institute of Textile Technology, Amritsar, Punjab 144011, India*

Correspondence should be addressed to A. K. Choudhary; choudharyak@nitj.ac.in

Academic Editor: Phillip W. Gibson

Many previous studies showed that seam appearance as well as performance depends on the interrelationship of fabrics, threads, and the stitch/seam selection and lastly on the sewing conditions, which include the needle size, needle thread tension, stitch density, and the appropriate operation and maintenance of the sewing machine. Present investigation elucidates the effect of blend composition, sewing thread size, and sewing needle parameters on garment seam characteristics (i.e., seam strength, seam strength efficiency, seam puckering, seam stiffness, and drape coefficient). The seam quality characterization is studied through the L9 orthogonal design methodology. In good quality apparels, compatibility of the seams with the functional requirement is very important for serviceability and life of the apparel. Reversibility as well as repairing of seams in the apparel is very limited in the condition of seam failure. Seam strength efficiency is higher for the uniform fiber matrix in the structure of fabrics. The polyester dominated suiting fabrics give minimum seam stiffness because polyester component has low flexural rigidity. The cotton dominated suiting fabrics have less seam puckering due to increase in fabric stiffness.

1. Introduction

The fabric quality influences not only the quality of the garment but also the ease with which a shell structure can be produced out with flat fabric. The specifications of fabrics for apparel manufacturing can be considered in terms of primary and secondary quality characteristics. The primary quality characteristics are static physical dimensions and secondary characteristics are the reactions of the fabric to an applied dynamic force. The apparel manufacturer is primarily interested in the secondary characteristics of the fabric and focus on the seam quality during the fabrication and production of apparel [1, 2]. On the other hand, apparel consumers are mainly paying attention to appearance, comfort, and wearability of fabric and evaluate seamquality based on the seam appearance and its mechanical intactness after wear and care procedures of apparel. The production of garments from high quality fabrics not only gives comfort to the wearer but also helps in the smooth working of manufacturing processes and leads to almost defect-free garments. Further, the primary function of a seam is to provide uniform stress transfer from one piece of fabric to another, thus preserving the overall integrity of the fabric assembly. For proper appearance, seam should not contain any defects including skipped stitches, unbalanced stitches, seam grin, distortion or unevenness or puckering, unsteadiness, improper drapeability, uneven seam density, and yarn severance or damage. The overall quality of a seam depends on its strength, elasticity, durability, stability, and appearance [3]. The seam characteristics parameters are seam strength, seam pucker, seam stiffness, seam appearance, and seam efficiency [4, 5]. According to many researchers [6, 7], there are various factors affecting the seam quality. These comprise sewing thread, sewing condition, and others.

A sewing thread is a strong, smooth, evenly spun, hard twisted and elastic, ply or cabled yarn treated by special finishing process to make it resistant to abrasion in its passage through the eye of the needle and through materials in seaming and stitching operation. Sewing thread may be

Table 1: Characteristics of medium weight suiting fabrics.

Samples number	Fabric blend composition (%)	Fabric thickness (mm)	Fabric weight (GSM)	Thread characteristics					
				Fractional cover		Thread density (per inch)		Yarn linear density (Ne)	
				Warp	Weft	Warp	Weft	Warp	Weft
1	Poly/Cot: 65/35	0.44	205	22.36	13.41	100	60	20	20
2	Poly/Cot: 15/85	0.42	210	22.36	13.5	100	54	20	16
3	Cotton: 100	0.40	215	22.36	13.5	100	54	20	16

constructed from staple yarn, continuous filament, or core spun yarns. The sewing thread plays a significant role for improving even though it usually represents much less than 1% by mass of a garment. The high strength of sewing thread always gives better functional performance of seam, namely, seam strength and seam strength efficiency. The different manufacturing parameters of sewing thread are type, ply, finish, twist, and size. The spun-polyester, 3-ply, normal twist, and standard finish sewing thread quality is widely used for all types of sewn fabrics [8]. However, the size of the sewing thread is the most crucial for the stability of seam in apparel, as the improper use of sewing thread size directly affects the seam quality of apparels. Rengasamy et al. [9] established that sewing thread type has a great influence on seam efficiency. He explained that seam efficiency of sewn fabric stitched with polyester textured thread lies in between those of fabric stitched with cotton and spun polyester threads, but more closer to fabric stitched with spun polyester threads. A few researchers conducted research on the effect of thread finishes on seam quality. In the course of garment manufacturing, for protective measures to prevent the thread strength reduction and/or breakage during sewing from the needle heat up, the different types of finishes can be applied over sewing thread according to the specific end use of apparel products. Similarly, Sundaresan et al. [10] found that sewing thread size and ply are the most influencing factors for the seam strength and seam puckering. Higher sewing thread size (Tex value) was subjected to greater friction during sewing operation, which ultimately reduces its strength leading to poor seam-strength and also greater seam puckering. On the contrary, more number of ply in a sewing thread helps to increase the strength of a sewing thread and leads to higher seam strength and/or seam efficiency with a less chance of seam slippage. But the diameter becomes high with more number of ply causing the structural jamming along the seam line, which in turn leads to seam puckering and change in draping behavior of apparel product due to increased stiffness. There is a trend in the apparel industry to select the lowest size of thread with the greatest possible strength for making a seam which is less visible and has good seam functional performance.

The fabric properties which affect the seam quality of apparel are cover factor, weight, thickness, strength, shrinkage, functional finishes, extensibility, bending rigidity, and shear rigidity, some of which form an integral part of low stress mechanical properties. The role of fibre types and its % in made-up fabric is very important for sewing performances and characteristics due to its effect on made-up fabric mechanical properties. In present experimental work, we have taken polyester-cotton blend and 100% cotton to see the influences of polyester and cotton fibre on stitching qualities and performances. The L9 orthogonal experimental design is taken into account to see the effect of blend composition, needle thread tension, needle size, and sewing thread linear density on the stitching performances like seam efficiency, seam stiffness, seam puckering, and drape coefficient of seam, as incorporating larger influencing factors and output effects through statistical analysis.

2. Materials and Methods

2.1. Materials. Commercially available three types of suiting fabrics of medium weight construction, varying in the blend composition of polyester and cotton components, are taken into consideration for this work. The details of suiting fabric are 210 g/m^2, 2/1 Twill, 100 ends/inch (EPI), and 60 picks/inch (PPI). The other characteristics of the fabrics are shown in Table 1.

The GSMs of fabrics are calculated with GSM cutter and electronic weighing balance. The ASTM-D105911 methodology is used for fabric yarn count determinations. The fabric thickness is measured under constant compressive load of 2 KPa with the thickness tester (Karl Schröder KG). The fabric yarn densities are measured according to ASTM-D3375 [11]. Commercially available polyester/cotton core spun sewing threads of different linear densities (40, 60, and 80 Tex) are used. These sewing threads are used in the experimental work due to their least reversible strain as compared to cotton sewing threads. The threads are characterized for actual linear density, number of plies, twist direction, and tensile properties. The twist per unit length is measured by direct counting method according to ASTM-D1423 [11], and the tensile testing of the threads is performed at a gauge length of 250 mm on Zwick tensile testing machine at 35 mm/minute (corresponding to 20 s breaking time), as per ASTM-D2256 [11]. The number of test of 30 is carried out and the error is less than 4% at 95% confidence level. The specifications of the sewing threads are shown in Table 2.

The most appropriate needle sizes are used in stitching works (i.e., Organ 11, 14, and 16) and the stitch density is kept constant at 10 stitches per inch (SPI). The details of the needle dimensions are given in Table 3.

2.2. Seam Preparation. All stitching work has been done on Juki industrial lockstitch sewing machine, Japan, at a speed of 4000 stitches/min and producing a balance seam.

Table 2: Specifications of sewing threads.

Sewing thread linear density (Tex)	Specifications of sewing threads			
	Twist direction (ply/single)	Tenacity (CN/Tex)	Breaking elongation (%)	Initial modulus (CN/Tex)
40	Z/S	44.73	25.78	264.45
60	Z/S	53.13	27.15	310.11
80	Z/S	61.14	29.14	442.1

Table 3: Needle dimensions.

Needle size	Shoulder diameter (mm)	Needle eye		
		Length (mm)	Width (mm)	Area (mm^2)
11	0.84	0.85	0.30	0.30
14	0.90	0.95	0.35	0.36
16	1.00	1.04	0.40	0.44

Table 4: L9 Orthogonal design of experiment.

Factors	Code	Levels		
		1	2	3
Blend composition (%)	A	P/C: 35/65	P/C: 15/85	Cotton100%
Needle thread tension	B	75	100	125
Needle size (Organ make)	C	11	14	16
Sewing thread linear density (Tex)	D	40	60	80

P: polyester; C: cotton.

Any wrinkled fabrics are ironed before preparation of the seam specimen. The seam specimens are prepared in such a way that the existing seams ran through the middle of each specimen. Stitch density is measured on the sewn fabrics by means of a magnifying glass and an accurately marked scale ASTM-D1683 [11]. Similar sewing thread is used as needle thread and the bobbin thread of machine for preparing the seams. During sample preparation the static bobbin thread tension is kept constant. The needle thread tension values are measured by using strain gauge type digital tension meter.

Table 5: Layout of the experiment in standard order.

Sl. no.	Fabric blend composition % (A)	Needle thread tension (B)	Needle size (C)	Sewing thread linear density (D)
1	1	1	1	1
2	1	2	2	2
3	1	3	3	3
4	2	1	2	3
5	2	2	3	1
6	2	3	1	2
7	3	1	3	2
8	3	2	1	3
9	3	3	2	1

2.3. Methods

2.3.1. Experimental Design. The individual and interactive effects of needle size, needle thread tension, and sewing thread linear density on seam quality critical dimensions of suiting fabrics are studied at three levels using L9 orthogonal design. The actual values of different parameters corresponding to coded levels are mentioned in Table 4.

Through the use of L9 orthogonal design, information on single linear effects, single quadratic effects, along with the two factor linear interaction can be obtained. In the present case, 4-factor, L9 orthogonal experimental design at three levels is chosen as it can adequately fulfill the present requirement with only 9 treatment (experiment) combinations wherein parameters can be kept at 3 levels. For effective statistical analysis and validity of inferences, the samples are prepared in a standard order; 1, 2, 3, 4, 5, 6, 7, 8, and 9 as shown in Table 5.

Conclusively, the testing samples are prepared by changing the parameters, that is, needle thread tension, needle size, and sewing thread linear density, at three levels as per L9 orthogonal design methodology (Table 5) on Juki industrial sewing machine.

The ANOVA technique is conducted to see the effect of individual factors on the seam quality dimensions. The three replicate measures (at different fabric positions from the whole fabric sample lot) are taken, and total degree of freedom is 26 (total combination $9 \times 3 = 27$; hence, degree of freedom is $27 - 1 = 26$). F-ratio and percentage contribution

Table 6: Seam quality characteristics. Response of nine experiments.

Experiment number	Factors				Seam stiffness (N)			Seam efficiency (%)			Drape coefficient (%)			Seam puckering (%)		
					(Blocks according to replicate measures)											
	A	B	C	D	1	2	3	1	2	3	1	2	3	1	2	3
1	1	1	1	1	2.00	1.50	2.00	87	86	85	57	55	56	23	21	28
2	1	2	2	2	1.50	1.50	2.00	81	80	79	53	50	52	48	55	59
3	1	3	3	3	1.50	1.50	2.00	78	79	77	61	62	60	68	73	70
4	2	1	2	3	3.00	2.50	3.00	88	86	89	77	74	75	75	77	72
5	2	2	3	1	3.00	3.50	3.00	63	64	62	70	71	68	37	47	47
6	2	3	1	2	2.50	2.50	3.50	77	76	75	74	73	72	70	62	57
7	3	1	3	2	3.00	3.50	5.00	97	97	96	68	66	66	50	50	50
8	3	2	1	3	3.50	3.00	3.00	79	80	78	68	65	66	69	59	64
9	3	3	2	1	2.50	3.00	3.50	89	87	90	71	68	68	31	33	36

A: blend composition; B: needle thread tension; C: needle size; D: sewing thread linear density (Tex value).

of different factors are studied. Finally regression equations (Table 8) relating the main factors with seam quality dimensions and response surface regression analysis are established. From the ANOVA analysis, % contribution of different factors is evaluated based on the following expression:

$$\text{contribution (\%)} = \frac{\left[\text{SS}_f - \left(\text{df}_f \times \text{MS}_e\right)\right]}{\text{SS}_t} \times 100, \quad (1)$$

where SS_f is sum of square of the factor, df_f is degree of freedom of the factor, MS_e is mean square of the error, and SS_t is total sum of squares.

Sewn fabrics are tested for seam stiffness on circular bend stiffness tester as per ASTM standard D 4032-94 [11]. The sample of 8 inch × 4 inch is cut from the fabric and sewn in warp direction with seam allowance of 1 inch; the short side of the specimen must be parallel to the machine (length) direction of the fabric. The air pressure control to the actuator is kept at 60 psi to confirm the tester plunger speed of 1.7 ± 0.3 s under no load conditions. Ten readings are carried out for each sample.

Sewn fabrics are tested for seam efficiency on Instron tensile testing machine at a speed of 305 mm/min and 75 mm gauge length as per ASTM-D1683. Ten tests are conducted for each sample. For seam strength testing, the sample of 8 inch × 4 inch is cut from the fabric and sewn in warp direction with seam allowance of 1 inch. Consider

$$\text{seam efficiency (\%)} = \frac{\text{seam tensile strength}}{\text{fabric tensile strength}} \times 100. \quad (2)$$

Sewn fabrics are tested for drape coefficient on drape meter. Most garments are sewn in the warp direction. Hence, experimental fabric seams are sewn in the warp direction. After seaming, samples are cut into 25 cm diameter round shapes with the seam positioned across the specimen diameter [12]. The drape coefficient can be calculated as:

$$\text{drape coefficient (\%)} = \frac{A_d - A_1}{A_2 - A_1} \times 100, \quad (3)$$

where A_d is the area of draped shadow, A_1 is the area of the disc, and A_2 is the area of the fabric.

Seam puckering has been calculated by measuring the fabric and seam thickness under constant compressive load of 2 KPa (ISO 9073-2 : 1997-02). Ten readings are carried out for each sample and puckering calculated by the following formula:

$$\text{seam puckering (i.e., thickness strain, \%)} = \frac{T_s - 2T}{2T} \times 100, \quad (4)$$

where T_s is the seam thickness in mm and T is fabric thickness in mm.

3. Results and Discussion

3.1. Study of the Seam Stiffness. It is observed from Table 6 that with the replicate measures seam stiffness is changing for all experimental runs. The main factor influencing the seam stiffness is fibre %, the percentage contribution being 67% (Table 7). It is also observed from Figure 1 that with the change in blend composition the seam stiffness increases. The increasing trend of seam stiffness with the change in blend percentage is due to the fact that the polyester fibre content goes on decreasing which has low bending stiffness in low load regions (flexural rigidity of polyester fibre is 0.30 mN·mm^2/tex^2). On the contrary, cotton dominated suiting fabrics has high flexural rigidity of 0.53 mN·mm^2/tex^2. So the polyester dominated suiting fabrics show less resistance when bent by the external force encountered during fabric manipulation in spreading and sewing. Hence, fibre types and % largely influence the seam stiffness (Table 7). However, the effect of sewing thread linear density on seam stiffness is statistically insignificant for medium weight suiting fabrics.

3.2. Study of Seam Strength Efficiency. Figure 2 showed the effect of blend composition and needle thread tension on the seam strength efficiency. As fabric strength increases, the seam strength efficiency decreases because it is inversely

TABLE 7: ANOVA results. F-value and contribution (%) of factors on seam characteristics.

Characteristics	Factors	F-value	% contribution
Seam stiffness (N)	Blend	35.92	67
	Needle thread tension	Not significant	Very small
	Needle size	Not significant	Very small
	Sewing size	Not significant	Very small
Seam efficiency (%)	Blend	308.3	32
	Needle thread tension	510.1	53
	Needle size	85.9	9
	Sewing thread linear density	48.86	5
Drape coefficient (%)	Blend	591.6	85
	Needle thread tension	55.6	7.8
	Needle size	Not significant	Very small
	Sewing thread linear density	31.9	4
Seam puckering (%)	Blend	21.19	10
	Needle thread tension	4.89	2
	Needle size	Not significant	Very small
	Sewing thread linear density	157.42	80

Not significant at 95% confidence level.

TABLE 8: Regression equations of seam quality characteristics.

Characteristics	Interactive factors	Regression equation	R^2
Seam stiffness (N)	A-D	$-0.44 + 2.30x + 0.44y - 0.41x^2 + 0.08xy - 0.166y^2$	0.76
Seam efficiency (%)	A-B	$153 - 33.27x - 50.61y + 9.16x^2 - 15xy + 11.5y^2$	0.98
	A-D	$88 - 32.44x + 17.33y + 9.16x^2 - 0.41xy - 3.83y^2$	0.97
Drape coefficient (%)	A-B	$28.41 + 49.9x - 14.92y - 10.80x^2 - 0.58xy + 4.22y^2$	0.99
Seam puckering (%)	A-D	$-50.9 + 54.4x + 41.81y - 11.6x^2 - 4xy - 3.91y^2$	0.96

A: blend composition; B: needle thread tension; C: needle size; D: sewing thread linear density.

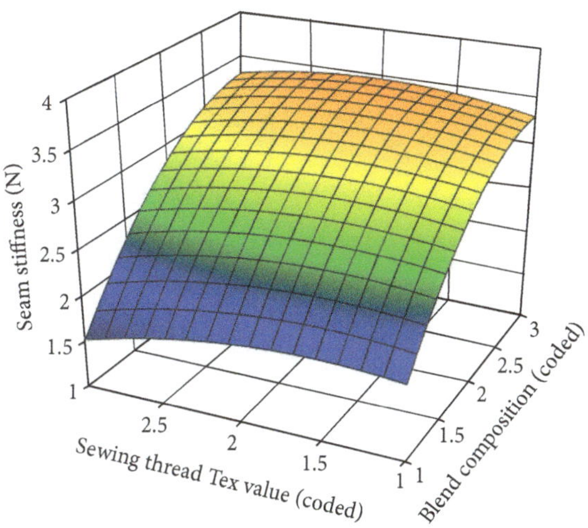

FIGURE 1: Effect of blend composition and sewing thread linear density (Tex value) on seam stiffness.

proportional to the fabric strength for a given sewing thread. The low seam strength efficiency for the polyester dominated suiting fabrics is attributed to the fact that the polyester based suiting fabrics have high extensibility (elongation at break—30%) and tenacity (45 gf/tex) as compared to cotton dominated fabrics (elongation at break—7%) and tenacity (35 gf/tex). Similarly, in seamed conditions this high strength and elongation is not desirable, hence, results in low seam efficiency for polyester dominated fabrics. The latter decreasing trend for the seam efficiency is due to the less uniform fibre matrix which may lead to fibre slippage and yarn failure. As the fibre matrix becomes uniform in cotton dominated fabrics (100% cotton), the seam efficiency increases as shown in Figure 2.

The needle thread tension has significant effect on seam efficiency, and the low needle thread tension during stitching improves the seam strength and, hence, seams efficiency considerably; the percentage contribution of needle thread tension is 53% (Table 7). This is due to some flexibility at the seam to improve the seam strength under sudden stress. However, at high needle thread tension, the fabric has been pulled at the seam leading to puckered, unstable seam resulting in low seam strength efficiency.

It has also been observed that the sewing thread linear density has considerable effect on the seam efficiency; the percentage contribution is 5% (Table 7). Seam efficiency increases with the sewing thread linear density. This is due to the fact that more number of fibres is incorporated in the coarser sewing thread resulting in high seam strength.

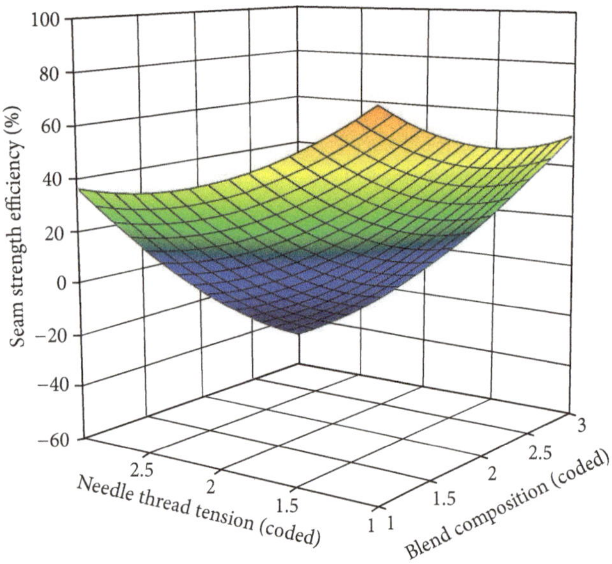

FIGURE 2: Effect of blend composition and needle thread tension on seam strength efficiency.

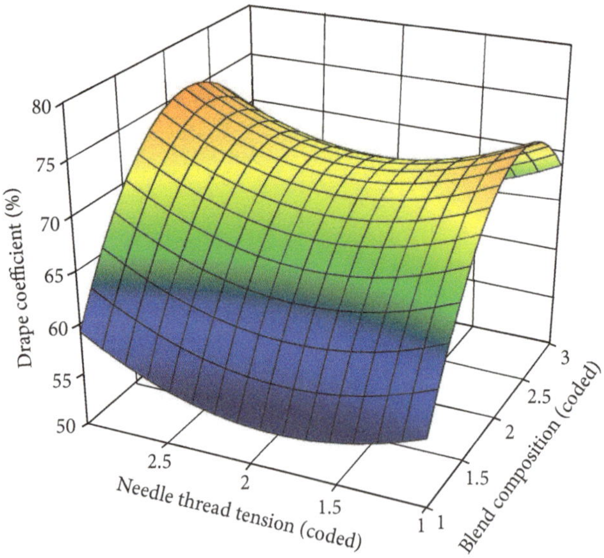

FIGURE 4: Effect of blend composition and needle thread tension on drape coefficient.

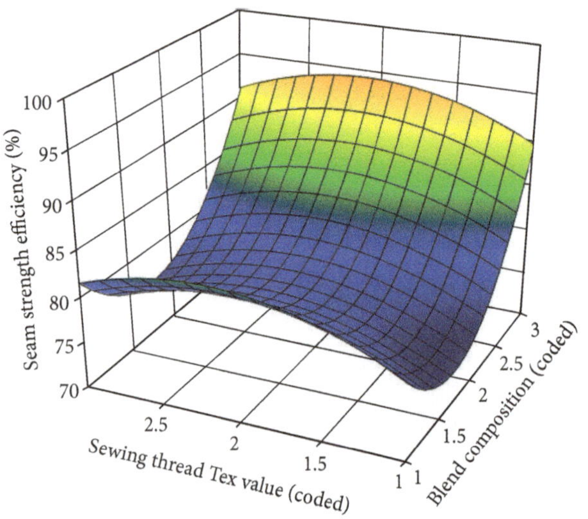

FIGURE 3: Effect of blend composition and sewing thread linear density (Tex value) on seam strength efficiency.

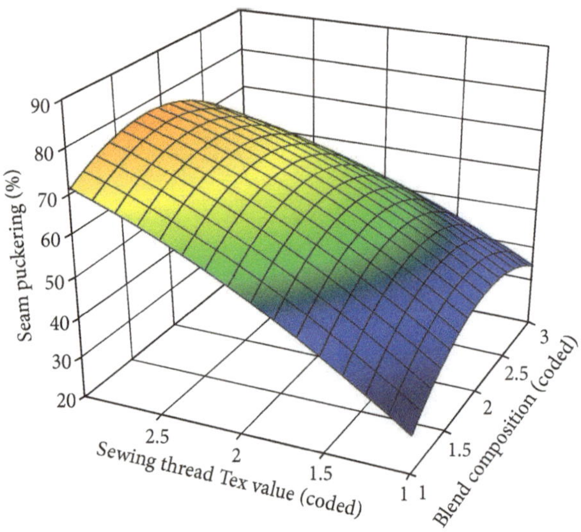

FIGURE 5: Effect of blend composition and sewing thread linear density (Tex value) on seam puckering.

Figure 3 showed also the effect of blend and sewing thread Tex value on seam strength efficiency. The seam efficiency first decreases and then increases considerably with change in blend composition.

The coarser sewing thread has better gripping with the bulkier and resilient polyester fibre component (density 1.40 g/cc) in the polyester suiting fabrics leading to high seam strength efficiency (Figure 3), but the later decreasing trend in seam efficiency is due to the reason of fabric structural jamming in cotton dominated fabrics leading to high fabric strength and low seam efficiency.

3.3. Study of the Draping Behavior. It is observed from Figure 4 that the blend composition of fabric has drastic effect on the drape coefficient (%); the percentage contribution is 85% (Table 7). The low-stress extensibility for P/C : 65/35 is high which determines the mobility of threads within the fabrics. Higher low-stress extensibility gives more wearing comfort, high drape ability, and consequently low drape-coefficient for polyester dominated suiting fabrics. As the cotton composition goes on increasing the drape coefficient increases in P/C : 15/85 since the bending stiffness of the cotton fibre is higher than the polyester fibre. The later decreasing trend of drape coefficient in cotton dominated suiting fabrics is due to the uniform fibre matrix in 100% cotton fabrics. However, the other factors needle thread tension and sewing thread linear density have negligible effect on draping behavior.

3.4. Study of the Seam Puckering. It is observed from Figure 5 that seam puckering is highly dependent on sewing thread linear density, the contribution being 80% (Table 7). This indicates that the coarser yarn causes the fabric structural jamming and leads to low flexibility at the seam causing seam puckering.

The polyester-based suiting fabrics with the extensible polyester component give higher extensibility in the low-load region; hence, the fabric dimensional stability is reduced which may lead to high seam puckering. The latter decreasing trend in cotton-based suiting fabric is due to the fact that the increase in fabric stiffness leads to the increase in in-plane compression resistance which causes the decrease in seam puckering value.

4. Conclusions

(i) The polyester dominated suiting fabrics give minimum seam stiffness because polyester component has low flexural rigidity.

(ii) Seam strength efficiency is high for the uniform fibre matrix in the textile structure.

(iii) At high needle thread tension, seam strength efficiency is low because fabric has been pulled leading to puckered and unstable seam.

(iv) The higher low-stress extensibility of polyester-cotton suiting fabrics P/C : 65/35 is high which gives high drapeability and wearing comfort.

(v) Seam pucker is most effective dimension for evaluating seam quality; bending stiffness has considerable effect on seam puckering.

(vi) Seam puckering and seam strength efficiency increase with the increase in sewing thread linear density

(vii) The cotton dominated suiting fabrics have less seam puckering due to increase in fabric stiffness which leads to increase in in-plane compression resistance.

References

[1] H. Carr and B. Latham, *The Technology of Clothing Manufacturing*, Blackwell Scientific, Oxford, UK, 1995.

[2] B. K. Behera and S. Sharma, "Low-stress behaviour and sewability of suiting and shirting fabrics," *Indian Journal of Fibre and Textile Research*, vol. 23, no. 4, pp. 233–241, 1998.

[3] K. P. S. Cheng and K. P. W. Poon, "Seam properties of woven fabrics," *Textile Asia*, vol. 33, no. 3, pp. 30–34, 2002.

[4] V. Dobilaite and M. Juciene, "The influence of mechanical properties of sewing threads on seam pucker," *International Journal of Clothing Science and Technology*, vol. 18, no. 5, pp. 335–345, 2006.

[5] D. H. Brain, "The prediction of strength of lockstitch seams in woven fabric," *Journal of Textile Institute*, vol. 61, no. 10, pp. 493–505, 1970.

[6] N. Tarafder, R. Karmakar, and M. Mondal, "The effect of stitch density on seam performance of garments stitched from plain and twill fabrics," *Man-Made Textiles in India*, vol. 50, no. 8, pp. 298–302, 2007.

[7] B. Meric and A. Durmaz, "Effect of thread structure and lubrication ratio on seam properties," *Indian Journal of Fibre and Textile Research*, vol. 30, no. 3, pp. 273–277, 2005.

[8] J. Ukponmwan, A. Mukhopadhyay, and K. N. Chatterjee, "Sewing threads," *Textile Progress*, vol. 30, no. 3-4, pp. 1–91, 2000.

[9] R. S. Rengasamy, V. K. Kothari, R. Alagirusamy, and S. Modi, "Studies on air-jet textured sewing threads," *Indian Journal of Fibre and Textile Research*, vol. 28, no. 3, pp. 281–287, 2003.

[10] G. Sundaresan, K. R. Salhotra, and P. K. Hari, "Strength reduction in sewing threads during high speed sewing in industrial lockstitch machine. Part II: effect of thread and fabric properties," *International Journal of Clothing Science and Technology*, vol. 10, no. 1, pp. 64–79, 1998.

[11] http://www.astm.org/.

[12] J. E. Booth, *Principles of Textile Testing*, CBS, 1996.

2

Moisture Transport for Reaction Enhancement in Fabrics

Phillip Gibson,[1] Heidi Schreuder-Gibson,[1] Pearl Yip,[1] Brendan Denker,[1] Hamid Benaddi,[2] Sa Wang,[3] Lev Bromberg,[3] and T. Alan Hatton[3]

[1] *U.S. Army Natick Soldier Research, Development and Engineering Center, Natick, MA 01760, USA*
[2] *Stedfast, Inc., Granby, QC, Canada J2G 3Y3*
[3] *Department of Chemical Engineering, Massachusetts Institute of Technology, Cambridge, MA 01239, USA*

Correspondence should be addressed to Phillip Gibson; phillip.gibson@us.army.mil

Academic Editor: Sambandam Anandan

The role of water in protective fabrics is critical to comfort and material performance. Excessive perspiration in clothing causes discomfort, and bound water can adversely affect the ability of carbon to adsorb chemicals. Yet the presence of water can also improve the moisture vapor transport of protective polymer films, and is essential for the hydrolytic destruction of nerve agents. Reported here are the findings of wicking and drying experiments conducted on various hydrophilic and hydrophobic cover fabrics that demonstrate the influence of wetting on permeation through fabrics. The influence of water content on reactive polymers capable of degrading nerve agent simulant is also discussed, and the importance of a novel "delivery system" for water to the reactive components through the use of a wicking fabric is introduced.

1. Introduction

Current chemical protective fabrics rely on layers of activated carbon to adsorb chemical agent vapors. A tightly woven outer hydrophobic and oleophobic shell fabric provides liquid protection and aerosol particle protection.

Another approach is to base protection on a semipermeable or reactive polymer membrane layer, which is permeable to water vapor, but which is also a barrier to chemical agents. The polymer membrane is usually laminated between an outer shell fabric and an inner fabric layer in contact with the skin. The outer fabric layer is treated to be hydrophobic and oleophobic. This allows water (rain), oils, and chemical agents to roll off of the fabric surface. However, the inner layer of these fabrics can be designed to pick up liquid sweat from the skin's surface. The liquid wicking characteristics of this inner laminated "comfort layer" fabric can affect the perceived comfort of the garment and the total amount of sweat vapor capable of being transmitted through the laminated fabric and membrane structure.

The research gap motivating the work presented in this study is the need to more fully understand how water amount and location affect reactive compounds present in fabrics and polymer membranes that are designed to hydrolytically decompose chemicals that are analogs for nerve agents. One such compound is the polymeric supernucleophilic pyridine catalyst called poly(butadiene-co-pyrrolidinopyridine) (polyBPP) [1–3], which functions to completely degrade the nerve agent surrogates, including the compound, diisopropyl fluorophosphate (DFP).

Water interactions are involved in the degradation kinetics of DFP and require that there must be a sufficient amount of water and complete contact between the DFP, water, and catalyst for rapid and complete destruction of the surrogate. The delivery of the liquid water from the interior of the protective fabric to the exterior regions containing the reactive compounds is an important process that has been designed into a layered fabric system comprised of a wicking comfort layer next to the skin, bonded to a polymer membrane capable of degrading DFP, and covered with a protective outer shell fabric that can be modified with polyBPP. Previous work on water interactions with chemical protective polymer membranes has shown that water content of the polymer layer is important for permeation of water and chemical

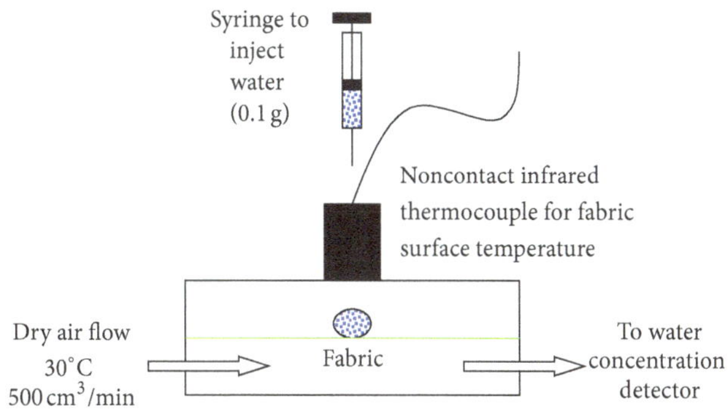

FIGURE 1: Test cell modified to accommodate an infrared thermocouple viewing fabric surface in the cell.

agent simulants [4–6]. Of particular interest is the increase in efficiency that the wicking comfort layer might contribute to the hydrolytic degradation of DFP within the reactive membrane. This paper focuses on fabric design features that promote efficient water interactions with reactive compounds within layered fabric systems.

2. Liquid Water Wicking/Transport

Novel drying experiments were developed to measure drying time by means of vapor flux from spreading liquids in a modified permeation test cell. Previous measurements of this type have been useful in distinguishing the difference in performance of single-layer fabrics that differ in their liquid wicking characteristics [7].

Measurement systems as shown in Figure 1 provide water vapor flux through the fabric as influenced by fabric surface treatments.

Fabrics can be treated to be fully wicking on both sides, partially wicking on one side and hydrophobic on the other side or fully hydrophobic on both sides. Figure 2 shows typical examples of water behavior in a single layer of fabric (hydrophobic on one side, and hydrophilic on the other) and the way that water wets the inner "comfort layer" of an activated carbon chemical protective garment, while it beads up on the outer protective shell fabric layer.

Figure 3 shows typical drying curves for a variety of single-layer fabrics that differ in their hydrophobic surface characteristics on one or both sides. It is clear that gradations of treatment can produce fabrics with transport behavior located anywhere on the spectrum between "fully wicking" and "fully repellent." Modern textile fabrication and finishing technology provides the ability to tailor fabric wicking and repellency to achieve specific performance requirements; however, it is unclear in some design applications what those performance targets should be.

Previous work on the wicking and wetting characteristics of fabrics showed that there are physiological and comfort consequences to the choice of fabric wicking characteristics [7]. The orientation of a fabric with wicking characteristics on one side and hydrophobic characteristics on the other is of particular importance.

Similar experiments on wicking and water transport were performed for a variety of more complicated fabrics. Included were several commercial polymer membrane laminates used in "breathable" rainwear. The behavior of these materials becomes much more complicated, since these are multilayer structures that usually include a solid polymer film that changes its transport behavior depending on how much water is in the polymer structure. The most interesting aspect of these measurements was the resulting observation that many of these materials had poor water vapor transport at low states of hydration but performed very well when exposed to liquid water at the same conditions. The improved performance seemed related to the presence of an inner wicking fabric layer that ensured liquid was transported through the fabric quickly to the polymer layer, where water was able to hydrate the membrane and quickly pass through to the other side.

An example of this increased water vapor transport for polymer membrane laminates exposed to liquid water is shown in Figure 4. A reactive polymer membrane laminate that had poor water vapor transport under steady-state vapor conditions showed improved performance approaching that of a fully wicking glass fiber filter layer (used as a control). The fabric surface temperature can be monitored during the wicking and drying process as another indicator of drying time, since the evaporation of the liquid causes a drop in surface temperature (Figure 4).

The measurement shown in Figure 4 was interesting, because it showed that a reactive polymer membrane laminate performed well at pushing water (sweat) from the skin surface through the wicking comfort layer to the reactive polymer membrane layer, where the water becomes available to aid in chemical reactions that degrade chemical warfare agents, as well as hydrating the membrane to improve water vapor transport. A more controlled study of the specific transport behavior of polymer membrane laminates, as opposed to single-fabric layers, was developed to understand these phenomena.

3. Materials and Methods

We chose a single-fabric system to study the effect of differential water and wicking performance of a "comfort layer" on the inner surface of a reactive polymer membrane laminate. The reactive polymer membrane layer is manufactured by Stedfast, Inc., under the trade name of ChemCAT. A knit comfort layer was chosen as the fabric to laminate to the inner surface of the membrane. The knit fabric is engineered to be hydrophilic on one side and hydrophobic on the other side. However, although the hydrophobic side is water repellent to some degree, it is capable of picking up liquid water (sweat) from the skin surface and transporting it through to the hydrophilic side, where it is retained. As a consequence, the hydrophobic side will wick sweat away from the skin and will have a very dry feel, since the liquid is segregated away from the skin surface in the outer hydrophilic layer.

By contrast, if the hydrophilic layer is against the skin, the sweat will be wicked into that layer, stay there against the skin,

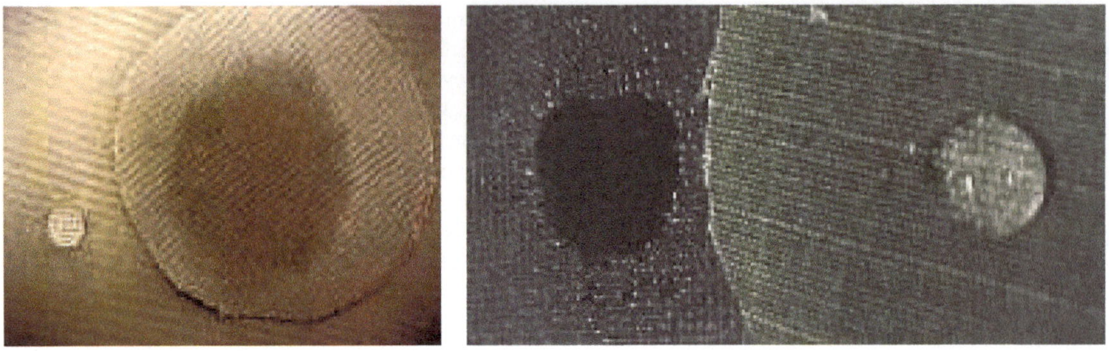

(a) (b)

FIGURE 2: Water on inner and outer surfaces of fabrics for drying experiments. (a) Single-layer fabric with hydrophilic treatment on inner surface and hydrophobic treatment on outer surface. (b) Standard carbon inner liner and outer shell fabric.

TABLE 1: Wicking test conditions.

(i) 0.1 g of Liquid water applied to top surface of fabric.
(ii) Test temperature of 30°C.
(iii) Gas flow rates of 2 L/min on both sides of fabric.
(iv) Humidity is fractional (0.80 is 80% r.h.).

Humidity on top	Humidity on bottom	Humidity gradient	Mean relative humidity
Constant humidity conditions			
0.0	0.0	N/A	0.0
0.4	0.4	N/A	0.40
0.8	0.8	N/A	0.80
Humidity gradient conditions			
0.4	0.0	0.4	0.2
0.6	0.2	0.4	0.4
0.8	0.4	0.4	0.6

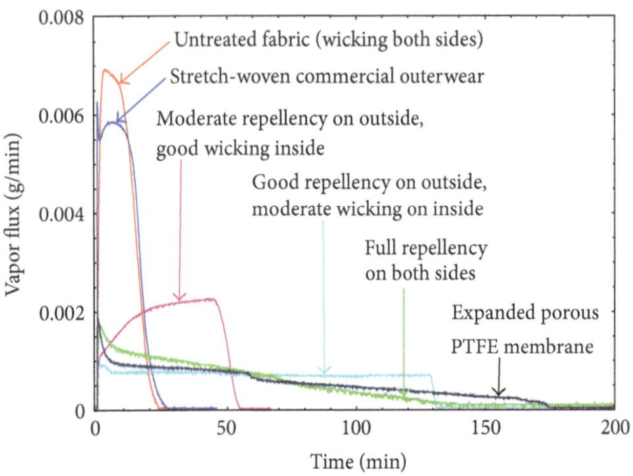

FIGURE 3: Typical drying behavior of fabrics with various wicking treatments on the inner and outer surfaces of a single layer of fabric [7].

and produce a much great sensation of wetness. The outer hydrophobic surface does not become wetted in this case.

Two variants of laminated fabric were produced with this hydrophobic/hydrophilic comfort layer: (1) the hydrophilic layer was oriented toward the skin surface (hydrophobic layer attached to the polymer membrane) and (2) the hydrophobic layer was oriented toward the skin surface (hydrophilic layer attached to the polymer membrane).

We supposed that the consequences of these two fabric orientations would be that, in case (1) where the hydrophilic layer was against the skin, the liquid water would be less available to the polymer membrane layer, since the water would not pass through the hydrophobic layer. For case (2), the hydrophobic fabric would pass water directly to the hydrophilic layer against the polymer membrane, where it would be quickly absorbed, producing a higher hydration state within the reactive polymer, enhancing transport and reactivity.

A wicking glass fiber filter layer was used as a control. This material was a borosilicate glass filter, with no resin binder or hydrophobic or oleophobic treatments. The glass fiber filter had high porosity and low air permeability and held liquid water within the pore spaces of the material with no preferential dependence of wicking through the material thickness. Figure 5 outlines the various material configurations used for the wicking study.

The test system shown in Figure 6 was modified slightly from that shown previously in Figure 1. Equal gas flows were

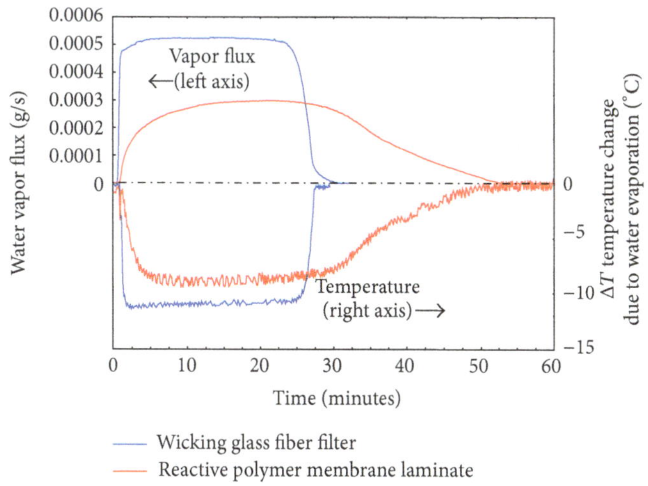

FIGURE 4: Simultaneous vapor flux and temperature drop measurements for a fully wicking layer and a reactive polymer laminate that includes a wicking fabric layer on the inner surface.

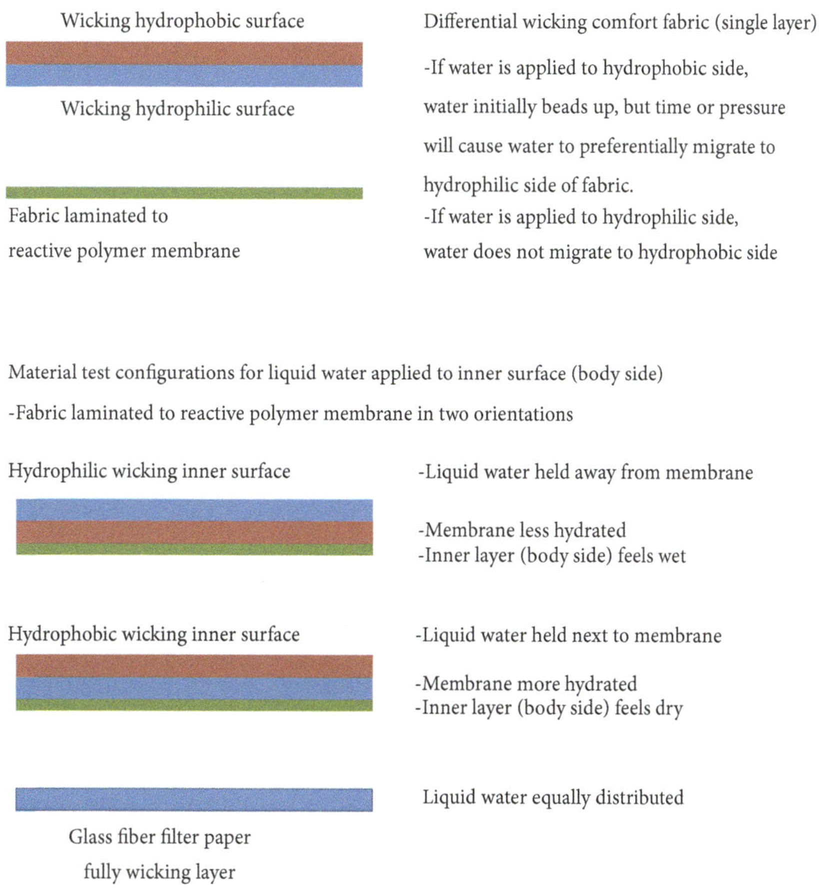

FIGURE 5: Material configurations for liquid water wicking and transport study.

used to sweep the top and bottom of the cells, and the humidity and flux was measured from both the top and bottom surfaces of the sample. This allowed the differences between the evaporative losses from the top surface to be compared directly to the water vapor flux that were actually transmitted through the fabric and polymer membrane layers.

The use of this configuration also allowed the sample to be equilibrated to a variety of humidity conditions, some of which included a gradient of humidity from the top to the bottom of the permeation cell (Table 1). This allowed direct comparison of the water vapor flux measured under vapor conditions to that obtained under the conditions where liquid

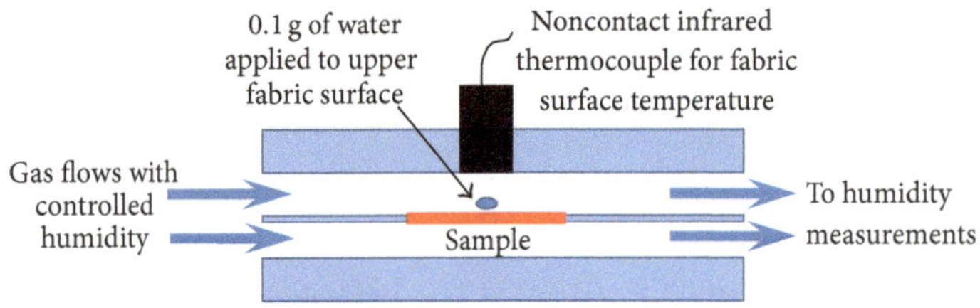

FIGURE 6: Experimental system of Figure 1 modified to add a ventilating top flow and simultaneous measurement of water vapor flux from the top and bottom flow channels of the permeation cell.

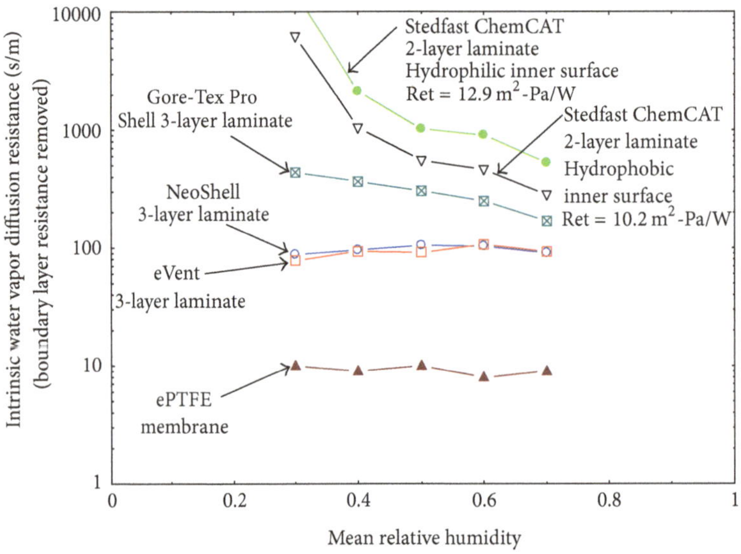

FIGURE 7: Water vapor diffusion resistance of ChemCAT laminates and several reference materials. Measured Ret values (ISO 11092) shown for ChemCAT laminates are consistent with the diffusion resistance values.

sweat is present. Further details on the instrumentation used in the permeation cell are available [4, 7].

Table 1 shows the wicking test conditions used for the study.

4. Results

The steady-state water vapor permeation results for the two laminates are shown in Figure 7. The results are shown in terms of water vapor diffusion resistance, where lower resistance implies better performance and comfort [4]. "Mean relative humidity" refers to the mean of the humidity on the two sides of the sample and is related to the equilibrium water content in the sample—a higher mean relative humidity implies higher water content in the polymer layer. The results shown in Figure 7 are consistent with many previous studies showing that the state of hydration and equilibrium water content in polymer membranes can have a large effect on the water vapor transport properties [8–16].

Figure 7 shows that the reactive polymer membrane laminates have high resistance to water vapor transport when compared to three high-performance commercial waterproof-breathable membrane laminates and a standard reference material (expanded polytetrafluoroethylene (ePTFE)). Although there are some differences in the measured properties of the two ChemCAT samples, we believe the differences are probably due to fabrication variations in the membrane and lamination method, rather than due to the orientation of the fabric layer. These commercial materials are similar in construction to the reactive polymer membrane laminates and serve as desirable performance goals for the reactive laminates. The "NeoShell" and the "eVent" laminates are composed of air permeable polymer membrane layers, while the Gore-Tex "Pro Shell" includes a thin solid polymer membrane layer much like the ChemCAT material.

Scanning electron micrographs of the functional polymer layer for each of the comparison materials are shown in Figure 8.

What are the consequences of the differing orientation of the laminated comfort layer upon the water vapor transport behavior of reactive polymer membrane laminates?

A representative case in Figure 9 shows that the reactive membrane laminate with the hydrophobic wicking inner surface has a very different water transport profile than the laminate with the reverse orientation (hydrophobic fabric side against the membrane). When the liquid water is held directly against the membrane in the hydrophilic layer, the water is transported quickly through the laminate, and a higher total amount is transported (area under the curve)

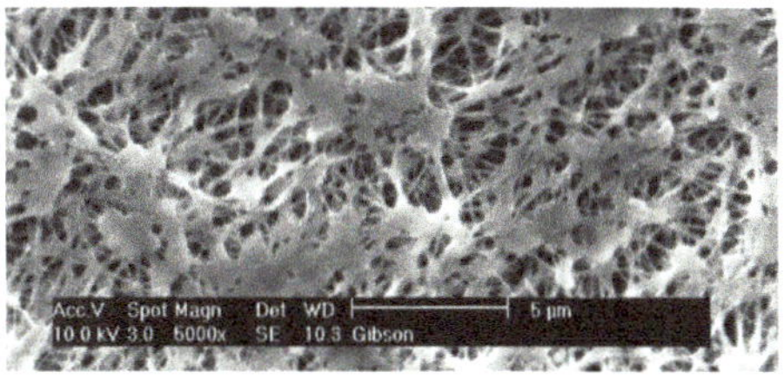

(a) eVent ePTFE film

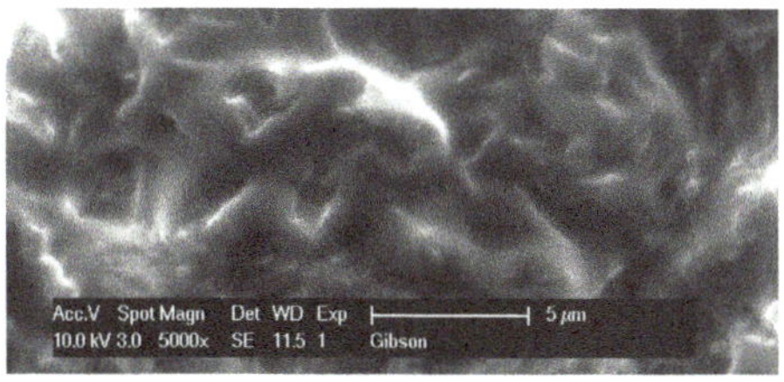

(b) Gore-Tex Membrane (coated side: continuous polymer film)

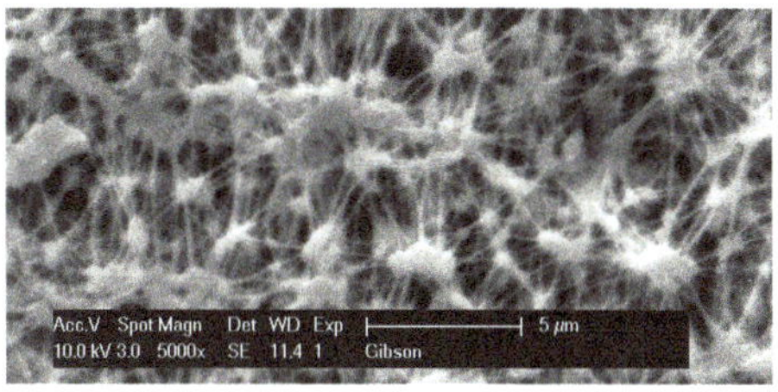

(c) Gore-Tex Membrane (uncoated side: ePTFE)

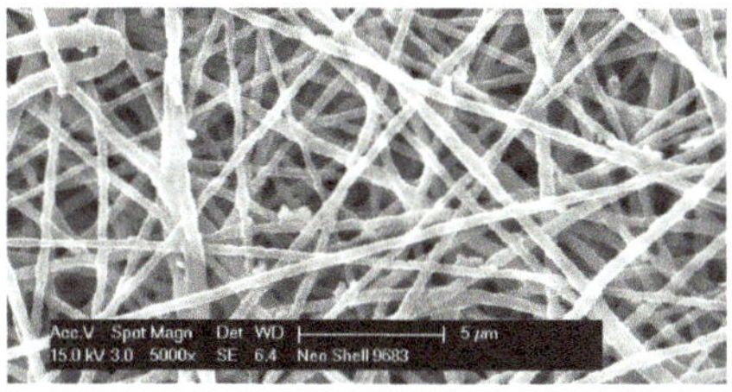

(d) NeoShell Electrospun Polyurethane Membrane

FIGURE 8: Scanning electron microscope images of typical membranes used in waterproof breathable fabric laminates. Images are of the membrane layer only (bonded supporting and protective textile layers: not shown).

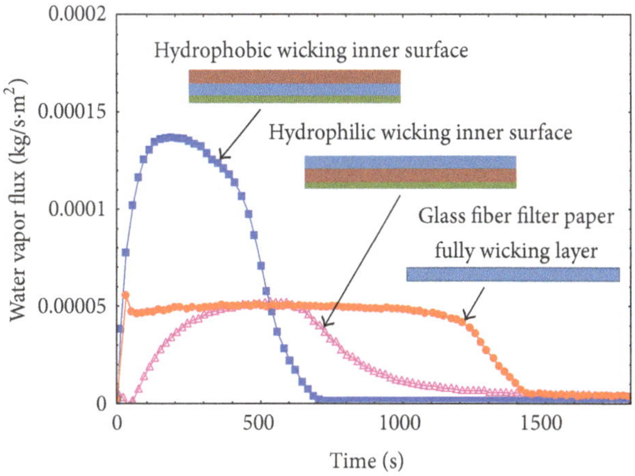

FIGURE 9: Water vapor transport from outer surface of fabric, following 0.1 g application of liquid water on inner surface of fabric. Sweep flow of 40% relative humidity on both sides of the fabric.

than is in the case for the laminate with the hydrophilic inner surface. The laminate with the hydrophilic inner surface transported about half the total amount of water as either the glass fiber filter or the laminate with the hydrophobic wicking inner surface. It is also interesting to note that although the membrane laminates showed poor water diffusion rates in the vapor diffusion test method, the transport behavior when the membranes are exposed to liquid water is much more efficient.

This can be further illustrated by looking at liquid water vapor transport rates when tested under a humidity gradient, as shown in Figure 10.

In this case, the fabrics are first equilibrated with a humidity of 0.8 (80% r.h.) over the top of the sample, and 0.4 (40% r.h.) across the bottom of the sample. Under these vapor transport conditions, the water vapor transport rate, of the reactive polymer membrane laminates are quite low and are an order of magnitude less than the water vapor transport rate through the reference glass fiber filter layer. This is also reflected in the water vapor transport curves presented previously in Figure 7, where the reactive polymer membrane laminates were shown to have much higher water vapor diffusion resistances than the commercial waterproof-breathable membrane laminates at all humidity conditions.

However, when the liquid water is applied to the inner comfort layer of the fabric, the water vapor transport rates increase in a similar fashion as was seen in Figure 9. Under these gradient conditions, the hydrophobic wicking layer is more successful at moving liquid water into the polymer membrane layer, increasing the water vapor transport rate quickly, and efficiently transferring water from one side of the laminate to the other. However, the water vapor transport rate does not exceed the performance of the reference glass fiber filter layer in this case but does achieve levels that are equivalent to the steady-state vapor transport rate of the glass fiber filter. The laminate with the hydrophilic wicking inner surface again lags behind the performance of hydrophobic wicking inner surface configuration because the liquid water is held away from the membrane. However, since the humidity in the flow in that upper portion of the cell is at 0.80 (80% r.h.) evaporative losses due to the ventilating flow are much less, and there is more time for the liquid water to be taken up by the polymer layer and transported to the other side. Consequently, the total amount of water transported through the membrane is equivalent for the two laminates. It just takes a longer time for the same amount of water to

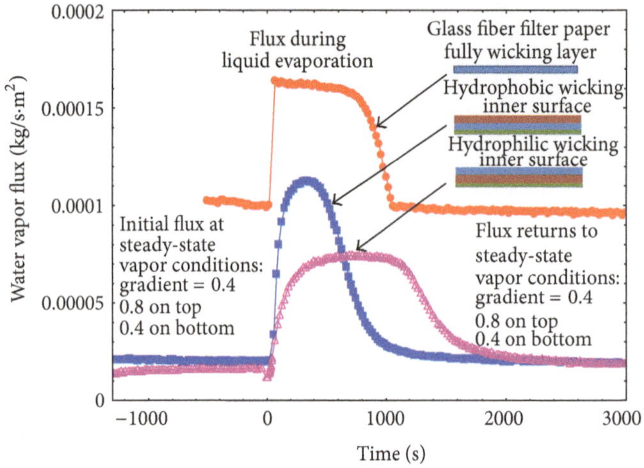

FIGURE 10: Water vapor transport under a humidity gradient of 0.4 (0.8 on top of the sample and 0.4 on the bottom of the sample).

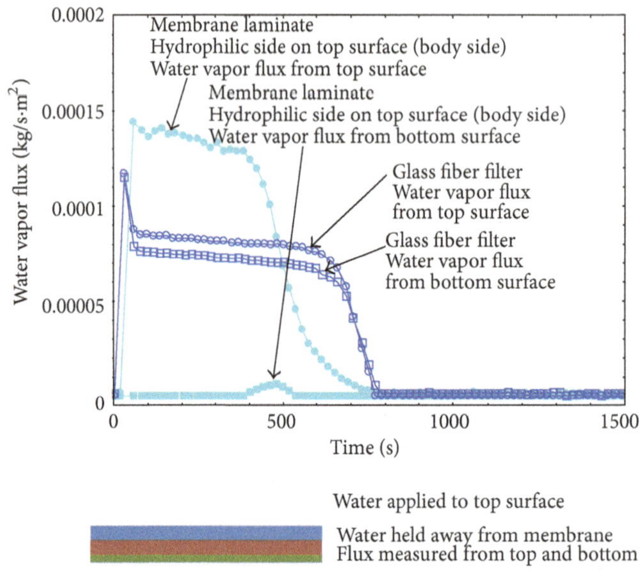

FIGURE 11: Water vapor transport from outer surface of fabric, following 0.1 g application of liquid water on inner surface of fabric. Sweep flow of 0% relative humidity on both sides of the fabric.

be transported through the laminate with the hydrophilic wicking inner surface.

The dual flow permeation test method also allows the directionality of water vapor permeation to be easily measured. Test conditions from Table 1 that use the same humidity on both sides of the sample should produce equivalent evaporation rates from a fully wicking sample. Both sides of the sample are wetted, and water evaporates equally well from both sides. However, we would expect that the reactive membrane laminate with the hydrophilic inner surface would have a very different behavior. The liquid water would be held in the hydrophilic layer and would mostly evaporate in the upper ventilating air flow, with much less diffusing through the polymer membrane to the other side of the permeation cell.

A good illustration of this directional permeation is obtained at the test condition of 0% humidity on both sides of the sample. This maximizes evaporation on the inner (upper side) of the test sample. Figure 11 shows that the glass fiber filter has equivalent water vapor flux from the top and bottom surfaces of the sample and that they both dry at the same rate.

The behavior of the laminate with the hydrophilic side towards the skin is dramatically different (Figure 11). Most of the liquid water is evaporated into the upper air flow. Only a tiny amount permeates through the polymer film, and only after a significant delay, and after most of the liquid water has already evaporated into the upper ventilating flow. It should also be noted that the polymer film is fully equilibrated to the humidity conditions of 0% r.h and contains very little water, resulting in a very low water vapor transport rate to begin with.

As in the previous examples, Figure 11 shows that when the hydrophilic layer is oriented towards the body, the layer fails to hydrate the polymer membrane effectively, and water vapor transport through the membrane is hindered.

Figures 9–11 illustrate the most important phenomena associated with the hydrophilic or hydrophobic orientation of the comfort layer on the reactive polymer membrane laminate. Other test conditions from Table 1 are consistent with these examples and support these general guidelines.

5. Conclusions

Permeation cell measurements demonstrated that the rate of liquid wicking was conveniently reflected in a measurable cooling effect on the fabric surface due to water evaporation. Results from vapor transport testing with wetting/nonwetting fabrics show that there are very large differences in drying time that can impact the rate at which the evaporating sweat from the body is transported through the fabric to the environment. Orienting a hydrophobic/hydrophilic fabric with the hydrophobic side towards the human skin surface and the hydrophilic surface towards a laminated reactive polymer membrane laminate is more efficient than the reverse orientation at picking up liquid from the skin surface and moving it to the polymer layer where it is available for hydration and chemical reaction.

Conflict of Interests

One of the authors, Dr. H. Benaddi, is the principal research scientist for the commercial entity, Stedfast, Inc., Canada, which manufactures the commercial material mentioned in the paper (Stedfast ChemCAT).

References

[1] S. Narang and S. Ventura, "Immobilized 4-dialkylaminopyridines: supernucleophilic chemical decontaminating agents," US Army Research Office Technical Report 23208.1-CH, 1987, http://handle.dtic.mil/100.2/ADA190999.

[2] S. Narang, S. Ventura, and R. Ramharack, "Polymeric dialkylaminopyridines as dupernucleophilic catalysts," in *Chemical Reactions on Polymers*, vol. 364 of *ACS Symposium Series*, chapter 6, pp. 72–80, 1988.

[3] S. Wang, L. Bromberg, H. Schreuder-Gibson, and T. Hatton, "Catalysis of organophophorous ester degradation by chromium (III) terephthlatalate (MIL-101) metal organic framework chelated to *N*, *N*-dimethylaminopyridine and related aminopyridines," *ACS Applied Materials and Interfaces*, vol. 5, no. 4, pp. 1269–1278, 2013.

[4] P. Gibson and H. Schreuder-Gibson, "Influence of hydration state on permeation testing and vapor transport properties of protective clothing layers," *Journal of Engineered Fibers and Fabrics*, vol. 4, no. 4b, pp. 11–20, 2009.

[5] J. Andzelm, J. Walker, H. Schreuder-Gibson, and P. Gibson, "Interaction of water and DFP with PVA-Am/PEI polymer blends," in *Proceedings of the Chemical and Biological Defense Physical Science and Technology Conference*, Dallas, Tex, USA, November 2009.

[6] H. Schreuder-Gibson, J. Walker, J. Owens et al., "The role of water in the deactivation of chemical and biological agents within fabrics," in *Proceedings of the Chemical and Biological Defense Physical Science and Technology Conference*, New Orleans, La, USA, November 2008.

[7] P. Gibson, "Water-repellent treatment on military uniform fabrics: physiological and comfort implications," *Journal of Industrial Textiles*, vol. 38, no. 1, pp. 43–54, 2008.

[8] P. Dolhan, "A comparison of apparatus used to measure water vapour resistance," *Journal of Coated Fabrics*, vol. 17, pp. 96–109, 1987.

[9] B. Farnworth, W. Lotens, and P. Wittgen, "Variation of water vapor resistance of microporous and hydrophilic films with relative humidity," *Textile Research Journal*, vol. 60, no. 1, pp. 50–53, 1990.

[10] P. Gibson, C. Kendrick, D. Rivin, M. Charmchi, and L. Sicuranza, "An automated water vapor diffusion test method for fabrics laminates, and films," *Journal of Coated Fabrics*, vol. 24, pp. 322–345, 1995.

[11] P. W. Gibson, "Water vapor transport and gas flow properties of textiles, polymer membranes and fabric laminates," *Journal of Coated Fabrics*, vol. 28, pp. 300–327, 1999.

[12] P. Gibson, D. Rivin, A. Berezin, and A. Nadezhdinskii, "Measurement of water vapor diffusion through polymer films and fabric/membrane laminates using a diode laser spectroscope," *Polymer-Plastics Technology and Engineering*, vol. 38, no. 2, pp. 221–239, 1999.

[13] P. W. Gibson, "Effect of temperature on water vapor transport through polymer membrane laminates," *Polymer Testing*, vol. 19, no. 6, pp. 673–691, 2000.

[14] D. Rivin, C. Kendrick, P. Gibson, and N. Schneider, "Studies of the Solubility and Transport Behavior of Water and Alcohols in Nafion," *Polymer*, vol. 81, p. 541, 1999.

[15] E. A. McCullough, M. Kwon, and H. Shim, "A comparison of standard methods for measuring water vapour premeability of fabrics," *Measurement Science and Technology*, vol. 14, no. 8, pp. 1402–1408, 2003.

[16] G. R. Lomax, "Breathable polyurethane membranes for textile and related industries," *Journal of Materials Chemistry*, vol. 17, no. 27, pp. 2775–2784, 2007.

3

Studies on Wicking Behaviour of Polyester Fabric

Arobindo Chatterjee and Pratibha Singh

Department of Textile Technology, National Institute of Technology, Jalandhar 144 011, India

Correspondence should be addressed to Arobindo Chatterjee; chatterjeea@nitj.ac.in

Academic Editor: Seshadri Ramkumar

This paper aims to investigate vertical wicking properties of polyester fabric based on change in sample direction and change in tension. Also experimental results are compared with theoretical results. Polyester fabric made out of spun yarn with four types of variation in pick density was used. Theoretical values of vertical wicking were calculated by using Lucas-Washburn equation and experimental results were recorded using strip test method. Maximum height reached experimentally in both warp way and weft way is more than that of the theoretical values. The maximum height attained by fabric experimentally in weft is more as compared to warp way. Vertical wicking increases with increase in tension. This paper is focused on wicking which plays a vital role in determining comfort and moisture transport behavior of fabric.

1. Introduction

Transport of water through textiles plays a very important role in deciding comfort, dyeing and finishing of textile fabrics, liquid filtration, and so forth. Transport of water takes place through the phenomenon of capillarity. Capillarity is the ability of liquids to penetrate into fine pores with wettable walls and be displaced from those with nonwettable walls [1]. Capillary wetting is due to the meniscus formed by fibres and yarns in warp and weft directions, especially in the interstitial area [2].

Wicking can only occur when a liquid wets fibres assembled with capillary spaces between them. The resulting capillary forces that arise from the wetting of the fabric surface due to pressure difference created by surface tension of the liquid across the curved liquid/vapor interface drive the liquid into the capillary spaces. Because capillary forces are caused by wetting, wicking is a result of spontaneous wetting in a capillary system [3]. Capillary forces are governed by the properties of the liquid, liquid-medium surface interactions, and geometric configurations of the pore structure in the medium.

During normal activity and in normal atmospheric condition, the heat produced in the body due to metabolism is liberated to the atmosphere by conduction, convection, and radiation and the body perspires in vapour form to maintain the temperature. However, while doing high level bodily activity and/or at higher atmospheric temperatures, heat production is very high and for transmitting this heat from the skin to the atmosphere the sweat glands produce liquid perspiration (insensible perspiration which is in vapour form and sensible perspiration which is in liquid form) [4]. To be in comfortable state, the clothing worn should allow both types of perspirations to transmit from the skin to the outside atmosphere. Moisture related properties influence the thermophysiological clothing comfort of the material [5].

Mathematical modeling of surface-tension-driven flow in yarns and fabrics can provide a way to develop an understanding of the liquid transport mechanism. The constituent yarns are responsible for the main portion of the wicking action, in capillary flow through textile fabrics [6, 7]. In many researches, the textile yarns were treated either as porous media [8–10], the liquid transport through which is described by Darcy's law [11], or as capillary tubes [12–15], the liquid flow through which can be modeled by Lucas-Washburn kinetics [16]. In the first case, the characteristic parameters, such as permeability, are difficult to quantify and are obtained empirically. In the second case, the effective radius of the capillary tube, the effective contact angle, and so forth are determined by fitting the experimental data [17].

TABLE 1: Details of sample used for experiment.

Composition	Type of yarn	EPI	PPI	Thickness (in mm)	GSM	Referred to as
100% polyester	Staple yarn	48	30	0.535	127	P1
100% polyester	Staple yarn	48	40	0.493	147	P2
100% polyester	Staple yarn	48	50	0.544	173	P3
100% polyester	Staple yarn	48	60	0.549	186	P4

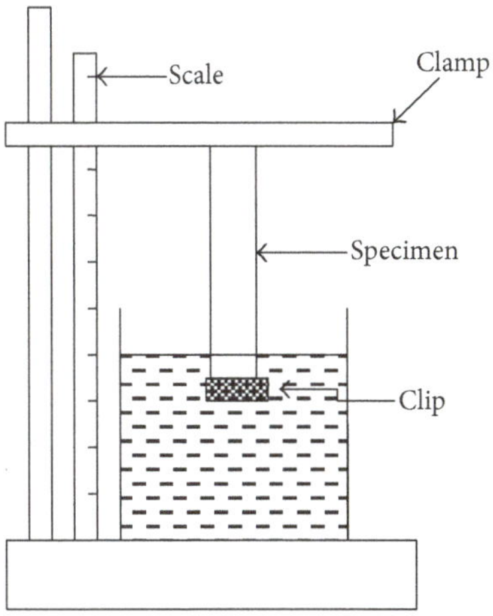

FIGURE 1: Vertical wicking apparatus.

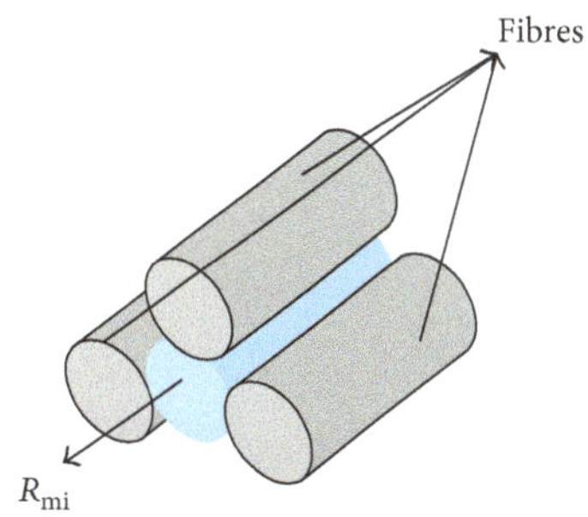

FIGURE 2: Microcapillary.

In a model developed for jersey knitting fabric macro- and microcapillary are considered which are collectively responsible for wicking in the fabric. The effective capillary radius is not calculated by fitting the experimental data; it is dependent on the geometrical conformation [18]. It was reported by various researchers that wicking depends on tortuosity of fabric [6, 7, 14, 19] and this factor was incorporated in single jersey knitted fabric [18].

Though this work focuses on the wickability, recently related properties such as moisture vapor transport have also attracted attention [20–22].

In this work, plain woven fabric is studied and tortuosity factor is considered. Direct method for calculation of macropore radius is given for plain weave.

2. Materials and Methods

2.1. Materials. Fabric was made out of 100% polyester spun yarn having 2/30's yarn count. The detailed specification of it is given in Table 1.

2.2. Methods

2.2.1. Vertical Wicking. The schematic diagram of the experimental setup for measurement of vertical wicking is shown in Figure 1. The fabric was tinted to facilitate visual tracking of the movement of water.

Specimens of 200 mm × 25 mm cut along warpwise and weftwise directions were prepared from these fabrics. The specimens were suspended vertically with their bottom ends dipped in a reservoir of distilled water. In order to ensure that the bottom ends of the specimens could be immersed vertically at a depth of 20 mm into the water, the bottom end of each specimen was clamped with a 3 g clip. To evaluate wicking performance at varying tensions 3 g weight was replaced by 12 g, 20 g, and 34 g weights. For kinetics of wicking heights, distance traveled by water on vertical strip was measured for every minute for the first 5 minutes and then readings were taken after every 5 minutes for 30 minutes.

The fabric wicking experiments were conducted in a standard atmosphere of 20 ± 2°C and 65 ± 2% relative humidity and the fabric was conditioned for 24 h before testing.

2.2.2. Capillary Rise in Fabric. Textile materials are hierarchical porous media. Fabric is composed of yarns which are running parallel to each other and yarns are made up of fibres or filaments oriented along the yarn axis. Capillary rise in fabric can be considered as the effect of capillary rise between yarns within a fabric and between fibres within a yarn constituting the fabric. The capillary formed between yarns may be termed as macrocapillary and capillary formed between fibres in yarn may be termed as microcapillary. Microcapillary, macrocapillary, and tortuosity were evaluated for theoretical calculation of capillary rise using Lucas-Washburn equation.

Microcapillary. Capillary rise between fibres (in a yarn) can be analysed like a flow in capillary tube of radius R_{mi} as shown in Figure 2.

The capillary rise of liquid is given by Washburn law [16] as follows:

$$\frac{dh}{dt} = \frac{(R_{mi}/\tau)^2}{8\eta h}\left(\frac{2\gamma_L\cos\theta}{R_{mi}} - \rho g h\right), \quad (1)$$

where dh/dt is the rate of change of capillary height with respect to time, R_{mi} is the microcapillary radius, τ is the tortuosity, η is liquid viscosity, h is the capillary height, γ_L is surface tension, $\cos\theta$ is the contact angle of water with the fibre, ρ is liquid density, and g is acceleration due to gravity.

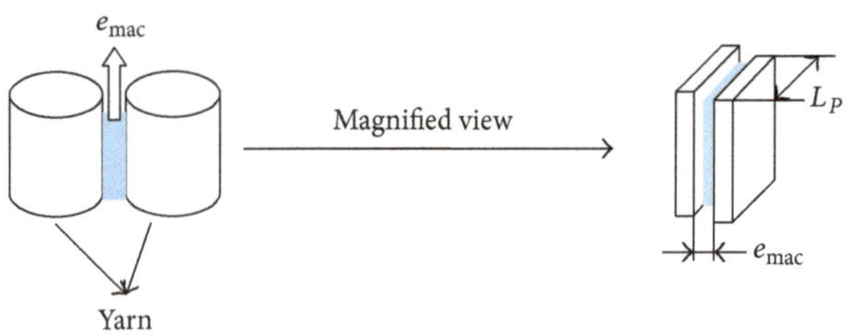

FIGURE 3: Macrocapillary.

At equilibrium the rate of change of height with respect to time is zero. That is, $dh/dt = 0$. Putting this value in (1) we get:

$$h_{\text{mic-eq}} = \frac{2\gamma_L \cos\theta}{R_{\text{mi}}\rho g}. \quad (2)$$

From (1), (3) can be obtained for further evaluation of kinetics of capillary rise as follows:

$$h = \frac{2\gamma_L \cos\theta}{R_{\text{mi}}\rho g}\left(1 - \frac{1}{e^{(tR_{\text{mi}}^2\rho g/\tau^2 8\eta)}}\right). \quad (3)$$

Microcapillary radius (R_{mi}) *can be calculated as* [18]

$$R_{\text{mi}} = \sqrt{\frac{t^2}{32n_s} - \frac{d_{\text{fiber}}^2}{8}}. \quad (4)$$

Macrocapillary. On fabric scale the capillary formed is between yarns as shown in Figure 3.

The equation describing the capillary kinetics of progression between two parallel plates, where ($L_p \gg e_{\text{mac}}$), is given by the equation of Poiseuille:

$$\frac{dh}{dt} = \frac{e_{\text{mac}}^2 \Delta P}{12\eta h}, \quad (5)$$

where $\Delta P = \Delta P_c - \rho g h$; ΔP_c is the difference in pressure related to the capillary forces (law of Laplace). In the case of two parallel plates with e_{mac} distance (as in two yarns) the Laplace law is

$$\Delta P_c = \frac{2\gamma_L \cos\theta}{e_{\text{mac}}}. \quad (6)$$

Thus the rate of change of height of liquid in macrocapillary is

$$\frac{dh}{dt} = \frac{(e_{\text{mac}}/\tau)^2}{12\eta h}\left(\frac{2\gamma_L \cos\theta}{e_{\text{mac}}} - \rho g h\right), \quad (7)$$

where dh/dt is the rate of change of capillary height with respect to time, e_{mac} is the macrocapillary radius, τ is the tortuosity, η is liquid viscosity, h is the capillary height, γ_L is surface tension, $\cos\theta$ is the contact angle of water with the fibre, ρ is liquid density, and g is acceleration due to gravity.

At equilibrium, the rate of change of height with respect to time is zero. That is, $dh/dt = 0$. Putting this value in (7), we get

$$h_{\text{mac-eq}} = \frac{2\gamma_L \cos\theta}{e_{\text{mac}}\rho g}. \quad (8)$$

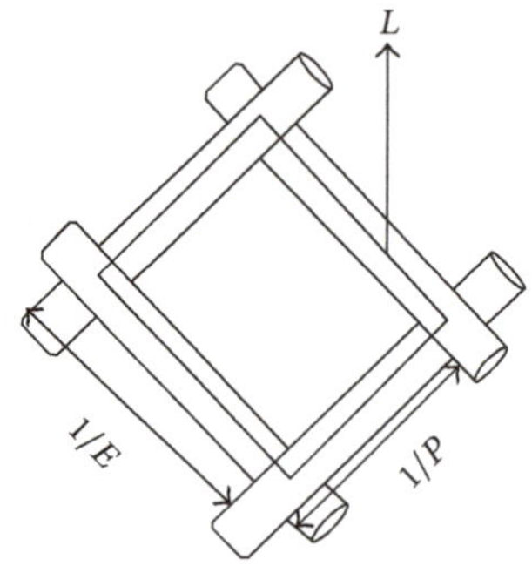

FIGURE 4: L in unit cell of plain weave.

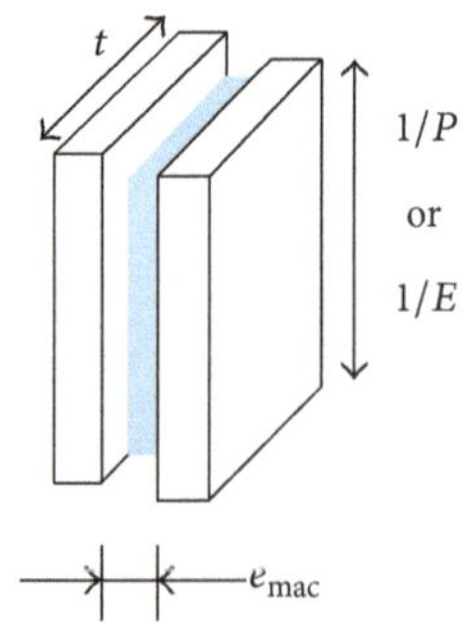

FIGURE 5: Parallelepiped.

From (7), (9) can be obtained for further evaluation of kinetics of capillary rise as follows:

$$h = \frac{2\gamma_L \cos\theta}{e_{\text{mac}}\rho g}\left(1 - \frac{1}{e^{(te_{\text{mac}}^2\rho g/\tau^2 12\eta)}}\right). \quad (9)$$

Macrocapillary radius (e_{mac}) *can be calculated as*

$$\text{Vacuum volume} = \text{Total volume} - \text{Yarn volume} \quad (10)$$

$$\text{Total volume} = \frac{t}{E \times P} \quad (11)$$

$$\text{Yarn volume} = \frac{\pi d_{\text{yarn}}^2 L}{4}, \quad (12)$$

where L is the length of yarn making one unit cell of plain weave (Figure 4).

Hence,

$$\text{Vacuum volume} = \frac{t}{E \times P} - \frac{\pi d_{\text{yarn}}^2 L}{4}. \quad (13)$$

The capillary rise between yarns (on fabric scale) can be regarded as equivalent to a flow between two distant parallel plates of capillary distance e_{mac}. The vacuum volume is equivalent to a parallelepiped volume having lengths $1/P$ (if it is warp way capillary) and $1/E$ (if it is weft way capillary), width e_{mac}, and thickness t (Figure 5):

$$\text{parallelepiped volume} = \frac{t \times e_{\text{mac warp}}}{P} \ (\text{warp way capillary}),$$

$$\text{parallelepiped volume} = \frac{t \times e_{\text{mac weft}}}{E} \ (\text{weft way capillary}). \quad (14)$$

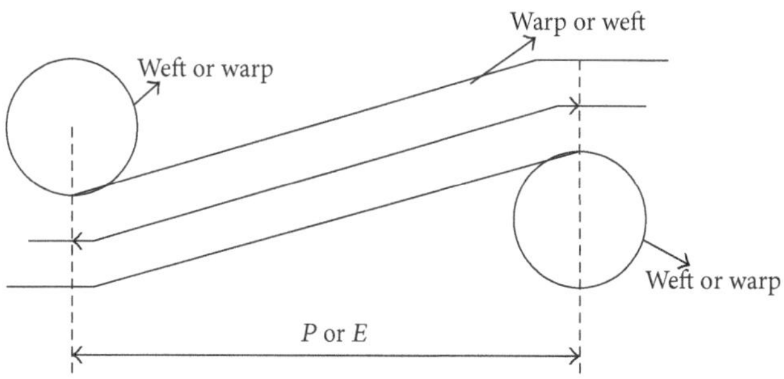

FIGURE 6: Geometry of the unit cell for a plain weave.

TABLE 2: Characteristics of distilled water at 20°C.

Parameters	Value
Density	998.29 kg/m^3
Dynamic viscosity	0.001003 Kg/m·s
Surface energy	72.5 mJ/m^2
Contact angle/polyester fibre	75°

Using (13) and (14), the macrocapillary radius is

$$e_{\text{mac warp}} = \frac{1}{E} - \frac{\pi d_{\text{yarn}}^2 LP}{4 \times t} \text{ (warp way capillary)},$$
$$e_{\text{mac weft}} = \frac{1}{P} - \frac{\pi d_{\text{yarn}}^2 LE}{4 \times t} \text{ (weft way capillary)}. \quad (15)$$

Tortuosity. The tortuosity illustrated in Figure 6 is defined as

$$\tau = \frac{L}{P \text{ or } E}, \quad (16)$$

where τ is the tortuosity, P and E denote picks per inch and ends per inch, respectively, and L is given by

$$L = (1 + C) \times (P \text{ or } E). \quad (17)$$

Here C denotes the crimp (it is measure of waviness in yarns) which can be calculated by taking out the yarns from the fabric and measuring crimp condition length and then applying force from both ends of the yarn and measuring the actual length after application of force [23].

Consider

$$C = \frac{\text{Actual length} - \text{Crimp condition length}}{\text{Crimp condition length}}. \quad (18)$$

From equations (16), (17), and (18) we get

$$\tau = (1 + C). \quad (19)$$

2.2.3. Liquid Characteristics. The characteristics of liquid used for carrying out wicking test are given in Table 2.

3. Results and Discussions

3.1. Experimental and Theoretical Vertical Wicking of Polyester. Figures 7 to 14 show the results of experimental and theoretical vertical wicking of polyester both warp and weft ways and the theoretical vertical wicking in micro- and macrocapillaries. It is seen from the figures that the maximum height reached in 1800 seconds experimentally for all the samples of polyester in both warp and weft ways is more than that of the theoretical values.

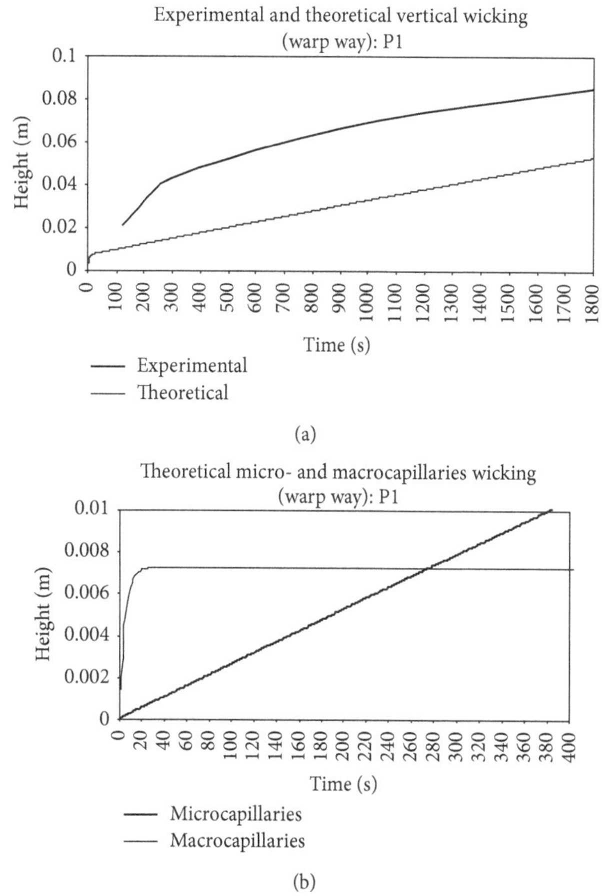

FIGURE 7: (a) Experimental and theoretical vertical wicking. (b) Theoretical micro- and macrocapillaries wicking (warp way, P1).

When the fabric strip is dipped in liquid reservoir, the smaller capillaries are filled first because of higher capillary pressure followed by larger capillaries. The mass of liquid retained in small capillaries is less as compared to large capillaries. Due to less capillary pressure in large capillaries the liquid advancement is less but the liquid retained by these capillaries acts as reservoir for small capillaries. Liquid reservoir is also present at the points where yarns are intersecting. So for wicking to take place, liquid placed at the bottom of fabric strip is the main reservoir; the liquid retained in large capillaries and at the intersection points of yarns act as mini reservoirs.

In theoretical calculation, it was assumed that the capillaries are uniform. But in actual fabric capillaries are not uniform or continuous between fibres and yarns. It may be assumed that in fabric a large number of small capillaries are joined by large number of mini reservoirs which act as source of liquid for these small capillaries. With time and gradual rise of liquid, the amount of liquid available in the

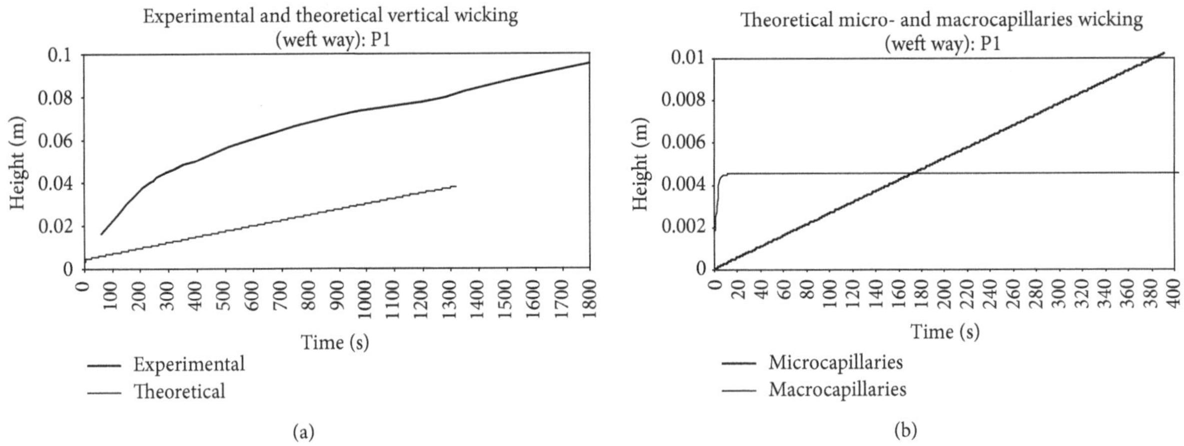

FIGURE 8: (a) Experimental and theoretical vertical wicking. (b) Theoretical micro- and macrocapillaries wicking (weft way, P1).

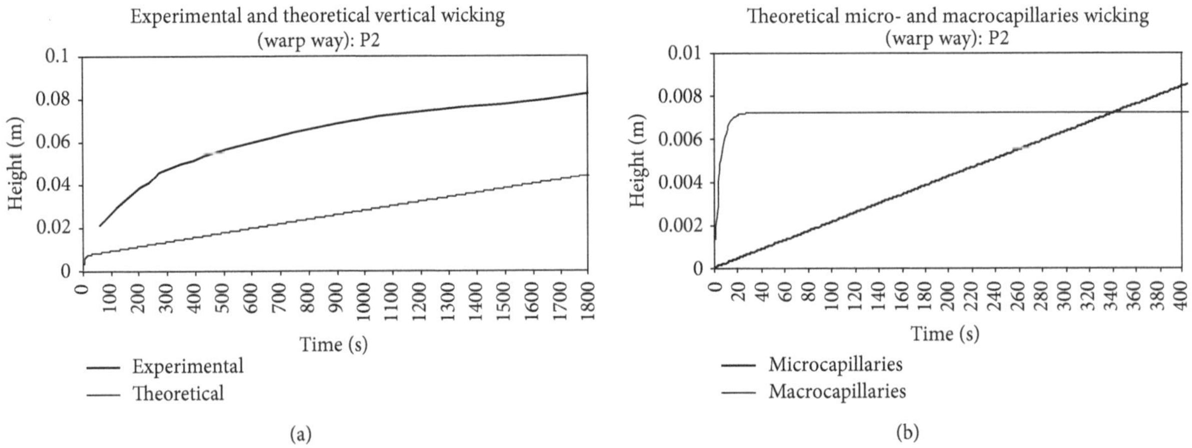

FIGURE 9: (a) Experimental and theoretical vertical wicking. (b) Theoretical micro- and macrocapillaries wicking (warp way, P2).

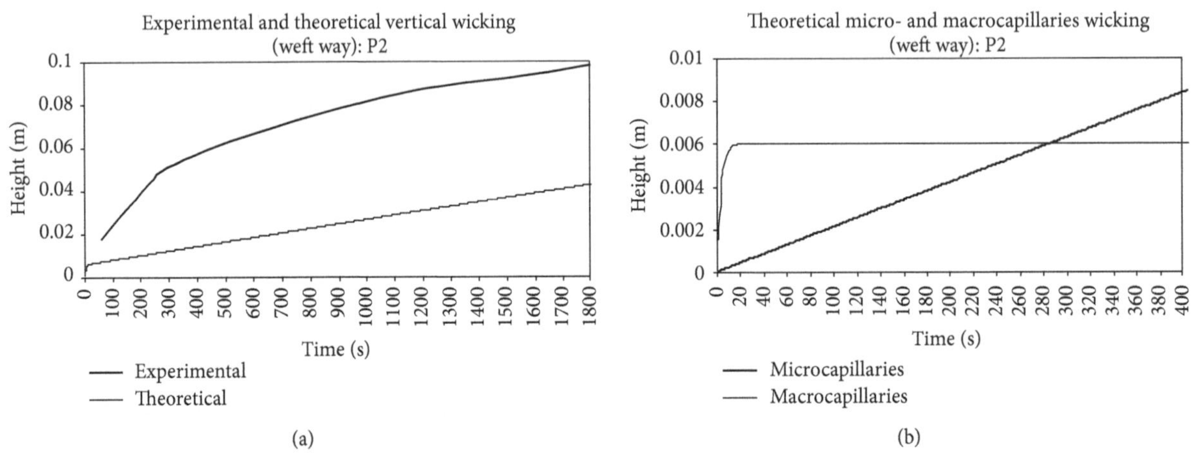

FIGURE 10: (a) Experimental and theoretical vertical wicking. (b) Theoretical micro- and macrocapillaries wicking (weft way, P2).

subsequent mini reservoirs keeps decreasing and when the pressure difference between the small capillaries and mini reservoirs ceases to exist, the equilibrium is achieved. Because of this reason, the experimental values are higher than the corresponding theoretical values.

3.2. Effect of Sample Direction for Vertical Wicking of Polyester. Samples are cut in warp way and weft way directions for vertical wicking test. Theoretical values are based on certain variables; change in those variables brings about change in maximum height reached in warp way and weft way

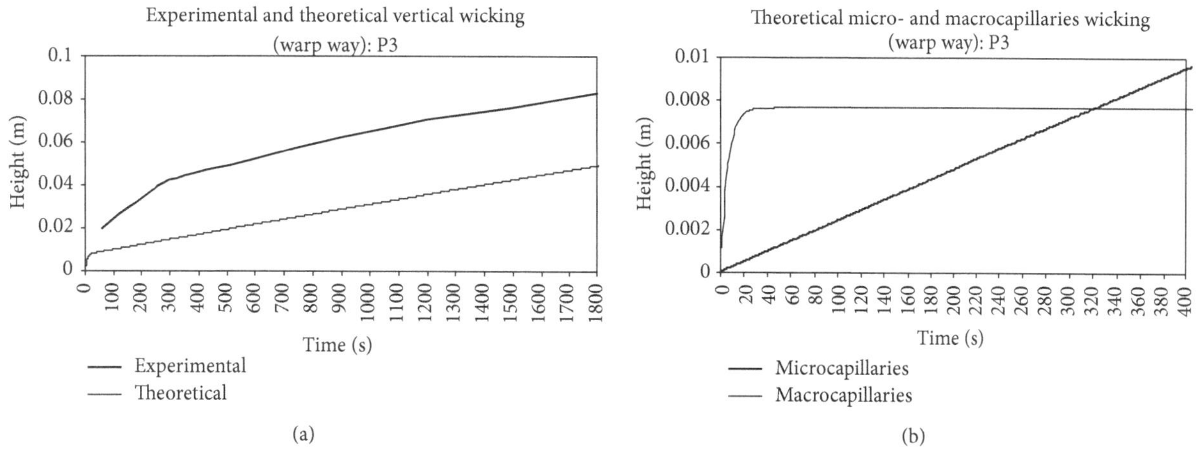

FIGURE 11: (a) Experimental and theoretical vertical wicking. (b) Theoretical micro- and macrocapillaries wicking (warp way, P3).

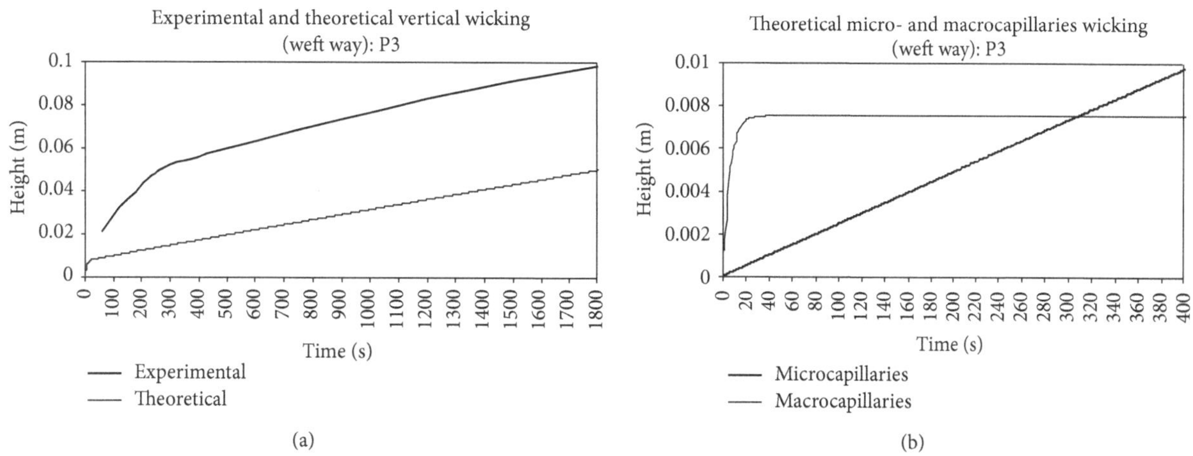

FIGURE 12: (a) Experimental and theoretical vertical wicking. (b) Theoretical micro- and macrocapillaries wicking (weft way, P3).

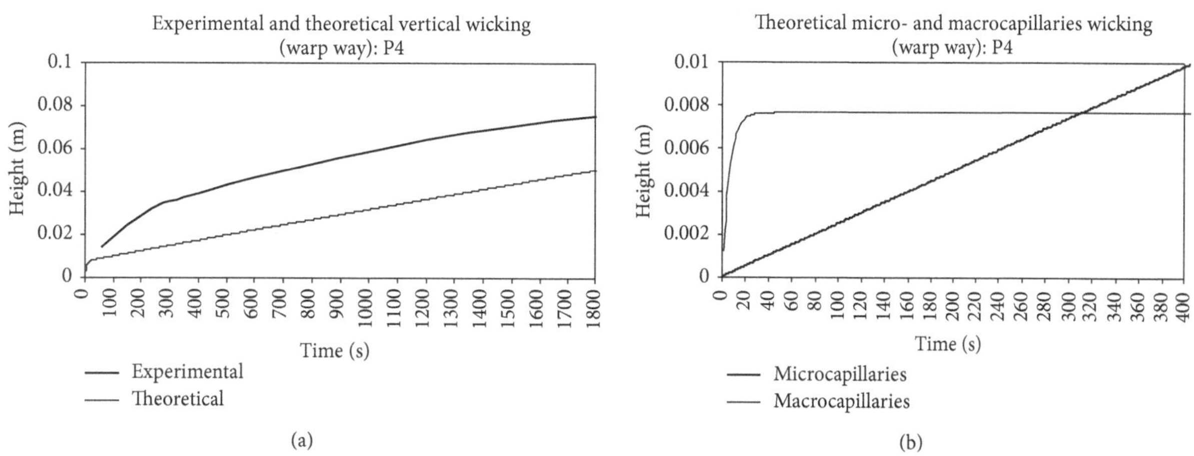

FIGURE 13: (a) Experimental and theoretical vertical wicking. (b) Theoretical micro- and macrocapillaries wicking (warp way, P4).

directions. From Table 3 it can be observed that for P1 and P2 fabric warp way wicking height is more as that of weft way wicking height but for P3 and P4 fabric the situation is reversed.

But for the experimental values for maximum height reached, it is observed that the maximum height reached in weft way is more as compared to maximum height reached in warp way in all cases of polyester samples (Table 3).

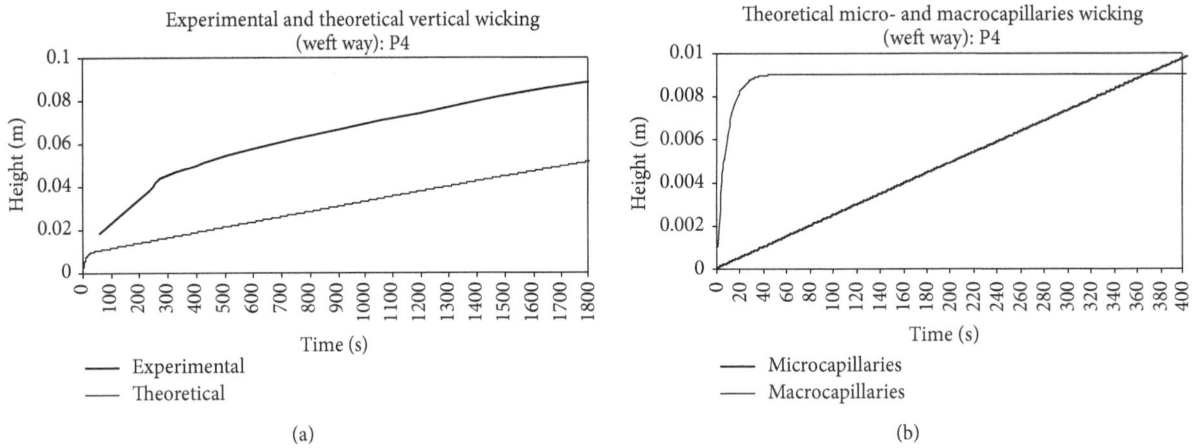

FIGURE 14: (a) Experimental and theoretical vertical wicking. (b) Theoretical micro- and macrocapillaries wicking (weft way, P4).

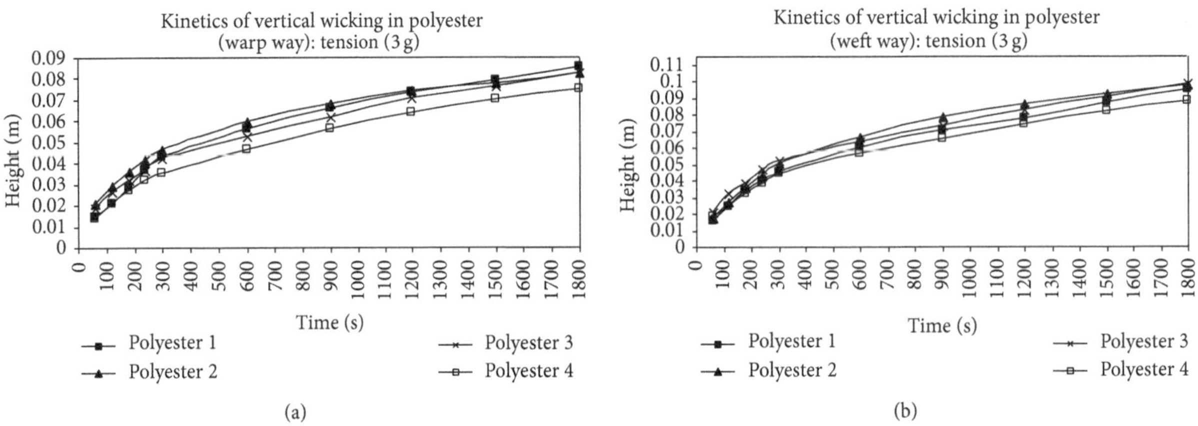

FIGURE 15: Kinetics of vertical wicking in polyester at 3 g, (a) warp way and (b) weft way.

Though the same yarn was used both as warp and as weft, a difference in warp way wicking and weft way wicking behaviour suggests that there has to be some difference in terms of size and distribution of capillaries in the respective directions. The difference in tension of warp and weft yarn during fabric formation may be responsible for difference in size and disposition of capillaries in two directions. Warp yarns, being in more stressed condition than weft yarns, have a more compact structure due to which the radius of the values of micro- and macrocapillaries formed will be less. This will slow down the rate of capillary rise and in a given time the height reached will be less. It is also evident from the experimental vertical wicking graphs (Figures 7, 8, 9, 10, 11, 12, 13, and 14) that the initial rate of wicking is greater in weft way than in warp way which supports the difference in size and disposition of capillary in warp way and weft way.

It may also be mentioned here that within the range investigated, irrespective of the ends per inch and picks per inch, the weft way wicking is more than that of the warp way wicking.

3.3. Effect of Change in Pick Density. As pick density increases, wicking height decreases. Figures 15, 16, 17, and 18 show the relation between pick density and wicking height. The results obtained are significant at 95% level. Theoretical calculation of effect of pick density on wicking height is also in line with experimental results and may be explained as follows.

(i) With increase in pick density fabric structure is becoming more compact, due to which thickness is increasing gradually which is influencing the microcapillary radius. Theoretically according to (4), thickness (t), number of fibres in yarn (n_s) and fibre diameter (d) can bring a change in micro capillary radius (R_{mi}). In this case thickness is the only variable and the other two factors are constant. So higher thickness value will give higher microcapillary radius. So, as pick density is increased micro capillary radius will increase (Table 1) and due to which capillary pressure will decrease which is responsible for wicking; the less the capillary pressure is, the less the wicking will be. Because of this theoretical reason, wicking height is reduced as pick density is increased. Theoretically, (15), change in pick density and thickness of fabric will affect the macrocapillary radius. With increase in pick density and thickness, the macrocapillary radius will

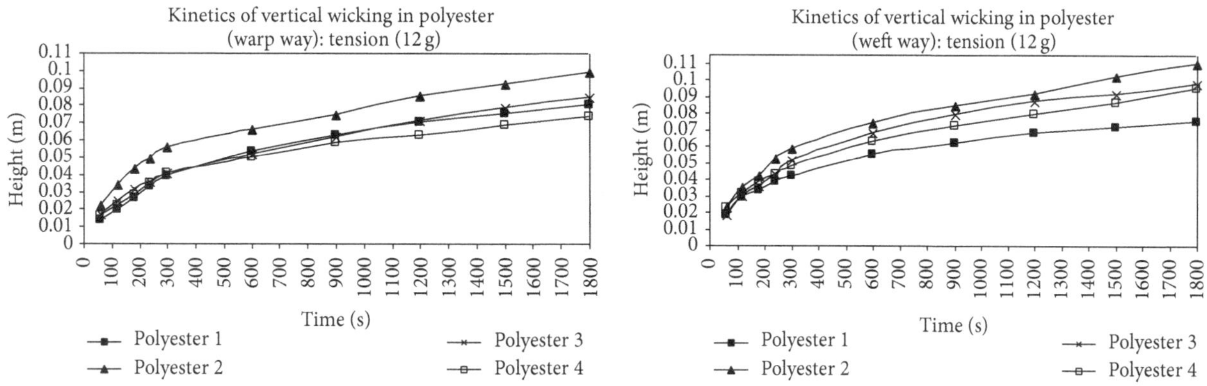

(a) (b)

FIGURE 16: Kinetics of vertical wicking in polyester at 12 g, (a) warp way and (b) weft way.

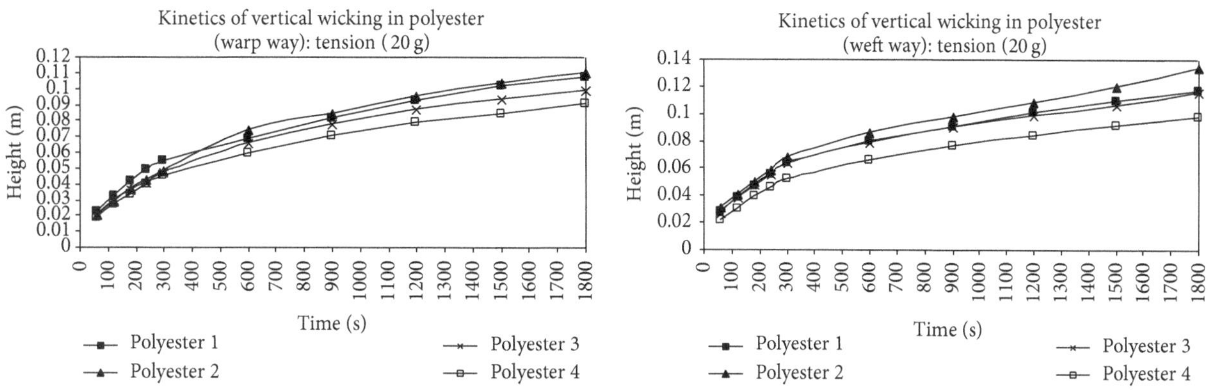

(a) (b)

FIGURE 17: Kinetics of vertical wicking in polyester at 20 g, (a) warp way and (b) weft way.

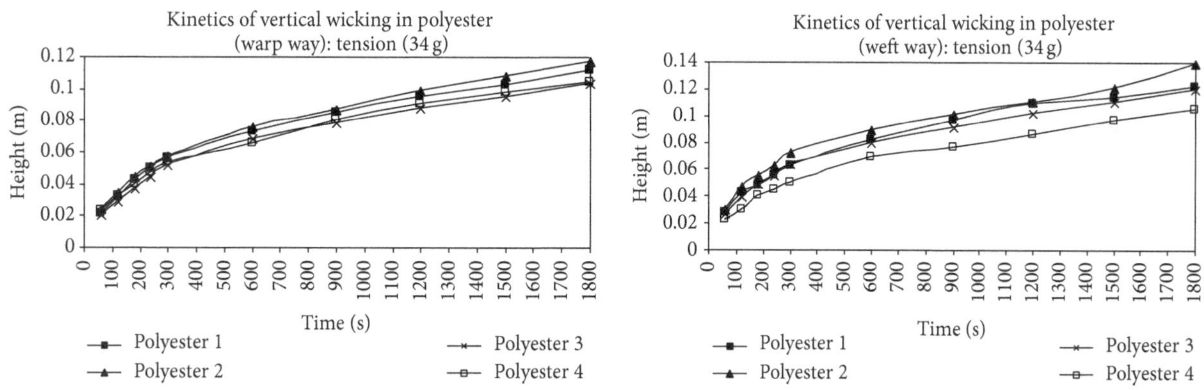

(a) (b)

FIGURE 18: Kinetics of vertical wicking in polyester at 34 g, (a) warp way and (b) weft way.

decrease (Table 4). Macrocapillaries are responsible for short term wicking less will be the capillary radius less would be the initial rate and hence less wicking height. Moreover, the liquid which is retained in these capillaries will decrease. Liquid retained in capillaries acts as reservoir and helps in increasing the wicking height. Due to this reason as we move on to higher pick density the wicking height is reduced.

(ii) With change in pick density crimp in yarn will change. The more the pick density is, the more the crimp will be there in yarn and the more the tortuosity will be (19). With an increase in the tortuosity of the capillaries, its wicking potential is reduced.

Fabric with pick density 30 does not follow the abovementioned trend. This may be due to the fact that in this fabric

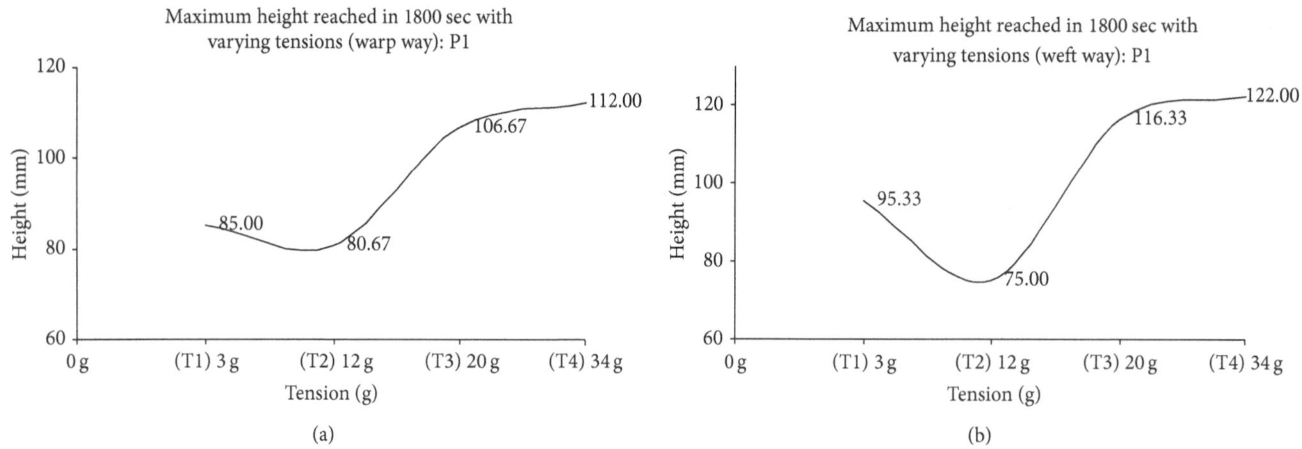

FIGURE 19: Maximum height reached with varying tension, (a) warp way and (b) weft way (P1).

TABLE 3: Maximum height reached in 1800 seconds (polyester).

	P1	P2	P3	P4
Theoretical values warp way (m)	0.053	0.044	0.049	0.051
Theoretical values weft way (m)	0.050	0.043	0.050	0.052
Experimental values warp way (m)	0.085	0.082	0.083	0.075
Experimental values weft way (m)	0.095	0.098	0.098	0.088

TABLE 4: Micro- and macrocapillary radius (polyester).

	P2	P3	P4
Microcapillary radius R_{mi} (m)	$5.037E-6$	$6.247E-6$	$6.359E-6$
Warp macrocapillary radius $e_{mac\ warp}$ (m)	$5.292E-4$	$5.000E-4$	$5.000E-4$
Weft macrocapillary radius $e_{mac\ weft}$ (m)	$6.350E-4$	$5.080E-4$	$4.230E-4$

TABLE 5: Thickness of polyester fabrics.

	P1	P2	P3	P4
Experimental thickness (m)	0.000535	0.000493	0.000544	0.000549
Theoretical thickness (m)	0.000526	0.000508	0.00053	0.000486

the stress generated at yarn intersection point is less due to smaller number of picks per inch and because of this the structure of yarns is closer to cylindrical structure. This assumption is also supported by the measured thickness values of the fabric and theoretical value of thickness which is calculated based on Pierce's Geometry (Table 5). The thickness of the fabric P1 is close to that of the fabric P3 both experimentally and theoretically.

In actual experiment the change in wicking height with variation in pick density may be attributed to the effect of the above-mentioned factors either individually or collectively.

3.4. Effect of Change in Tension on Polyester. Wicking height increases with increase in tension in P2, P3, and P4 fabrics (Figure 15). Due to presence of twist in yarn, when tension was applied on fabric, the fibres inside the yarn will become compact (due to lateral pressure developed) and hence more parallel along yarn axis. Because of this dimension of microcapillary radius is decreasing, so when strip with tension attached at bottom is immersed in water liquid will wick in these small capillaries formed because of higher capillary pressure developed since capillary pressure is inversely related to micro capillary radius.

At higher tension, dimensional change in capillaries is negligible as is observed in Figures 19, 20, 21, and 22 that from 20 g tension to 34 g tension. As at higher tension or after critical tension micro and macro capillaries available for wicking to take place are becoming constant.

Behaviour of P1 fabric with increase in tension is different from other fabrics. In Figure 22 maximum wicking height at 12 g tension is less than 3 g. As 3 g load is less to bring any change in arrangement of fibres inside yarn and in yarn itself. Due to smaller number of yarns in warp (EPI 48), and weft (PPI 30) way, structure of yarn in fabric is more similar to cylindrical structure. And when tension is increased to 12 g the cylindrical structure becomes flat due to application of load; hence, micro- and macrocapillaries that existed are distributed and may lead to increase in capillary radius due to which less wicking is observed in this fabric at 12 g tension. From 12 g to 20 g and 34 g this fabric is behaving in the same way as that of P2, P3, and P4.

4. Conclusions

Wicking takes place through microcapillaries and macrocapillaries in fabric. It is difficult to analyse the interconnection between micro- and macrocapillaries, so kinetics for microcapillary and macrocapillary was studied at different wicking

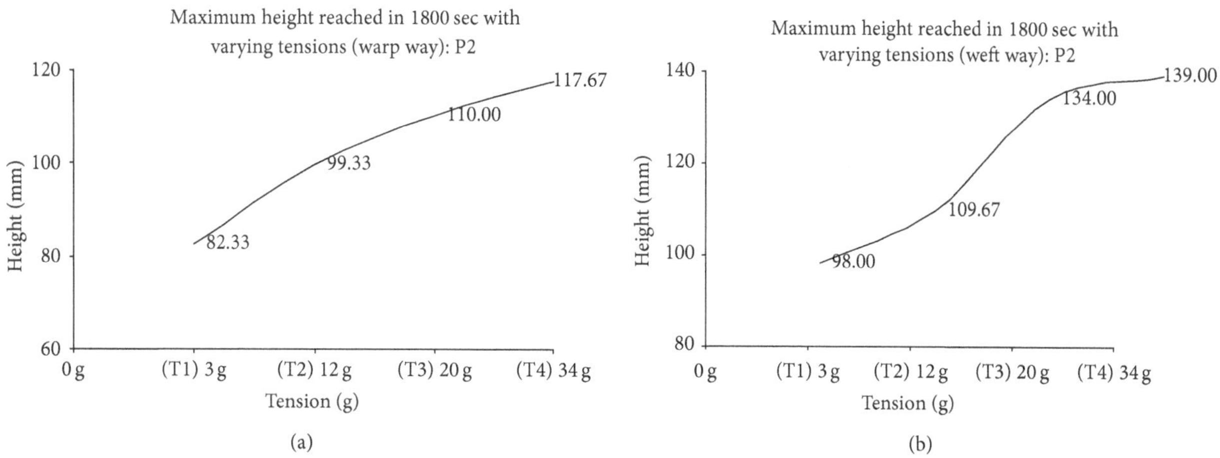

FIGURE 20: Maximum height reached with varying tension, (a) warp way and (b) weft (P2).

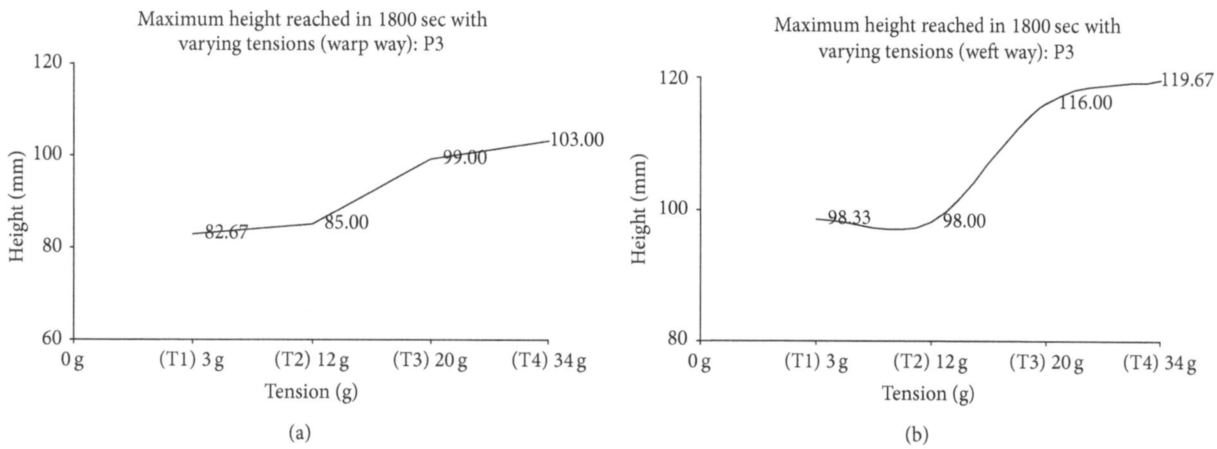

FIGURE 21: Maximum height reached with varying tension, (a) warp way and (b) weft way (P3).

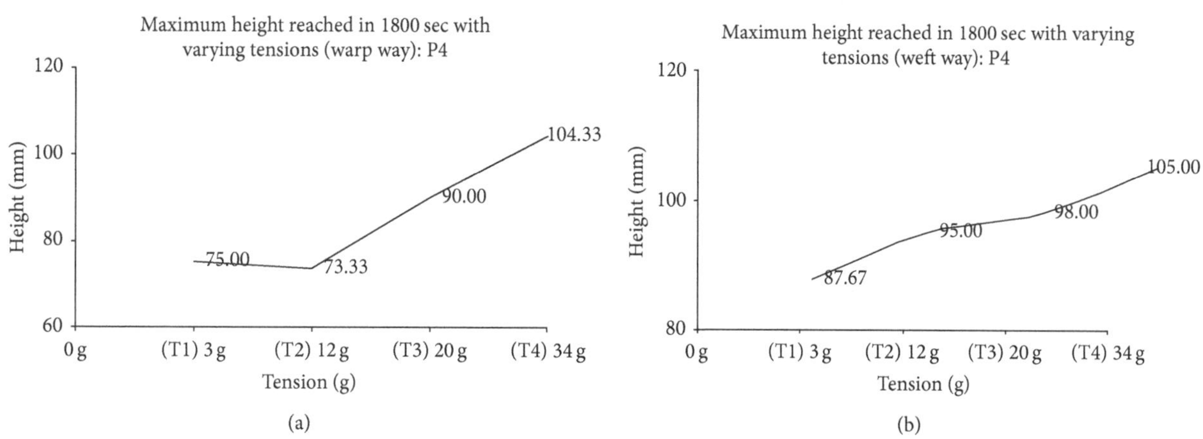

FIGURE 22: Maximum height reached with varying tension, (a) warp way and (b) weft way (P4).

moments by using Lucas-Washburn equation. Macrocapillaries are responsible for short term wicking and microcapillaries are responsible for long term wicking reaching maximum height with slow diffusion rate. Tension influences the results of the wicking test. With increase in tension wicking height increases and after a critical tension no further change is observed. Actual structure of textile is more complex than an idealized assembly of cylinder, so exact prediction is difficult. Further refinement of the equations is necessary.

Conflict of Interests

The authors declare that there is no conflict of interests regarding the publication of this paper.

References

[1] B. V. Zhmud, F. Tiberg, and K. Hallstensson, "Dynamics of capillary rise," *Journal of Colloid and Interface Science*, vol. 228, no. 2, pp. 263–269, 2000.

[2] D. Knittel and E. Schollmeyer, "Notes on future developments for textile finishing processes," *The Journal of The Textile Institute*, vol. 91, part 3, pp. 166–167, 2000.

[3] E. Kissa, "Wetting and wicking," *Textile Research Journal*, vol. 66, no. 10, pp. 660–668, 1996.

[4] K. C. Parsons, *Human Thermal Environments*, Taylor & Francis, London, UK, 1993.

[5] B. Das, A. Das, V. Kothari, R. Fanguiero, and M. D. Araujo, "Moisture flow through blended fabrics—effect of hydrophilicity," *Journal of Engineered Fibers and Fabrics*, vol. 4, no. 4, pp. 20–28, 2009.

[6] N. R. S. Hollies, M. M. Kaessinger, and H. Bogaty, "Water transport mechanisms in textile materialsl—part I: the role of yarn roughness in capillary-type penetration," *Textile Research Journal*, vol. 26, pp. 829–835, 1956.

[7] N. R. S. Hollies, M. M. Kaessinger, B. S. Watson, and H. Bogaty, "Water transport mechanisms in textile materials—part II: capillary-type penetration in yarns and fabrics," *Textile Research Journal*, vol. 27, pp. 8–13, 1957.

[8] S. C. Amico and C. Lekakou, "Mathematical modelling of capillary micro-flow through woven fabrics," *Composites A*, vol. 31, no. 12, pp. 1331–1344, 2000.

[9] S. C. Amico and C. Lekakou, "Axial impregnation of a fiber bundle—part 1: capillary experiments," *Polymer Composites*, vol. 23, no. 2, pp. 249–263, 2002.

[10] S. C. Amico and C. Lekakou, "Axial impregnation of a fiber bundle—part 2: theoretical analysis," *Polymer Composites*, vol. 23, no. 2, pp. 264–273, 2002.

[11] P. K. Chatterjee, *Absorbent Technology*, Elsevier Scientific, New York, NY, USA, 1985.

[12] Y. K. Kamath, S. B. Hornby, H. D. Weigmann, and M. F. Wilde, "Wicking of spin finishes and related liquids into continuous filament yarns," *Textile Research Journal*, vol. 64, no. 1, pp. 33–40, 1994.

[13] A. B. Nyoni and D. Brook, "Wicking mechanisms in yarns—the key to fabric wicking performance," *Journal of the Textile Institute*, vol. 97, no. 2, pp. 119–128, 2006.

[14] A. Perwuelz, P. Mondon, and C. Caze, "Experimental study of capillary flow in yarns," *Textile Research Journal*, vol. 70, no. 4, pp. 333–339, 2000.

[15] A. Perwuelz, M. Casetta, and C. Caze, "Liquid organisation during capillary rise in yarns—influence of yarn torsion," *Polymer Testing*, vol. 20, no. 5, pp. 553–561, 2001.

[16] E. W. Washburn, "The dynamics of capillary flow," *Physical Review*, vol. 17, no. 3, pp. 273–283, 1921.

[17] T. Liu, K.-F. Choi, and Y. Li, "Wicking in twisted yarns," *Journal of Colloid and Interface Science*, vol. 318, no. 1, pp. 134–139, 2008.

[18] B. Sofien, F. Faten, and B. Sassi, "Capillary rise in macro and micro pores of jersey knitting structure," *Journal of Engineered Fibres and Fabrics*, vol. 3, no. 3, pp. 47–54, 2008.

[19] H. Ito and Y. Muraoka, "Water transport along textile fibers as measured by an electrical capacitance technique," *Textile Research Journal*, vol. 63, no. 7, pp. 414–420, 1993.

[20] S. S. Ramkumar, A. Purushothaman, K. D. Hake, and D. D. Mcalister III, "Relationship between cotton varieties and moisture vapor transport of knitted fabrics," *Journal of Engineered Fibers and Fabrics*, vol. 2, no. 4, 2007.

[21] S. Irandoukht and A. Irandoukht, "Development of the predictive models for the fabric water vapor resistance," *Journal of Engineered Fibers and Fabrics*, vol. 6, no. 2, pp. 40–49, 2011.

[22] S. Lee and S. K. Obendorf, "Statistical modeling of water vapor transport through woven fabrics," *Textile Research Journal*, vol. 82, no. 3, pp. 211–219, 2012.

[23] B. P. Saville, *Physical Testing of Textiles*, The Textile Institute, Woodhead, Cambridge, UK, 2000.

4

Colour Fading of Textile Fabric by Plasma Treatment

H. F. Cheung, C. W. Kan, C. W. M. Yuen, J. Yip, and M. C. Law

Institute of Textiles and Clothing, The Hong Kong Polytechnic University, Hung Hom, Kowloon 852, Hong Kong

Correspondence should be addressed to C. W. Kan; tccwk@inet.polyu.edu.hk

Academic Editor: Phillip W. Gibson

Colour fading of a reactive dye (C.I. Reactive Blue 19) dyed textile fabric was performed by atmospheric pressure plasma (APP) treatment with the use of plasma jet. Under the APP treatment condition of treatment time = 5 sec/mm; ignition power = 160 W; oxygen concentration = 1%; jet distance = 3 mm, significant colour-fading effect was achieved. For comparison purpose, the reactive dye dyed textile fabric was subjected to conventional enzymatic colour-fading process. Experimental results revealed that the APP-induced colour-fading effect was comparable with conventional enzymatic colour-fading process.

1. Introduction

Recently, textile manufacturers start fading textile products which become very popular amongst young customers [1]. Therefore, companies have been trying in recent years to develop various techniques to improve the visual aspect of fabrics especially the faded looks [2–4]. The conventional technologies involve creating designs by fading the colour in certain areas of textile fabric using the processes such as sanding, sand blasting, brushing, prewashing, rinsing, stone washing, sand washing, snow washing, stone washing with enzymes, and bleaching [5, 6]. Although desired fading effect could be achieved by such methods, the following problems were encountered: (i) difficulty in application and time consuming due to problem in work flow; (ii) decrease in the wear resistance of the product; (iii) inability to create standard and reproducible designs; (iv) successful application of designs is not possible on all textile surfaces; (v) inability to create required nuances in shading; (vi) inability to produce identical fading effect on both sides of the products; (vii) loss of quality [7–9]. In addition, the production of faded looks in fabric using conventional technologies involves large amount of water and most of which being highly contaminated by chemical products used in the process. Also the time-consuming and old-fashioned processes are not suitable for mass production and increase the production cost [7–9]. In order to cope with these problems, a novel system that does not have the drawbacks involved in the conventional technologies should be adapted for treating the textile fabrics. The plasma treatment, being a dry treatment, would be an alternative to conventional technologies. This paper will investigate the use of atmospheric pressure plasma treatment for fading colour from textile fabrics.

2. Experimental

2.1. Materials. 100% cotton single jersey-knitted fabric was used (fabric weight = 220 g/m^2; yarn count = 32 S/2). The cotton fabric was dyed with C.I. Reactive Blue 19 under industrial condition. After dyeing, the cotton fabric was conditioned at 65 ± 2% relative humidity and 21 ± 1°C for 24 hours prior to all experiments.

2.2. Atmospheric Pressure Plasma (APP) Treatment. APP colour-fading treatment was conducted by an atmospheric pressure plasma jet (APPJ, Surfx Technologies LLC, CA, USA) with a rectangular nozzle. After glow plasma was generated by radio frequency of 13.56 kHz. Helium (He) and oxygen (O_2) were used as carrier and reactive gas, respectively. Plasma operation parameters were set as jet distance = 3 mm, oxygen concentration = 1%, treatment time = 5 sec/mm (treating 1 mm of fabric with 5 seconds), and ignition power = 160 W. Schematic diagram of the experimental setup was shown in Figure 1.

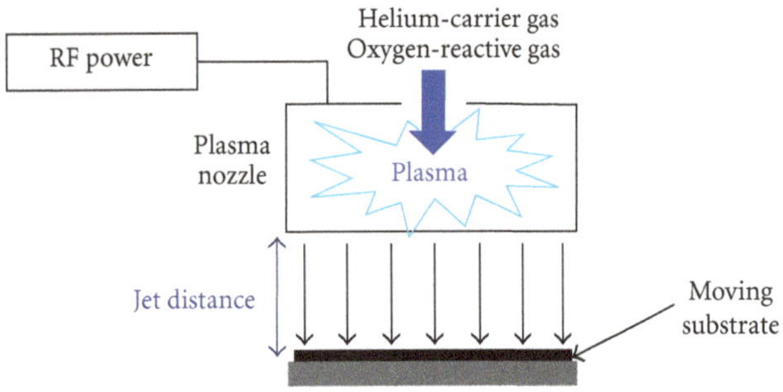

FIGURE 1: Schematic diagram of APP treatment.

2.3. Enzymatic Colour-Fading Process. Knitted fabric was cut into size of 22 cm × 22 cm and was conditioned at 65 ± 2% relative humidity and 21 ± 1°C for at least 24 hours prior to enzyme treatment. A neutral enzyme (Lava Cell NNM, DyStar) with activity 130 EGU/g was used for treating the fabric. Enzyme treatment was carried out using a Launder-Ometer (Atlas Material Testing Technology LLC, Chicago, USA) which was rotated at a speed of 40–45 rpm. The fabric was treated with 0.1% (on weight of fabric) enzyme with a liquor ratio of 50 : 1 at 50°C for 30 minutes. The pH value of the enzyme treatment was set at 7.5 with buffer solution, and mechanical agitation was simulated by using 50 stainless steel balls of 6 mm diameter. After enzyme treatment, the fabrics were rinsed with deionized water to wash away abraded fibres attached on fabrics. The fabrics were then washed thoroughly with nonionic detergent (1 g/L) with a liquor ratio 50 : 1 at 80°C for 10 minutes to deactivate the enzyme action and washed away unfixed dye from fabric surface. After that, the fabrics were rinsed with running water, squeezed to remove excessive water, and were dried completely in an oven at a temperature of 90°C. After drying, the fabrics were placed for conditioning at 65 ± 2% relative humidity and 21 ± 1°C for at least 24 hours before further evaluations.

2.4. Measurement of Weight Change. The weight of fabric samples before and after treatment was measured by Shimadzu BX300 meter after conditioning at 21 ± 1°C and relative humidity of 65 ± 2% for 24 hours. The percent change of weight was calculated, after five measurements, as in the following:

$$\text{Weight change } (\%) = \frac{W - W_o}{W_o} \times 100, \tag{1}$$

where W_o is the initial weight, and W is the weight of the fabric sample after treatment.

2.5. Measurement of Colour. Colour properties of samples were measured by Macbeth Color Eye 7000A spectrophotometer. The conditions for measurement were specular excluded, large aperture with 10° observer, and illuminant D_{65}. The samples were measured twice, and colour yields (expressed by K/S value) were obtained from the reflectance curve which were calculated according to (2). Finally, the sum value of K/S (K/S Sum) was obtained by individual K/S values over the visible spectrum ranging from wavelength of 400 nm to 700 nm. The higher the K/S Sum value, the higher the colour yield would be

$$\frac{K}{S} = \frac{(1 - R)^2}{2R}, \tag{2}$$

where K = absorption coefficient, depending on concentration of the colorant; S = scattering coefficient, caused by the dyed substrate; and R = reflectance of the coloured sample.

In addition, CIE colour coordinates, that is, L^* (lightness and darkness), a^* (redness and greenness), b^* (yellowness and blueness), and ΔE (total colour difference) were measured.

TABLE 1: Weight change of different samples.

Fabric sample	Percentage weight change (%)*
APP-induced colour fading	−5.0
Enzymatic colour fading	−10.0

*Negative value refers to weight loss.

3. Results and Discussion

3.1. Weight Change. Table 1 shows that percentage of weight changes after different treatments. The negative value in percentage weight change indicates that weight loss after treatment. In the enzymatic colour-fading, the action of enzyme and mechanical agitation will abrade fibre surface, and cotton fibre would be released resulting in weight reduction. As a result, after enzyme treatment, weight loss was observed because of the enzymatic hydrolysis which could not be avoided, but a high degree of enzymatic hydrolysis was not expected [10]. The hydrolysis is required to obtain the desired effect of colour removal. On the other hand, weight loss was also observed in the APP-induced colour-fading process. During APP treatment, the fibre surface was subjected to localised ablation, and only surface etching would be resulted. On the whole, APP treatment introduces lower percentage weight loss than the enzyme treatment because the APP treatment affects only the surface of the material and no effect on the bulk properties of the materials.

3.2. Colour Properties. The K/S Sum value is related to the concentration of the colorant in the fabric. K/S Sum values of different fabrics are shown in Table 2. From the results, it is noted the differently treated fabrics have lower K/S Sum values than the untreated fabric. Generally speaking, the higher the K/S Sum value, the higher the colour yield will be, and it can conclude that a paler shade was obtained after different treatments. The APP-treated fabric has a paler shade than the untreated fabric. The reason is that during the APP treatment, the oxygen plasma would oxidize the dyes in the fabric surface leading to a colour-fading effect [11]. However, in the case of enzymatic colour-fading treatment, the enzyme in the aqueous medium can penetrate effectively into the interior part of fabric. The enzymatic hydrolysis induced by enzyme in the fabric would be occurred [12, 13]. More fibres on fabric are hydrolysed by enzyme, and the weaken fibres

TABLE 2: Colour properties.

Fabric sample	K/S Sum	L^*	a^*	b^*	$\Delta E^{\#}$
Untreated	435.63	22.94	1.66	−3.20	—
APP-induced colour fading	387.23	25.55	1.49	−2.63	2.68
Enzymatic colour fading	383.95	25.38	1.72	−3.78	2.51

$^{\#}\Delta E = [(\Delta L^*)^2 + (\Delta a^*)^2 + (\Delta b^*)^2]^{1/2}$.
$\Delta L^* = L^*_{\text{treated fabric}} - L^*_{\text{untreated fabric}}$.
$\Delta a^* = a^*_{\text{treated fabric}} - a^*_{\text{untreated fabric}}$.
$\Delta b^* = b^*_{\text{treated fabric}} - b^*_{\text{untreated fabric}}$.

are further removed by abrasion of fabric and mechanical agitation. Thus, more dye particles are removed with the cotton fibres. As a result, the enzyme-treated fabric gives a paler shade than the APP-treated fabric.

CIE colour coordinates include colour qualities in terms of L^* (lightness and darkness), a^* (redness and greenness), and b^* (yellowness and blueness) of the fabrics are shown in Table 2. The higher the L^* value is, the lighter the sample is. The L^* values of differently treated fabrics were increased which indicates that the fabric lightness was increased after both APP and enzymatic colour-fading treatments. Since APP treatment is physically a surface etching process, the surface-protruding fibre in the fabric surface would be removed together with the dye leading to a smooth surface. In the case of enzyme, there are two factors contributing the increase in lightness induced by enzyme treatment. The first factor is that the enzyme action gives reactive-dyed cotton fabric an aged appearance (i.e., biostoning), while the second factor is to give cotton fabric a revived appearance by removing the surface fibre (i.e., depilling or biopolishing). Both factors contribute to change the lightness of the enzyme-treated cotton fabric. During the action of enzyme and mechanical agitation, the depilling or biopolishing effect occurs at the fabric surface, and consequently the surface fibres were removed [14]. At the same time, the action of enzyme on the fibre surface would increase the smoothness and evenness. The smooth surface induced by enzyme would enhance the regular reflection of light from the fabric surface, and therefore the lightness was increased.

The a^* value represents redness and greenness of the fabric sample. The more positive the value of a^* is, the redder the shade of the sample will be. On the contrary, the value of a^* being towards negative indicates the greenness of the shade of the sample. It is noted that APP-treated fabric have a reduction in a^* value which indicates that fabric will turn greener than the untreated fabric. During the APP treatment, thermal oxidation effect may occur in the fabric surface leading to certain degree of yellowness [15]. The thermal induced yellow colour together with the original blue colour in the fabric surface would result in increased greenish effect. On the contrary, the a^* value of enzyme-treated fabric increased slightly which indicates the redder shade was obtained after enzyme treatment. During the enzyme treatment, the hydrolysis of the cellulose would occur, and blue reactive dye would be released from the fibre and re-deposited on the fabric surface leading to a further colouring of fabric surface which may be termed as backstaining [16]. As a result, the a^* value increased.

In general, the b^* value describes the yellowness and blueness of a sample. The higher the positive value of b^* is, the more will be the yellowish of the sample. On the contrary, negative value of b^* refers to the bluish colour of the sample. It is revealed that the APP treatment increases the b^* value because under the effect of oxygen plasma not only the reactive dye is removed due to the impact of active plasma particles (ions, free radicals, etc) with fibre [11], but also the thermal oxidation in fabric surface would induce certain degree of yellowness [15]. As a result, a yellowish shade would be imparted after APP treatment. With the use of enzyme treatment, the b^* values of enzyme-treated fabric decrease further, that is, the b^* values of the enzyme-treated fabrics become more negative which indicates that the enzyme treatment can increase the bluish shade in the fabric. During the enzyme treatment, the hydrolysis of cellulose would occur, and the blue dye would be released from the fibre and redeposited on the surface of the fabric leading to backstaining [16]. As a result, b^* value decreased.

Table 2 shows the ΔE values of differently treated fabrics. ΔE value contains the information of colour depth, shade, and the hue of a sample. ΔE value is calculated by using the CIE L^*, a^*, and b^* values with the equation $\Delta E = (\Delta L^{*2} + \Delta a^{*2} + \Delta b^{*2})^{1/2}$ where ΔL^*, Δa^*, and Δb^* values were the difference between L^*, a^*, and b^* values of a pair of colour standard and sample. The larger the ΔE value, the greater will be the colour difference between this pair of colour standard and sample. It was noted after the APP and enzyme treatments that colour difference was appeared between the control and the treated fabrics. This colour difference was expected because after the APP and enzyme treatment, changes in $^*L$, a^*, and b^* values were noted, and these three values would contribute the total colour difference values. However, it is interesting to note that the APP-induced and enzymatic colour-fading process obtain similar ΔE values. This finding can support the argument to use APP treatment as an alternative to conventional enzyme treatment to create colour-fading effect on textile fabric.

4. Conclusion

In the present study, the dyed cotton fabrics were subjected to APP-induced colour-fading process, and the colour-fading effect was compared with conventional enzymatic colour-fading process. Experimental results revealed that APP treatment could achieve similar colour-fading effect when compared with conventional enzymatic colour-fading process without much loss in fabric weight and with shorter treatment time. Therefore, it could be concluded that with the careful control of the APP process parameters, APP treatment could be flexible as an alternative to conventional chemical-based colour-fading process to be used in textile industry.

Acknowledgment

Authors would like to thank the financial support from the The Hong Kong Polytechnic University for this work.

References

[1] A. Card, M. A. Moore, and M. Ankeny, "Performance of garment washed denim blue jeans," *AATCC Review*, vol. 5, no. 6, pp. 23–27, 2005.

[2] M. Sariisik, "Use of cellulases and their effects on denim fabric properties," *AATCC Review*, vol. 4, no. 1, pp. 24–29, 2004.

[3] A. Cavaco-Paulo, "Mechanism of cellulase action in textile processes," *Carbohydrate Polymers*, vol. 37, no. 3, pp. 273–277, 1998.

[4] N. Ozdil, E. Ozdooan, and T. Oktem, "Effects of enzymatic treatment on various spun yarn fabrics," *Fibres and Textiles in Eastern Europe*, vol. 11, no. 4, pp. 58–61, 2003.

[5] R. Mori, T. Haga, and T. Takagishi, "Bending and shear properties of cotton fabrics subjected to cellulase treatment," *Textile Research Journal*, vol. 69, no. 10, pp. 742–746, 1999.

[6] A. S. Aly, A. B. Moustafa, and A. Hebeish, "Bio-technological treatment of cellulosic textiles," *Journal of Cleaner Production*, vol. 12, no. 7, pp. 697–705, 2004.

[7] C. Q. Yang, W. Zhou, G. C. Lickfield, and K. Parachura, "Cellulase treatment of durable press finished cotton fabric: effects on fabric strength, abrasion resistance, and handle," *Textile Research Journal*, vol. 73, no. 12, pp. 1057–1062, 2003.

[8] A. Cavaco-Paulo and M. Jose Rios, "Mechanical properties of cellulase-treated fabrics analyzed," *American Dyestuff Reporter*, vol. 86, no. 9, pp. 20–28, 1997.

[9] M. Montazer and A. S. Maryan, "Application of laccases with cellulases on denim for clean effluent and repeatable biowashing," *Journal of Applied Polymer Science*, vol. 110, no. 5, pp. 3121–3129, 2008.

[10] W. Rakowski, "Plasma treatment of wool today. Part 1: fibre properties, spinning and shrinkproofing," *Journal of the Society of Dyers and Colourists*, vol. 113, no. 9, pp. 250–255, 1997.

[11] M. Ghoranneviss, B. Moazzenchi, S. Shahidi, A. Anvari, and A. Rashidi, "Decolorization of denim fabrics with cold plasmas in the presence of magnetic fields," *Plasma Processes and Polymers*, vol. 3, no. 3, pp. 316–321, 2006.

[12] B. Dhurai and V. Natarajan, "Effect of enzyme softening of cotton hosiery yarns on knittability and dimensional properties of weft knitted fabrics," *Indian Journal of Fibre and Textile Research*, vol. 32, no. 2, pp. 241–247, 2007.

[13] C. W. Kan, C. W. M. Yuen, S. Q. Jiang, W. S. Tung, and S. Y. Cheng, "Influence of enzymatic treatment on the properties of linen," *Journal of Applied Polymer Science*, vol. 104, no. 1, pp. 286–289, 2007.

[14] K. J. Vishnu Vardhini and N. Selvakumar, "Biopolishing enzymes and their influence on the properties of cotton materials—a review," *Colourage*, vol. 53, no. 12, pp. 45–52, 2006.

[15] C. W. Kan, C. W. M. Yuen, and W. Y. Tsoi, "Using atmospheric pressure plasma for enhancing the deposition of printing paste on cotton fabric for digital ink-jet printing," *Cellulose*, vol. 18, no. 3, pp. 827–839, 2011.

[16] C. W. Kan and W. Y. Wong, "Color properties of cellulase-treated cotton denim fabric manufactured by torque-free ring spun yarn," *Textile Research Journal*, vol. 81, no. 9, pp. 875–882, 2011.

5

Ultraviolet Protection by Fabric Engineering

Mukesh Kumar Singh[1] and Annika Singh[2]

[1] *Uttar Pradesh Textile Technology Institute (Formerly Known as GCTI), Kanpur 208001, India*
[2] *Institute of Biosciences and Biotechnology, CSJMU, Kanpur, India*

Correspondence should be addressed to Mukesh Kumar Singh; mukesh70ster@gmail.com

Academic Editor: Anindya Ghosh

Background. The increasing emission of greenhouse gases has evoked the human being to save the ozone layer and minimize the risk of ultraviolet radiation (UVR). Various fabric structures have been explored to achieve desired ultraviolet protection factor (UPF) in various situations. *Objective.* In this study, the effect of various filament configurations like twisted, flat, intermingled, and textured in multifilament yarns on fabric in different combinations is assessed in order to engineer a fabric of better ultraviolet protection factor (UPF). *Methods.* In order to engineer a fabric having optimum UV protection with sufficient comfort level in multifilament woven fabrics, four different yarn configurations, intermingled, textured, twisted, and flat, were used to develop twelve different fabric samples. The most UV absorbing and most demanding fibre polyethylene terephthalate (PET) was considered in different filament configuration. *Results.* The combinations of intermingled warp with flat, intermingled, and textured weft provided excellent UVR protection comparatively at about 22.5 mg/cm^2 fabric areal density. The presence of twisted yarn reduced the UV protection due to enhanced openness in fabric structure. *Conclusion.* The appropriate combination of warp and weft threads of different configuration should be selected judiciously in order to extract maximum UV protection and wear comfort attributes in multifilament woven PET fabrics.

1. Introduction

Prolong and frequent exposure of human being against sun causes different dermatological problems [1]. The short-term exposure to ultraviolet radiation (UVR) causes sunburn and in medical science it is erythema [2, 3]. Prolong sunburn leads to photoageing of skin and results in terms of both nonmelanoma and melanoma skin cancer [4]. The specific band of UV radiation (100–400 nm) can be classified into three groups on the basis of wavelength. The UVA region of light belongs to span of wavelength 315–400 nm while the UVB radiation comes in the range of 280–315 nm [5]. The region 100–280 nm is extremely dangerous and belongs to UVC radiation. Divinely, nature has stratospheric ozone layer in atmosphere to absorb UVB and UVC and to block it to reach on earth surface [6]. The intensity of UVR is the highest in Australia and some part of eastern and southern Europe, so protection of adolescents and children (skin thickness is very less) than adults and outdoor workers is very essential [7].

When UVR strikes on textile surfaces, part of it can be transmitted, absorbed, and reflected by the textiles as shown in Figure 1. Different textile fibres have the tendency to absorb a part of UVR and convert it into a different energy form. Another part of radiation is reflected or scattered by the fibr itself and this may be a part of transmitted radiation, because another part of UVR transmits directly through the fabric via interstices between fibres and yarns in the fabrics [8, 9]. Growing evidence of unusual depletion of stratospheric ozone has encouraged the different research group globally to take new research programme to focus on different aspects of UVR. The Australian radiation laboratory (ARL) is credited to track back the research of UVR protection in 1980s [10]. It is evident that the clothing provides UV protection up to some extent. Clothing is associated with several factors which governs the effectiveness of UVR protection. In addition to the fibrousmaterial, the fabric properties like fabric sett, fibre blend composition, tightness of weave, thickness, and areal density also affects the UPF of resultant fabrics [11–15]. Wet processing of textiles (bleaching, usage of UV

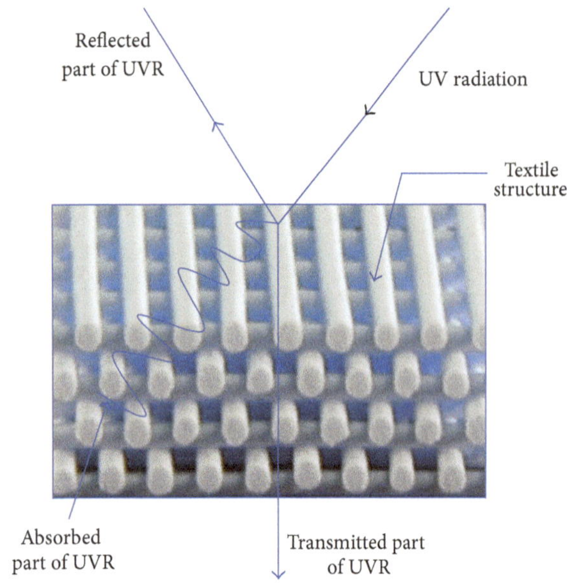

Figure 1: UV radiation and textile structure.

Table 1: UPF rating and protection grades.

Protection grade	Rating	UPF range	Transmission (%)
Good	No value	20–29	5.0–3.4
Very good	30	30–40	3.3–2.5
Excellent	40	>40	<2.5

absorber, and coloration) also influences the UVR protection function of textiles and some refurbishment activities like wetness, stretch, heat, or chemical treatments [16–21]. The combined effect of these parameters (fibre, yarn, fabric, and wet processing treatment) complicates the subject of UVR protection by clothing. The physicochemical type of fibrous material and fabric openness are other driving parameters to decide the UVR protection of clothing. It has been found that fibres containing conjugated aromatic system like PET are more effective for UVR absorption [9, 14, 22]. The cellulosic fibres having no double bond in their molecular backbone have low UVR absorption capacity, exhibiting comparatively low UPF of textiles made thereof. Natural colour pigment and waxes in natural fibres act as UV absorber. Naturally coloured cotton has higher UV absorption capacity. Different group of scientists conducted that UVR transmission occur through textile through interstices between fibre and yarns and this has been supported by a reduced free space between yarns in the fabric after subsequent washing and consequent increase in ultraviolet protection factor (UPF). In spite of the established fact that yarn structure can influence the UPF of fabric by altering the openness of fabric. Most of the research work was concerned about the influence of yarn linear density but there is no systematic research associated with the UV transmission behaviour of fabric composed of multifilament yarns of different filament configurations [23, 24]. Various UPF ratings and grades are compiled in Table 1.

In this research the efforts were focused to evaluate the effect of filament configuration on UV protection of plain woven fabric for suiting applications. In order to avoid the influence of bleaching dyeing and subsequent finishing treatments, the present research was performed in scoured and desized state of fabrics.

2. Purpose of Study

Various yarn configurations are used to produce woven fabrics which influence the yarn and fabric compactness in different ways. Hence, to understand the role of various filament configurations like intermingled, textured, flat, and twisted a systematic and scientific study is concluded in this paper. Findings of this work will definitely helpful to produce PET multifilament yarns based woven fabric of higher UV protection.

3. Quantitative Assessment of Fabrics for Ultraviolet Protection Factor (UPF)

Spectrophotometer was used to assess the percent transmission at wavelength intervals upto 5 nm in the 290–400 nm spectral span. The UPF is calculated by the following equation [25]:

$$\text{UPF} = \frac{\sum_{\lambda=290}^{\lambda=400} e(\lambda) \cdot s(\lambda) \cdot \Delta(\lambda)}{\sum_{\lambda=290}^{\lambda=400} e(\lambda) \cdot s(\lambda) \cdot \tau(\lambda) \cdot \Delta\lambda}, \quad (1)$$

where $E(\lambda)$ is the spectral weighting function of erythemal action spectra, $S(\lambda)$ spectral irradiation for appropriate solar radiation spectrum ($W\,\text{m}^{-2}\,\text{nm}^{-1}$), $\tau(\lambda)$ spectral transmittance through specimen, and $\Delta(\lambda)$ appropriate wavelength measuring interval (nm).

4. Materials and Methods

Intermingled, textured, and flat yarns (220/78/00) and twisted yarn (220/78/680) were used to produce 12 different fabric samples. Various combinations of warp and weft considered for this study are shown in Table 2 along with fabric code. The textured yarn was not used as warp because it is not practically feasible to use this yarn as warp. The intermingled yarn was characterized in terms of nip frequency. The nips were counted by placing a free length of yarn on stagnant water surface. The filaments were spread out and then nips become visible. Finally nips were counted and expressed in nips per meter. The typical diagrammatic conformations of four different filament configurations are shown in Figures 2(a) to 2(d).

All warp beams were produced by direct warping method on Tsudakoma TW/N warping machine. Total numbers of ends were kept 1200. Parallel creel was used with ceramic tensioner by keeping the tension 14 to 18 grams. Flat warps were sized on Tsudakoma KS200 sizing machine with 5% poly vinyl alcohol (PVA) and 0.25% antistatic to reduce the static charge generating tendency of PET filament. Twelve fabric samples were manufactured using various combinations of warp and weft of flat, twisted, intermingled, and textured configuration on Nissan NW-155 Water Jet looms under identical operational conditions to keep fabric areal

Table 2: Fabric manufacturing details.

Fabric detail (warp/weft)	Sample code	Warp versus weft combinations	Ends per cm (EPC)	Picks per cm (PPC)	Fabric areal density (mg/cm^2)	Fabric thickness (mm)
FI/I	S1	Intermingled versus intermingled	32	26	22.21	0.61
FI/Tx	S2	Intermingled versus textured	32	26	22.61	0.67
FI/F	S3	Intermingled versus flat	32	26	22.11	0.62
FI/Tw	S4	Intermingled versus twisted	32	26	22.58	0.58
FTw/I	S5	Twisted versus intermingled	32	26	22.18	0.57
FTw/Tx	S6	Twisted versus textured	32	26	22.38	0.64
FTw/F	S7	Twisted versus flat	32	26	22.06	0.58
FTw/Tw	S8	Twisted versus twisted	32	26	22.95	0.60
FF/I	S9	Flat versus intermingled	32	26	22.02	0.59
FF/Tx	S10	Flat versus textured	32	26	22.14	0.62
FF/F	S11	Flat versus flat	32	26	22.15	0.56
FF/Tw	S12	Flat versus twisted	32	26	22.17	0.58

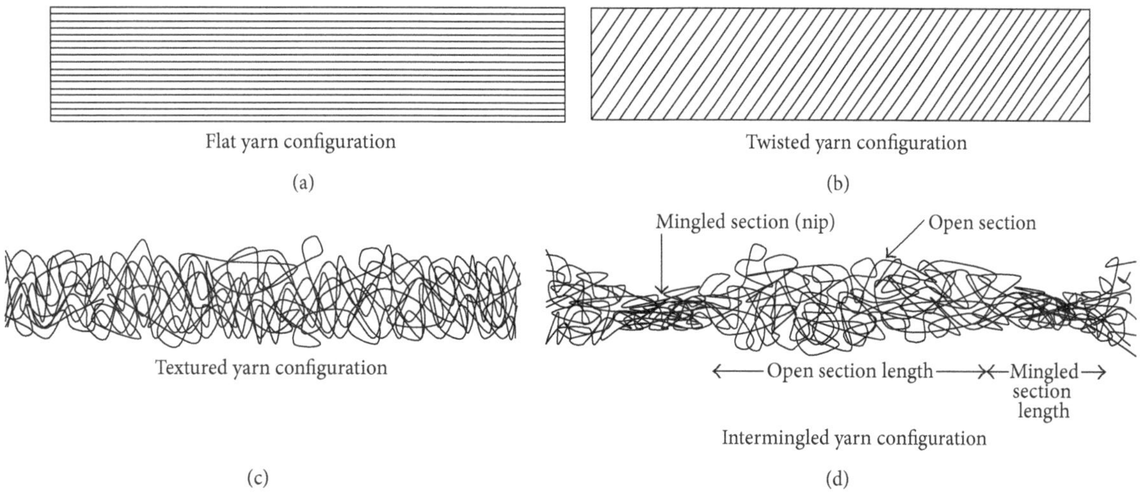

Figure 2: Constituent filament conformations in (a) flat, (b) twisted, (c) textured, and (d) intermingled yarns.

density 22 mg/cm^2. The details of fabric constructions are given in Table 2.

The woven fabrics were scoured and desized in Star Hisaka Jet Dyeing Machine by keeping temperature 90°C for 60 minutes and used 2 gpl LISAPOL N nonionic detergent and 2 gpl Finocon-FBOL as desizing agent. After scouring and desizing, fabric samples were passed through Tex Print Stenter at 40 meters per minute speed by keeping the temperature 180°C. The fabric areal density was maintained at about 120 gram per square meter.

For tensile testing, multifilament yarns were mounted on Instron tensile testing machine (Instron 4200) as per ASTM D-3822-01 with special jaw to prevent any slippage during loading by taking specimen length 500 mm. The jaw speed was adjusted to ensure the yarn breaking time 20 ± 2 sec as per ASTM D-3822-01. The tenacity at maximum, percent strain, and Young's modulus were obtained based on 64 readings. Air permeability was measured by the TEXTEST FX 3300 air permeability tester. The testing was carried out using test area 5.08 cm^2 and test pressure 100 pa. An average of 30 observations is reported here. UVF of various fabric samples is measured by using EN 13758-1 : 2002.

5. Results

Tensile attributes of four different yarns of different configurations are shown in Table 3. The tensile and boiling water response of all four yarns having intermingled, textured, flat, and twisted configurations are shown in Table 3. The tensile and boiling behaviour of yarns having different configuration clearly indicates the effects of filament configuration. The difference in strength and strain % is not considerable. The filament configuration mainly affects the initial modulus of the yarns as clearly shown in Table 3.

Open porosity is an established governing factor to influence the fabric transmission behaviour, such as heat flow, air, and moisture permeability. The fabric porosity depends on fabric construction, thickness, and surface profile. All twelve

Table 3: Tensile behaviour of different configured yarns.

Yarn sample	Max. tensile strength (cN/Tex)	Strain (%)	Initial modulus (cN/Tex)	Boiling water shrinkage (%)
Intermingled	25.53	16.24	486.6	7.1
Textured yarn	26.12	17.97	688.8	3.8
Flat	26.92	15.05	660.7	7.5
Twisted	27.24	14.32	591.6	1.8

fabric samples were woven under identical conditions as mentioned in the second table. The differences in basic manufacturing characteristics originated from four basic configurations of yarn flat, twisted, intermingled and textured. The change in filament configuration in multifilament yarn results in varying porosity as well as thickness. The thickness of fabric samples woven with textured weft was slightly increased and this may be attributed to the highly bulky configuration of textured yarns. The filament configuration influences the openness of fabric structure which is evident from air permeability data (Table 4). The fabric openness based on the total volume of interconnected voids both inside and between the yarns. The four filament yarn configurations were used in this study varying in their openness index drastically. So, it is difficult to measure the actual yarn density because of the fact that yarn is never uniform in terms of its diameter although the packing index of various multifilament yarns varies with change in yarn configurations. At higher twist level yarn surface becomes more irregular and prevents a close packing between the yarns namely higher porosity. The mechanism of intermingled yarn formation consisted mainly of loops and nodes while textured yarn consisted of wavy structure of constituent filaments. The structural features of intermingled, textured, and twisted yarns decided the compactness of the woven structure. The differences in the yarn surface characteristic are clearly identified in pictorial model Figure 2 and the results suggested that these differences influenced the transmission behaviour of woven fabrics.

In order to perceive the effect of multifilament yarn configuration on porosity, the air permeability of the woven structures was measured since it is considered that majority air flow takes place between yarn interstices. As per the results regarding air permeability (Table 4) of woven fabrics of nearly same areal density produced from high twist yarn (S8) was the most permeable. The flat yarn without any twist enabled the least permeability in fabric sample S11 while the fabric sample designated as S6 enabled highest air permeability.

Evidently, the compactness of the flat yarn structure reduces the interyarn stices wherever they found. In consequence of that the fabric sample S11 (FF/F) consists of both flat warp and weft exhibited lowest air permeability amongst different fabric samples. It looks here that at least two elements are opposed to each other: fabric geometry and filament configuration in multifilament yarns or in otherworld size of pores and number of pores per unit area, as similarly concluded by Stankovic et al., 2009, in case of knitted fabrics [26]. Gooding et al., (1964) [27] also similarly inferred that the increased number of pores per unit area would reduce the air permeability due to higher air drag force to air flow due to increase available specific area inside the total number of pores of reduced size. The uneven surface of intermingled, textured, and flat yarn configurations may also have been a reason for closing the interyarn stices in woven fabric and diminishing the air permeability.

The results obtained by UV protection analysis given in Table 4. It is a general consideration that the majority of interyarn stices and the macroporosity of yarns and fabrics are the major factors influencing the UV transmission through fabrics. There is an old concept that UV transmission is not possible through yarns [8, 10, 11], but recent research inferred that UV transmission is possible through yarns. The UV protection analysis of different fabrics produced from yarns having different configurations suggested that the yarn structure playing a decisive role to decide the UV protection by fab rics. The fabric sample S11 consists of flat warp and flat weft exhibited the lowest air permeability due to the lowest openness and permitted the lowest amount of UVA and UVB radiation to be passed through. The fabric sample S8 consists of twisted warp and weft thread recognized by highest air permeability that translated in terms of highest UVA and UVB transmittance. The fabric sample S1 consists of both warp and weft yarn intermingled configuration (FI/I) exhibited higher UVA and UVB protection than S2 (FI/Tx), that is, fabric made with intermingled warp and textured weft. This may be attributed to the fact that textured configuration has comparatively higher open structure than intermingled configuration in which nodes remain present at regular interval. The UVA and UVB transmittance reduced when weft was flat with intermingled warp in S3 (FI/Tw). This may be attributed again to yarn configuration, that is, flat yarns are most compact among the four considered configurations (intermingled, textured, flat, and twisted). The UV protection again reduced as twisted yarn used as weft. This may be attributed to the fact that twisted yarns leave higher interyarn space than flat yarn [26]. The twisted warp was used with different weft thread of varying configurations from S5 to S8; the UV protection followed the trend S7 > S5 > S6 > S8.

The above results and trend showed that yarn configuration is an important variable to play with UV protection of multifilament woven PET fabrics. The flat yarn with zero twist offered most compact structure which was again configured with least air permeability in S11 fabric samples. This may be attributed to the fact that each filament in zero twist flat yarn is free to tie in free space wherever they get and in consequence of that maximum compactness constituted. The intermingled yarn configuration reached at second position to achieve compactness in fabric structure followed by textured configuration of multifilament PET yarns. This may be explained on the basis of ideal geometry of textured configuration (Figure 2). The textured configuration consists

TABLE 4: Porosity, transmission, and ultraviolet protection factor (UPF) values.

Sample code	Air permeability ($cm^3/cm^2/s$)	Moisture vapour transmittance (%)	Transmission (%)			UPF
			UVA	UVB	UVC	
S1	5.128	67.45	0.83	0.86	0.89	40+
S2	12.92	71.92	0.92	0.95	0.99	40+
S3	3.02	64.94	0.78	0.80	0.80	40+
S4	8.35	69.21	0.87	0.90	0.94	40+
S5	75.45	68.24	1.12	1.17	1.20	39.8
S6	151.04	71.04	1.85	1.99	2.00	33.6
S7	64.17	66.16	1.01	1.08	1.10	40+
S8	129.44	70.16	1.65	1.73	1.78	37.4
S9	2.32	64.53	0.75	0.71	0.69	40+
S10	33.41	69.28	0.99	1.00	1.00	40+
S11	1.46	57.72	0.682	0.693	0.685	40+
S12	12.11	68.63	0.956	0.998	1.006	40+

TABLE 5: Bending and compression behavior of fabric samples.

Fabric code	B	LC	WC
S1	0.045	0.693	0.037
S2	0.039	0.845	0.027
S3	0.244	0.749	0.038
S4	0.051	0.774	0.029
S5	0.048	0.933	0.021
S6	0.055	0.929	0.025
S7	0.212	0.911	0.022
S8	0.072	0.881	0.021
S9	0.079	0.929	0.032
S10	0.043	0.042	0.025
S11	0.383	0.843	0.054
S12	0.047	0.845	0.026

of wavy filament structure, offers higher UV transmission through yarn itself. The twisted yarn is very compact in its structure and leaves maximum porosity as evident by air permeability data in case of fabric sample S8 consists of both twisted warp and weft.

The moisture vapour permeability % is vary important attribute of wearing comfort follow a typical trend and that is, Tw > Tx > I > F (i.e moisture vapour permeability is highest for twisted configurations and lowest for flat configurations. These results indicated that interfibre space and interyarn stices both contribute to decide the moisture vapour transmissibility of fabric samples.

The bending and compression behaviour of fabric samples exhibited that yarn configuration and appropriate combination of warp and weft threads of different structure are decisive factor to engineer it for better fabric hand and compiled in Table 5. The twisted yarn configuration exhibited highest bending rigidity. The bending rigidity of S3 having intermingled warp and flat weft showed higher bending rigidity. Fabric samples S7 (consist of twisted warp and flat weft) and S11 (consist flat warp and flat weft) also showed higher bending toughness as shown in Table 5.

The compression behaviour of fabric is among the prime contributors to decide the sensorial comfort of the fabric samples (Kawabata et al., 1983) [28]. The linearity in compression (LC), compression energy (WC), and resiliency of compression (RC) data (Table 5) for different fabric samples reflect the influence of yarn configuration on compression behaviour of fabric samples. The compression energy of fabric sample S2 is lowest which indicates the lowest resistance against fabric compression while the S11 (F F/F) consists of both flat warp and flat weft threads exhibited highest compression energy, namely indication of difficult to compress.

6. Conclusion

In conclusion, yarn configuration proved to be the remarkable determinant of the UV protection characteristics of woven fabrics through the effectiveness of filament configuration, that is, fibre packing in the multifilament yarn. The appropriate combination of warp and weft threads of different configuration should be selected in order to produce the maximum UV protection and other wear comfort properties of multifilament woven PET fabrics. The mentioned research provides guidelines for engineering multifilament woven PET fabrics to achieve adequate UPF on the basis of yarn configuration of warp and weft threads. In order to achieve ideal UV protective clothing, fabric engineering may prove its crucial role.

Abbreviations

SPF: Sun protection factor
UVR: Ultraviolet radiation reaching the earth's surface, ultraviolet radiation (UVR).

References

[1] D. Saravanan, "UV protection textile materials," *Autex Research Journal*, vol. 7, no. 1, pp. 53–62, 2007.

[2] S. Dornelles, J. Goldim, and T. Cestari, "Determination of the minimal erythema dose and colorimetric measurements as

indicators of skin sensitivity to UV-B radiation," *Photochemistry and Photobiology*, vol. 79, no. 6, pp. 540–544, 2004.

[3] M. K. Singh, "Sun protective clothing," *Asian Textile Journal*, vol. 14, no. 1-2, pp. 91–97, 2005.

[4] I. Algaba, A. Riva, and P. C. Crews, "Influence of fiber type and fabric porosity on the UPF of summer fabrics," *AATCC Review*, vol. 4, no. 2, pp. 26–31, 2004.

[5] European Standard EN 13758-1, "Textiles-Solar Ultraviolet Protective Properties-Part-I: Methods of test for Apparel Fabrics," 2002.

[6] C. R. Roy, H. P. Gies, and S. Toomey, "The solar UV radiation environment: measurement techniques and results," *Journal of Photochemistry and Photobiology B*, vol. 31, no. 1-2, pp. 21–27, 1995.

[7] M. Pailthorpe, "Textile and sun protection: the current situation," *Australasian Textiles*, vol. 14, no. 6, pp. 54–66, 1994.

[8] J. Alvarez and B. Lipp-Symonowicz, "Examination of the absorption properties of various fibres in relation to UV radiation," *Autex Research Journal*, vol. 3, no. 2, pp. 72–77, 2003.

[9] T. Gambichler, A. Avermaete, A. Bader, P. Altmeyer, and K. Hoffmann, "Ultraviolet protection by summer textiles. Ultraviolet transmission measurements verified by determination of the minimal erythema dose with solar-simulated radiation," *British Journal of Dermatology*, vol. 144, no. 3, pp. 484–489, 2001.

[10] T. Gambichler, K. L. Hatch, A. Avermaete, P. Altmeyer, and K. Hoffmann, "Influence of wetness on the ultraviolet protection factor (UPF) of textiles: in vitro and in vivo measurements," *Photodermatology Photoimmunology and Photomedicine*, vol. 18, no. 1, pp. 29–35, 2002.

[11] C. A. Wilson and A. V. Parisi, "Protection from solar erythemal ultraviolet radiation—simulated wear and laboratory testing," *Textile Research Journal*, vol. 76, no. 3, pp. 216–225, 2006.

[12] M. Khazova, J. B. O'Hagan, and K. L. Grainger, "Assessment of sun protection for children's summer 2005 clothing collection," *Radiation Protection Dosimetry*, vol. 123, no. 3, pp. 288–294, 2007.

[13] H. P. Gies, C. R. Roy, G. Elliott, and W. Zongli, "Ultraviolet radiation protection factors for clothing," *Health Physics*, vol. 67, no. 2, pp. 131–139, 1994.

[14] S. Davis, L. Capjack, N. Kerr, and R. Fedosejevs, "Clothing as protection from ultraviolet radiation: which fabric is most effective?" *International Journal of Dermatology*, vol. 36, no. 5, pp. 374–379, 1997.

[15] I. E. S. Clark, K. J. L. Grainger, J. L. Agnew, and C. M. H. Driscoll, "Clothing protection measurements," *Radiation Protection Dosimetry*, vol. 91, no. 1–3, pp. 279–281, 2000.

[16] D. G. Stanford, K. E. Georgouras, and M. T. Pailthorpe, "The effect of laundering on the sun protection afforded by a summer weight garment," *Journal of the European Academy of Dermatology and Venereology*, vol. 5, no. 1, pp. 28–30, 1995.

[17] A. V. Parisi, M. G. Kimlin, L. Mulheran, L. R. Meldrum, and C. Randall, "Field-based measurements of personal erythemal ultraviolet exposure through a common summer garment," *Photodermatology Photoimmunology and Photomedicine*, vol. 16, no. 3, pp. 134–138, 2000.

[18] K. Hoffmann, K. Kaspar, T. Gambichler, and P. Altmeyer, "In vitro and in vivo determination of the UV protection factor for lightweight cotton and viscose summer fabrics: a preliminary study," *Journal of the American Academy of Dermatology*, vol. 43, no. 6, pp. 1009–1016, 2000.

[19] S. Q. Wang, A. W. Kopf, J. Marx, A. Bogdan, D. Polsky, and R. S. Bart, "Reduction of ultraviolet transmission through cotton T-shirt fabrics with low ultraviolet protection by various laundering methods and dyeing: clinical implications," *Journal of the American Academy of Dermatology*, vol. 44, no. 5, pp. 767–774, 2001.

[20] J. H. Xin, W. A. Daoud, and Y. Y. Kong, "A new approach to UV-blocking treatment for cotton fabrics," *Textile Research Journal*, vol. 74, no. 2, pp. 97–100, 2004.

[21] M. Gorensek and F. Sluga, "Modifying the UV blocking of polyester fabric," *Textile Research Journal*, vol. 74, no. 6, pp. 469–474, 2004.

[22] P. C. Crews, S. Kachman, and A. G. Beyer, "Influences on UVR transmission of undyed woven fabrics," *Textile Chemist and Colorist*, vol. 31, no. 6, pp. 17–26, 1999.

[23] G. Hustvedt and P. C. Crews, "Textile technology: the ultraviolet protection factor of naturally-pigmented cotton," *Journal of Cotton Science*, vol. 9, no. 1, pp. 47–55, 2005.

[24] A. Riva, I. Algaba, and M. Pepió, "Action of a finishing product in the improvement of the ultraviolet protection provided by modal and modal sun fabrics: modelisation of the effect," *Fibers and Polymers*, vol. 8, no. 2, pp. 205–211, 2007.

[25] J. M. Menter and K. L. Hatch, "Clothing as solar radiation protection," *Current Problems in Dermatology*, vol. 31, pp. 50–63, 2003.

[26] S. B. Stankovic, D. Popovic, G. B. Poparic, and M. Bizjak, "Ultraviolet protection factor of gray-state plain cotton knitted fabrics," *Textile Research Journal*, vol. 79, no. 11, pp. 1034–1042, 2009.

[27] A. C. Goodings, "Air flow through textile fabrics," *Textile Research Journal*, vol. 34, no. 8, pp. 713–724, 1964.

[28] S. Kawabata, "Recent development in the objective measurement of apparel fabric handle, quality and physical properties," in *Objective Evaluation of Apparel Fabrics*, R. Postle, S. Kawabata, and M. Niwa, Eds., pp. 15–28, Textile Machinery Society of Japan, Osaka, Japan, 1983.

6

A Comparative Study between Physical Properties of Compact and Ring Yarn Fabrics Produced from Medium and Coarser Yarn Counts

Ghada Ali Abou-Nassif

Fashion Design Department, Design and Art Faculty, King Abdul Aziz University, Jeddah, Saudi Arabia

Correspondence should be addressed to Ghada Ali Abou-Nassif; dr_ghada2013@yahoo.com

Academic Editor: Arunangshu Mukhopadhyay

In compact spinning, fibers in drawing stage are condensed using air suction, which results in the produced yarns which significantly differ from ring spun yarns in their physical and mechanical properties. This study compares between physical properties of compact and ring yarn fabrics woven from coarser and medium yarn counts. The experimental results were statistically analyzed using ANOVA. The findings of this study revealed that woven fabrics produced from compact spun yarns are superior to those produced from ring spun yarns with respect to breaking strength, breaking elongation, abrasion resistance, tearing strength, and air permeability.

1. Introduction

Compact spinning, which is a modified version of ring spinning, produces a novel yarn structure, and the development of compact spinning has set new standards in yarn structure [1, 2]. However, compact spinning is oriented to better fiber utilization and the high quality rather than higher productivity [3].

In conventional ring spinning, the zone between the nip line of the pair of delivery rollers and the twisted end of the yarn is called the "spinning triangle." This represents the critical weak spot of the ring spinning process. In this zone, the fiber assembly contains no twist. The edge fibers splay out from this zone and make little or no contribution to the yarn strength. Furthermore, the edge fibers lead to the familiar problem of yarn hairiness [4, 5].

In compact spinning, the fibers which have left the drafting system are guided via the perforated drums or lattice aprons over the openings of the suction slots. Following the air flow, the fibers move sideways and are consequently condensed. This condensing has such a favorable effect on the ratio of the width of the condensed fiber to yarn diameter that the spinning triangle is nearly eliminated. When spinning without a spinning triangle, almost all the fibers are incorporated into the yarn structure under the same tension. As the twist insertion takes place very close to the nip line, even short fibers can take up tension. This results in increased strength, as more fibers contribute to the yarn strength. The utilization degree of fibers can thus be increased [6–11].

According to previous studies, mechanical compact spinning significantly improves yarn tensile properties and reduces its hairiness [12, 13]. Until now there are many studies about the comparison of the conventional ring and compact yarns properties [14–18].

Most researches operate on the comparison between compact and ring yarn woven fabrics focused on fabrics made from finer yarns. The objective of this study emphasizes the comparison between such fabrics woven from medium and coarser yarns.

2. Experimental Work

2.1. Material. Two sets of fabrics were woven from three weft yarn counts, namely, 20, 24, and 30 Ne. These weft yarns were spun from Egyptian cotton of Giza 86 (2.5% span length, 30 mm; fineness, 4.5 micronair; tenacity, 32.7 g/tex;

and breaking elongation, 6%). The weft yarn samples were spun on ring and compact spinning machines with twist factor 4 for all yarn counts. The yarn samples were spun on Reiters' ring frame using conventional and rotorcraft compact spinning mode (ROCOS). Experimental ring and compact spun yarns were separately used to produce plain woven fabrics on Air-jet weaving machine.

The Air-jet weaving machine has the following particulars:

(i) machine model: Toyoda 1998,

(ii) shedding mechanism: electronic dobby,

(iii) number of harness frames: 6,

(iv) machine running speed: 650 rpm,

(v) weave structure: plain 1/1,

(vi) fabric width: 160 cm,

(vii) warp width: 165 cm,

(viii) filling density: 60 picks/inch,

(ix) warp density: 66 ends/inch,

(x) weft yarn counts: 20, 24, and 30/1 Ne,

(xi) warp yarn count: 24/1 Ne.

2.2. Laboratory Testing. Since the variation of the fabric samples was in the weft yarns and in spinning type, all tests were performed in the weft direction. Before testing, all fabric samples were conditioned for 24 hours under the standard atmospheric conditions (20 ± 2°C temperatures, 65 ± 2% relative humidity).

Breaking strength and elongation of all test specimens were measured on an Instron tensile strength tester model 4411 according to ASTM standard D 2256. The tests were carried out using 150 mm gauge length at an extension rate 300 mm/min. Twenty observations were taken for each fabric sample and then averaged. Air permeability tests were conducted on air permeability tester according to ASTM standard D737-96. Before testing, tension was applied on the fabric specimen to eliminate the wrinkle. Recorded air flow was divided by the exposed area of the test specimen. Exposed area of fabric samples was 10 cm^2 under the pressure drop of 5 inch of water column. The average of twenty readings was taken for each woven fabric sample.

The abrasion resistances of the fabrics samples were tested on the Martindale pilling and abrasion tester at 9 kPa pressure according to BS EN ISO 12947-2. A circular specimen (38 mm), mounted in a specimen holder and subjected to a defined load, is rubbed against an abrasive medium. The evaluation of the abrasion resistance is determined from the mass loss of specimen after specific amount (10,000 cycles) of rubs. Tearing strength of the woven fabric samples was determined by using Elmendorf tearing tester in accordance with ASTM standard test method D 1424. A template was used to cut fabric strips of 100±2 mm length and 63±0.15 mm width. The critical dimension, that is, the distance to be torn, was taken as 43 ± 0.15 mm and the observation was recorded from the scale of the tester.

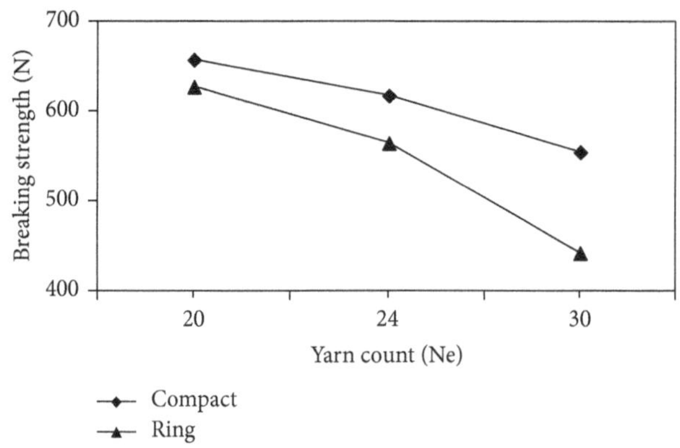

FIGURE 1: Effect of yarn counts and fabric type on the woven fabric breaking load.

3. Results and Discussion

3.1. Breaking Load. The breaking load test results are shown in Figure 1 and the statistical results of the breaking load property are given in Table 1. The statistical analysis proved that spinning type and yarn count have a significant influence on the woven fabrics' breaking load at 0.01 significance level. It is shown that the weft yarn count was related inversely to the woven fabrics' breaking load. As the yarn count increases, the fabrics' breaking load decreases. The negative effect of weft yarn count on the breaking strength of the woven fabric may be due to the lower tensile strength of the constituent yarns with the increase in English count of the weft yarns.

The results also indicated that the fabrics produced with compact spun yarns have better tensile strength results compared to the fabrics produced with conventional ring spun yarns. The difference between the tensile strength of the compact and ring yarn fabrics is more pronounced in the case of finer yarns, that is, yarn count 30 Ne. However, in coarser yarns, compact spun yarns show a slight increase in tensile results compared to conventional ring spun yarns. The statistical analysis revealed that the average values of breaking load of compact yarn woven fabrics are 657, 616, and 554 cN for weft yarns of counts 20, 24, and 30 Ne, respectively, whilst in the case of ring yarns woven fabrics the mean breaking load values of fabrics woven from weft yarn counts 20, 24, and 30 Ne are 627, 566, and 441 cN, respectively. The superior breaking strength of compact yarn fabrics over ring yarn ones may be ascribed to the exploitation of all fibers in the compact yarn cross-section which in turn increases the tensile properties of such yarns and then increases the breaking load of compact yarn fabrics.

3.2. Breaking Elongation. According to the statistical analysis' results listed in Table 2, the effect of weft yarn count and spinning type is significant on breaking elongation of woven fabrics under study at significance level 0.05. The breaking elongation of ring and compact yarn fabrics against the weft yarn counts is plotted in Figure 2. As seen from this figure, the breaking elongation of ring and compact yarn fabrics is inversely related to the weft yarn counts. As the weft

Table 1: Analysis of variance results for the effects of spinning type and yarn count on fabric breaking load.

Source of variation	SS	df	MS	*F*	*P* value	*F* crit
Yarn count	107.0126	2	53.50632	1433.845	0.000697	19.00003
Spinning type	36.35882	1	36.35882	974.3318	0.001025	18.51276
Error	0.074633	2	0.037317			
Total	143.4461	5				

Table 2: Analysis of variance results for the effects of spinning type and yarn count on fabric breaking elongation.

Source of variation	SS	df	MS	*F*	*P* value	*F* crit
Yarn count	72.33333	2	36.16667	31	0.03125	19.00003
Spinning type	28.16667	1	28.16667	24.14286	0.039012	18.51276
Error	2.333333	2	1.166667			
Total	102.8333	5				

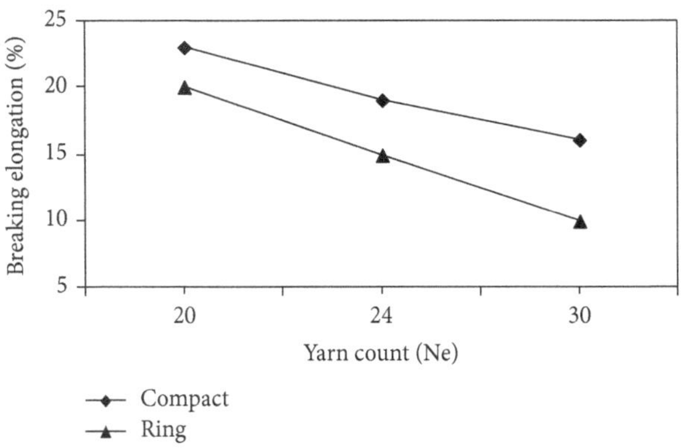

Figure 2: Effect of yarn counts and fabric type on the woven fabric breaking elongation.

yarn counts increase the breaking elongation of both fabrics decreases. The inverse effect of yarn counts on fabric breaking elongation can be attributed to the decrease in yarn diameter with the increase of its count, which in turn leads to the reduction of breaking elongation of the constituent yarns, which reflected in lower fabric breaking elongation.

It is also observed that breaking elongation of compact yarn fabrics is higher than those produced from ring yarns. It is also apparent that the difference between breaking elongation of such fabrics is more pronounced at finer weft yarns. At coarser weft yarns the breaking elongation of compact yarn fabrics and that of ring yarn fabrics are close to each other. The statistical results revealed that the mean values of breaking elongation of compact and ring yarn fabrics are 23, 19, and 16% and 20, 15, and 10% for weft yarn counts 20, 22, and 24 Ne, respectively. The higher breaking elongation of compact yarn fabrics in comparison with ring yarn ones may be attributed to the higher elongation of compact yarns because of its exploitation of all fibers in the yarn cross section, which in turn leads to higher elongation of compact yarn fabrics.

3.3. Tearing Strength. Tearing strength is one of the important aspects of finished fabrics. It refers to the rupture of a fabric progressively along a line thread by thread. Tearing strength mainly depends on fiber, yarn, and fabric characteristics. The analysis of variance for the effects of weft yarn count and spinning type was listed in Table 3. From this table it is noticed that weft yarn count and spinning type have a significant influence on tearing strength of the woven fabrics at 0.05 significance level. The effect of weft yarn count on tearing strength of both compact and ring yarn fabric was shown in Figure 3. It is shown that weft yarn count has a negative effect on the tearing strength of both types of fabrics. As the weft yarn count increases the tearing strength decreases. It is also observed that the fabrics made from compact spun yarns are more tear resistant than those made of ring spun yarns. The tearing resistance of these fabrics is found to be higher in weft direction. The mean values of tearing strength of compact yarn fabrics are 4200, 3900, and 3600 gm for weft yarn counts 20, 24, and 30 Ne, respectively. The average values of tearing strength of ring yarn fabrics were found to be 3900, 3510, and 3020 gm for weft yarn counts 20, 24, and 30 Ne, respectively.

3.4. Abrasion Resistance. Abrasion is the mechanical deterioration of fabric components by rubbing them against another surface. Abrasion ultimately results in the loss of performance characteristics, such as strength, but it also affects the appearance of a fabric. The abrasion resistance of textile materials is affected by many factors (e.g., fiber fineness, yarn count, yarn type, and weave) in a very complex and as yet little understood manner.

In this study abrasion resistance of compact and ring yarn fabrics was characterized by the percent mass loss of the fabrics after 10000 rubs. As the mass loss increases the abrasion resistances decrease. The analysis of variance for the effects of spinning type and weft yarn count on the abrasion resistance of the woven fabrics under study was tabulated in Table 4. From this table it is noticed that spinning type and weft yarn count have a profound effect on fabric abrasion resistance at 0.05 significance level.

The effect of weft yarn count on percent mass loss of compact and ring yarn fabrics is illustrated at Figure 4. From this figure a decreasing trend is detected confirming

TABLE 3: Analysis of variance results for the effects of spinning type and yarn count on fabric tearing strength.

Source of variation	SS	df	MS	F	P value	F crit
Yarn count	548433.3	2	274216.7	26.84013	0.035919	19.00003
Spinning type	268816.7	1	268816.7	26.31158	0.035968	18.51276
Error	20433.33	2	10216.67			
Total	837683.3	5				

TABLE 4: Analysis of variance results for the effects of spinning type and yarn count on fabric tearing strength.

Source of variation	SS	df	MS	F	P value	F crit
Yarn count	39	2	19.5	39	0.025	19.00003
Spinning type	13.5	1	13.5	27	0.035099	18.51276
Error	1	2	0.5			
Total	53.5	5				

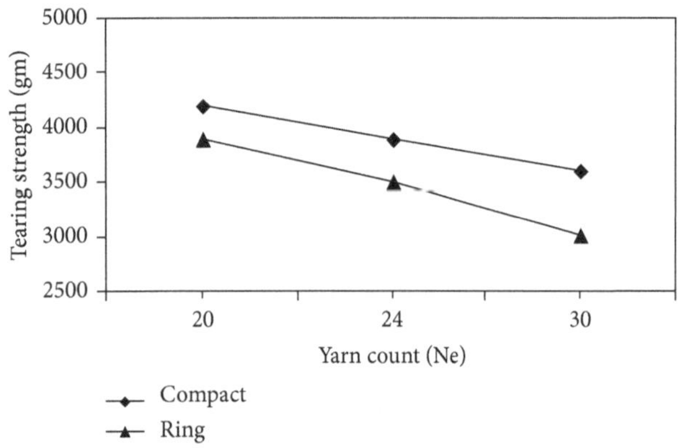

FIGURE 3: Effect of yarn counts and fabric type on the woven fabric tearing strength.

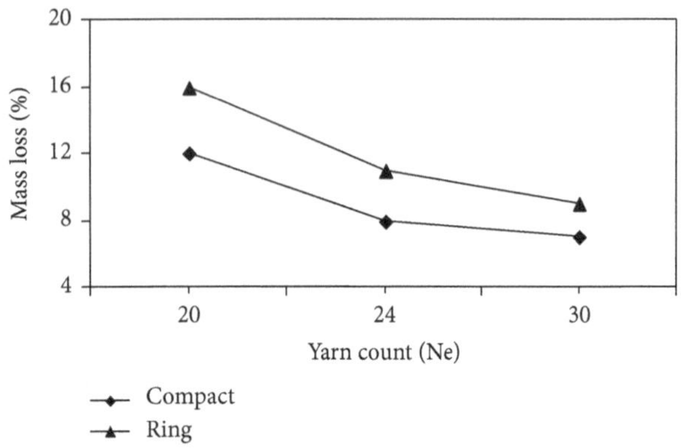

FIGURE 4: Effect of yarn counts and fabric type on the woven fabric abrasion resistance.

that as the weft yarn count increases the percent mass loss of the woven fabrics decreases. It was also clearly showed that woven fabrics made from compact yarn showed less mass loss after 10,000 cycles than the conventional ring based fabric. This means that compact yarn fabrics are more resistant to abrasion than those made from ring spun yarns. The statistical analysis proved that the percent mass loss of compact yarn fabrics made from yarn counts 20, 24, and 30 Ne is 12%, 8%, and 7%, respectively, whereas in the case of ring yarn fabrics the percent mass loss was 16%, 11%, and 9% for yarn counts 20, 24, and 30 Ne, respectively. Compact yarns based fabrics are expected to have lower friction values because of their uniform and less hairy structures and as a result higher abrasion resistance is expected. In contrast to the previous results, the difference between abrasion resistance of compact and ring yarn fabrics is more pronounced at coarser yarns.

3.5. Air Permeability. Air permeability is mainly affected by two parameters, porosity and fabric thickness. Since the fabric structure is the same for all fabric samples, plain weave, the fabric thickness will only affect the air permeability of the woven fabrics. Fabric thickness depends mainly on the yarn diameter, which in turn relies on yarn counts. Thus the only parameters which could affect the air permeability in this study are the yarn counts and yarn type.

The statistical analysis for the effects of spinning type and weft yarn count on the air permeability of the woven fabrics under study was tabulated in Table 5. From this table it is noticed that spinning type and weft yarn count have a significant influence on air permeability of such fabrics at 0.05 significance level.

The effects of weft yarn count on air permeability of compact and ring yarn fabrics are depicted in Figure 5. It can be seen from this figure that the relation between yarn count and air permeability of the woven fabrics under study is a direct relationship. As the filling yarn count increases the air permeability of compact and ring yarn fabrics increases. As the yarn count increases the fabric thickness decreases, which in turn leads to passage of much air through fabric cross section. It is also shown that woven fabrics made from compact spun yarns are more permeable than those made from ring spun yarns. This is because the compact spinning process helps for compacting yarn structure and thus leads to finer yarn cross section which in turn leads to more spaces between yarns in the fabric cross section.

The difference between the air permeability of compact and ring yarn fabrics is more pronounced in the case of

TABLE 5: Analysis of variance results for the effects of spinning type and yarn count on fabric air permeability.

Source of variation	SS	df	MS	F	P value	F crit
Yarn count	330.3333	2	165.1667	39.64	0.024606	19.00003
Spinning type	160.1667	1	160.1667	38.44	0.025042	18.51276
Error	8.333333	2	4.166667			
Total	498.8333	5				

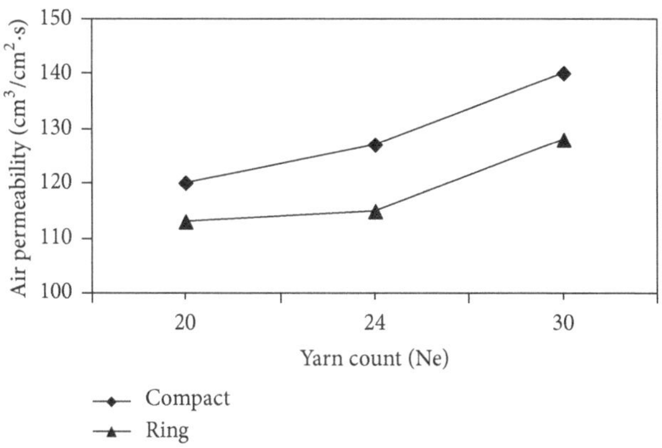

FIGURE 5: Effect of yarn counts and fabric type on the woven fabric air permeability.

medium yarn counts, while at coarser yarns, the air permeability of compact yarn fabrics and that of ring yarn fabrics are close to each other as shown from Figure 5. The statistical analysis proved that the average values for air permeability of compact yarn fabrics are 120, 127, and 140 (cm^3/cm^2·sec) for weft yarn counts 20, 24, and 30 Ne, respectively. The corresponding air permeability values for ring yarn counts are 113, 115, and 128 (cm^3/cm^2·sec) for counts 20, 24, and 30 Ne, respectively.

4. Conclusions

Due to the difference in the yarn structure between compact and ring spun yarns, the physical and mechanical properties of fabrics woven from them will also be different. In this study physical and mechanical properties of fabrics woven from compact and ring spun yarns at coarser and medium counts were studied. The conclusion can be drawn as follows.

(i) There is a significant difference between the physical properties of compact and ring yarn fabrics in relation to breaking strength, breaking elongation, tearing strength, abrasion resistance, and air permeability.

(ii) It was found that compact yarn fabrics were superior to ring yarn fabrics in all these properties because of their compact and fineness compared to corresponding ring yarns.

(iii) The difference between compact and ring yarn fabrics was more pronounced at finer counts, whereas for coarser ones the properties of such fabrics are comparable approximately, and any one of the spinning systems can replace the other.

Conflict of Interests

The author declares that there is no conflict of interests regarding the publication of this paper.

References

[1] P. Artzt, "Short staple spinning on the way to new yarn structures and better raw material utilisation," *International Textile Bulletin*, vol. 45, no. 4, pp. 16–23, 1999.

[2] H. Stalder, "New spinning process ComforSpin," *Melliand International*, no. 6, pp. 22–25, 2000.

[3] U. Meyer, *Compact Yarns: Innovation as a Sector Driving Force*, Melliand International, 2000.

[4] G. Egbers, "The future of spinning and weaving," *Melliand International*, no. 3, pp. 34–36, 1999.

[5] W. Kampen, "The advantages of condensed spinning," *Milliand International*, no. 4, pp. 58–59, 2000.

[6] P. Artzt, "Possibilities of improving the efficiency of compact spinning," *Melliand English*, no. 3, pp. 19–21, 2002.

[7] K. P. S. Cheng and C. Yu, "A study of compact Spun Yarns," *Textile Research Journal*, vol. 73, no. 4, pp. 345–349, 2003.

[8] A. Hellwig, "Compact yarn in weaving—a new fashion or a must for high-quality fabrics," *Melliand English*, vol. 83, no. 6, pp. 84–85, 2002.

[9] A. Olbrich, "The Air-Com-Tex 700 condenser ring spinning machine," *Melliand International*, vol. 6, pp. 26–29, 2000.

[10] J. Smekal, "Air-Com-Tex 700 for compact spinning yarns," *Melliand International*, vol. 7, no. 1, pp. 18–19, 2001.

[11] M. A. Saad and A. A. Almetwally, "Spinning techniques vs. yarn properties," *Textile Asia*, vol. 39, no. 7, pp. 35–41, 2008.

[12] Y. Beceren and B. U. Nergis, "Comparison of the effects of cotton yarns produced by new, modified and conventional spinning systems on yarn and knitted fabric performance," *Textile Research Journal*, vol. 78, no. 4, pp. 297–303, 2008.

[13] S. Ganesan, A. Venkatachalam, and V. Subramaniam, "Fibre migration in compact spun yarns: part II—mechanical compact yarn," *Indian Journal of Fiber & Textile Research*, vol. 32, no. 2, pp. 169–172, 2007.

[14] J. R. Dash, S. M. Ishtiaque, and R. Alagirusamy, "Properties and processibility of compact yarns," *Indian Journal of Fibre and Textile Research*, vol. 27, no. 4, pp. 362–368, 2002.

[15] T. Jackowski, D. Cyniak, and J. Czekalski, "Compact cotton yarn," *Fibres and Textiles in Eastern Europe*, vol. 12, no. 4, pp. 22–26, 2004.

[16] M. Nikolić, Z. Stjepanovič, F. Lesjak, and A. Štritof, "Compact spinning for improved quality of ring-spun yarns," *Fibres and Textiles in Eastern Europe*, vol. 11, no. 4, pp. 30–35, 2003.

[17] S. Mavruz and R. T. Ogulata, "Statistical investigation of properties of ring and compact yarns and knitted fabrics made of these kinds of yarns," *Tekstil ve Konfeksiyon*, vol. 18, no. 3, pp. 197–205, 2008.

[18] A. A. Almetwally and M. M. Salem, "Comparison between mechanical properties of fabrics woven from compact and ring spun yarns," *Autex Research Journal*, vol. 10, no. 1, pp. 35–40, 2010.

7

Dyeing Studies with *Eucalyptus*, Quercetin, Rutin, and Tannin: A Research on Effect of Ferrous Sulfate Mordant

Rattanaphol Mongkholrattanasit,[1] Charoon Klaichoi,[2] Nattadon Rungruangkitkrai,[3] Nattaya Punrattanasin,[4] Kamolkan Sriharuksa,[4] and Monthon Nakpathom[5]

[1] *Department of Textile Chemistry Technology, Faculty of Industrial Textiles and Fashion Design, Rajamangala University of Technology Phra Nakhon, 517 Nakhonsawan Road, Kwang Suan Chitladda, Dusit District, Bangkok 10300, Thailand*
[2] *Department of Textile Product Design, Faculty of Industrial Textiles and Fashion Design, Rajamangala University of Technology Phra Nakhon, 517, Nakhonsawan Road, Kwang Suan Chitlada, Dusit District, Bangkok 10300, Thailand*
[3] *Department of Textile Science, Faculty of Agro-Industry, Kasetsart University, 50 Phaholyotin Road, Chatuchak District, Bangkok 10900, Thailand*
[4] *Department of Textile Science and Technology, Faculty of Science and Technology, Thammasat University, Rangsit Center, Pathumthani 12121, Thailand*
[5] *National Metal and Materials Technology Center, National Science and Technology Development Agency, Pathumthani 12120, Thailand*

Correspondence should be addressed to Rattanaphol Mongkholrattanasit; rattanaphol.m@rmutp.ac.th

Academic Editor: Nabil Ibrahim

Natural dyes from *Eucalyptus* leaf extract, quercetin, rutin, and tannin were applied to silk fabric by pad-batch and pad-dry techniques under different conditions. Ferrous sulfate was used as a mordant. The dyeing properties were evaluated by measuring *K/S* and *CIELAB* values. In addition, the different fastness properties were evaluated. The effect of dyes at different concentration levels with respect to their colour strength was also studied.

1. Introduction

Natural dyes are comprised of dyes and pigments that are obtained from various parts of plants including roots, bark, leaves, flowers, and fruit [1]. The major applications are coloring of food, leather, wood, and natural fibers like wool, silk, cotton, and flax. Natural dyes may have a wide range of shades. Dyeing with natural dyes, however, normally requires the use of mordants, which are metallic salts of aluminum, iron, chromium, and copper, among others, to ensure a reasonable color fastness to sunlight and washing [2]. The metal ions of these mordants can act as electron acceptors for electron donors to form coordination bonds with the dye molecules, making them insoluble in water [3]. Lately, there has been increasing interest in natural dyes, as the public is becoming more aware of the ecological and environmental problems related to the use of synthetic dyes. The use of natural dyes cuts down significantly on the amount of toxic effluent resulting from synthetic dye processes. Natural dyes have also been used for printing and dye-sensitized solar cells [4–7]. Alongside their aesthetic qualities, natural dyes may offer other benefits, being antibacterial, deodorizing, and UV protective [8–14]. The use of natural dyes in textile applications is growing rapidly, reflecting the strict environmental standards being established in many countries, and the concern about the health hazards associated with synthetic dyes, for example, the recent ban on the use of azo dyes by the European Union, has also increased the scope for the use of natural dyes [15].

Eucalyptus is one of the most important sources of natural dye, yielding several yellowish-brown colorant. *Eucalyptus* has ample natural tannins and polyphenols ranging from 10% to 12% [16]. The major coloring component of *Eucalyptus* bark is quercetin, which is also an antioxidant. It has been used as a food dye with strong antioxidant properties [17]. *Eucalyptus* leaves contain up to 11% of the major components of tannin

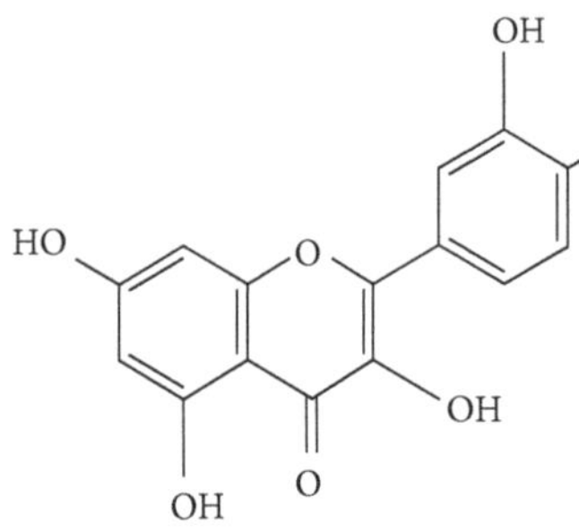

Quercetin (C.I. 75670)

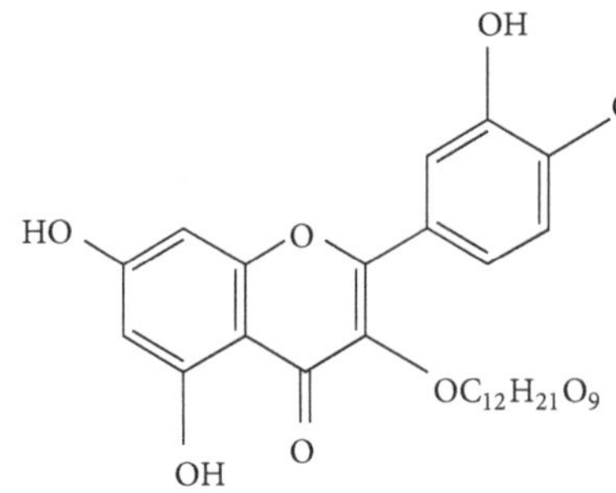

Rutin (C.I. 75730)

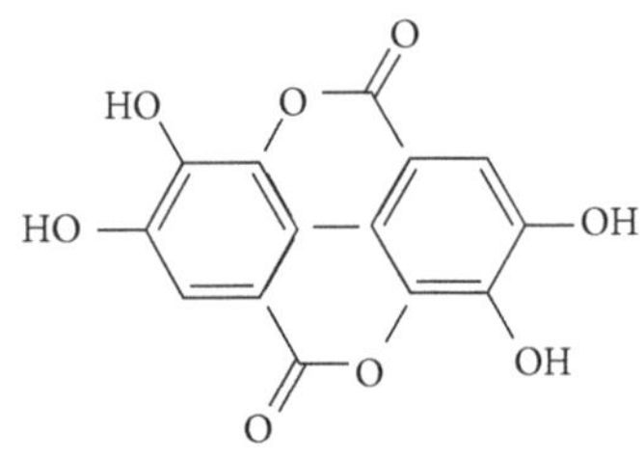

Ellagic acid (C.I. 75270)

FIGURE 1: Color composition of *Eucalyptus* leaf extract dye.

(gallic acid and ellagic acid) with flavonoids (quercetin, rutin, etc.) as minor substances [18–20]. Tannins and flavonoids are considered to be very useful substances during the dyeing process because of their ability to fix dyes within fabrics. The structures of the important coloring components of *Eucalyptus* leaves are given in Figure 1.

Vegetable tannin extracts contain a variety of amorphous materials including polyphenolic tannins of large relative molecular mass (M_r), such as hydrolysable gallotannin, and tannic acid, as well as non-tannins, such as flavones and gums [21]. Because tannins have a large M_r and are water-soluble phenolic compounds, they undergo typical phenolic reactions, notably the chelation of metal ions. Tannins have been used in textiles for several thousands of years, as exemplified by the dyeing of cotton and silk with dyewoods, in which the tannin is "fixed" by a metal salt (e.g., $CuSO_4$) employed as a mordant [22]. Burkinshow [23] used tannins as mordants to increase the uptake of cationic dyes (e.g., mauvein) onto cotton by firstly applying tannin to the cotton, followed by "fixing" of tannin by the application of metal (Fe, Al, Cu, Pb, or Sn) salts.

Flavonoids (polyphenolic pigments) are widely present in plants. Rutin (3,3′,4′,5,7-pentahydroxyflavone-3-rham-no-glucoside) and quercetin (3,3′,4′,5,7- pentahydroxyl-flavone) are phenolic compounds derived from hydroxyl substitutions on a flavone chromophore. Flavone-based compounds are known to form stable complexes with metal cations (Fe, Cu, Al, and Cr) [24]. Flavonoids and tannins are two of the most interesting natural phenolic compounds.

Pure quercetin, rutin, and tannin were also used as colorants in this study in order to compare their dyeing characteristics using *CIELAB* and *K/S* values as well as their fastness properties with those of *Eucalyptus* leaves. $FeSO_4{\cdot}7H_2O$ mordant was used in this work. Our interest lies in the color that these compounds are imparted when added to silk fabric by pad-dry and pad-batch dyeing techniques.

2. Experimental

2.1. Materials and Chemicals. The *Eucalyptus* leaves (*Eucalyptus camaldulensis*) used in this study were collected in Thailand. Quercetin dihydrate, 98% purity ($C_{15}H_{10}O_7{\cdot}2H_2O$, $M_W = 338.80$), rutin hydrate, 95% purity ($C_{27}H_{30}O_{16}{\cdot}xH_2O$, $M_W = 610.52$), and tannin Ph. Eur. 5 ($C_{76}H_{52}O_{46}$, $M_W = 1701.20$) were purchased from Sigma, Fluka and Lachner, respectively. Plain weave scoured and bleached silk fabric was used. The mordant used was ferrous(II) sulfate heptahydrate ($FeSO_4{\cdot}7H_2O$). The anionic wetting agent, Altaran S8 (sodium alkylsulfate), and anionic soaping agent, Syntapon ABA, were supplied by Chemotex Děčin, Czech Republic.

2.2. Instruments. The mordanting and dyeing were carried out in a two-bowl padding mangle machine (Mathis, type number HVF.69805). A spectrophotometer (Datacolor 3890) was used to measure the color strength. The color strength, in terms of K/S values, was calculated using the Kubelka-Munk equation, $K/S = (1 - R)^2/2R$, where R is reflectance of the dyed fabric, K is the sorption coefficient, and S is the scattering coefficient.

2.3. Dye Extraction from Eucalyptus Leaves. Fresh *Eucalyptus* leaves were dried in sunlight for one month and crumbled using a blender. They were then used as the raw material for dye extraction, which was achieved by the reflux technique (70 g of crumbled *Eucalyptus* leaves was mixed with 1 L of distilled water and refluxed for 1 h). The suspension was then filtered and the dye solution was separated into two parts: one

Table 1: Color value of silk fabric dyed with eucalyptus leaf extract and quercetin dyes.

Type of dye	Concentration	Pad-batch					Pad-dry				
		K/S	L^*	a^*	b^*	Dyed sample	K/S	L^*	a^*	b^*	Dyed sample
Eucalyptus	1 g/L dye (without mordant)	0.59	86.5	2.0	−3.6		0.58	86.4	2.1	−3.8	
	5 g/L dye (without mordant)	0.71	85.8	1.4	−1.4		0.66	86.2	1.5	−1.2	
	1 g/L dye + 5 g/L $FeSO_4$	1.54	56.9	2.1	−1.8		1.48	60.9	2.3	−1.6	
	5 g/L dye + 5 g/L $FeSO_4$	1.75	69.1	2.4	−1.2		1.63	58.2	1.9	−1.1	
Quercetin	1 g/L dye (without mordant)	1.17	85.4	−2.2	6.0		1.09	86.3	−1.6	5.1	
	5 g/L dye (without mordant)	1.58	86.3	−3.4	10.2		1.38	87.0	−3.0	8.2	
	1 g/L dye + 5 g/L $FeSO_4$	4.55	46.9	3.5	20.1		4.30	51.9	1.7	15.2	
	5 g/L dye + 5 g/L $FeSO_4$	8.03	36.7	1.4	14.9		7.61	39.8	2.9	16.8	

for evaporating with a rotary evaporator and one for dyeing. The crude dye extract was used for obtaining the standard calibration curve. The dilution of the *Eucalyptus* leaf extract gives a relatively clear solution with a linear dependence on the concentration absorbance and an absorption peak (λ_{max}) at 262 nm [25]. The concentration of 20 g/L was calculated from a standard curve of concentrations versus absorbance at the 262 nm wavelength.

2.4. Mordanting and Pad Dyeing. A simultaneous padding process was used in this research. To study the effect of dye concentration, two concentrations of the *Eucalyptus* leaf extract, quercetin, rutin, and tannin dyes were chosen: 1 and 5 g/L. Ferrous sulfate mordant was used at concentration of 5 g/L for each dye concentration, and 1 g/L of an anionic wetting agent was added to the dye solution. The pH of the dyeing solution was adjusted to 4 using an acetic acid. This pH was adopted from the previous studies [12, 26, 27]. The fabric was then immersed in the dye solution at room temperature and padded on a two-bowl padding mangle at 80% pick up. For the pad-dry method, the samples were dried at 90°C for 3 min. For the cold pad-batch method, the padded fabric was

TABLE 2: Color value of silk fabric dyed with rutin and tannin dyes.

Type of dye	Concentration	Pad-batch					Pad-dry				
		K/S	L^*	a^*	b^*	Dyed sample	K/S	L^*	a^*	b^*	Dyed sample
Rutin	1 g/L dye (without mordant)	0.71	87.5	0.05	0.1		0.65	87.4	0.8	−1.6	
	5 g/L dye (without mordant)	0.83	86.1	−0.4	2.0		0.76	87.0	−0.1	1.1	
	1 g/L dye + 5 g/L $FeSO_4$	4.10	60.1	1.5	20.0		2.30	69.4	3.3	19.7	
	5 g/L dye + 5 g/L $FeSO_4$	4.73	56.4	1.2	19.2		3.00	63.1	2.3	19.0	
Tannin	1 g/L dye (without mordant)	0.58	86.2	2.0	−3.4		0.56	86.7	2.4	−4.1	
	5 g/L dye (without mordant)	0.78	84.7	0.9	−1.0		0.67	86.8	1.3	−1.0	
	1 g/L dye + 5 g/L $FeSO_4$	2.04	53.0	2.3	−1.8		1.77	54.5	2.1	−2.1	
	5 g/L dye + 5 g/L $FeSO_4$	2.96	40.1	2.9	−4.4		2.81	41.6	4.1	−6.4	

rolled on a glass rod and a plastic sheet wrapped around the rolled fabric. Fabric was then kept at room temperature for 24 hours. The samples were washed in 1 g/L of the soaping agent at 80°C for 15 min and air-dried at room temperature.

2.5. Evaluation of Color Strength and Fastness Properties. The color strength (*K/S*) and *CIELAB* of the dyed samples were evaluated using a spectrophotometer (Datacolor 3890). All samples showed a λ_{max} value of 400 nm. The color fastness to washing, light, and rubbing of the dyed samples was determined according to ISO 105-C06 A1S: 1994, ISO 105-B02: 1994, and ISO 105-X12: 2001, respectively.

3. Results and Discussion

3.1. Effect of Dyeing on CIELAB and K/S Values. The color value results are presented in Tables 1 and 2. Silk fabrics dyed with *Eucalyptus* leaf extract and tannin dye showed a pale yellowish-grey shade, while those dyed with ferrous sulfate showed a dark greyish-brown color. Silk fabric dyed with quercetin without a mordant had a yellowish-green color, while those mordanted with ferrous sulfate produced a dark yellowish-brown shade. Silk substrates dyed with rutin gave a pale yellowish-green color, while those dyed with ferrous sulfate had a yellowish-brown color.

Table 3: Color fastness to washing at 40°C (ISO 105-C06 A1S: 1994).

Dyeing and mordanting conditions	Color change	Color staining of multifibers					
		Acetate	Cotton	Nylon	Polyester	Acrylic	Wool
Pad-batch							
Eucalyptus	4-5	4-5	4-5	4-5	4-5	4-5	4-5
Eucalyptus + Fe	4	4-5	4	4-5	4-5	4-5	4
Quercetin	4-5	4-5	4-5	4-5	4-5	4-5	4-5
Quercetin + Fe	4	4-5	4	4	4-5	4-5	4
Rutin	4-5	4-5	4-5	4-5	4-5	4-5	4-5
Rutin + Fe	4	4-5	4-5	4-5	4-5	4-5	4-5
Tannin	4-5	4-5	4-5	4-5	4-5	4-5	4
Tannin + Fe	4-5	4-5	4-5	4-5	4-5	4-5	4-5
Pad-dry							
Eucalyptus	4-5	4-5	4-5	4-5	4-5	4-5	4-5
Eucalyptus + Fe	4	4-5	4-5	4-5	4-5	4-5	4
Quercetin	4-5	4-5	4-5	4-5	4-5	4-5	4-5
Quercetin + Fe	4	4-5	4	4-5	4-5	4-5	4-5
Rutin	4-5	4-5	4-5	4-5	4-5	4-5	4-5
Rutin + Fe	4	4-5	4	4-5	4-5	4-5	4-5
Tannin	4-5	4-5	4-5	4-5	4-5	4-5	4-5
Tannin + Fe	4	4-5	4	4-5	4-5	4-5	4

Note: Fe: $FeSO_4$.

Table 4: Color fastness to light (ISO 105-B02: 1994).

Dyeing and mordanting conditions	Color change	
	Pad-batch	Pad-dry
Eucalyptus	3-4	3-4
Eucalyptus + $FeSO_4$	3-4	3-4
Quercetin	2	2
Quercetin + $FeSO_4$	3-4	3-4
Rutin	3	3
Rutin + $FeSO_4$	3-4	3-4
Tannin	4	4
Tannin + $FeSO_4$	4-5	4-5

From Tables 1 and 2, it is clear that the color shade of the fabrics dyed with tannin (a major constituent of *Eucalyptus* leaves) is colorimetrically and visually observed to be very similar to that one using *Eucalyptus* leaf extract dye. The colors obtained with the different dyes vary in their tone due to the fact that when the dyes (*Eucalyptus* leaf extract, quercetin, rutin, and tannin) are combined with ferrous sulfate to form dye-ferrous complexes, different shades are attained.

Tables 1 and 2 also show the color strength (K/S) values of silk fabric dyed with *Eucalyptus* leaf extract, quercetin, rutin, and tannin, respectively. It can be observed that the K/S values increase with an increase in dye concentration. Silk fabric dyeing by pad-batch technique showed higher K/S values than pad-dry technique. This is because the pad-batch method exposes the fibers to the dye molecules for a longer time period, promoting greater dyeability.

It can be concluded that the successful dyeing of silk fabrics with *Eucalyptus* leaf extract dye, quercetin, rutin, and tannin is due to the formation of ferrous coordination complexes. Ferrous sulfate readily chelated with the dyes. As the coordination number of ferrous sulfate is 6, some coordination sites remain unoccupied when they interact with the fiber, which allows functional groups, such as amino and carboxylic acids, on the fiber to occupy these unoccupied sites. Thus ferrous sulfate can form a ternary complex on one site with the fiber and on another site with the dye [28].

3.2. Color Fastness Properties. The fastness ratings of silk fabric dyed with and without mordants at a dye concentration of 5 g/L and ferrous sulfate (5 g/L) are presented in Tables 3–5. Table 3 indicates that the washing fastness rating of silk fabric dyed with *Eucalyptus* leaf extract, quercetin, rutin, and tannin is very good (4 to 4-5). A probable explanation for the good fastness property is that tannin and flavonoids (quercetin and rutin) can form metal chelates with the ferrous mordant. Hence, after mordanting, the tannin and flavonoids are insoluble in water, which improves the washing fastness.

As seen in Table 4, a light fastness in the range of 3-4 (fair) can be observed in the silk fabrics, except for that dyed with quercetin without mordant, whose rating was 2 (poor). This is attributed to the fact that the presence of 3-hydroxy groups in quercetin reduces the light fastness due to lower photostability [29]. Silk dyed with tannin with or without a mordant was in the range of 4 to 4-5 (good).

From Table 5, very good (4-5) dry rubbing fastness can be observed in silk fabric dyed with *Eucalyptus* leaf extract, quercetin, rutin, and tannin, except for those mordanted with ferrous sulfate, whose ratings were 3 to 4 (fair to

TABLE 5: Color fastness to rubbing (ISO 105-X12: 2001).

Dyeing and mordanting conditions	Pad-batch				Pad-dry			
	Warp direction Color staining		Weft direction Color staining		Warp direction Color staining		Weft direction Color staining	
	Dry	Wet	Dry	Wet	Dry	Wet	Dry	Wet
Eucalyptus	4-5	4-5	4-5	4-5	4-5	4-5	4-5	4-5
Eucalyptus + Fe	3	2-3	3	2-3	3	2-3	3	2-3
Quercetin	4-5	4-5	4-5	4-5	4-5	4-5	4-5	4-5
Quercetin + Fe	3-4	3-4	3-4	3-4	3-4	3-4	4	3-4
Rutin	4-5	4-5	4-5	4-5	4-5	4-5	4-5	4-5
Rutin + Fe	4	4	4	4	4	4	4	4
Tannin	4-5	4-5	4-5	4-5	4-5	4-5	4-5	4-5
Tannin + Fe	3	2-3	3	2-3	3	2-3	3	2-3

Note: Fe: $FeSO_4$.

good). However, the fabrics dyed with *Eucalyptus* and tannin mordanted with ferrous sulfate show a wet rubbing rating of 2 to 3 (poor to fair). This is attributed to a difference in the extent to which the low aqueous soluble ferrous-tannate complexes were able to diffuse within the dyed fiber. The large molecular size complex that was formed within the dyeing bath could be anticipated to display very low diffusional behavior and, therefore, to deposit mostly on the periphery of the dyed fiber, resulting in a low rubbing fastness [21].

4. Conclusions

Silk fabrics dyed with *Eucalyptus* leaf extract, quercetin, rutin, and tannin using the pad-batch method show higher color strength than those dyed with the pad-dry technique. Tannins are considered as a main colorant in dyeing processes because of the shade similarities of *Eucalyptus* leaves. The color fastness to washing showed very good results, whereas the color fastness to rubbing was fair to good, except for silk fabric dyed with tannin mordanted with *Eucalyptus* with ferrous sulfate, where ratings were poor when subjected to wet rubbing. The light fastness rating of the silk fabric mordanted with ferrous sulfate showed a fair to good result, but in the case of the silk fabric dyed with quercetin without mordant, the fastness rating was poor. The application of natural dyes to silk fabric by the pad-batch technique can be considered to be an effective ecooption; hence this technique could be considered to be the most suitable for small scale industry and cottage dyeing.

References

[1] R. L. M. Allen, *Colour Chemistry*, Nelson, London, UK, 1971.

[2] S. M. Robertson, *Dyes From Plants*, Van Nostrand Reinhold, New York, NY, USA, 1973.

[3] R. Mongkholrattanasit, J. Kryštůfek, J. Wiener, and J. Studničková, "Properties of wool and cotton fabrics dyed with eucalyptus, tannin and flavonoids," *Fibres and Textiles in Eastern Europe*, vol. 19, no. 2, pp. 90–95, 2011.

[4] K. Wongcharee, V. Meeyoo, and S. Chavadej, "Dye-sensitized solar cell using natural dyes extracted from rosella and blue pea flowers," *Solar Energy Materials and Solar Cells*, vol. 91, no. 7, pp. 566–571, 2007.

[5] N. M. Gómez-Ortíz, I. A. Vázquez-Maldonado, A. R. Pérez-Espadas, G. J. Mena-Rejón, J. A. Azamar-Barrios, and G. Oskam, "Dye-sensitized solar cells with natural dyes extracted from achiote seeds," *Solar Energy Materials and Solar Cells*, vol. 94, no. 1, pp. 40–44, 2010.

[6] M. Rekaby, A. A. Salem, and S. H. Nassar, "Eco-friendly printing of natural fabrics using natural dyes from alkanet and rhubarb," *The Journal of the Textile Institute*, vol. 100, no. 6, pp. 486–495, 2009.

[7] R. Agarwal, N. Pruthi, and S. J. S. Singh, "Effect of mordants on printing with Marigold flowers dye," *Natural Product Radiance*, vol. 6, no. 4, pp. 306–309, 2007.

[8] O. Habbal, S. S. Hasson, A. H. El-Hag et al., "Antibacterial activity of *Lawsonia inermis* Linn (Henna) against *Pseudomonas aeruginosa*," *Asian Pacific Journal of Tropical Biomedicine*, vol. 1, no. 3, pp. 173–176, 2011.

[9] N. Reddy, S. Han, Y. Zhao, and Y. Yang, "Antimicrobial activity of cotton fabrics treated with curcumin," *Journal of Applied Polymer Science*, vol. 127, no. 4, pp. 2698–2702, 2012.

[10] Y. H. Lee, "Dyeing, fastness, and deodorizing properties of cotton, silk, and wool fabrics dyed with coffee sludge (*Coffea arabica L.*) extract," *Journal of Applied Polymer Science*, vol. 103, no. 1, pp. 251–257, 2007.

[11] Y. H. Lee, E. K. Hwang, Y. J. Jung, S. K. Do, and H. D. Kim, "Dyeing and deodorizing properties of cotton, silk, wool fabrics dyed with amur corktree, *Dryopteris crassirhizoma, Chrysanthemum boreale, Artemisia* extracts," *Journal of Applied Polymer Science*, vol. 115, no. 4, pp. 2246–2253, 2010.

[12] R. Mongkholrattanasit, J. Kryštůfek, J. Wiener, and M. Viková, "UV protection properties of silk fabric dyed with eucalyptus leaf extract," *The Journal of the Textile Institute*, vol. 102, no. 3, pp. 272–279, 2011.

[13] D. Grifoni, L. Bacci, G. Zipoli, L. Albanese, and F. Sabatini, "The role of natural dyes in the UV protection of fabrics made of vegetable fibres," *Dyes and Pigments*, vol. 91, no. 3, pp. 279–285, 2011.

[14] A. K. Sarkar and R. Dhandapani, "Study of natural colorants as antibacterial agents on natural fibers," *Journal of Natural Fibers*, vol. 6, no. 1, pp. 46–55, 2009.

[15] V. Sivakumar, J. Vijaeeswarri, and J. L. Anna, "Effective natural dye extraction from different plant materials using ultrasound," *Industrial Crops and Products*, vol. 33, no. 1, pp. 116–122, 2011.

[16] S. Ali, N. Nisar, and T. Hussain, "Dyeing properties of natural dyes extracted from eucalyptus," *The Journal of the Textile Institute*, vol. 98, no. 6, pp. 559–562, 2007.

[17] P. S. Vankar, V. Tiwari, and J. Srivastava, "Extracts of steam bark of eucalyptus globules as food dye with high antioxidant properties," *Electronic Journal of Environmental, Agricultural and Food Chemistry*, vol. 6, no. 11, pp. 2550–2556, 2007.

[18] L. Chapuis-Lardy, D. Contour-Ansel, and F. Bernhard-Reversat, "High-performance liquid chromatography of water-soluble phenolics in leaf litter of three *Eucalyptus* hybrids (Congo)," *Plant Science*, vol. 163, no. 2, pp. 217–222, 2002.

[19] E. Conde, E. Cadahia, Garcia-Vallejo, and M. C. :, "Low molecular weight polyphenols in leaves of *Eucalyptus camaldulensis. E. globules* and *E. rudis*," *Phytochemical Analysis*, vol. 8, no. 4, pp. 186–193, 1997.

[20] E. Conde, E. Cadahia, M. C. Garcia-Vallejo, Fernandez de Simon, and B. :, "High ressure liquid chromatographic analysis of polyphenols in leaves of *Eucalyptus cmadulensis. E. globules* and *E. rudis*: proanthocyanidins, ellagitannins and flavonol glycosides," *Phytochemical Analysis*, vol. 8, no. 2, pp. 78–83, 1997.

[21] S. M. Burkinshaw and N. Kumar, "A tannic acid/ferrous sulfate aftertreatment for dyed nylon 6,6," *Dyes and Pigments*, vol. 79, no. 1, pp. 48–53, 2008.

[22] S. M. Burkinshaw and N. Kumar, "The mordant dyeing of wool using tannic acid and $FeSO_4$, part 1: initial findings," *Dyes and Pigments*, vol. 80, no. 1, pp. 53–60, 2009.

[23] S. M. Burkinshaw, *Chemical Principles of Synthetic Fibre Dyeing*, Chapman and Hall, Glasgow, UK, 1995.

[24] C. Septhum, V. Rattanaphani, and S. Rattanaphani, "UV-Vis spectroscopic study of natural dyes with alum as a mordant," *Suranaree Journal Science Technology*, vol. 14, no. 1, pp. 91–97, 2007.

[25] E. A. Yarosh, T. I. Gigoshvili, and M. D. Alaniya, "Chemical composition of eucalyptus jumanii cultivated in the humid georgian subtropics," *Chemistry of Natural Compounds*, vol. 37, no. 1, pp. 86–87, 2001.

[26] R. Mongkholrattanasit and T. Vitidsant, "Dyeing and colour fastness properties of silk and cotton fabrics dyed with eucalyptus leaves extract," in *Proceedings of the 6th International Conference Textile Science (TEXSCI '07)*, pp. 285–286, Liberec, Czech Republic, 2007.

[27] R. Mongkholrattanasit, W. Wongphakdee, and C. Sirikasemlert, "Dyeing and colour fastness properties of silk and cotton fabrics dyed with eucalyptus bark extract," *RMUTP Research Journal*, vol. 1, no. 1, pp. 41–49, 2007.

[28] S. D. Bhattacharya and A. K. Shah, "Metal ion effect on dyeing of wool fabric with catechu," *Coloration Technology*, vol. 116, no. 1, pp. 10–12, 2000.

[29] S. V. Bhat, B. A. Nagasampagi, and M. Sivakumar, *Chemistry of Natural Products*, Narosa, New Delhi, India, 2006.

8

Effects of Different Mordants on Silk Fabric Dyed with Onion Outer Skin Extracts

Mohammad Gias Uddin

Department of Textile Engineering, Faculty of Engineering, Ahsanullah University of Science and Technology, Dhaka 1208, Bangladesh

Correspondence should be addressed to Mohammad Gias Uddin; giasdtt@gmail.com

Academic Editor: Phillip W. Gibson

At present, a higher demand is put towards the use of natural dyes due to increased awareness of the environmental and health hazards associated with the synthesis and use of synthetic dyes. This research was conducted using onion outer skins as a potential source of natural plant dyes. In this study, extraction of dye was carried out in aqueous boiling method. Premordanting technique was followed using different mordants, namely alum, ferrous sulphate, tin, tannic acid, tartaric acid, and their combinations on silk fabric. Fabric samples dyed without using any mordant were then compared with the dyed samples pretreated with the mordants. The range of colors developed on dyed materials was evaluated by measuring the color values with respect to K/S values and color coordinates. It was concluded that the color values were found to be influenced by the addition of mordants, and thus different fashion hues were obtained from the same amount of dye extract using different mordants. Ferrous sulphate was found as the most influential mordant. ΔE_{cmc} values between unmordanted (Reference dyed) and metallic mordanted fabric samples were found higher than those between unmordanted and nonmetallic mordanted fabric samples. The dyed samples were evaluated for color fastness to washing, light, drycleaning, rubbing, and perspiration. The color fastness properties were found to be satisfactory and improved in many cases. From the fastness results, it was obvious that these dyes can also be applied on silk fabric without using any mordant if required.

1. Introduction

Natural dyes are noncarcinogenic, produce soothing and exclusive color to the textiles, and have better biodegradability and higher compatibility with the environment [1–6]. Among the natural dyes, plant dyes are potential sources because of their easy availability and abundant nature. Onion outer skins are the most commonly discarded household and commercial food waste which can be used as dyes for coloring natural textile materials. These dyes, which are known as pelargonidin (3,5,7,4-tetrahydroxyanthocyanidin) [7] (see Figure 1), work like acid dyes that can dye the protein fibers at high efficiency [8]. The amount of pelargonidin was found to be 2.25% in certain solvent extraction process using soxhlet apparatus [7]. Due to presence of four hydroxy groups (Auxochrome groups) pelargonidin exhibits good dyeing properties for dyeing of natural fibers [7, 9].

However, with the increased interest in natural plant dyes a greater importance was given to using mordant. Most of the natural dyes require some sort of mordant to set permanently in any fiber. Metal ions of mordants act as electron acceptors for electron donors to form coordination bonds with the dye molecules, making them insoluble in water [10]. This leads to improve dye uptake and retention, which result in a greater depth of shade and color fastness properties [11]. Common mordants are alum, chrome, stannous chloride, copper sulphate, ferrous sulphate, and so forth [12]. Chrome and copper are considered to be harmful for human skin as objectionable heavy metals beyond a certain limit of their presence [13]. Organic substances that have been used as mordants are mud, blood, and cow dung [14]. Also to reduce the effect of some toxic mordants, natural mordants such as tannins are recommended, as they can provide fabrics with characteristics similar to those provided by synthetic mordants [15].

Again, depending on the metal character of mordant, the complex formation not only strengthens dyestuff fixation on the substrate but also changes the color of dyeing. In

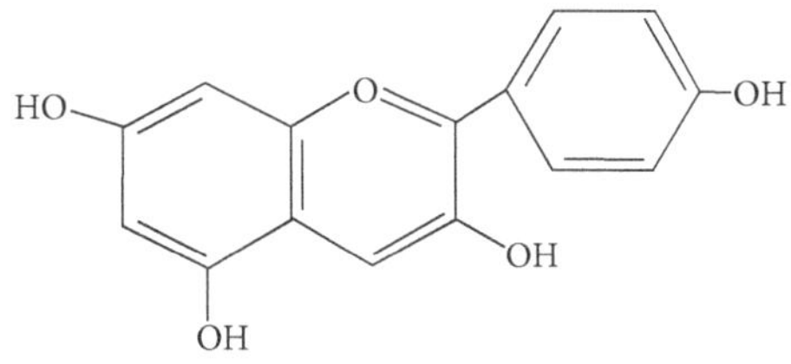

Figure 1: Chemical structure of pelargonidin.

some cases the resulting change in shade can be seen as an opportunity to steer color in a wider range. Different types of mordants and their combination on the same natural dye may darken, brighten, or drastically alter the color of the dye-mordant solution and influence the shade of the final product [16] which may be a desired effect as well as an unwanted phenomenon [17]. So, final color, its brilliance, and color fastness properties are not only dependent on the dye itself but are also determined by varying concentration and suitable application of mordants.

2. Materials and Methods

2.1. Materials. The raw silk fabric (22 denier warp, 22 denier weft, 132 ends/inch, 132 picks/inch, and 22 g/m^2 fabric area density) purchased from Sopura Silk Ltd., Dhaka, Bangladesh, was used in this study.

2.2. Methods

2.2.1. Extraction Process. The onion skins were dried under direct sunlight and grinded into small units with the help of grinding machine. The wastages are removed using a fine strainer and finally weight was taken. After drying, crushing, and straining, the weight of 1 kg skins was found to be 805 gm. The dried and crushed skins were shown in Figure 2.

The color component was extracted from the skins in aqueous extraction process. It was done by extracting fixed quantity of crushed skins under pH 5 with a material to liquor ratio of 1 : 10 (weight of crushed skins in g; amount of water in mL) at boil for 60 min. In each process of extraction, the mixture was cooled down and finally the dye extracts were filtered accurately.

2.2.2. Degumming. The raw silk fabric was degummed in an aqueous soap solution (15 g/L soap) at pH 9. The material to liquor ratio during the treatment was maintained at 1 : 50. The temperature was gradually raised to 80°C and run for 60 minutes. The degummed fabric was washed with 2 g/L detergent at 65°C for 10 min.

2.2.3. Bleaching. Then, the fabric was treated with 35% hydrogen peroxide (3 mL/L) solution, maintaining a material to liquor ratio of 1 : 50 at pH 9 and temperature 60°C for 60 minutes followed by wash with 2 g/L detergent at 65°C for 10 min.

2.2.4. Mordanting. Premordanting was carried out on silk fabric using 5% (on fabric weight) of ferrous sulphate, alum, stannous chloride, tannic acid, and tartaric acid mordants individually and using four different combinations of mordants such as ferrous sulphate-alum (2.5% + 2.5%), ferrous sulphate-alum-tin (2% + 2% + 1%), alum-tin (2.5% + 2.5%), and alum-tin-tannic acid (2% + 2% + 1%) at 60°C for 60 minutes keeping material to liquor ratio 1 : 30. Again, cream of tartar (CT) was used as a mordant assistant with alum and stannous chloride separately.

2.2.5. Dyeing. Dyeing was carried out in exhaust method with the dye extracts as per standard dyeing conditions recommended for silk fabric [18] at 80°C for 60 minutes under pH 5, keeping material to liquor ratio 1 : 50 in a IR sample dyeing machine. Opticid PSD (1.5 g/L) was used as a buffering agent in the extracted dye liquor.

2.2.6. Color Yield of Dyed Fabrics. Dyed textile samples were analyzed by measuring the reflectance curve between 350 and 750 nm by the reflectance spectrophotometer with illuminant D_{65} at 10° observer. The minimum of the curve ($R_{\min}$) was used to determine the ratio of light absorption (K) and scatter (S) via the Kubelka-Munk function using [19]

$$\left(\frac{K}{S}\right)_{\text{Dyed}} = \frac{(1 - R_{\min})^2}{2R_{\min}}. \quad (1)$$

2.2.7. Color Coordinates of Dyed Fabrics. The color coordinates of the dyed samples were determined by a dual beam reflectance spectrophotometer using color-matching software (Datacolor International, USA) and the three parameters L^*, a^*, and b^* were determined in the CIELab system. In addition, chroma (C^*) and hue (h) values were determined from the spectrophotometer readings.

2.2.8. Color Difference (ΔE_{cmc}). Mordanted and unmordanted dyed samples were measured in the mentioned spectrophotometer and ΔE_{cmc} value was determined to show the color difference between them taking unmordanted dyed sample as standard.

2.2.9. Color Fastness. Washing fastness test was carried out in ISO 105 C02 method, light fastness in ISO 105 B02 method, dry cleaning fastness in ISO 105 D01 method, rubbing fastness in ISO 105 X12 method, and perspiration fastness in ISO 105 E04 method.

3. Results and Discussion

3.1. Dyed Samples. The use of mordants produced different shades on silk dyed with onion extracts which were shown in Table 1.

3.2. Color Measurements of Dyed Fabrics. The results of color measurements of the dyed silks were shown in Table 2.

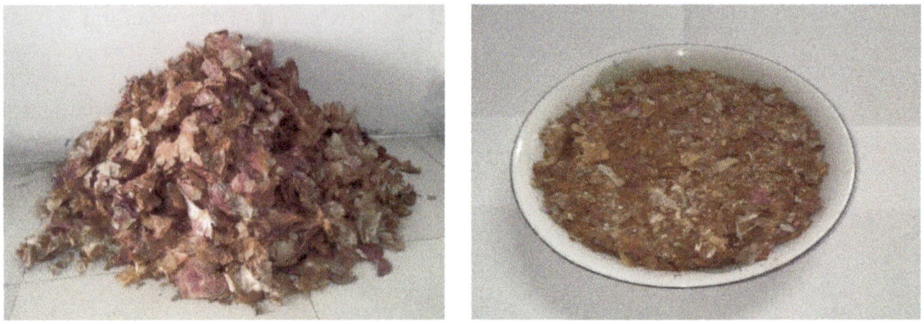

(a) Dried onion skins (b) Crushed onion skins

FIGURE 2: Dried and crushed onion skins.

TABLE 1: Shades of dyed silk.

Mordant	No mordant (Reference dyed)	Ferrous sulphate	Ferrous sulphate-alum	Ferrous sulphate-alum-tin
Shade	(1) Camel	(2) Otter	(3) Dusky dull green	(4) Bone brown
Mordant	Alum	Alum + CT	Alum + Tin	Tin
Shade	(5) Amber gold	(6) Oak buff	(7) Honey yellow	(8) Sandstorm
Mordant	Tin + CT	Tannic acid (TA)	Alum-Tin-TA	Tartaric acid
Shade	(9) Clay	(10) Toast	(11) Honey gold	(12) Toasted nut

3.2.1. Color Yield of Dyed Fabrics. K/S value of the unmordanted dyed sample was found to be 11.235. In the mordanting method, mordant resulted in improved color yield of the dyed fabrics, except tin-CT and alum-CT. Ferrous sulphate as a mordant significantly increased the color yield of silk. The K/S value was found to be 15.284 using ferrous sulphate which showed the relative surface color strength value of 136.04% considering the unmordanted dyed sample as standard. Besides, using ferrous sulphate-alum as a combination, the maximum color strength 146.94% (K/S = 16.509) and using ferrous sulphate-alum-tin as a combination the color strength 131.93% (K/S = 14.822) was found.

Among ferrous sulphate, alum, tin, tannic acid, and tartaric acid mordants, when these were used individually in a single mordanting process on silk, the order of color yield was found to be ferrous sulphate > tartaric acid > tin > tannic acid > alum, that is, color yield gradually decreased when approached from ferrous sulphate to alum. Again, among the four different combinations of mordants, that is, ferrous sulphate-alum, ferrous sulphate-alum-tin, alum-tin, and alum-tin-tannic acid, the order in the combination process was found to be ferrous sulphate-alum > ferrous sulphate-alum-tin > alum-tin > alum-tin-tannic acid.

The increase in K/S values due to mordanting reveals that dye molecules are capable of forming metal complex with the positively charged metals. In case of silk fibers, dye anions and metal cations have strong attraction towards positively charged amino and negatively charged carboxyl groups, respectively. Hence, they enter the fiber and form ionic bonding between dye and fiber and metal and fiber and finally with dye and metal ions. The dye-metal chellates thus produced also form coordinate bonds with the uncharged amine ($-NH_2$) groups of silk. Again, one molecule of dye can form a bond with one site of fiber molecule. But one molecule of mordant can form bonds with two or more molecules of dyes. As a result, when the mordant molecule binds on to fiber it holds two molecules of dye with it. Therefore, using mordant, the color yield was increased.

It is known that some transition metal ions can bond strongly with natural dye molecules due to their ability to form coordination complexes and thus produce deep color on the fabric. Iron salts such as ferrous sulphate as transition metal mordant form a large number of complexes with

TABLE 2: Color yield, color coordinates, and color difference of the dyed fabrics.

Mordant type	K/S at λ_{max}	Color strength%	Color coordinates					ΔE_{CMC}
			L^*	a^*	b^*	C^*	h^0	
No mordant (Reference dyed)	11.235	100	63.19	15.53	25.43	29.69	58.92	—
Fe	15.284	136.04	37.54	4.23	18.61	19.09	77.19	15.25
Fe + Alum	16.509	146.94	39.16	4.56	21.11	21.59	77.82	15.08
Fe + Alum + Tin	14.822	131.93	50.76	8.45	25.41	26.78	71.60	9.60
Alum	12.980	115.53	55.81	9.72	34.92	36.25	74.45	12.25
Alum + CT	9.3829	83.51	60.58	10.88	26.62	28.76	67.77	5.92
Alum + Tin	13.411	119.37	63.12	15.22	33.79	37.06	65.76	6.29
Tin	14.188	126.28	63.24	15.62	29.38	33.27	62.00	2.82
Tin + CT	9.0721	80.75	65.94	13.67	26.73	30.02	62.92	2.92
Tannic acid (TA)	13.795	122.79	65.93	14.22	27.04	30.56	62.26	2.56
Alum + Tin + TA	11.719	104.31	57.88	11.85	31.28	33.45	69.26	7.86
Tartaric Acid	14.484	128.93	60.60	18.45	28.23	33.72	56.84	2.71

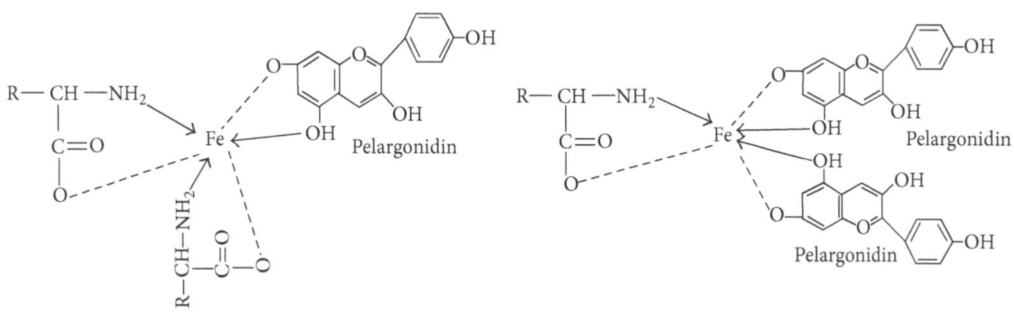

FIGURE 3: Possible structures of pelargonidin with $FeSO_4$ on silk.

the dye molecules, mostly octahedral ones with coordination number 6. As a result, some coordination sites remain unoccupied when they interact with the fiber and at that time functional groups such as amino and carboxylic groups on the silk fiber can occupy these unoccupied sites. Thus, ferrous sulphate salts can form a ternary complex on one site with the fiber and in the other site with the dye. This resulted in higher dye uptake as well as shade change due to mordanting with ferrous sulphate. On the other hand, aluminium and tin salts formed weak coordination complexes with the dye. This tends to form quite strong bonds with the dye molecule but not with the fiber. Thus, they block the dye and reduce its interaction with the fiber. This is the reason behind the lower K/S values in case of aluminium and tin than those obtained from ferrous sulphate salts [13, 20–23].

The possible structures of complexes with pelargonidin dye molecules and ferrous sulphate or alum in premordanting of silk were shown in Figures 3 and 4, respectively [9].

In contrast, as a mordant assistant cream of tartar (CT) decreased the color yield of dyed fabrics when used with alum or tin.

3.2.2. Effect of Mordanting on Color Characteristics of Dyed Silk. The ΔL^* values of the dyed silk samples as shown in Figure 5 indicated that ferrous sulphate mordant resulted in the highest increase in color depth (ΔL^* values were lower), followed by alum when single mordant was considered. Darkness was increased 40.59% using ferrous sulphate mordant in comparison to the reference dyed sample. This could be attributed to the affinity of mordant with color and fabric.

Among the combined applications of mordants, in all cases, it was observed that the highest darkness achieved by ferrous sulphate was reduced while alum was used with ferrous sulphate and further reduced while tin was used with the combination of ferrous sulphate and alum. In contrast, tin-CT and tannic acid produced lighter shades but when tannic acid was used with alum and tin in combination, the color depth was found to be increased. Besides, cream of tartar (CT) with alum or tin brought changes in the lightness scale of the dyed fabrics. The test results showed that CT increased L^* value when used along with alum or tin than their individual use.

From the Δa^* values of dyed samples as shown in Figure 6, it was clear that the addition of ferrous sulphate mordant increased the greenness quality 72.76% of the reference samples. On the contrary, tartaric acid improved 18.80% in redness value from the corresponding reference sample.

The Δb^* values of the samples as shown in Figure 7 indicated that ferrous sulphate and ferrous-alum mordanted samples were bluer than the reference sample. In other cases, yellowness of dyed fabrics was increased. Significant

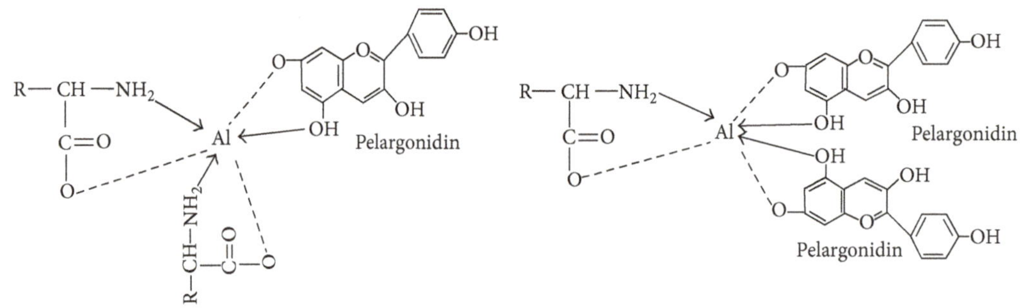

FIGURE 4: Possible structures of pelargonidin with alum on silk.

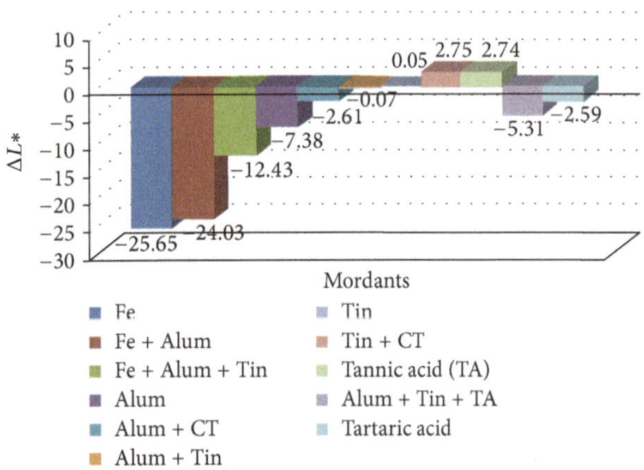

FIGURE 5: ΔL^* values of dyed silk.

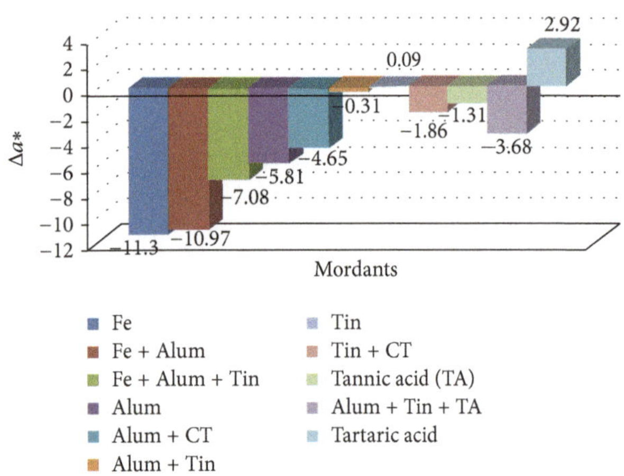

FIGURE 6: Δa^* values of dyed silk.

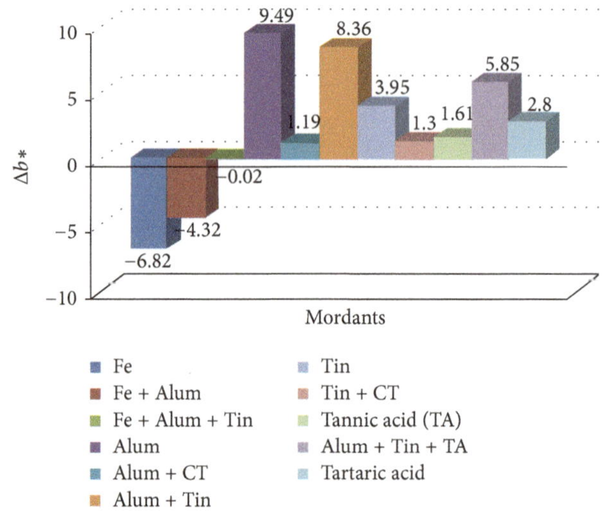

FIGURE 7: Δb^* values of dyed silk.

yellowness was increased in case of alum, alum-tin, and alum-tin-tannic acid combinations which were 37.32%, 32.87%, and 23%, respectively, over the reference dyed sample.

The color saturation values (C^*) were found least in ferrous sulphate mordanted samples (19.09) whereas the values were found maximum in case of alum-tin (37.065) and alum (36.25) mordanted samples. In addition, the hue angles lie within 56.84° to 77.82°, so all of the dyed samples were closer to yellowish shade than the reddish.

3.2.3. Color Difference (ΔE_{CMC}). The most striking results were found in case of ferrous sulphate mordanted samples. Higher color difference was noticeable between reference and ferrous sulphate mordanted samples and the color difference reduced from ferrous sulphate to alum and then alum to tin in case of metallic mordants as shown in Table 2. Again, the ΔE_{CMC} value between reference and nonmetallic mordanted fabric sample was found lower than that between reference and metallic mordanted fabric sample.

3.3. Fastness Results

3.3.1. Washing Fastness. The results of washing, light, and rubbing fastness of the dyed fabrics were shown in Table 3.

The unmordanted dyed silk showed rating 4. Using mordants and combinations, the color change ratings were found within 3/4 to 5, where rating 5 (Excellent) was found using tin, tin-CT, tannic acid, and tartaric acid mordants. The ratings were found 4 in case of using alum, alum-CT, and alum-tin-tannic acid. So it can be said that overall ratings of color change were good. As wash fastness of dye is influenced by the rate of diffusion of dye molecules and state of dyes inside the fiber, it can be explained that they have a tendency to aggregate inside the fiber. Thus, their molecular size is

Table 3: Washing, light, and dry cleaning fastness of dyed fabrics.

Mordant type	Washing fastness							Light fastness	Dry cleaning	
	CC*	CS*								
		Di-acetate	Bleached cotton	Polyamide	Polyester	Acrylic	Wool	(CC)	CC	CS
No mordant (Reference dyed)	4	5	5	5	5	5	5	4	4/5	5
Fe	3/4	5	5	5	5	5	4	6	5	5
Fe + Alum	3	5	5	5	5	5	4	6	4	5
Fe + Alum + Tin	3/4	5	5	5	5	5	4	7	4/5	5
Alum	4	5	5	5	5	5	5	4	4	5
Alum + CT	4	5	5	5	5	5	5	4	4/5	5
Alum + Tin	3	5	5	5	5	5	4	3	4.5	5
Tin	5	5	5	5	5	5	5	3	5	5
Tin + CT	5	5	5	5	5	5	5	3	4/5	5
Tannic acid (TA)	5	5	5	5	5	5	5	3	4/5	5
Alum + Tin + TA	4	5	5	5	5	5	5	4	4/5	5
Tartaric acid	5	5	5	5	5	5	5	3	4/5	5

CC*: color change; CS*: color staining.

increased which exhibits good wash fastness. In addition, in case of mordanted samples, complexing with mordant has the effect of insolubilizing the dye, making it color fast.

Moreover, this can be explained that the good fastness to washing for the sample dyed without mordant was due to the affinity of coloring component through H-bonding and van der Waals forces.

On the other hand, the color staining ratings were found from 4 to 5 for all the dyed fabrics, except when ferrous sulphate and its combinations were used as mordants. There were very slight staining observed on the adjacent wool fiber of the multifiber fabrics in case of ferrous sulphate and its combination as well as alum-tin samples where ratings were found 4 and almost no staining on the other fibers of the multifiber fabric.

3.3.2. Light Fastness. As shown in Table 3, the unmordanted dyed fabric showed rating 4 and this was improved to 6 using ferrous sulphate mordant. The rating was further improved to 7 using ferrous sulphate-alum-tin combination. In other cases, in comparison to unmordanted sample, rating was either unchanged or decreased by 1 grade.

As ferrous sulphate has more affinity towards dye and fiber compared to tin or alum, it can be said that ferrous sulphate can bind with more dye molecules than tin or alum. The fabrics mordanted with ferrous sulphate, alum, or tin may have the same number of dye molecules destroyed during exposure to light. But as the fabrics mordanted with ferrous sulphate had deeper shades due to bonding with more number of dye molecules, it seemed to fade less compared to the fabric mordanted with tin or alum.

It has been clearly shown that ferrous sulphate mordanted samples dyed with the extracts showed excellent light fastness. This happened due to the formation of a complex with transition metal which protected the chromophore from photolytic degradation. The photons which were sorbed by the chromophoric group dissipated their energy by resonating within the six member ring formed, thus protecting the dye.

3.3.3. Dry Cleaning Fastness. The dry cleaning fastness ratings as shown in Table 3 for both color change and color staining to the perchloroethylene solution were found within 4 to 5. Only in case of color change rating of alum and ferrous sulphate-alum, the rating was 4 and in all other cases it was within 4/5 to 5. In most cases, it can be stated that the pelargonidin dyes were almost completely resistant to perchloroethylene solution.

3.3.4. Rubbing Fastness. From the comparison carried out between the staining ratings of dry and wet rubbing fastness as shown in Figure 8, it was noticed that dry rubbing fastness was found slightly better than wet rubbing fastness.

All the staining ratings of dry rubbing fastness were ranged within 4 to 5. Besides the overall better dry rubbing fastness, the wet rubbing fastness of the ferrous sulphate-alum-tin and alum-tin-tannic acid combinations was 3 and in other cases it was ranged within 3/4 and 5.

3.3.5. Perspiration Fastness. The results of color fastness to perspiration test are shown in Table 4. In most of the cases, the color change ratings found for acid perspiration were higher than the corresponding ratings for alkaline perspiration. But the reverse trend was found in case of alum-CT and tin-CT mordanted samples. Equal ratings for color change of both acid and alkaline perspiration tests were achieved in case of ferrous sulphate-alum-tin, alum-tin-tannic acid, and tartaric acid mordanted samples.

Among the staining results, significant to moderate staining (rating 2/3) was noticeable only in three cases on wool fiber of the multifiber fabrics under acidic perspiration (i.e., ferrous sulphate, ferrous sulphate-alum-tin, and alum-tin

Table 4: Perspiration fastness of dyed fabrics.

Mordant type	Color change		Color staining to adjacent multifiber fabric											
			Di-acetate		Bleached cotton		Polyamide		Polyester		Acrylic		Wool	
	Ac*	Ak*	Ac	Ak	Ac	Ak	Ac	Ak	Ac	Ak	Ac	Ak	Ac	Ak
No mordant (Reference dyed)	4/5	4	3	3/4	4	4	4	4/5	3	4	3	3/4	4	4
Fe	3/4	3	3/4	4/5	2/3	4	3	4/5	3	4	3	4	2/3	4
Fe + Alum	4	3/4	4	4/5	3	3	3/4	3/4	3/4	3/4	3/4	3/4	3/4	3
Fe + Alum + Tin	3/4	3/4	4	4/5	2/3	3/4	3	4	3	4	3	4	2/3	3/4
Alum	4	3/4	4	4	4/5	4/5	4/5	4/5	4/5	4/5	4	4	4/5	4/5
Alum + CT	3/4	4/5	4/5	4	4/5	4/5	4/5	4/5	4/5	4/5	4	4	4/5	4/5
Alum + Tin	4	3/4	3/4	4/5	3	3	3	3/4	3	3/4	3	3/4	2/3	3
Tin	3/4	3	4	3/4	4/5	4	4/5	4	4/5	3/4	4	3	4/5	4/5
Tin + CT	4	4/5	3/4	3/4	4/5	4/5	3/4	4/5	3/4	4/5	4/5	3	4/5	4
Tannic acid (TA)	4/5	3/4	4/5	2/3	4/5	3/4	4/5	4	4	3	4	2/3	4/5	3/4
Alum + Tin + TA	4/5	4/5	4	4/5	4/5	5	4/5	4/5	4/5	4/5	4	3/4	4/5	4/5
Tartaric acid	4	4	3/4	3/4	4	4	4	4/5	3/4	3	3	3	4/5	4

Ac*: ratings when treated under acidic perspiration solution, while Ak*: under alkaline perspiration solution.

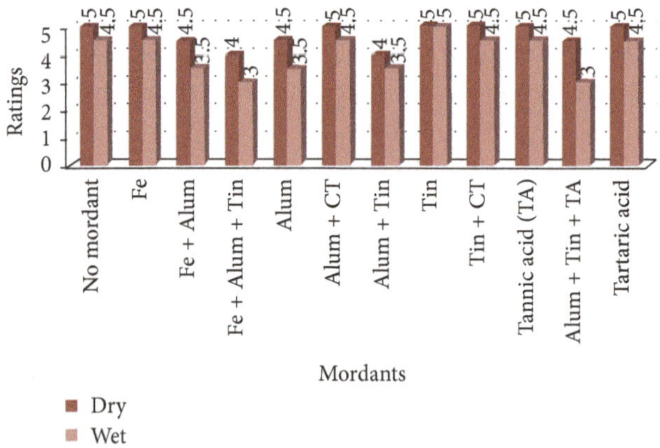

Figure 8: Rubbing fastness results of dyed fabrics.

mordanted samples), in one case on the acrylic fiber under alkaline condition (i.e., tannic acid mordanted sample), one case of on di-acetate fiber under alkali perspiration (tannic acid mordanted sample), and two cases on bleached cotton under acidic perspiration (ferrous sulphate and ferrous-alum-tin mordanted sample). These are the cases that affected the ratings most. All other staining results were found to be satisfactory.

Test data of perspiration indicated that there were changes noticeable in the dyed samples after treating under the perspiration solutions, and this may be due to several factors, such as the ionization of some of the dyes during alkaline perspiration testing or some dyes may detach from the substrate from the dye-fiber bond (between the skin dye and the silk fiber) during acidic perspiration treatments.

4. Conclusions

This study demonstrated that onion outer skins can be used as a potential source of dyes for dyeing of silk fabric. This gave the chance to produce different fashion hues on silk fabric from the same dye extract of onion skins using metallic and nonmetallic mordants and their combinations. It was concluded that the color values with respect to K/S and L^*, a^*, b^* values were found to be influenced by the addition of mordants. Ferrous sulphate was found as the most influential mordant with respect to color values. Mordants improved K/S values of the dyed fabrics except a few cases when compared to unmordanted dyed fabrics. Using ferrous sulphate, alum, tin, tannic acid, and tartaric acid as a single mordant on silk, the order of color yield found was ferrous sulphate > tartaric acid > tin > tannic acid > alum. Again, among the combinations ferrous sulphate-alum, ferrous sulphate-alum-tin, alum-tin, and alum-tin-tannic acid, the order found was ferrous sulphate-alum > ferrous sulphate-alum-tin > alum-tin > alum-tin-tannic acid. Fastness properties were found satisfactory and improved in many cases. Light fastness rating was improved from 4 to 7 using ferrous sulphate-alum-tin combination. Furthermore, there is a tremendous scope to use the extracted dyes on silk fabric without using any mordant. The plant dyes extracted from onion skins can be considered as a copartner with synthetic dyes for dyeing of silk fabric. The areas where synthetic dyes, fixatives, and other additives are imported and are, therefore, relatively expensive; the plant dyes especially from this type of waste materials can offer an attractive alternative.

Conflict of Interests

The author of the paper has declared that he has no conflict of interests.

Acknowledgment

The author would like to express his gratitude to Professor Dr. Md. Zulhash Uddin (Dean, Faculty of Textile Chemical Processing Engineering and Applied Science, Bangladesh

University of Textiles) due to his technical cooperation in this work.

References

[1] E. Tsatsaroni, M. Liakopoulou-Kyriakides, and I. Eleftheriadis, "Comparative study of dyeing properties of two yellow natural pigments—effect of enzymes and proteins," *Dyes and Pigments*, vol. 37, no. 4, pp. 307-315, 1998.

[2] T. Bechtold, A. Turcanu, E. Ganglberger, and S. Geissler, "Natural dyes in modern textile dyehouses—how to combine experiences of two centuries to meet the demands of the future?" *Journal of Cleaner Production*, vol. 11, no. 5, pp. 499–509, 2003.

[3] P. Saravanan and G. Chadramohan, "Dyeing of silk with ecofriendly natural dye obtained from Barks of *Ficus religiosa* L," *Universal Journal of Environmental Research and Technology*, vol. 1, no. 3, p. 268, 2011.

[4] L. G. Angelini, A. Bertoli, S. Rolandelli, and L. Pistelli, "Agronomic potential of *Reseda luteola* L. as new crop for natural dyes in textiles production," *Industrial Crops and Products*, vol. 17, no. 3, pp. 199–207, 2003.

[5] Z. M. Win and M. M. Swe, "Purification of the natural dyestuff extracted from Mango bark for the application on protein fibers," *World Academy of Science, Engineering and Technology*, vol. 22, p. 536, 2008.

[6] S. Waheed and A. Alam, "Studies of some natural dyes," *Journal of the Chemical Society of Pakistan*, vol. 26, no. 3, pp. 255–263, 2004.

[7] S. Chandravanshi and S. K. Upadhyay, "Interaction of natural dye (*Allium cepa*) with ionic surfactants," *Journal of Chemistry*, vol. 2013, Article ID 685679, 6 pages, 2013.

[8] F. M. Tera, K. E. Elnagar, and S. M. Mohamed, "Dyeing and light fastness properties of onion scale dye on different fabric types for conservative applications," *Journal of Textile and Apparel, Technology and Management*, vol. 7, no. 3, pp. 1–6, 2012.

[9] A. Önal, "Extraction of dyestuff from onion (*Allium cepa* L.) and its application in the dyeing of wool, feathered-leather and cotton," *Turkish Journal of Chemistry*, vol. 20, no. 3, pp. 194–203, 1996.

[10] R. Mongkholrattanasit, J. Kryštůfek, J. Wiener, and M. Vikov, "Dyeing, fastness, and uv protection properties of silk and wool fabrics dyed with eucalyptus leaf extract by the exhaustion process," *Fibres and Textiles in Eastern Europe*, vol. 86, no. 3, pp. 94–99, 2011.

[11] H. Böhmer, *Koekboya: Natural Dyes and Textiles: A Colour Journey from Turkey to India and Beyond*, Remhöb, Ganderkesee, Germany, 2002.

[12] S. S. Kulkarni, U. M. Bodake, and G. R. Pathade, "Extraction of natural dye from chili (*Capsicum Annum*) for textile coloration," *Universal Journal of Environmental Research and Technology*, vol. 1, pp. 58–63, 2011.

[13] A. C. Dweck, "Natural ingredients for colouring and styling," *International Journal of Cosmetic Science*, vol. 24, no. 5, pp. 287–302, 2002.

[14] D. Cardon, *Natural Dyes: Sources, Tradition, Technology and Science*, Archetype Publications, London, UK, 2007.

[15] G. Arroyo-Figueroa, G. M. L. Ruiz-Aguilar, G. Cuevas-Rodriguez, and G. G. Sanchez, "Cotton fabric dyeing with cochineal extract: influence of mordant concentration," *Coloration Technology*, vol. 127, no. 1, pp. 39–46, 2011.

[16] B. Glover, "Are natural colorants good for your health? Are synthetic ones better?" *Textile Chemist and Colorist*, vol. 27, no. 4, pp. 17–20, 1995.

[17] T. Bechtold and R. Mussak, Eds., *Handbook of Natural Colorants*, John Wiley & Sons, West Sussex, UK, 2009.

[18] *Clariant Manual. Recommendations for Dyeing Silk*, Sandoz Chemicals, Basel, Switzerland.

[19] R. Mcdonald, Ed., *Color Physics for Industry*, Society of Dyers and Colourists, Bradford, UK, 2nd edition, 1997.

[20] S. D. Bhattacharya and A. K. Shah, "Metal ion effect on dyeing of wool fabric with catechu," *Coloration Technology*, vol. 116, no. 1, pp. 10–12, 2000.

[21] M. Rekaby, A. A. Salem, and S. H. Nassar, "Eco-friendly printing of natural fabrics using natural dyes from alkanet and rhubarb," *The Journal of the Textile Institute*, vol. 100, no. 6, pp. 486–495, 2009.

[22] R. Mongkholrattanasit and N. Punrattanasin, "Properties of silk fabric dyed with Eucaliptus, Quercetin, Rutin and Tannin using padding techniques," in *Proceedings of the RMUTP International Conference on Textiles & Fashion*, Bangkok, Thailand, July 2012.

[23] F. A. Cotton and G. Wilkinson, *Advanced Inorganic Chemistry: A Comprehensive Text*, John Wiley & Sons, New York, NY, USA, 3rd edition, 1972.

9

Dyeing of Silk with Anthocyanins Dyes Extract from *Liriope platyphylla* Fruits

Huayin Wang,[1] Ping Li,[2] and Wenlong Zhou[1]

[1] *Key Laboratory of Advanced Textile Materials and Manufacturing Technology (Ministry of Education), Zhejiang Sci-Tech University, Hangzhou, Zhejiang 310018, China*
[2] *Yancheng Institute of Industry Technology, Yancheng, Jiangsu 224005, China*

Correspondence should be addressed to Ping Li; yiitliping@163.com

Academic Editor: Joanne Yip

A new source of natural anthocyanins dyes, from *Liriope platyphylla* fruit, is proposed. This paper analyzes the dye extracts, the primary color components of the extracts, the color features of the extracts under different pH conditions, and their application in silk dyeing. The research shows that, nine anthocyanins are found in *L. platyphylla* fruits by analyzing the results of the HPLC/DAD, MS, and MS/MS spectra. The five major anthocyanins related to delphinidin, petunidin, and malvidin derivatives take up 91.72% of total anthocyanin contents. The color of the solution is red under acidic condition (pH < 3.0) and stays in yellow under alkaline condition with pH values above 7.0. The dye extracts applied to silk fabric with mordant free dyeing show different color under different pH conditions, changing between purple, blue, green, and yellow. However, the dyed colors is light and the dyeing rate is low. Metal mordant such as Sn in chelation enhances the dye depth and improves the fastness of the dyed silk fabrics, especially in silk fabrics dyed by premordanting and metamordanting.

1. Introduction

Natural dyes are generally environment friendly and have many advantages over synthetic dyes. In recent years, the applications of natural dyes are researched due to their biodegradability and higher compatibility with the environment [1–3]. Anthocyanins, with their common structure shown in Figure 1, are flavonoids commonly found in flower petals, fruits, and leaves and produce orange, red, violet, and blue colours [4, 5]. Earlier studies show that anthocyanins are especially abundant in bilberry, strawberry, raspberry, grape skin, blackberry, and so on [6–10]. Interests in these water-soluble pigments have been increased substantially during the last decade because of the increasing evidence demonstrating potential therapeutic effects. Researches also show that anthocyanins have strong oxidation resistance, inhibit the growth of cancerous cells, inhibit inflammation, and have antiobesity effects. The therapeutic properties of anthocyanins have been reviewed and proved [11–14]. Despite the great application potentials that anthocyanins represent for food, pharmaceutical, and cosmetic industries, their application in textile is limited as the pigments lack affinity for the fiber and cannot be sustained during washing and therefore requires further investigation.

In general, anthocyanins show colour variation caused by pH or metal chelation changes. As natural dye used in textile, it is important to study the color coordinates of anthocyanin dye extracts and the dyed fabrics influenced by the pH values of dyeing bath, the metal mordant, and the mordanting processes, which would help to understand the basic color effects of the dyed fabrics using anthocyanin extracts as natural dyes.

L. platyphylla is an evergreen perennial, widely distributed in China, India, Japan, and Korea. Its dried tubers have been commonly used as traditional herbal medicine to cure acute and chronic inflammation and cardiovascular diseases including thrombotic diseases [15, 16]. Various phytochemical and biochemical investigations on tubers of *L. platyphylla* have been reported [17–19]. However, up to now, few reports can be found on analysis and applications of the fruits of *L. platyphylla*. In this work, the analysis of the anthocyanin components of the *L. platyphylla* fruit extracts was

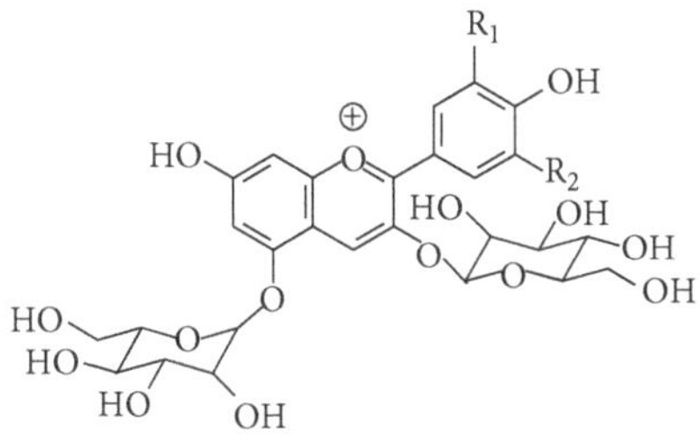

R_1	R_2	Anthocyanin	Aglycon
H	H	Pelargonin	Pelargonidin
OH	H	Cyanin	Cyanidin
OCH_3	H	Peonin	Peonidin
OH	OH	Delphin	Delphinidin
OCH_3	H	Petunin	Petunidin
OCH_3	OCH_3	Malvin	Malvidin

FIGURE 1: Structure of common anthocyanins.

FIGURE 2: Fruits used for extraction.

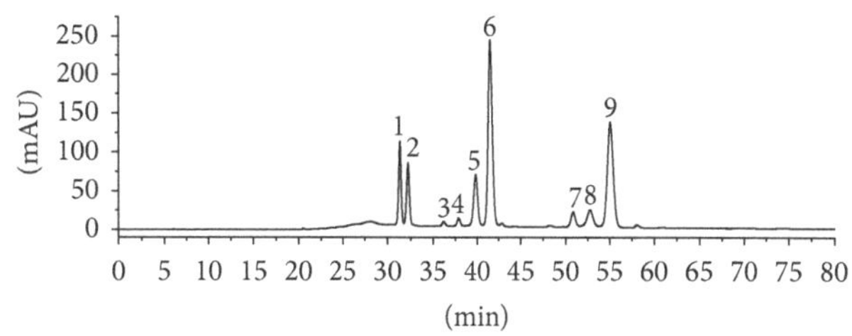

FIGURE 3: HPLC separation of anthocyanins in the fruits of *L. platyphylla*.

performed with HPLC/DAD, MS, and MS/MS spectra. The color coordinates of the extracts and silk fabrics dyed by the extracts were analyzed with CIE1964-XYZ and CIE1976LAB of the Commission Internationale de L'Éclairage.

2. Experiments

2.1. Chemicals. The methanol, $SnCl_2$, and hydrochloric acid were analytical-grade reagents obtained from Tianjin Kemiou Chemical Reagent Co., Ltd. (China). Trifluoroacetic acid (TFA) was analytical-grade reagent obtained from Aladdin Chemistry Co., Ltd. (China).

2.2. Plant Material. Fruits of *L. platyphylla* were harvested in Zhejiang Sci-Tech University, Hangzhou, Zhejiang, China, in January 2013 and the samples (Figure 2) were packed immediately and stored in a freezer at −20°C until processed.

One criterion of the fruit is its moisture content, which is evaluated first for the fresh fruits. Moisture content is the percentage of water contained in a material. 100 g of the fresh fruits was taken and kept in an oven at 100°C for 24 h and weighted afterward. Triplicate assays were applied. The moisture content (%) of the fresh fruits is 50.2 ± 8.6%.

2.3. Extraction of Anthocyanin Dyes. The fruits were crushed with a mortar and weighed for preparation. Three typical methods are used in this paper, which are mainly differentiated by the solvents used. All the three methods extracted anthocyanins by putting the fruits into a curtain solvent, stirring them for 10 min, and keeping them 24 h at room temperature in darkness without stirring [30, 31]. The mixture was then filtered through a Buchner funnel under vacuum. The solid residue was washed with the same solvent until a clear solution was obtained. The combined filtrates were dried using a rotary evaporator at 55°C or 60°C, which was chosen individually for each solvent for the best extraction results. The concentrate was dissolved in distill water and the solution obtained was used for dyeing. The first method uses methanol containing 0.1% (V/V) of concentrated hydrochloric acid with 55°C for drying and the second method uses solvent $CH_3OH/TFA/H_2O$ 49.5/0.5/50 v/v/v with 60°C for drying, while the third method uses methanol containing 5.0% (W/V) of concentrated citric acid with 55°C for drying.

2.4. Total Anthocyanins Content Measurement. Total anthocyanins content (TAC) was normally determined using pH differential method and expressed as cyanidin-3-glucoside (Cy-3-glc) equivalents, according to the following formula [32, 33]:

$$c\left[\mathrm{mg/L}\right] = \frac{A \times M_w \times \mathrm{DF} \times 1000}{\varepsilon_M \times d}, \qquad (1)$$

where A is the absorbance value in visible region, equaling $(A_{\lambda\max} - A_{700})_{\mathrm{pH1.0}} - (A_{\lambda\max} - A_{700})_{\mathrm{pH4.5}}$, which corresponds to sodium acetate-acetic acid buffer solution when pH = 4.5 and potassium chloride-hydrochloric acid buffer solution when pH = 1.0 to dilute the anthocyanin extracts; M_w is the molecular weight (g/mol) of Cy-3-glc (449.2 g/mol in the analysis); DF is the dilution factor; ε_M is the molar extinction coefficient (26,900 for Cy-3-glc); and d is the path length of the cuvette (1 cm in the analysis). The experiments were performed in triplicate assays.

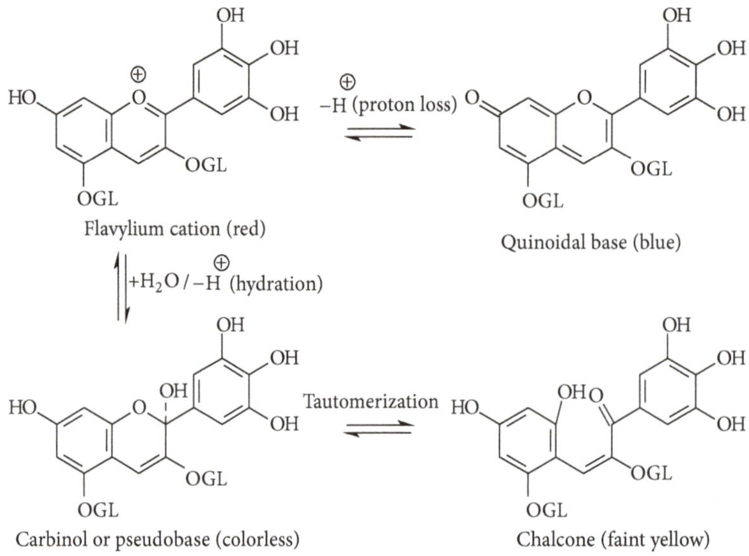

FIGURE 4: Structure change of anthocyanin with pH in aqueous solution.

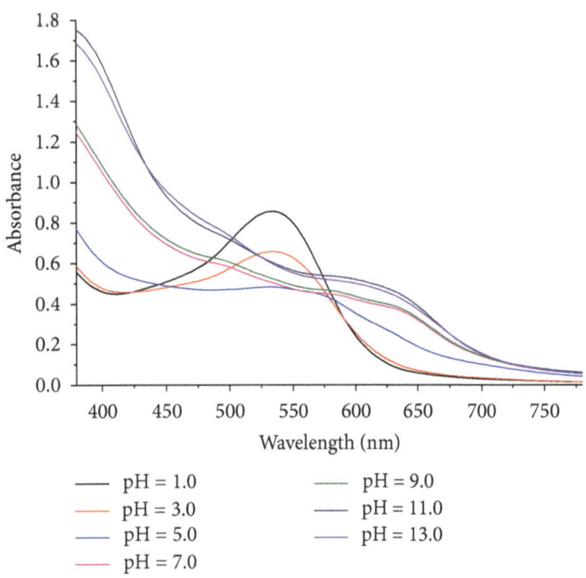

FIGURE 5: Visible spectra of *L. platyphylla* fruit extracts at several pH values.

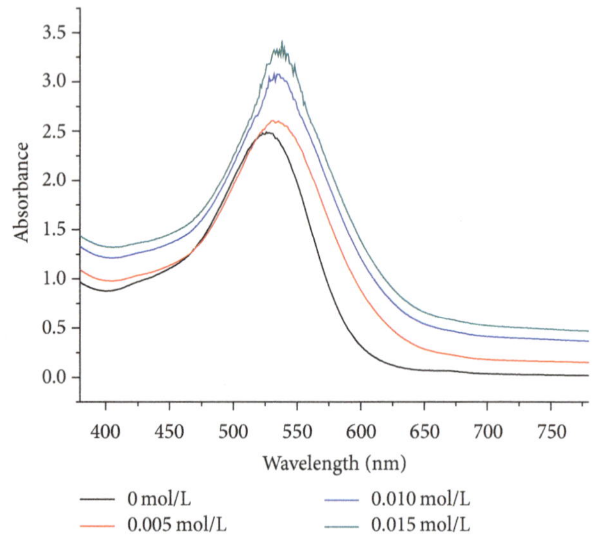

FIGURE 6: Visible spectra of Sn chelated anthocyanins extracts in buffer solutions at pH 1.0.

TABLE 1: Total anthocyanin contents by different acids (mg/100 g fresh weight; mean ± SD of triplicate assays).

Extracting acid	TAC
0.1% HCl	474.45 ± 25.67
5% citric acid	615.84 ± 34.21
0.5% TFA	535.37 ± 28.75

2.5. HPLC/DAD Analysis. HPLC/DAD experiments were performed using the HPLC machine (1260 Infinity, Agilent Technologies, CA, USA). For analysis, the separations of anthocyanins were carried out on Agilent Zorbax SB-C18 column (4.6 × 250 mm, 5 μm) (Agilent Technologies, CA, USA) equipped with a C18 guard column (Agilent Technologies, CA, USA). The column oven temperatures were set to 35°C. The mobile phase consisted of 5% formic acid (solvent A) and 100% CH_3CN (solvent B). For the separation of anthocyanin extracts on SB-C18 column (4.6 × 250 mm, 5 μm), the elution conditions were set as follows: isocratic elution 0% B, 0–10 min; linear gradient from 0% B to 10% B, 10–20 min; 14% B, 20–60 min; 100% B, 60–80 min; flow rate 0.5 mL min^{-1}; and injection volumes 10 μL. The detected wavelength was 520 nm and all spectra peaks from 250 nm to 640 nm were recorded.

2.6. HPLC/MS and HPLC/MS/MS Analysis. HPLC/MS and HPLC/MS/MS experiments were performed using the same HPLC machine (Agilent Technologies) equipped with a mass

TABLE 2: Identification of anthocyanins in *L. platyphylla* fruit extracts.

Peak	RT (min)	MS	MS^2	UV-Vis data				Tentative identification	References
				$\lambda_{vis,max}$	$\lambda_{acyl,max}$	$A_{440}/A_{vis,max}$	$A_{acyl,max}/A_{vis,max}$		
1	31.330	465	303	524	346	28.74	9.51	Dp 3-glu	[20–22]
2	32.217	611	465/303	525	347	28.18	9.51	Dp 3-rut	[23]
3	35.882	479	317	515	332	32.47	6.84	Pt 3-gal	[21, 24]
4	37.867	595	449/287	517	325	31.13	18.13	Cy 3-rut	[21, 23, 25–27]
5	39.807	479	317	525	345	27.99	11.19	Pt 3-glu	[21, 22, 26, 28]
6	41.460	625	479/317	527	347	27.38	9.97	Pt 3-rut	[23]
7	50.835	609	463/301	518	331	31.97	8.77	Pn 3-rut	[23, 24, 29]
8	52.762	493	331	527	348	27.96	11.48	Mv 3-glu	[20–22]
9	55.024	639	493/331	528	348	26.86	10.57	Mv 3-rut	[23]

RT: retention time; Dp: delphinidin; Cy: cyanidin; Pt: petunidin; Pn: peonidin; Mv: malvidin; glu: glucoside; rut: rutinoside; gal: galactoside.

spectrometer (LCQ FLEET, Thermo Fisher Scientific, PA, USA) and electrospray ionization (ESI). The HPLC column and HPLC conditions for the mobile gradient were the same as those in HPLC/DAD experiments. The MS analytical parameters were as follows: polarity, positive; ion source, turbo spray (ESI); capillary temperature, 350°C; capillary voltage, 60 V; spray voltage, 4.5 KV; and full-scan range, from m/z 400–1200. MS^2 scan of the most abundant ion uses relative collision energy of 20%.

2.7. Mordanting. The weighed silk fabrics were treated with tin metal salts for premordanting, metamordanting, and postmordanting. The amount of mordant used was 5% (o.w.f.). The fabrics were mordanted at liquor ratio of 50 : 1. In premordanting, fabrics were immersed in the mordant solution at 60°C for 30 min. In metamordanting, fabrics were immersed in a dyeing bath containing both mordant and extracted dyes for 120 min. In postmordanting, fabrics were immersed in a dyeing bath containing only extracted dyes for 120 min and only mordant at 60°C for 30 min, separately. Due to the light sensitivity of the mordants, fabrics after mordanting should be dyed immediately.

2.8. Dyeing. The weighted silk fabric was dyed at a liquor ratio of 50 : 1. 20 g of dye concentrates was dissolved in 1000 mL deionized water for a piece of fabric weighing 20 g. The fabric was kept in dye bath for 120 min at 60°C. The pH value of the dyeing solution was adjusted with dilute solutions of hydrochloric acid or sodium hydroxide. The dyed fabric was rinsed with cold water and finally dried at room temperature in shade.

2.9. Color Analysis. Colorimetric methods are very useful in characterizing the color properties of pigments. Tristimulus colorimetry has been widely applied. An absorbance spectrum ($\Delta\lambda = 1$ nm) of anthocyanin dye solutions was recorded with a UV-VIS double-beam Spectrophotometers TU-1950 (Beijing Purkinje General Instrument Co., Ltd., Beijing, China). D65 standard illuminant and 10° standard observer

TABLE 3: Color coordinates of the *L. platyphylla* fruit extract at several pH values.

pH	L^*	a^*	b^*	C^*	h
1.0	62.21	45.84	0.80	45.84	1.00
3.0	65.82	33.99	3.69	34.19	6.20
5.0	67.56	12.06	7.09	13.99	30.47
7.0	64.10	4.29	21.13	21.56	78.53
9.0	63.04	4.74	21.47	21.99	77.54
11.0	58.10	3.49	29.67	29.87	83.29
13.0	58.59	4.79	32.04	32.39	81.50

were considered in the calculation. Reference blank measurements were made with the cuvette filled with distilled water. The whole visible spectrum (380–780 nm) was recorded. The CIELAB parameters (L^*, a^*, b^*, C^*, and h) were determined following the CIE1964-XYZ and CIE1976LAB of the Commission Internationale de L'Éclairage.

Color values of dyed fabrics were evaluated by means of color strength (K/S), color differences (ΔE_{cmc}), and other CIELAB values (L^*, a^*, b^*, C^*, and h). With 10 mm sample diameter (geometry d/10°), CIELAB values were calculated for illuminant D65 on Datacolor SF600 spectrophotometer with software data match Color Tools 3.1 (Datacolor International, NJ, USA). Five measurements were made on each sample and variations in the percentage reflectance values over range of 400–700 nm were recorded.

In the uniform space CIELAB, L^* corresponds to the brightness (100 = white, 0 = black), a^* to the red-green coordinate (positive sign = red, negative sign = green), and b^* to the yellow-blue coordinate (positive sign = yellow, negative sign = blue). Chroma (C^*) is the quantitative attribute of colourfulness. Hue (h) is the attribute according to which colors are defined as reddish, greenish, and so forth. The total colour difference (ΔE^*) between two samples is estimated using the following expression:

$$\Delta E^* = \left[(\Delta L^*)^2 + (\Delta a^*)^2 + (\Delta b^*)^2\right]^{1/2}, \quad (2)$$

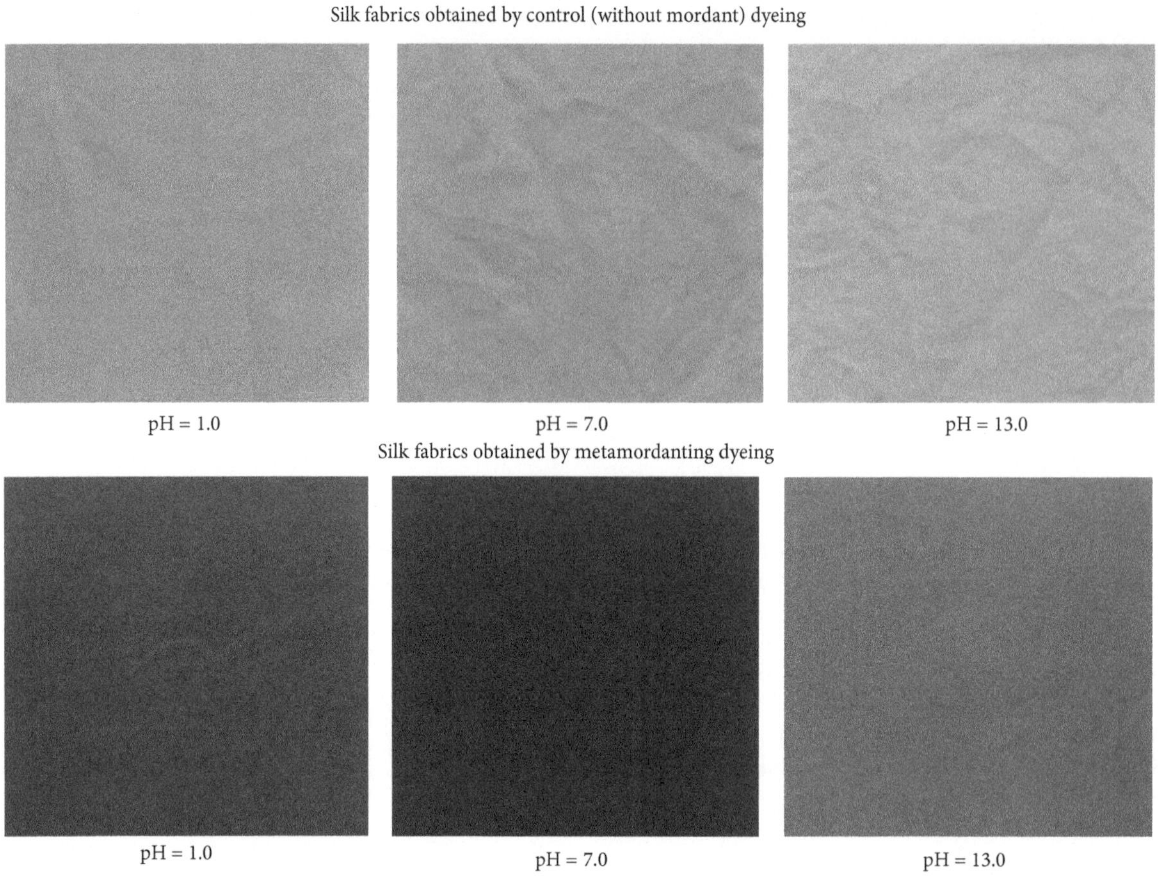

FIGURE 7: Some colour samples obtained by metamordanting dyeing and control (without mordant) dyeing.

TABLE 4: Color coordinates of the Sn chelated anthocyanins extracts in buffer solutions at pH 1.0.

No.	Concentration of Sn^{2+} (mol/L)	λ_{max} (nm)	L^*	a^*	b^*	C^*	h
1	0	526	46.90	69.19	24.84	73.52	19.75
2	0.005	531	29.72	56.47	−2.30	56.52	357.66
3	0.010	532	20.74	46.53	−5.05	46.80	353.80
4	0.015	534	16.91	42.43	−6.83	42.97	350.85

TABLE 5: Silk dyeing with *L. platyphylla* fruit extracts at several pH.

pH	L^*	a^*	b^*	C^*	h^*	ΔE^*
1.0	58.76	16.79	−7.59	18.43	335.66	40.21
3.0	69.26	8.88	0.41	8.89	2.65	26.14
5.0	70.68	0.25	1.95	1.97	82.64	22.94
7.0	71.93	−0.66	7.65	7.68	94.93	22.17
9.0	74.66	−0.17	12.73	12.73	90.76	21.29
11.0	74.34	1.13	13.42	13.47	85.19	21.95
13.0	80.92	2.94	18.76	18.99	81.08	20.48

where ΔE^* indicates the total color differences between dyed fabric and undyed original silk fabric. The color indicators of the undyed silk fabric are $L^*(93.6)$, $a^*(-0.28)$, $b^*(3.00)$, $C^*(3.01)$, and $h^*(95.42)$.

TABLE 6: Silk dyeing with *L. platyphylla* fruit extracts with tin mordanting at several pH by premordanting.

pH	L^*	a^*	b^*	C^*	h^*	ΔE^*
1.0	45.1	15.24	−15.57	21.79	314.39	54.2
3.0	43.76	11.92	−23.95	26.75	296.47	57.96
5.0	36.62	10.41	−24.43	26.56	293.08	64.13
7.0	42.49	5.81	−19.41	20.26	286.65	56.14
9.0	51.61	2.01	−12.77	12.93	278.93	44.91
11.0	64.14	2.14	4.82	5.28	66.07	29.62
13.0	72.98	0.17	9.57	9.57	88.98	21.65

2.10. Fastness Testing of Dyed Fabrics. The dyed silk fabrics were tested according to AATCC standard methods. The specific tests were color fastness to light, AATCC 16-2004;

TABLE 7: Silk dyeing with *L. platyphylla* fruit extracts with tin mordanting at several pH by metamordanting.

pH	L^*	a^*	b^*	C^*	h^*	ΔE^*
1.0	55.5	13.29	−16.16	20.92	309.42	44.75
3.0	55.7	8.19	−15.93	17.91	297.2	43.2
5.0	55.52	6.34	−17.93	19.05	289.44	43.97
7.0	43.73	6.4	−22.01	22.92	286.2	56.19
9.0	52.84	−0.72	−18.32	18.34	267.76	46
11.0	56.7	4.41	−5.91	7.37	306.72	38.24
13.0	63.02	2.61	2.43	3.57	42.98	30.72

TABLE 8: Silk dyeing with *L. platyphylla* fruit extracts with tin mordanting at several pH by postmordanting.

pH	L^*	a^*	b^*	C^*	h^*	ΔE^*
1.0	67.46	1.07	−1.93	2.21	299.03	26.63
3.0	71.78	−4.92	0.24	4.92	177.21	22.47
5.0	72.86	−5.22	0.36	5.24	176.01	21.47
7.0	73.49	−5.3	5.91	9.01	99.46	19.85
9.0	78.72	−0.62	11.19	11.21	93.18	16.98
11.0	76.84	−0.66	11.89	11.91	93.19	18.97
13.0	85.93	−0.86	16.75	16.77	92.95	15.76

color fastness to crocking, AATCC 8-2007; color fastness to washing, AATCC 61-2009; and color fastness to perspiration, AATCC 15-2009. The colors of the specimens were rated by referencing to the Gray Scale for Color Change. The staining on each undyed original fabric was rated by means of the Gray Scale for Staining.

3. Results and Discussion

3.1. Effect of Acids on Total Anthocyanin Content. Anthocyanin from the fruits was extracted using 0.1% HCl, 5% citric acid, or 0.5% TFA. The total anthocyanin contents extracted from the fruits using different acids are shown in Table 1. The results show that the citric acid gives better yield and deeper color, which produce 29.8% and 15.0% more anthocyanins than 0.1% HCl and 0.5% TFA, respectively.

3.2. Characterization of Anthocyanins in the Fruit Extracts. Figure 3 shows the HPLC chromatograms of anthocyanin extracts carried on C18 columns. In Figure 3, nine different anthocyanins from the fruits are identified. Peaks 1, 2, 5, 6, and 9, eluted at 31.330 min, 32.220 min, 39.807 min, 41.460 min, and 55.024 min, indicate the major anthocyanins and take up 9.32%, 8.06%, 9.51%, 36.54%, and 28.29% of the total anthocyanin content, respectively. The four others (peaks 3, 4, 7, and 8) take up total 8.28%, each accounting for 0.5% to 4% of the total anthocyanin content.

By combining the results of DAD, MS, and MS/MS spectra, the nine different anthocyanins extracted from the fruit of *L. platyphylla* are summarized in Table 2. Among the nine different anthocyanins, delphinidin 3-glucoside (Dp 3-glu), delphinidin 3-rutinoside (Dp 3-rut), cyanidin-3-glucoside (Cy-3-glu), petunidin 3-glucoside (Pt 3-glu), petunidin 3-rutinoside (Pt 3-rut), malvidin 3-glucoside (Mv 3-glu), and malvidin 3-rutinoside (Mv 3-rut) were confirmed by previous reported NMR spectroscopy data [34].

3.3. Effect of pH Values on Anthocyanin Dye Colors. Anthocyanins show great susceptibility toward pH variation, being more stable in acidic media at low pH values than in alkaline solutions at high pH value. The ionic nature of anthocyanins enables the changes of the molecule structures according to the prevailing pH values and results in different colors and hues at different pH values. In acidic aqueous solution, anthocyanins exist in the form of four main equilibrium species (Figure 4) [35]: red flavylium cation, blue quinoidal base, colorless carbinol or pseudobase, and yellowish chalcone. Under acidic conditions (pH < 2.0), the anthocyanins exist primarily in the form of red flavylium cation. Increasing the pH value causes fast loss of the proton and produces quinoidal base forms, blue or violet. At the same time hydration of flavylium cation occurs and the carbinol or pseudobase is generated, which slowly reaches equilibrium and produces the chalcone in faint yellow. The relative amounts of above four forms of anthocyanins at the equilibrium condition vary according to pH values.

The visible spectra of *L. platyphylla* fruit extracts at different pH values are shown in Figure 5. The values of the color indexes are described in Table 3. Figure 5 shows that, with the increasing of pH value, the maximum visible absorption peak moves towards the long wavelength. The absorbance decreases first and then increases. Table 3 shows that the color of the solution is red under strong acidic condition (pH < 3.0) and stays in yellow under alkaline condition with pH value above 7.0.

As seen in Table 3, the lightness index L^* is inversely correlated to the color contents in the solution. When under acidic condition (pH < 3.0), the index L^* of the *L. platyphylla* fruit dye solution is stable, the pigment mainly presents in the form of red anthocyanin molten salt, and the a^* is positive and relatively high. When pH increases to above 3.0, the pigment gradually transforms from colored substances into colorless substances, which are mainly colorless methanol false alkali. When pH increases to above 5.0, the value of L^* starts to increase and the solution color becomes deepened, transforming from red anthocyanin molten salt to blue quinone alkali. With the further increasing of the pH value, the solution color becomes lighter, since the blue quinone alkali or colorless methanol false base structure transforms to light yellow chalcone structure. A similar conclusion can be obtained from the analysis of a^* and b^* values. The C^* value reflects the singularity of each pigment structure in the solution. The more complex the pigment structure is, the lower the value of C^* is. Based on the data in Table 3, the C^* value is higher under strong acid (pH < 3.0) or strong base (pH > 11.0), which suggests the chemical structures of pigments in these two solutions are simple. They are mainly in the form of anthocyanin molten salt at pH < 3.0 and chalcone structure at pH > 11.0, which is consistent with the previous analysis.

TABLE 9: Fastness properties of dyed silk fabrics with *L. platyphylla* fruit extracts.

Dyeing methods	pH value of dyeing liquor	WF[a]	Perspiration		$Crock_{dry}$	$Crock_{wet}$	LF[c]
			CC[b]	Staining			
Control (without mordant)	1.0	2	2	1.5	2.5	2	1.5
	3.0	2.5	2	1.5	2.5	2	1.5
	5.0	2.5	2	1.5	2.5	2.5	1.5
	7.0	2.5	2	1.5	2.5	2.5	1.5
	9.0	2	2	2	2.5	2.5	1.5
	11.0	2.5	2.5	2	2.5	2.5	1.5
	13.0	2.5	2.5	2	2.5	2.5	2
Premordanting	1.0	3.5	3	3.5	3.5	3.5	2.5
	3.0	3.5	3	3.5	3.5	3.5	4
	5.0	4	3.5	4	4	4	3.5
	7.0	4	3.5	3.5	4	3.5	3.5
	9.0	3.5	3.5	3.5	4	3.5	3.5
	11.0	4	3.5	3.5	4.5	3.5	3.5
	13.0	4.5	3.5	4	4.5	3.5	3.5
Metamordanting	1.0	3.5	3	3	3.5	3.5	3.5
	3.0	3.5	3.5	3.5	3.5	3.5	3.5
	5.0	4	3.5	3.5	3.5	4	4
	7.0	4	4	3.5	3.5	3.5	3.5
	9.0	3.5	3.5	4	3.5	3.5	4
	11.0	4	3.5	3.5	4	3.5	4
	13.0	4	3.5	4	4	3.5	3.5
Postmordanting	1.0	3.5	3.5	3.5	3.5	3.5	3.5
	3.0	3.5	3.5	3.5	3.5	3.5	4
	5.0	4	3.5	3.5	4	4	4
	7.0	4	4	3.5	4	3.5	4
	9.0	4	3.5	3.5	4	3.5	3.5
	11.0	4	4	4	4	3.5	3.5
	13.0	4	4	4	4	3.5	3.5

[a]Washing fastness, [b]color change, and [c]light fastness.

3.4. Metal and Anthocyanin Chelation. Metals have been commonly used to stabilize the color of cyanidin, delphinidin, and petunidin, which have more than one free hydroxyl group in the chromane ring and are capable of metal chelation [36, 37]. The most common metals in anthocyanin complexes are tin (Sn), copper (Cu), iron (Fe), aluminum (Al), and magnesium (Mg) [38]. The main characteristic of anthocyanins and anthocyanindins with *o*-di-hydroxyl groups in the B ring (Cy, Dp, and Pt) is their ability to form metal-anthocyanin complexes [35, 39]. The resultant formations of metal-anthocyanin complexes were suggested by the sample colors as shown by L^*, a^*, b^*, C^*, and h. Visible spectra of Sn chelated *L. platyphylla* fruits extracts at pH 1.0 are shown in Figure 6. The color coordinates are shown in Table 4. As shown in Figure 6, with the increasing of the concentrations of Sn^{2+}, the bathochromic maximum absorption wavelength of the *L. platyphylla* fruits shifts with $\Delta\lambda = 5$–8 nm, the absorbance increases significantly, and the hyperchromic change is obvious.

Based on the CIELAB color system in Table 4, with the increasing of the Sn^{2+} concentration, the color indicators such as L^*, a^*, b^*, and C^* of the solution show a decreasing trend. The color of the solution changes from red to blue violet as shown in Table 4, which suggests that the Sn^{2+} forms a violet complex with the red flavylium cation.

3.5. Dyeing of Silk by the Extracts. The silk was dyed using the *L. platyphylla* fruit extracts without mordanting, with tin premordanting, metamordanting or postmordanting at different pH values. The corresponding CIELAB values of dyed silk are shown in Tables 5, 6, 7, and 8, respectively. Some colour samples are shown in Figure 7. As concluded from the pigment color indicators, the dyed silk fabric without mordant dyeing shows different color results under different pH conditions, such as purple, blue, green, and yellow. The dyed color is light and the dyeing rate is low. The color is comparable deep with pH = 1.0 and becomes lighter with the increasing of the pH values.

Compared to silk fabrics dyed without mordanting in Table 5, silk fabrics dyed with stannous chloride mordant has deeper dyed color as shown in Table 6 to Table 8. The total color difference ΔE^* is increased obviously. The color of the fabric dyed with mordants appears to be red-purple, blue-purple, blue, green, yellow, and so forth.

Different pH values of dyeing solutions cause differences not only in hue color and ΔE^* values but also in L^*, a^*, b^*,

C^*, and h^* values. In case of mordant dyeing, the color depth of dyed silk fabric with stannous chloride postmordanting is low and the dyeing results are not satisfactory. When the dyed fabric is under postmordant processing, part of the anthocyanins transfers from fabric to the dyeing bath.

The washing fastness, perspiration fastness, crocking fastness, and light fastness were investigated with the results shown in Table 9. The results are satisfactory for $SnCl_2$ mordanted fabrics due to the special chelation of anthocyanin color moieties and stannous.

4. Conclusion

The anthocyanin extracted from *L. platyphylla* fruits as a source of natural dyes has been utilized for the first time for silk fabrics dyeing to my best knowledge. In principal components analysis, nine anthocyanins from the fruits were identified by analyzing the results of the HPLC/DAD, MS, and MS/MS spectra. The five major anthocyanins related to delphinidin, petunidin, and malvidin derivatives take up 91.72% of total anthocyanin contents. In terms of acids used for extraction, 5% citric acid gave better yield and deeper color compared with 0.1% HCl and 0.5% TFA. In silk fabrics dyeing, the fruit pigment under mordant free condition presents different color results, changing between purple, blue, green, and yellow. However, from the point of the dyed color, the color is light and the dyed rate is low. The metal mordant such as tin in chelation with the extracts greatly enhances the dye depth and improves the fastness properties of the dyed silk fabrics, especially for silk fabrics dyed with premordanting and metamordanting.

Conflict of Interests

The authors have declared that they have no conflict of interests.

References

[1] T. Bechtold, A. Turcanu, E. Ganglberger, and S. Geissler, "Natural dyes in modern textile dyehouses—how to combine experiences of two centuries to meet the demands of the future?" *Journal of Cleaner Production*, vol. 11, no. 5, pp. 499–509, 2003.

[2] L. G. Angelini, A. Bertoli, S. Rolandelli, and L. Pistelli, "Agronomic potential of *Reseda luteola* L. as new crop for natural dyes in textiles production," *Industrial Crops and Products*, vol. 17, no. 3, pp. 199–207, 2003.

[3] E. Tsatsaroni, M. Liakopoulou-Kyriakides, and I. Eleftheriadis, "Comparative study of dyeing properties of two yellow natural pigments—effect of enzymes and proteins," *Dyes and Pigments*, vol. 37, no. 4, pp. 307–315, 1998.

[4] R. L. Jackman, R. Y. Yada, M. A. Tung, and R. A. Speers, "Anthocyanins as food colorants—a review," *Journal of Food Biochemistry*, vol. 11, no. 3, pp. 201–247, 1987.

[5] Ø. M. Andersen and M. Jordheim, "Anthocyanins," in *Encyclopedia of Life Sciences (ELS)*, John Wiley & Sons, Chichester, UK, 2010.

[6] Q. Du, G. Jerz, and P. Winterhalter, "Isolation of two anthocyanin sambubiosides from bilberry (*Vaccinium myrtillus*) by high-speed counter-current chromatography," *Journal of Chromatography A*, vol. 1045, no. 1-2, pp. 59–63, 2004.

[7] A. Kirca, M. Özkan, and B. Cemeroğlu, "Storage stability of strawberry jam color enhanced with black carrot juice concentrate," *Journal of Food Processing and Preservation*, vol. 31, no. 5, pp. 531–545, 2007.

[8] Q. Tian, R. M. Aziz, G. D. Stoner, and S. J. Schwartz, "Anthocyanin determination in black raspberry (*Rubus occidentalis*) and biological specimens using liquid chromatography-electrospray ionization tandem mass spectrometry," *Journal of Food Science*, vol. 70, no. 1, pp. C43–C47, 2005.

[9] X. Wang, H. Tong, F. Chen, and J. D. Gangemi, "Chemical characterization and antioxidant evaluation of muscadine grape pomace extract," *Food Chemistry*, vol. 123, no. 4, pp. 1156–1162, 2010.

[10] P. Murapa, J. Dai, M. Chung, R. J. Mumper, and J. D'Orazio, "Anthocyanin-rich fractions of blackberry extracts reduce UV-induced free radicals and oxidative damage in keratinocytes," *Phytotherapy Research*, vol. 26, no. 1, pp. 106–112, 2012.

[11] D. Ghosh, "Anthocyanins and anthocyanin-rich extracts in biology and medicine: biochemical, cellular, and medicinal properties," *Current Topics in Nutraceutical Research*, vol. 3, no. 2, pp. 113–124, 2005.

[12] J. Kong, L. Chia, N. Goh, T. Chia, and R. Brouillard, "Analysis and biological activities of anthocyanins," *Phytochemistry*, vol. 64, no. 5, pp. 923–933, 2003.

[13] M. A. Lila, "Anthocyanins and human health: an in vitro investigative approach," *Journal of Biomedicine and Biotechnology*, vol. 2004, no. 5, pp. 306–313, 2004.

[14] T. K. McGhie and M. C. Walton, "The bioavailability and absorption of anthocyanins: Towards a better understanding," *Molecular Nutrition and Food Research*, vol. 51, no. 6, pp. 702–713, 2007.

[15] S. B. Choi, J. D. Wha, and S. Park, "The insulin sensitizing effect of homoisoflavone-enriched fraction in *Liriope platyphylla* Wang et Tang via PI 3-kinase pathway," *Life Sciences*, vol. 75, no. 22, pp. 2653–2664, 2004.

[16] S. Hong, H. Jeong, H. Chung et al., "An herbal formula, Herbkines, enhances cytokines production from immune cells," *Journal of Ethnopharmacology*, vol. 98, no. 1-2, pp. 149–155, 2005.

[17] M. H. Kwak, J. E. Kim, I. S. Hwang et al., "Quantitative evaluation of therapeutic effect of Liriope platyphylla on phthalic anhydride-induced atopic dermatitis in IL-4/Luc/CNS-1 Tg mice," *Journal of Ethnopharmacology*, vol. 148, no. 3, pp. 880–889, 2013.

[18] G. Li, W. Ra, J. Park et al., "Developing EST-SSR markers to study molecular diversity in *Liriope* and *Ophiopogon*," *Biochemical Systematics and Ecology*, vol. 39, no. 4–6, pp. 241–252, 2011.

[19] Y.-C. Tsai, S.-Y. Chiang, M. El-Shazly et al., "The oestrogenic and anti-platelet activities of dihydrobenzofuroisocoumarins and homoisoflavonoids from *Liriope platyphylla* roots," *Food Chemistry*, vol. 140, no. 1-2, pp. 305–314, 2013.

[20] Y. S. Oh, J. H. Lee, S. H. Yoon et al., "Characterization and quantification of anthocyanins in grape juices obtained from the grapes cultivated in Korea by HPLC/DAD, HPLC/MS, and HPLC/MS/MS," *Journal of Food Science*, vol. 73, no. 5, pp. 378–389, 2008.

[21] J. C. Mi, L. R. Howard, R. L. Prior, and J. R. Clark, "Flavonoid glycosides and antioxidant capacity of various blackberry,

blueberry and red grape genotypes determined by high-performance liquid chromatography/mass spectrometry," *Journal of the Science of Food and Agriculture*, vol. 84, no. 13, pp. 1771–1782, 2004.

[22] M. Fanzone, F. Zamora, V. Jofré, M. Assof, C. Gómez-Cordovés, and Á. Peña-Neira, "Phenolic characterisation of red wines from different grape varieties cultivated in Mendoza province (Argentina)," *Journal of the Science of Food and Agriculture*, vol. 92, no. 3, pp. 704–718, 2012.

[23] E. Alexandra Pazmio-Durán, M. M. Giusti, R. E. Wrolstad, and M. B. Glória, "Anthocyanins from banana bracts (*Musa X paradisiaca*) as potential food colorants," *Food Chemistry*, vol. 73, no. 3, pp. 327–332, 2001.

[24] L. Cabrita and Ø. M. Andersen, "Anthocyanins in blue berries of Vaccinium padifolium," *Phytochemistry*, vol. 52, no. 8, pp. 1693–1696, 1999.

[25] W. Huang, S. Zhang, G. Qin, W. Le, and J. Wu, "Isolation and determination of major anthocyanin pigments in the pericarp of *P. Communis* L. cv. 'Red Du Comices' and their association with antioxidant activity," *African Journal of Agricultural Research*, vol. 7, pp. 3772–3780, 2012.

[26] Y. Lu, L. Y. Foo, and H. Wong, "Isolation of the first C-2 addition products of anthocyanins," *Tetrahedron Letters*, vol. 43, no. 37, pp. 6621–6624, 2002.

[27] E. Sariburun, S. Şahin, C. Demir, C. Türkben, and V. Uylaşer, "Phenolic content and antioxidant activity of raspberry and blackberry cultivars," *Journal of Food Science*, vol. 75, no. 4, pp. C328–C335, 2010.

[28] C. Qin, Y. Li, W. Niu, Y. Ding, X. Shang, and C. Xu, "Composition analysis and structural identification of anthocyanins in fruit of waxberry," *Czech Journal of Food Sciences*, vol. 29, no. 2, pp. 171–180, 2011.

[29] M. Dueñas, J. J. Pérez-Alonso, C. Santos-Buelga, and T. Escribano-Bailón, "Anthocyanin composition in fig (*Ficus carica* L.)," *Journal of Food Composition and Analysis*, vol. 21, no. 2, pp. 107–115, 2008.

[30] M. T. Escribano-Bailón, C. Alcalde-Eon, O. Muñoz, J. C. Rivas-Gonzalo, and C. Santos-Buelga, "Anthocyanins in berries of Maqui (Aristotelia chilensis (Mol.) Stuntz)," *Phytochemical Analysis*, vol. 17, no. 1, pp. 8–14, 2006.

[31] M. P. Kähkönen, J. Heinämäki, V. Ollilainen, and M. Heinonen, "Berry anthocyanins: isolation, identification and antioxidant activities," *Journal of the Science of Food and Agriculture*, vol. 83, no. 14, pp. 1403–1411, 2003.

[32] T. Sukwattanasinit, J. Burana-Osot, and U. Sotanaphun, "Spectrophotometric method for quantitative determination of total anthocyanins and quality characteristics of Roselle (Hibiscus sabdariffa)," *Planta Medica*, vol. 73, no. 14, pp. 1517–1522, 2007.

[33] J. Lee, R. W. Durst, and R. E. Wrolstad, "Determination of total monomeric anthocyanin pigment content of fruit juices, beverages, natural colorants, and wines by the pH differential method: collaborative study," *Journal of AOAC International*, vol. 88, no. 5, pp. 1269–1278, 2005.

[34] J. H. Lee and M. Choung, "Identification and characterisation of anthocyanins in the antioxidant activity-containing fraction of Liriope platyphylla fruits," *Food Chemistry*, vol. 127, no. 4, pp. 1686–1693, 2011.

[35] A. Castañeda-Ovando, M. de Lourdes Pacheco-Hernández, M. Elena Páez-Hernández, J. A. Rodríguez, and C. A. Galán-Vidal, "Chemical studies of anthocyanins: a review," *Food Chemistry*, vol. 113, pp. 859–871, 2009.

[36] T. Goto, H. Tamura, T. Kawai, T. Hoshino, N. Harada, and T. Kondo, "Chemistry of metalloanthocyanins," *Annals of the New York Academy of Sciences*, vol. 471, pp. 155–173, 1986.

[37] Y. Osawa, "Copigmentation of anthocyanins," in *Anthocyanins as Food Colors*, P. Markakis, Ed., pp. 41–68, Academic Press, New York, NY, USA, 1982.

[38] R. N. Cavalcanti, D. T. Santos, and M. A. A. Meireles, "Non-thermal stabilization mechanisms of anthocyanins in model and food systems—an overview," *Food Research International*, vol. 44, no. 2, pp. 499–509, 2011.

[39] G. A. Ellestad, "Structure and chiroptical properties of supramolecular flower pigments," *Chirality*, vol. 18, no. 2, pp. 134–144, 2006.

10

Heat Generation by Polypyrrole Coated Glass Fabric

A. M. Rehan Abbasi,[1,2] J. Militky,[1] and J. Gregr[1]

[1] *Technical University of Liberec, Studentska 2, 46117 Liberec, Czech Republic*
[2] *Department of Textile Engineering, BUITEMS, 87100 Quetta, Pakistan*

Correspondence should be addressed to A. M. Rehan Abbasi; rehan_abbaci@hotmail.com

Academic Editor: Shuichi Tanoue

Vapor deposition technique was employed to coat polypyrrole (PPy) on glass substrate using $FeCl_3$ as oxidant and p-toluenesulfonic acid ($^{-}$OTs) as doping agent. The Joule heating effect of PPy coated E-glass fabric was studied by supplying various DC electric fields. The coated fabric exhibited reasonable electrical stability, possessed medium electrical conductivity and was effective in heat generation. An increase in temperature of conductive fabric subjected to constant voltage was observed whereas decrease in power consumption was recorded. Thickness of PPy coating on glass fibers was analyzed by Laser confocal microscope and scanning electron microscope.

1. Introduction

In the late 1970s, MacDiarmid, Heeger, and Shirakawa discovered how to get polymers conducting electricity [1]. The first material becoming an intrinsically conducting polymer (ICP) was polyacetylene, after a doping with iodine. The announcement of this discovery quickly reverberated around scientific communities, and intensity of the research for other conducting polymers magnified dramatically [1–3]. A new generation of polymers was then developed, exhibiting the electrical and optical properties of metals or semiconductors, at the same time retaining the attractive mechanical properties and processing advantages of polymers. Intrinsically, conducting polymers were immediately seen as a new route to mimic metallic conductivity, besides the well-known approach to insert conductive fillers into an inherently insulating resin, or to coat a plastic substrate with a conductive metal solution [4].

Ability to apply conducting polymers to any fiber type and textile structure provides unlimited flexibility in the design of conductive fabrics. Each fiber on the surface of the textile is coated by a thin layer of conducting polymer. As a result of chemical polymerization in the presence of a fabric, a continuous thin film of conducting polymer is formed on the fabric surface. Current will flow through these fabrics without any wiring through them and the heat generated is not affected by holes or cuts in the fabric as the entire surface is electrically conductive.

Wide ranging modulation of electrical properties of conducting textiles leads to various other applications. Electrical conductivity is sensitive to external stimuli such as temperature and pressure, suggesting applications in the area of sensors.

Kinetics of degradation of conductivity of chemically and electrochemically synthesized polypyrrole at elevated temperatures has been thoroughly investigated. However, only a few brief reports have been published on the aspect of heat generation. In one of these reports, low temperature radiant heating panels were produced by PPy coating, and effect of heat generation was studied, but their applications are still limited due to comparatively low tensile strength of the fabric. The aim of this work is to develop heating panels by coating PPy on glass fabric for commercial and industrial applications, specially suggested for space heating systems at low cost. The reason for choosing glass fiber is its higher tensile strength compared to other fibers of the same price.

2. Methodology

2.1. Sample Preparation. PPy was coated on E-glass fabric having specifications given in Table 1. The glass fabric was first immersed in aqueous solution of $^{-}$OTs and $FeCl_3$ as

TABLE 1: Specification of the E-glass fabric used for experiment.

Properties	Index
Weave	Plain
Fineness (tex)	
Warp	1250
Weft	1250
Warp/cm	3.67
Weft/cm	3.5
Cover factor (SI)	
Warp	13
Weft	12.4
Areal density (g·m^{-2})	900
Thickness (mm)	1.1
Fabric porosity (ϕ)	0.678

an oxidant in molar ratio 1 : 2, respectively, at the temperature of 10°C and squeezed with rubber rollers padding machine with 100% pickup.

Fabric porosity can be calculated by

$$\phi = 1 - \left(\frac{\rho_{\mathrm{ta}}}{\rho_{\mathrm{f}}}\right). \quad (1)$$

Here, ρ_{fa} is the density of fabric that can be evaluated through dividing areal density by the thickness, whereas ρ_{f} is the known density of fiber.

As the amount of PPy depends upon the oxidizing agent present in the system, different specimens were developed by taking different concentrations of $FeCl_3$ as 0.3, 0.4, 0.5, and 0.6 mol/L keeping dopant to oxidizing agent ratio as 1 : 2 constant.

If released to air, a vapor pressure of 8.35 mmHg at 25°C indicates pyrrole exists solely as a vapor in the ambient atmosphere. Therefore, fabric was then immediately exposed to vapors of pyrrole monomer for 8 hours. Later, fabric was rinsed with ethanol and distilled water to remove excess of monomer and biproducts. Finally, it was dried in air at 25 ± 4°C for 24 hours. The formation of the conductive polymer coating on the textile substrate starts from deposited oligomers, which form islands in the initial stage and finally grow and cover the surface of the substrate completely [5].

For evaluation of electrical properties of fabrics, two parameters were selected, namely, the surface and volume resistivity. These parameters can be calculated from the information of surface and bulk resistivity, the dimensions of the electrodes, and/or the thickness of the sample. For this purpose, concentric electrode system was used conforming to DIN EN 1149-1, EN 100 015, EN 61340-5-1, together with Agilent 53131A digital multimeter. Measurements of electrical resistance were performed in the climatic conditions t = 20 ± 2°C and φ = 40 ± 2%, which is in accordance with the standard ČSN 80 0059.

The morphology and thickness of PPy coating on glass fiber was characterized by scanning electron microscope (SEM) Carl Zeiss Ultra plus after gold coating and Laser confocal microscope in the Technical University of Liberec, Czech Republic.

2.2. Temperature Measurements. For the purpose of evaluating heat generation by polypyrrole coated glass fabric samples, 5 V, 10 V, 15 V, and 20 V DC electromotive forces by DC voltage supplier NZ-2229.2 from Statron, Czech Republic, were applied via stainless steel clamp type electrodes. Meanwhile DC current passing through the substrate was measured by connecting Agilent 53131A digital multimeter in series. The rise in temperature was determined by the infrared thermal camera by Fluke together with k-type thermocouples in connection with ART data acquisition module. The infrared thermal camera was helpful to figure the homogeneity of the specimen out.

3. Results

Oxidation of the pyrrole monomer in the presence of the substrate causes polymerization to take place on the fabric surface, resulting in a thin and smooth conductive coating, whilst some of the polymer particles formed on the PPy coating deposited on the surface at prolonged reaction time and at higher monomer concentration, which can be seen as clumps of PPy deposition on the surface and shown as inhomogeneity in the PPy coating, as shown in Figure 1. The morphology of the glass surface after coating with PPy shown in Figure 2 explains that each fiber in the fabric structure is covered with PPy depending upon the concentration. The average thickness of coating of PPy on E-glass fiber was measured as 1.165 ± 0.05 μm.

The average thickness was also measured with the help of Laser confocal microscopy and was found to be around 1.372 μm as shown in Figure 2, which also proves the homogeneity and smoothness of PPy coating on glass fibers.

The weight gain (Δm) by the glass fabric after coating follows a linear function of concentration of $FeCl_3$ (C) as revealed in Figure 3, and this relation can be expressed mathematically as (2) with coefficient of determination R^2 = 0.962:

$$\Delta m = 9.0442 \cdot C. \quad (2)$$

The presence of PPy on the E-glass fabric specimen was characterized by Fourier transform infrared spectroscopy (FTIR) as shown in Figure 4. The strong absorbance of control specimen at 958 cm^{-1} shows the occurrence of Si–O–Si and Si–OH in E-glass structure. The absorption at 1546 and 1452 cm^{-1} assigned to ring vibration of C=C and C=N as well as at 1168 cm^{-1} for N–C vibration confirms the presence of pyrrole ring of PPy. Absorptions at 1303 and 1035 cm^{-1} confirm in-plane vibration of =C–H, whereas absorption at 788 cm^{-1} shows out-plane vibration of =C–H.

Sharp increase in temperature of fabric specimen was observed in first 3 min with the applied voltage which was then stabilized after reaching a plateau, as shown in Figure 5, and it shows the ease with which heat could be generated by employing a reasonable DC voltage.

Power density P_D (volume specific power) is the amount of power (time rate of energy transfer) per unit volume. It is then also called volume power density which is expressed as (W/m^3). Volume power density is sometimes an important

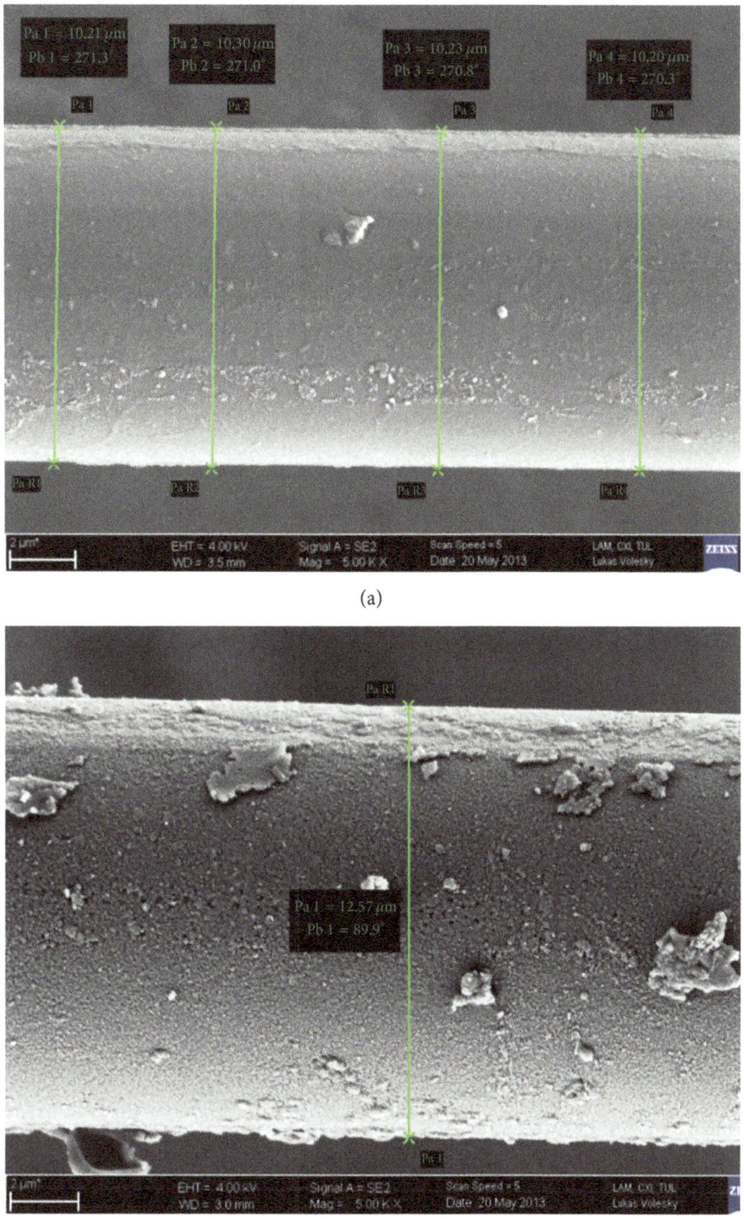

(a)

(b)

FIGURE 1: SEM micrographs: (a) E-glass control and (b) E-glass coated with PPy using $FeCl_3$ 0.6 mol/L.

consideration where space is constrained. In this study, power density P_D is extremely linear function of voltage supplied to the PPy coated glass fabric as illustrated in Figure 6 and expressed as

$$\dot{e}_{gen} = \frac{\dot{E}_{gen}}{v_{element}} = P_D = 82.58V - 245.83. \quad (3)$$

Here, $\dot{e}_{gen}$ shows the power generated per unit volume, $\dot{E}_{gen}$ the quantity of heat energy, and $v_{element}$ the volume of the heating element which in our case is fabric.

However, temperature rise in fabric sample at equilibrium (after 3 min) follows exponential function of voltage with $R^2 = 0.9717$, that can be written as

$$T = T_0 e^{0.0795V}. \quad (4)$$

Here, T is the temperature in (°C) measured at equilibrium at applied DC V, and T_0 is the temperature of fabric at ambient conditions, which was recorded as 22.3°C for this experiment.

4. Conclusion

For the specimen the rate of change of temperature has two distinct phases: an initial sharp rise during the first 100 s or so, followed by a leveling off to plateau at approximately 300 to 400 s. During the first phase, the electrical energy input to

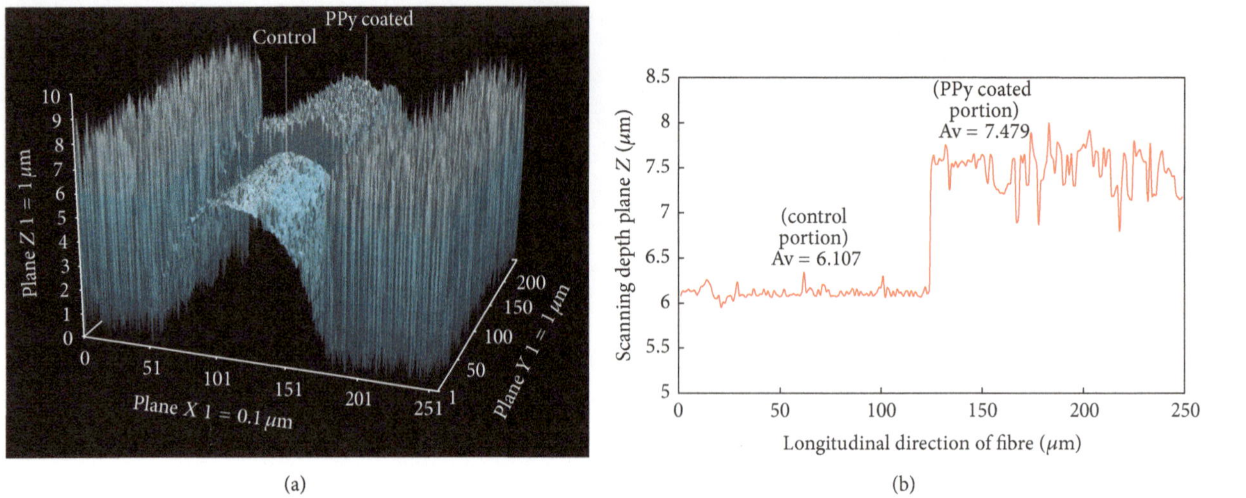

FIGURE 2: The surface morphology of the PPy coating using $FeCl_3$ 0.6 mol/L studied by Laser confocal microscope (a) describes the surface morphology and (b) comparison of thickness.

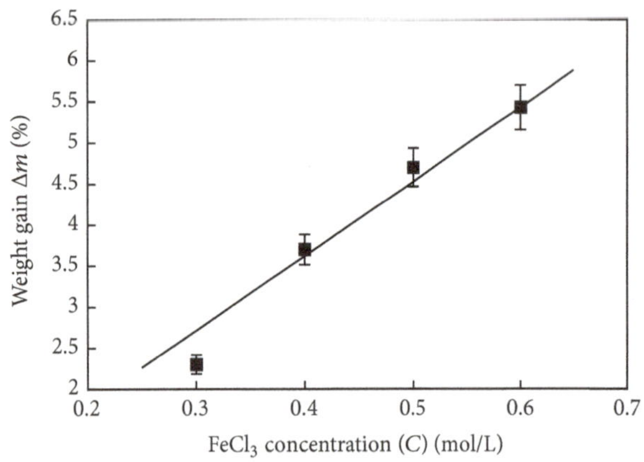

FIGURE 3: The dependence of weight gain by glass fabric after coating by PPy.

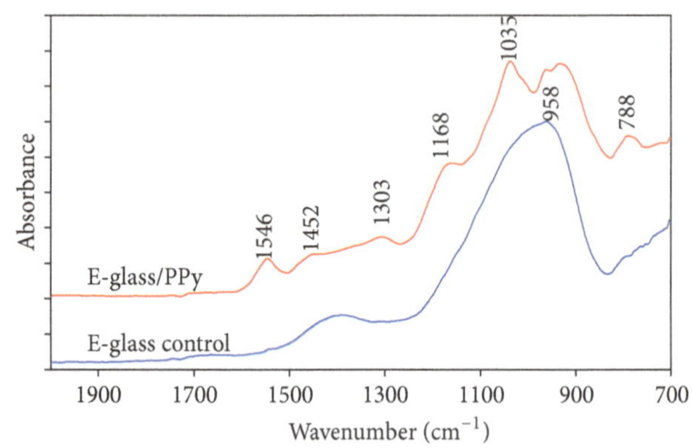

FIGURE 4: FTIR absorbance curves of E-glass control and E-glass coated with PPy.

the samples exceeds the combined thermal energy losses in the forms of convection, radiation, and conduction and hence, the temperature rises. As the temperature rises the heat losses increase until they are balanced with the heat generated and the temperature of the fabric stops rising. This is the stable temperature of the fabric under the applied voltage.

The electrical power supplied to the specimen that ultimately converts into thermal power follows a linear dependence on the DC voltage supplied, whereas rise in temperature of the specimen is a power function of supplied voltage.

The estimated cost to produce PPy coated E-glass fabric at laboratory scale was roughly calculated as less than Euro 1.15 per m^2, and this kind of heat generating industrial textile can find its practical application as heating systems for small and medium spaces as well as for the drying of wet cement in civil works.

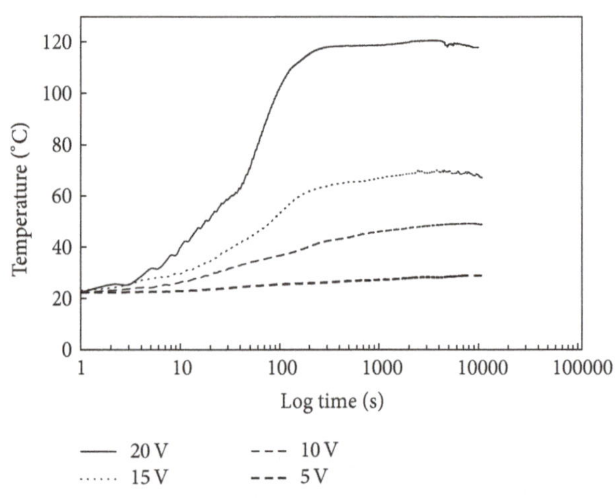

FIGURE 5: Escalation of temperature of E-glass fabric coated by PPy using 0.6 mol/L $FeCl_3$ with increase in DC voltage V and time.

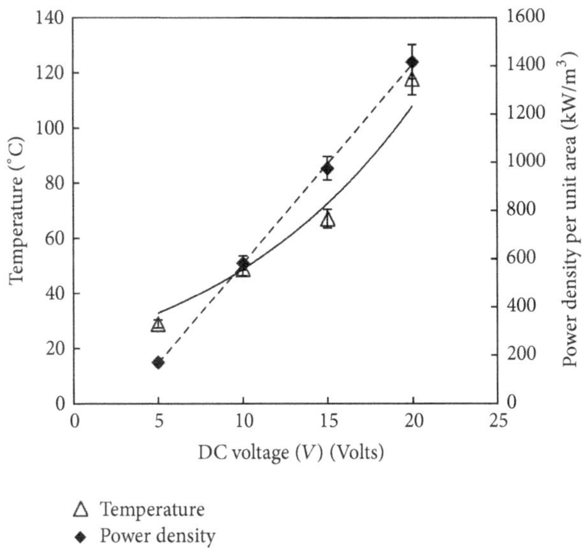

FIGURE 6: Rise in power density and temperature of sample D4 as a function of DC voltage V.

Acknowledgment

This study was supported by student grant scheme (SGS) in Technical University of Liberec, Czech Republic.

References

[1] Y. Cao, P. Smith, and A. J. Heeger, "Counter-ion induced processibility of conducting polyaniline," *Synthetic Metals*, vol. 57, no. 1, pp. 3514–3519, 1993.

[2] C. K. Chiang, C. R. Fincher, Y. W. Park et al., "Electrical conductivity in doped polyacetylene," *Physical Review Letters*, vol. 39, no. 17, pp. 1098–1101, 1977.

[3] A. G. MacDiarmid, "Polyaniline and polypyrrole: where are we headed?" *Synthetic Metals*, vol. 84, no. 1–3, pp. 27–34, 1997.

[4] J. Edenbaum, Ed., *Plastics Additives and Modifiers Handbook*, Van Nostrand Reinhold, New York, NY, USA, 1992.

[5] T. Lin, L. Wang, X. Wang, and A. Kaynak, "Polymerising pyrrole on polyester textiles and controlling the conductivity through coating thickness," *Thin Solid Films*, vol. 479, no. 1-2, pp. 77–82, 2005.

11

A Study on Herbal Finish to Prevent Bed Sore Using *Mangifera indica* and Triphala Dried Fruit

Kiruthika Deivasigamani,[1] Siva Kumar Kolandaivel,[2] and Kavitha Krishnamoorthi[1]

[1] *Kumaraguru College of Technology, Coimbatore 641049, India*
[2] *Angel College of Engineering and Technology, Tirupur 641665, India*

Correspondence should be addressed to Kavitha Krishnamoorthi; s_kavishna@yahoo.co.in

Academic Editor: Ajoy K. Sarkar

"Bed sores" owe their name to the observation that patients who were bedridden and not properly repositioned would often develop ulcerations or sores on their skin, typically over bony prominences. These bed sores, which result from prolonged pressure, are also called "decubitus ulcers," "pressure sores," "skin breakdown," and "pressure ulcers." They are associated with adverse health outcomes and high treatment costs. This study focuses on developing herbal finish to prevent bed sores. For this purpose, functional finishing was utilized to produce a bedsore preventing fabric. Plain weave cotton fabric was dyed by using herbal extracts, dried, and then cured. Two samples in different ratios (1 : 2 and 2 : 1) of two herbal extracts (*Mangifera indica* and Triphala dried fruits) were prepared with and without binder (chitosan). Air permeability, wicking ability, Drop Test for absorbency, and antibacterial properties were examined for those samples. Based on the results, fabric dyed with *Mangifera indica* and Triphala dried fruits in the ratio of 2 : 1 with chitosan is determined as the most promising combination for the final product.

1. Introduction

The skin is a large barrier organ that protects the human body from environmental hazards (heat, cold, chemicals, mechanical forces, etc.) and maintains the integrity of the body, whereas the clothing system provides extra layer(s) of barrier to enhance the aesthetic, thermophysiological, and sensorial comfort of the wearer [1]. However, direct contact and interactions between textiles and skin may cause reactions, even damage or diseases [2] (see Figure 1).

Bed sores can result from one period of sustained pressure on the skin; however, most bed sores probably occur as a result of repeated incidents of blood flow interruption without adequate time for recovery. In fact, low pressure endured for long periods of time is believed to be more significant in producing pressure ulcers than higher pressure for shorter durations. The role that textiles play in the prevention and formation of these injuries is examined.

2. Materials and Methodology

2.1. Fabric Particulars. Plain weave cotton fabric with 60s yarn count, EPI: 88, and PPI: 86.

2.2. Herb Particulars. See Table 1.

2.3. Methodology. The selected herbs were dried in sunshade and grinded. Methanol extract of the herbs was obtained and tested for antibacterial and wound healing property. Chitosan was then diluted with citric acid. Scoured and bleached fabric was padded, dried, and cured with and without binder. Fabric was then tested for its antibacterial and wound healing property, washing fastness, air permeability, wicking ability, and water absorbency. Finally its results were discussed (see Figure 2).

2.4. Finishing Agents. Methanol extract of *Mangifera indica* and Triphala dried fruits (*Terminalia chebula*, *Terminalia bellerica*, and *Emblica officinalis*) has the antibacterial and

TABLE 1

SI number	Botanical name	Common name	Chemical constituents
(1)	*Mangifera indica* [3]	Mango (leaf)	C-Glucoside xanthone mangiferin, tannins, alkaloids, glycosides, terpenoids, flavonoids, amino acids, proteins, euxanthin acid, and euxanthon.
(2)			Triphala dried fruits
	(a) *Terminalia chebula* [4]	*Chebulic myrobalan*	Tannins, chebulagic acid, chebulinic acid, ellagic acid, a tannin terchebin, an ellagitannin terchebulin, syringic acid, and gallic acid.
	(b) *Terminalia bellerica* [5]	*Belleric myrobalan*	Hexahydroxydiphenic acid ester, β-sitosterol, gallic acid, ellagic acid, ethyl gallate, galloyl glucose, chebulagic acid, mannitol, glucose, galactose, and rhamnose.
	(c) *Emblica officinalis* [6]	Indian gooseberry/amla	Organic acids—ascorbic acid, vitamin, tannin; bioflavonoids—quercetin, kaempferol; polyphenols—gallic acid; cytokinins—zeatin.

TABLE 2

Samples	M1	M2	C1	C2
Mangifera indica extract	2%	1%	2%	1%
Triphala dried fruits extract	1%	2%	1%	2%
Chitosan diluted with citric acid (fabric dipped for 30 min)	*	*	6%	6%
MLR ratio (fabric dipped for 2 hrs)	1:20	1:20	1:20	1:20
Drying (for 5 min)	80°C	80°C	80°C	80°C
Curing (for 2 min)	150–180°C	150–180°C	150–180°C	150–180°C

* denotes that samples M1 and M2 are not treated with Chitosan.

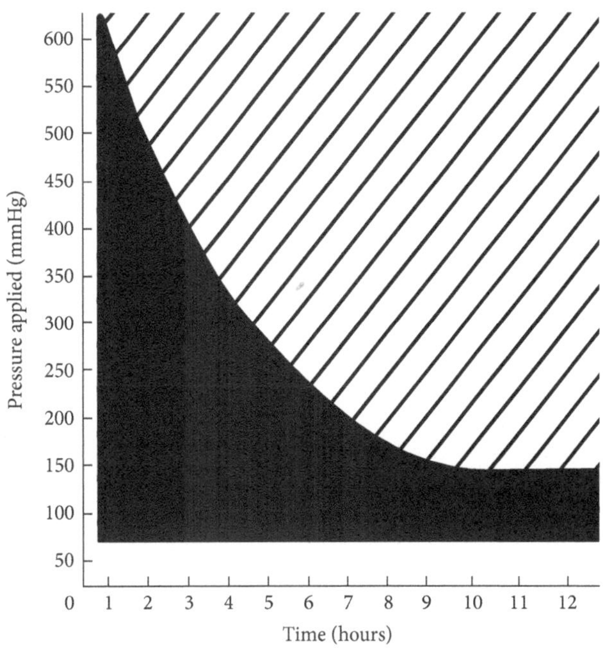

FIGURE 1: Time versus pressure curve for tissue breakdown in a canine model [2].

wound healing property. 25 grams of herbal powder is dissolved in 100 mL of methanol and then it is concentrated into 50 mL using Soxhlet apparatus to obtain the extract.

2.4.1. Recipe. See Table 2.

FIGURE 2: Herbal extract along with binder.

3. Results and Discussion

3.1. Wickability. Fabric specimen is cut into the size of $12'' \times 1''$ and hung over the stand. A beaker is filled with tinted distilled water. The lower end of the fabric strip is made to touch the water in such a way that half an inch height of the fabric is kept immersed inside the water. To keep the fabric strip in straight configuration a load of 10 g of weight is applied to the bottom of it. The rise of the water level in the fabric is measured after 10 minutes.

The wickability of treated and untreated plain weave cotton fabric samples was assessed by time for wetting. The vertical wicking test results are shown the table.

Table 3: Wickability result of the treated and untreated fabric samples.

SI number	Type of the sample	Wicking height in centimeters
1	Bleached fabric	5
2	Sample 1-M1	7.5
3	Sample 2-M2	7.5
4	Sample 3-C1	9
5	Sample 4-C2	9.5

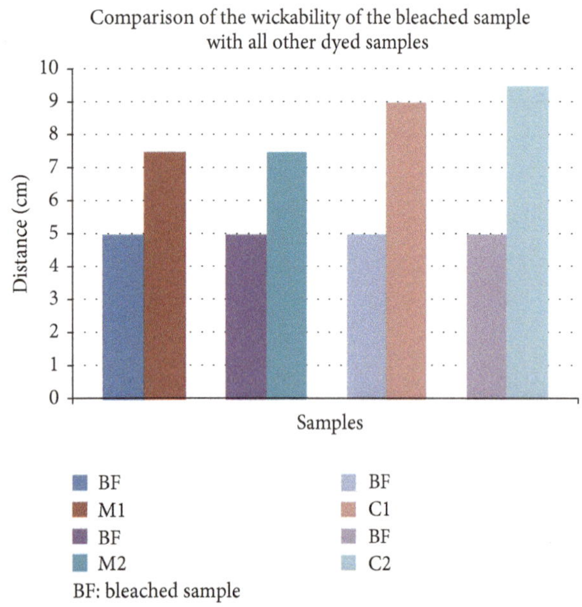

Figure 3: Comparison of wickability results of the treated and untreated samples.

3.1.1. Wickability Test Results. The wicking property of the finished fabric has been improved for all the samples when compared with the bleached fabric. Wickability of the samples treated with chitosan is highly increased (see Table 3 and Figure 3).

3.2. Drop Test. It is used to measure the time taken for the fabric to absorb a drop of water. The absorption rate of fabric changes as per the thickness of the fabric, weave structure, and finishes applied to the fabric.

3.2.1. Drop Test Results. The time taken to absorb a drop of water is decreased in the finished samples when compared with the bleached sample. This shows that the absorbency is increased after finishing. Comparing to all the four samples sample 3-C1 showed good absorbency (see Table 4 and Figure 4).

3.3. Air Permeability. It is the volume of air measured in cubic cm passed per second through 1 cm^2 for the fabric at a pressure one cm head of water. It is measured by using air permeability tester. The air permeability of a fabric is a measure of how well it allows the passage of air through it.

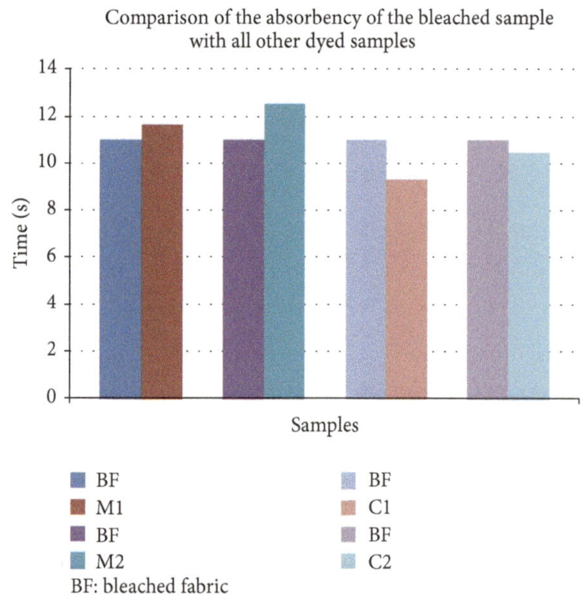

Figure 4: Comparison of Drop Test results of the treated and untreated fabric samples.

3.3.1. Air Permeability Test to Various Fabrics. Air permeability is an important factor in the performance of textile materials used to provide an indication of the breathability of coated fabrics. The air permeability of treated and untreated plain weave of cotton fabric samples was assessed. Consider

$$\text{Air permeability} = \frac{1000 * r}{3600 * A}, \quad (1)$$

where r = rotameter reading and $A = 4\,\text{cm}^2$ (testing area).

3.3.2. Air Permeability Test Results. Air permeability of the plain weave cotton fabric is high in all the treated samples compared to the untreated samples (see Table 5 and Figure 5).

3.4. Antibacterial Test

3.4.1. Results of Antibacterial Activity of the Herbal Extracts (Kirby-Bauer Disk Diffusion Susceptibility Test). Mangifera indica and Triphala dried fruits have antibacterial activity against *Staphylococcus aureus* (see Table 6 and Figure 6).

3.4.2. Results of Antibacterial Activity of Textile Fabric. Only sample 1-M1 and sample 3-C1 have antibacterial activity against *Staphylococcus aureus* (see Table 7 and Figure 7).

3.5. Wound Healing Test: Angiogenic Activity of the Herbal Extract and Herbal Based Textile Fabric. The herbal extract of Triphala dried fruits containing compounds with angiogenesis modulating properties showed strong angiogenic activity in CAM (chorioallantoic membrane) treated with herbal extract, by increasing the size and number of blood vessels as compared to the control.

The herbal based treated cloth C1 showed better angiogenic activity in CAM. The size and number of blood

TABLE 4: Drop Test result for the treated and untreated fabric samples.

SI number	Name of the samples	Time taken to absorb a drop of water in seconds			
		Trial 1	Trial 2	Trial 3	AVG
1	Bleached fabric	11.36	10.91	10.84	11.03
2	Sample 1-M1	11.25	11.29	12.50	11.68
3	Sample 2-M2	12.62	12.89	12.08	12.53
4	Sample 3-C1	9.56	9.03	9.33	9.30
5	Sample 4-C2	10.36	10.37	10.67	10.46

TABLE 5: Average value evaluation of air permeability.

SI number	Name of the samples	Rotameter reading			Air permeability	Air permeability
		Trial 1	Trial 2	AVG	(cc/sec/cm)^2	(cc/sec/cm)^2/60
1	Bleached fabric	242	304	273	18.9	0.315
2	Sample 1-M1	376	344	360	25	0.416
3	Sample 2-M2	450	410	430	29.8	0.496
4	Sample 3-C1	360	370	365	25.3	0.422
5	Sample 4-C2	480	490	485	33.6	0.561

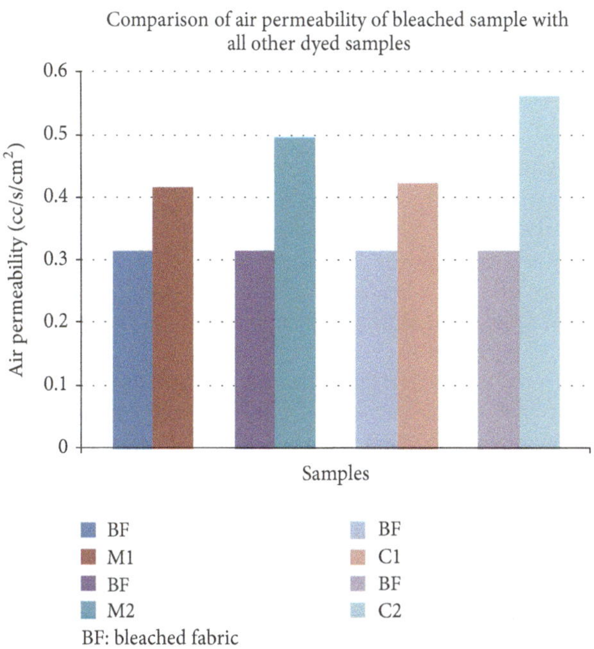

FIGURE 5: Average value evaluation of air permeability.

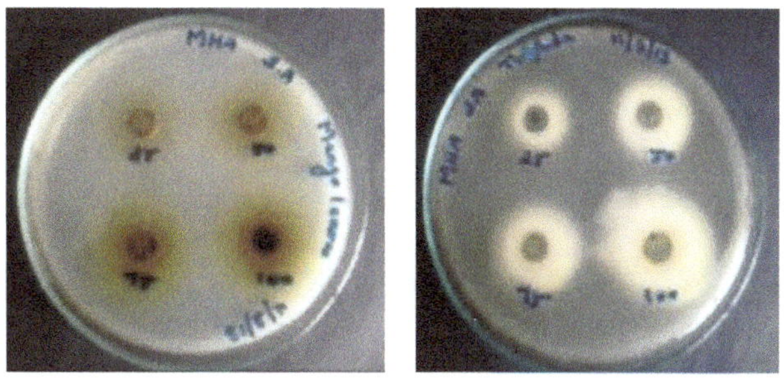

FIGURE 6: Antibacterial activity of the herbal extract.

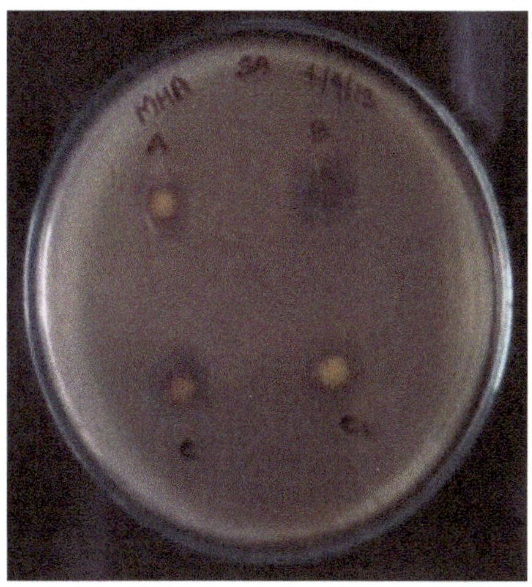

FIGURE 7: Antibacterial test of the treated samples.

vessels were high compared to control egg treated with saline (see Table 8 and Figure 8).

3.6. Washing Fastness Test

3.6.1. Results of Washing Fastness Test for Its Antibacterial Property. Only sample C1 has the washing fastness of antibacterial property for up to 3 washes, which shows that chitosan acts as a binder up to certain level (see Table 9).

3.6.2. Results of Washing Fastness Test for Its Wound Healing Property. Only sample C1 has the washing fastness for wound healing property for more than 10 washes, which shows that chitosan acts as a binder up to certain level (see Table 10).

Table 6: Antibacterial activity of the herbal extract.

Plant extract	*Staphylococcus aureus*				*Pseudomonas aeruginosa*			
Concentration	25 μL	50 μL	75 μL	100 μL	25 μL	50 μL	75 μL	100 μL
Mangifera indica	Nil	Nil	6 mm	11 mm	Nil	Nil	8 mm	12 mm
Triphala dried fruits	11 mm	13 mm	15 mm	21 mm	10 mm	12 mm	14 mm	16 mm

Table 7: Antibacterial activity of textile fabric coated with the herbal extract.

SI number	Name of the samples	*Staphylococcus aureus*	*Pseudomonas aeruginosa*
1	Sample 1-M1	3 mm	Nil
2	Sample 2-M2	Nil	Nil
3	Sample 3-C1	11 mm	Nil
4	Sample 4-C2	Nil	Nil
	Control (chloramphenicol)	**24 mm**	**10 mm**

Table 8: Angiogenic activity of herbal extract and herbal based textile fabric.

Sample name	Number of blood vessel formation
Triphala dried fruits	12
M1	Nil
M2	Nil
C1	16
C2	Nil
Control	8

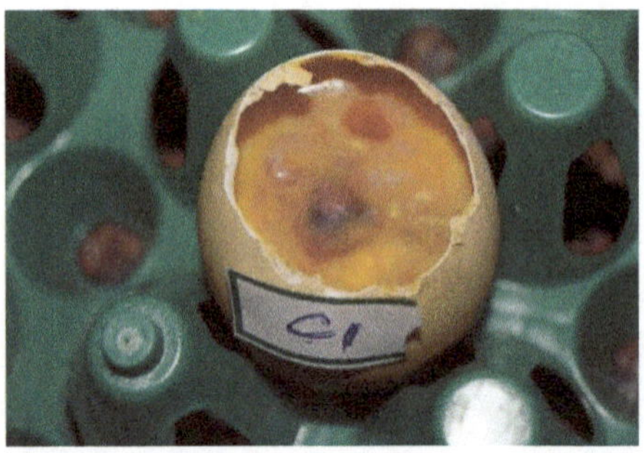

Figure 8: Wound healing test of the treated sample C1.

Table 9: Antibacterial activity of fabric coated with the herbal extract (for washing fastness).

SI number	Name of the sample	*Staphylococcus aureus*
1	C1, up to 3 washes	3 mm
2	C1, up to 5 washes	Nil
	Control (chloramphenicol)	24 mm

4. Conclusion

In this work, herbal based antibacterial finish was applied to the cotton fabric. Extracts of *Mangifera indica* and Triphala dried fruit were applied to the fabric in two different ratios with and without the binder chitosan.

Based on the tests conducted, the obtained results were shown in the tables and figures. According to the results, the treated samples M1 and C1 showed good antibacterial property and sample C1 showed excellent wound healing property. The samples treated with chitosan have high wickability and water absorbency properties. Air permeability in the bleached fabric is low and it has been increased in all the treated samples. Washing fastness result shows that the fabric treated with chitosan shows antibacterial property for up to 3 washes and wound healing property for more than 10 washes.

So it is concluded that sample C1 has much better properties for preventing bed sore. The herbal extract of *Mangifera indica* shows good antibacterial property when applied to the fabric compared to the Triphala dried fruits and chitosan has good binding property. And it is suggested that this fabric could be used for preventing bed sore and also for other applications where antibacterial finish is required. And also it gives cost-wise benefit to people. In addition, it is environment friendly.

Table 10: Wound healing activity of fabric coated with the herbal extract (for washing fastness).

Sample name	Number of blood vessel formation
C1, up to 3 washes	14
C1, up to 5 washes	13
C1, up to 10 washes	10
Control	8

Conflict of Interests

The authors declare that there is no conflict of interests regarding the publication of this paper.

References

[1] N. Bergstrom, A. Bennett, C. E. Carlson et al., *Pressure Ulcer Treatment: A Critical Practice Guideline*, AHCPR, 1995.

[2] N. C. Petersen and S. Bittmann, "The epidemiology of pressure sores," *Scandinavian Journal of Plastic and Reconstructive Surgery*, vol. 5, no. 1, pp. 62–66, 1971.

[3] N. Neelima, M. Sudhakar, M. B. Patil, and B. V. S. Lakshmi, "Anti-ulcer activity and HPTLC analysis of *Mangifera indica* L.

leaves," *International Journal of Pharmaceutical and Phytopharmacological Research*, vol. 1, no. 4, pp. 146–155, 2012.

[4] W. D. Schindler and P. J. Hauser, *Chemical Finishing of Textile*, Woodhead, 2004.

[5] A. K. Meena, A. Yadav, U. Singh et al., "Evaluation of physicochemical parameters on the fruit of *Terminalia bellirica* roxb," *International Journal of Pharmacy and Pharmaceutical Sciences*, vol. 2, supplement 2, p. 97, 2010.

[6] N. K. Savala, P. N. Haritha, and M. L. Narasu, "Haematological studies of Emblica officinalis formulation on wistar rats," *International Journal of Medical and Health Sciences*, vol. 1, no. 2, pp. 29–34, 2012.

12

A Design Tool for Clothing Applications: Wind Resistant Fabric Layers and Permeable Vents

Phillip Gibson,[1] Michael Sieber,[1] Jerry Bieszczad,[2] John Gagne,[2] David Fogg,[2] and Jintu Fan[3]

[1] *U.S. Army Natick Soldier Research, Development, and Engineering Center, Natick, MA 01760, USA*
[2] *Creare, Inc., Hanover, NH 03755, USA*
[3] *Cornell University, Ithaca, NY 14850, USA*

Correspondence should be addressed to Phillip Gibson; phillip.gibson@us.army.mil

Academic Editor: Jiri Militky

A computational clothing design tool is used to examine the effects of different clothing design features upon performance. Computational predictions of total heat and mass transfer coefficients of the clothing design tool showed good agreement with experimental measurements obtained using a sweating thermal manikin for four different clothing systems, as well as for the unclothed bare manikin. The specific clothing design features examined in this work are the size and placement of air-permeable fabric vents in a protective suit composed primarily of a fabric-laminated polymer film layer. The air-permeable vents were shown to provide additional ventilation and to significantly decrease both the total thermal insulation and the water vapor resistance of the protective suit.

1. Introduction

Windproof and wind resistant clothing layers are important in applications such as cold weather clothing and military chemical and biological protective ensembles. Completely windproof clothing provides protection from heat loss due to air penetration in cold conditions but can also become very uncomfortable when the wearer is working hard and generating a lot of heat and sweat. The same situation arises in protective clothing—some ventilation through the fabric layers can be very helpful in mitigating heat stress and extending wear time. Design tools that allow simultaneous assessment of factors such as wind penetration, heat transfer, moisture transport, and permeation of toxic substances can assist in developing new clothing systems that strike a good balance between protection and comfort.

We have shown previously that the wind resistance and aerosol filtration properties of clothing layers can be controlled by varying the deposition of electrospun fiber membranes onto textile substrates [1], as shown in Figure 1.

Since nanofiber layers make it possible to control air flow resistance over several orders of magnitude without affecting other clothing properties, we are interested in the resulting consequences of using these materials in new clothing designs. Electrospun elastomeric polyurethane membranes are now used in commercial outerwear fabrics such as "NeoShell" by Polartec, Inc., and take advantage of this ability to "tune" air flow through the fabric without significantly affecting other properties such as thermal insulation or water vapor diffusion (breathability). These materials may be used for either an entire clothing item, such as jacket or pants, or incorporated into permeable "vents," where the more permeable fabric in certain areas of the garment provides increased ventilation.

2. Approach

A protective clothing design tool under development by Creare, Inc., for the U.S. military provides a physics-based computational framework for iterative modification

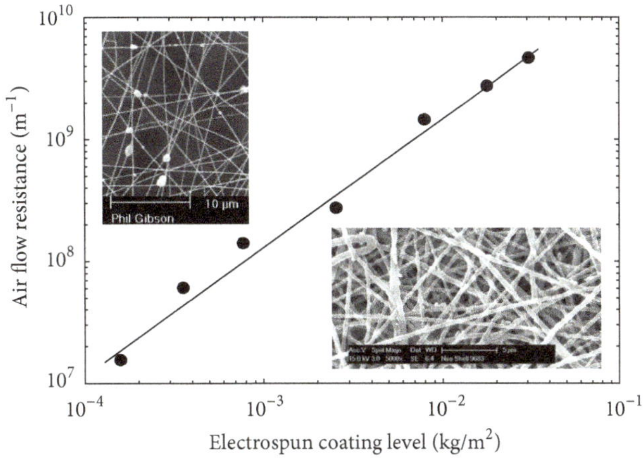

FIGURE 1: Air flow resistance of fabric layer modified by electrospun fiber coating level [1].

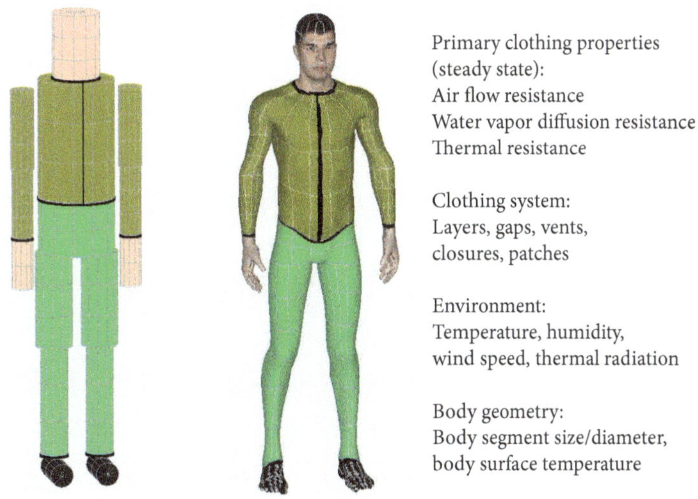

Cylinders mapped to more realistic clothed image

FIGURE 2: Typical IP SPM simulation.

and assessment of protective clothing systems [2]. Creare's Individual Protection System Performance Model (IP SPM) is built on a foundation of advanced computational fluid dynamic and experimental results, but the software itself is purposefully simplified to allow nonexpert users to modify clothing designs and assess the consequences of various clothing aspects such as fit, closures, interfaces, material properties, and layering, upon comfort and protection. The IP SPM model treats the clothed human body as an assemblage of fabric-covered cylinders, although the graphical user interface maps clothing layers onto a more anatomically correct human figure, as shown in Figure 2. Computational models that include more detailed human and clothing geometry are possible [3] and in fact are used as some of the guides for the IP SPM software, but too much detail has been found to add an unneeded level of complexity for a user-friendly design tool, especially for nonexpert users.

In this study we focus upon the use of the IP SPM tool to assess the relative importance of changing the air flow resistance of fabric layers or permeable vent sections upon the calculated total heat transfer resistance of the clothing system. Several recent experimental papers have discussed the trade-offs between fabric air flow resistance, clothing vents, and the interaction with air flow due to external wind or body motion [4–6].

3. Results and Discussion

3.1. Comparison of IP SPM Predictions with Thermal Manikin Measurements. The Creare IP SPM can be configured to simulate a thermal manikin, which is an important tool in clothing comfort research. Figure 3 shows baseline comparison results for the IP SPM thermal manikin calculated overall heat transfer coefficients as compared to some empirical correlations obtained with humans [7] and thermal manikins [8], as well as computational and analytical results from heated cylinders [9].

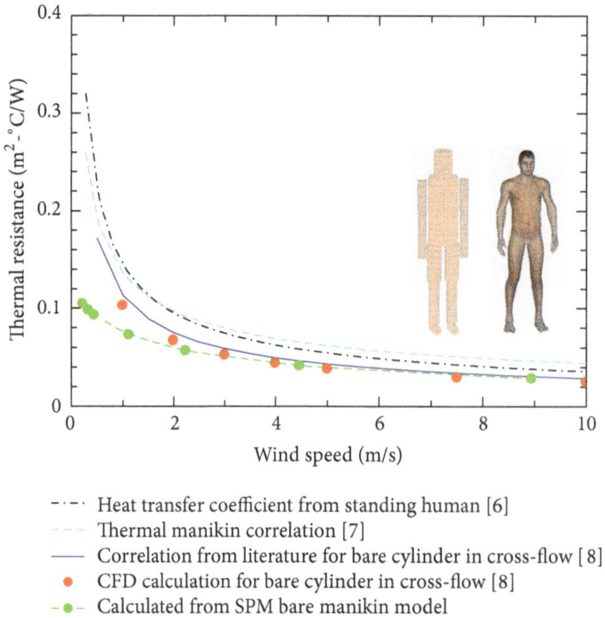

FIGURE 3: Overall heat transfer coefficient of IP SPM thermal manikin as a function of wind speed.

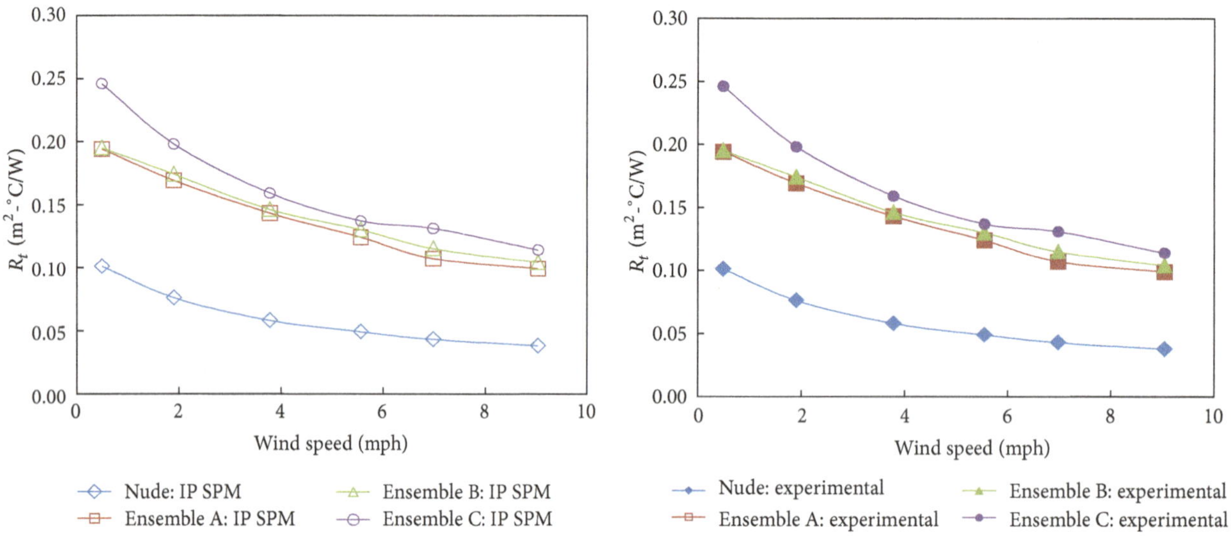

(a) IP SPM

(b) Thermal manikin

FIGURE 4: Comparison of thermal insulation for the nude manikin and three clothing ensembles: (a) IP SPM model results versus (b) experimental thermal manikin results.

Figure 3 shows an underprediction of bare manikin values at low wind speeds when radiative heat transfer is included in the IP SPM model; when this is removed from the calculation (h_{rad} of 4.5 W/m^2-°C), the IP SPM results show good agreement with the convective-only heat transfer calculations from [8] shown in Figure 3.

IP SPM predictions can be compared to experimental thermal manikin results. Inputs to the IP SPM model require clothing thermal resistance, water vapor diffusion resistance, and air flow resistance for the fabric and some information on the fit (space between the fabric and the body surface), tightness of the closures and seams, and the extent of coverage of the fabric over different sections of the body. Figure 4 shows a comparison of IP SPM predictions to experimental results obtained by a thermal manikin [10, 11] for both the bare (nude) condition and for three separate clothing ensembles on the manikin.

As shown in Figure 4, the performance rankings between the ensembles were correctly replicated by the IP SPM predictions, for both the overall thermal resistance (shown

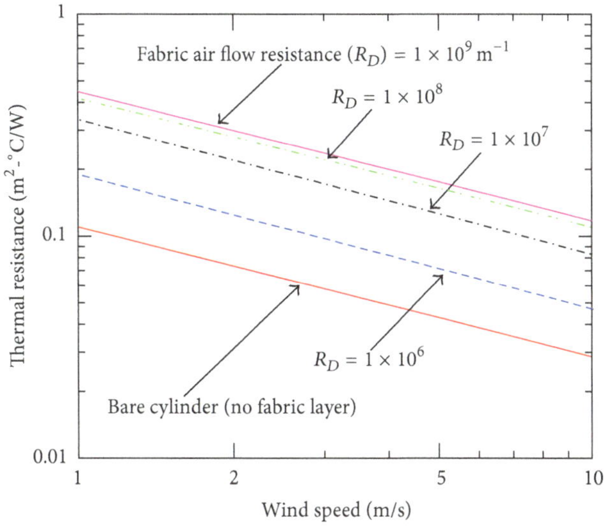

FIGURE 5: Overall heat transfer resistance of fabric-covered cylinders in cross-flow conditions at various wind speeds [9].

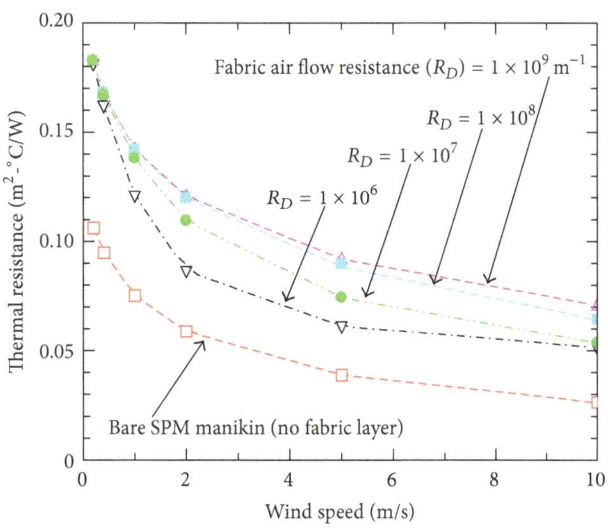

FIGURE 6: Individual protection system performance model (IP SPM) predictions for variable air flow resistance of fabric layer.

in Figure 3) and evaporative resistances (not shown). This is an important result because it indicates that in situations where the exact experimental configurations are not known or defined, the IP SPM can still be used to determine relative garment rankings for thermal performance using nominal configuration values (e.g., for garment fit).

3.2. Effect of Fabric Air Flow Resistance on Clothing System Thermal Resistance. The IP SPM purposefully simplifies the clothing/body model to be an assemblage of fabric-covered cylinders. For a practical design tool, it has been found that faithful reproduction of body geometry, along with details of fabric folds, drape, and fit, can quickly make a computational tool such as this only usable by dedicated experts. Previous computational simulations of fabric-covered cylinders as an analog to the human body [9] examined the effect of varying only the fabric air flow resistance (all other properties kept constant) and calculated the overall heat and mass transfer coefficient, as shown in Figure 5.

The IP SPM allows a very similar computation to be carried out for a more realistic clothing and body geometry. The IP SPM is configured to simulate a set of wind-resistant clothing (jacket and pants) that only varies the single property of air flow resistance. A similar relationship thermal resistance as a function of external wind speed as was shown in Figure 5 is shown in Figure 6 for the IP SPM model. Air flow resistance is more important at high wind speeds, and once the fabric becomes "wind-proof," any further increase in air flow resistance does not significantly affect heat transfer.

An example of the use of the IP SPM to examine some simple clothing design variations is shown in Figure 7. Suit A and suit B vary only in the air flow resistance of the jacket

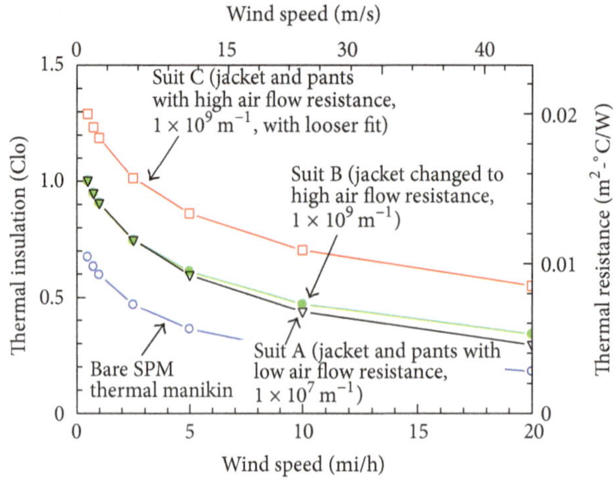

FIGURE 7: Use of IP SPM to examine effect of fabric air flow resistance on overall thermal insulation.

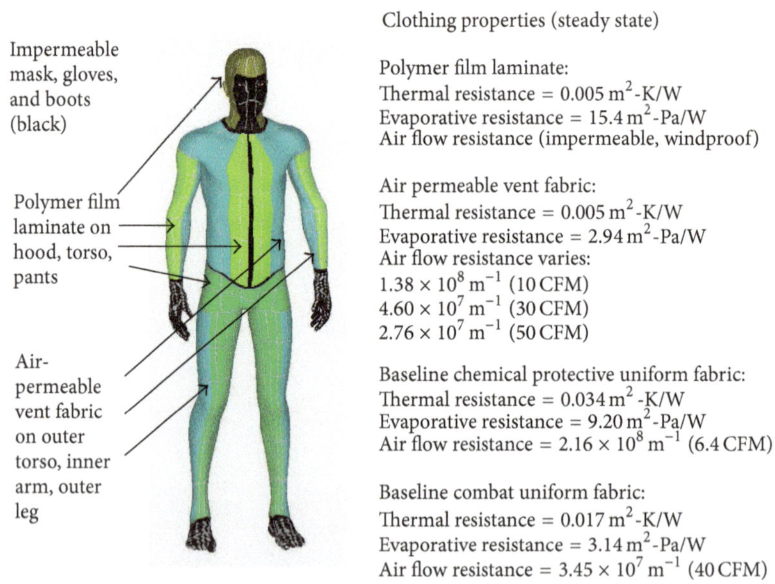

FIGURE 8: IPM SPM simulation to vary air flow resistance of permeable fabric vents in a membrane suit.

(causing slightly more heat loss at higher air flow conditions for the less-permeable jacket). Suit C is wind-resistant, and also has a much looser fit which traps more air under the clothing, resulting in increased thermal insulation properties for that configuration.

3.3. Effect of Permeable Fabric Vents on Thermal and Water Vapor Resistance. An example of a more detailed design study was performed with the IP SPM to examine the effect of incorporating permeable fabric vents into a chemical-protective suit that is primarily composed of a solid polymer membrane that provides chemical protection, yet allows some water vapor transport for comfort. It is anticipated that such a complete membrane-based suit will not allow enough ventilation by itself but must incorporate a venting strategy of some kind. One solution would be to incorporate separate vent panels composed of a more air-permeable fabric into the membrane-based suit.

Figure 8 shows the basic IP SPM setup for this simulation. Large air-permeable fabric vents are incorporated into the protective suit, and the vent fabric air flow resistance is varied in a systematic way. The suit also incorporates a protective hood, mask, boots, and gloves, which are impermeable to both air and water vapor. For comparison purposes, results are also modeled for the polymer film laminate suit with no vents incorporated into the design and two other baseline cases of a suit made up of a standard chemical protective uniform fabric and a standard combat uniform fabric. The additional case of the nude manikin simulation was also modeled, to provide a reference point for how much each clothing system decreases thermal and evaporative transfer from the human body.

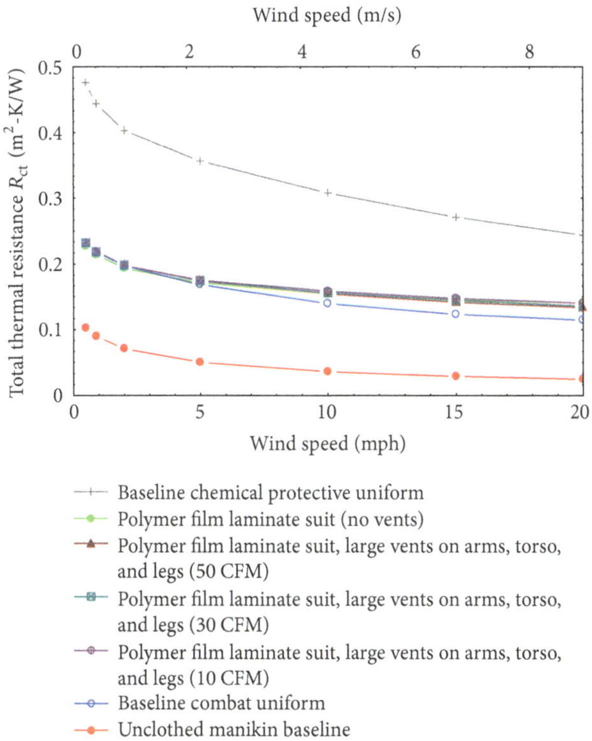

FIGURE 9: Overall IP SPM manikin thermal resistance.

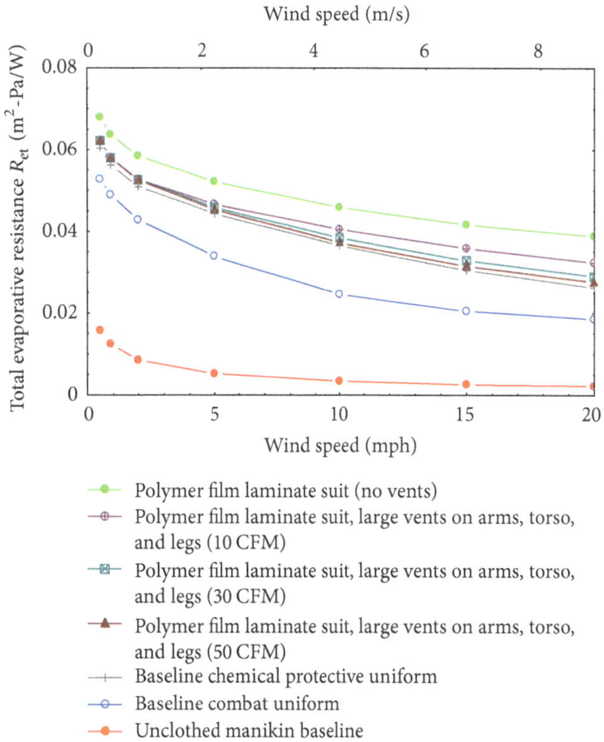

FIGURE 10: Overall IP SPM manikin evaporative resistance.

Figures 9 and 10 show the simulation result for these cases described in Figure 8. Figure 9 shows the overall thermal resistance of the IP SPM thermal manikin, over a range of air flows from 0.5 to 20 mph. For the IP SPM manikin thermal resistance, all the designs incorporating the polymer film laminate (vented or unvented) showed very little differences, since the thermal resistance of the fabric itself is minimal. However, there are large differences between the polymer film laminate suit and the baseline chemical protective uniform, which is a much thicker material, with a higher measured thermal resistance, which impedes dry heat transfer from the body to the environment. The baseline combat uniform shows a lower thermal resistance at higher wind speeds, as a consequence of its lower air flow resistance (higher air permeability), as opposed to the polymer film laminate fabric, which is impermeable to wind. For overall IP SPM manikin thermal resistance, the large air-permeable fabric vents on the arms, torso, and legs do not make an appreciable difference to the thermal comfort of this suit design.

Figure 10 shows the overall evaporative resistance of the IP SPM manikin, over a range of air flows from 0.5 to 20 mph. For the evaporative case, there are significant differences between the different suit designs and fabrics. It is still the case that none of the polymer film laminate suit designs approach the performance of either the baseline combat uniform or the nude manikin, but the incorporation of air-permeable fabric vents does favorably impact the performance of the polymer film laminate suit. The unvented polymer film laminate suit has the highest evaporative resistance of any of the suits modeled. As the vent fabric air permeability increases, the polymer film laminate suit becomes progressively better, and the design incorporating fabric vents with an air flow resistance of $2.8 \times 10^7\ m^{-1}$ (50 CFM) has the best performance, approaching the evaporative resistance of the baseline chemical protective uniform, which is entirely air-permeable.

4. Conclusions

Simplified clothing design tools would greatly enhance our ability to iteratively vary clothing design features and examine the consequences for comfort and protection. Further verification and validation with experimental thermal manikin measurements will be required for a variety of clothing ensembles, so that results predicted with the model can be used as a reliable design tool in the future.

Conflict of Interests

Jerry Bieszczad, John Gagne, and David Fogg are employed by the commercial entity, Creare, Inc., which developed the computational software mentioned in the paper.

Acknowledgment

Funding for this work was provided by the United States Defense Threat Reduction Agency.

References

[1] P. Gibson, H. Schreuder-Gibson, and D. Rivin, "Transport properties of porous membranes based on electrospun nanofibers," *Colloids and Surfaces A: Physicochemical and Engineering Aspects*, vol. 187-188, pp. 469–481, 2001.

[2] J. Bieszczad, "Physics-based modeling and test methods for improving aerosol protection and reducing thermal burden of IPE," in *Proceedings of the Chemical and Biological Defense Science and Technology Conference*, Las Vegas, Nev, USA, November 2011.

[3] P. Gibson, J. Barry, R. Hill, P. Brasser, M. Sober, and C. Kleijn, "Computational modeling of clothing performance," in *Thermal and Moisture Transport in Fibrous Materials*, N. Pan and P. Gibson, Eds., pp. 542–559, Woodhead Publishing, Cambridge, UK, 2006.

[4] M. P. Morrissey and R. M. Rossi, "The effect of wind, body movement and garment adjustments on the effective thermal resistance of clothing with low and high air permeability insulation," *Textile Research Journal*, vol. 84, no. 6, pp. 583–592, 2014.

[5] Y. Ke, G. Havenith, X. Zhang, X. Li, and J. Li, "Effects of wind and clothing apertures on local clothing ventilation rates and thermal insulation," *Textile Research Journal*, vol. 84, no. 9, pp. 941–952, 2014.

[6] M. P. Morrissey and R. M. Rossi, "The influence of fabric air permeability on the efficacy of ventilation features," *International Journal of Clothing Science and Technology*, vol. 25, no. 6, pp. 440–450, 2013.

[7] U. Danielsson, "Windchill and the risk of tissue freezing," *Journal of Applied Physiology*, vol. 81, no. 6, pp. 2666–2673, 1996.

[8] A. V. M. Oliveira, A. R. Gaspar, S. C. Francisco, and D. A. Quintela, "Convective heat transfer from a nude body under calm conditions: assessment of the effects of walking with a thermal manikin," *International Journal of Biometeorology*, vol. 56, no. 2, pp. 319–332, 2012.

[9] P. Gibson, "Modeling heat and mass transfer from fabric-covered cylinders," *Journal of Engineered Fibers and Fabrics*, vol. 4, no. 1, pp. 1–8, 2009.

[10] J. Fan and X. Qian, "New functions and applications of Walter, the sweating fabric manikin," *European Journal of Applied Physiology*, vol. 92, no. 6, pp. 641–644, 2004.

[11] J. Fan and Y. S. Chen, "Measurement of clothing thermal insulation and moisture vapour resistance using a novel perspiring fabric thermal manikin," *Measurement Science and Technology*, vol. 13, no. 7, pp. 1115–1123, 2002.

13

Evaluation of Durability to Laundering of Triclosan and Chitosan on a Textile Substrate

Anupama Sargur Ranganath[1] and Ajoy K. Sarkar[2]

[1] *Department of Design and Merchandising, Colorado State University, Fort Collins, CO 80523, USA*
[2] *Textile Development and Marketing, Fashion Institute of Technology, State University of New York, NY 10001-5992, USA*

Correspondence should be addressed to Ajoy K. Sarkar; ajoy_sarkar@exchange.fitnyc.edu

Academic Editor: Sambandam Anandan

In a hospital environment, textile substrates have been implicated as a vector in the transmission of disease. To mitigate the harmful spread of disease via textile substrates, an effective measure is treatment of textiles with antimicrobial agents. The current investigation compares one of the most widely used chemically synthesized antimicrobial agents, triclosan with chitosan, a naturally occurring antimicrobial agent. For the study, samples of a common polyester/cotton textile used in hospital settings were treated with triclosan and chitosan based antimicrobial agents. Following treatment, the samples were analyzed for their effectiveness against *Staphylococcus aureus* and *Escherichia coli* using standard AATCC Test Methods. The efficacy of the treatment to laundering was then evaluated by subjecting the treated samples to 50 washings and repeating the tests against the challenge microorganisms. Data obtained were statistically analyzed at a 95% confidence interval. Results showed that before laundering both triclosan and chitosan treated samples were extremely effective as antimicrobial agents. After laundering, chitosan was less effective against *E.coli* but maintained efficacy against *S. aureus*. The effectiveness of triclosan was not adversely affected after the laundering treatments.

1. Introduction

The Centers for Disease Control and Prevention, USA, estimate that approximately 1.7 million healthcare associated infections (HAIs) and 98,987 associated deaths occur each year on account of infection-causing bacteria. Hence, the control of infections has been identified as the most important target by the United States Department of Health and Human Services [1]. Textile substrates have been implicated as one of the vectors of transmission of infections. Fortuitously, the use of antimicrobial textiles is an effective deterrent against the spread of infection-causing bacteria [2].

Triclosan, a chlorinated phenolic compound, is a derivative of a diphenyl ether compound. Triclosan is most widely used in various medical and consumer care products. Triclosan belongs to *leaching* or controlled release category of antimicrobial agents [2, 3]. The mechanism of action of triclosan on a textile substrate is illustrated in Figure 1. The figure depicts a treated fiber containing a reservoir of antimicrobial agent. When the chemical agent ions are released outside the fiber they destroy microbes present in the vicinity. Triclosan inhibits the growth of microbes by using an electrochemical mode of action to penetrate and disrupt the cell wall of microbes [2, 4]. It is probable that, with eventual usage and repetitive laundering of the treated material, the reservoir in the fiber will get depleted and lose its antimicrobial efficacy [5]. As well, via launderings the ions may enter the water stream. Triclosan has been reported to cause environmental problems due to its nonbiodegradability. Other studies have reported increased resistance to triclosan by certain strains of *S. aureus* and other bacteria. The U.S. Food and Drug Administration is currently engaged in an ongoing scientific and regulatory review of triclosan [6].

Chitin, a poly(β-(1-4)-N-acetyl-D-glucosamine), is a natural polysaccharide. Chitin is synthesized by many living organisms. It is the most abundantly found polymer second only to cellulose. When chitin is acetylated to at least about 50%, it is called chitosan [7]. Chitosan belongs to the category of *nonleaching* or the bound type antimicrobial agents, wherein the antimicrobial agent is fixed to the fiber surface via

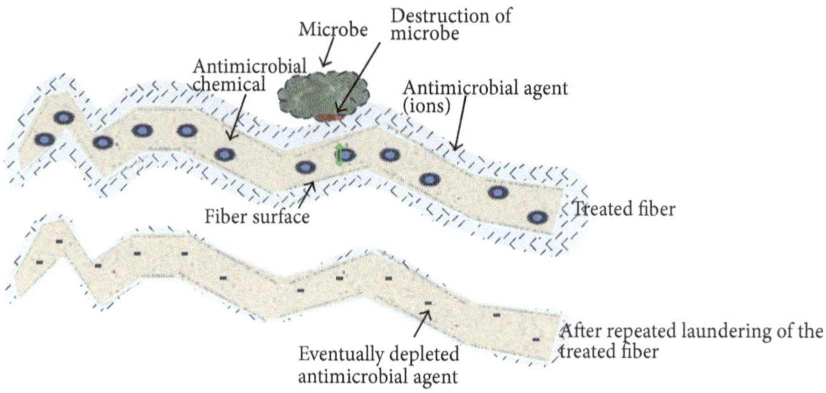

FIGURE 1: Mechanism of action of triclosan.

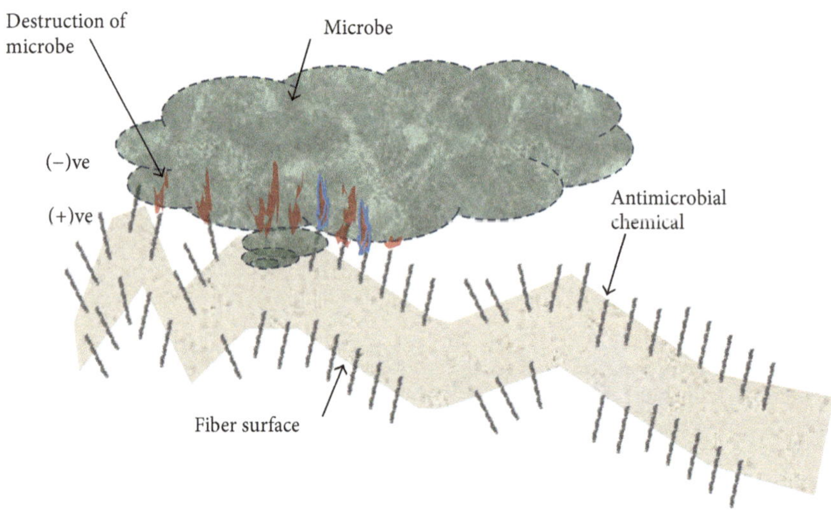

FIGURE 2: Mechanism of action of chitosan.

covalent bonds and is able to destroy the microbes that come in contact with the fiber [2]. The mechanism of action of chitosan on a textile substrate is shown in Figure 2. Mechanism studies suggest that the positively charged chitosan interacts with negatively charged residues at the cell wall of fungi or bacteria. The interaction changes cell permeability and causes the leakage of intracellular substances [8, 9]. Other studies suggest that the formation of the polymeric substance around the bacterial cell prevents the nutrients from entering the cell [10].

Based on a literature review of antimicrobial textiles, it was found that a majority of the research conducted to date has focused on synthesizing and evaluating uniquely distinct antimicrobial agents on different textile substrates with the purpose of proving their effectiveness against microbes. Very few studies have concentrated on comparing the durability to laundering and antimicrobial efficacy of different agents on a specific substrate against target challenge microorganisms. Therefore, the endeavor of the present research was to evaluate the durability to laundering of two such agents, triclosan and chitosan. Secondly, as mentioned before, there are rising concerns with regard to triclosan, a chemically synthesized agent. The study would consequently also shed light on whether chitosan, a naturally derived antimicrobial agent, could serve as a potential replacement for synthetic antimicrobial agents.

2. Materials and Methods

2.1. Materials. The antimicrobial agent based on triclosan was a halogenated phenol with the chemical constitution 5-chloro-2-(2,4-dichlorophenoxy) phenol. A 6% chitosan based formulation was the other antimicrobial agent. The plain weave fabric treated with the two antimicrobial agents had the following characteristics: fiber content: 65% polyester/35% cotton; weight: 154 g/m^2 (ASTM D3776); fabric count: 158; and thickness: 0.017 inches (ASTM D1777-96). The treated substrates were challenged against *Staphylococcus aureus* (ATCC 6538, PML Microbiologicals, Inc.), Gram positive bacteria, and *Escherichia coli* (ATCC 8739, PML Microbiologicals, Inc.), Gram negative bacteria.

2.2. Methods. The purpose of the study was to compare the antimicrobial efficacy of triclosan and chitosan when applied to a substrate using manufacturers' recommended concentration and treatment protocols. Adhering to the recommendations ensured optimum performance of the two agents. Hence, the treatment methods for triclosan and chitosan are different as enumerated below.

2.2.1. Triclosan Treatment Method. The substrate was washed to remove dust and other impurities. Exhaust procedure was adopted to treat the substrate with triclosan. The concentration of triclosan used was 4% on weight of fabric (owf) with a material to liquor ratio (MLR) of 1 : 50. As per the MLR calculations, required amount of water was measured in a treatment bath and heated to 50°C. Triclosan was added to the bath and pH was maintained between 4 and 6. The substrate was then added to the bath and temperature increased to 120°C. Treatment was continued at this temperature for 60 minutes. Finally fabrics were rinsed and air-dried.

2.2.2. Chitosan Treatment Method. The substrate was washed to remove dust and other impurities. Treatment of the substrate with a chitosan based antimicrobial agent was done by the pad-cure method. In this method, the chitosan based antimicrobial was diluted with 20 parts of water for 1 part of chitosan. The diluted solution was considered as the stock solution. The optimum concentration for the treatment was 2% on weight of the stock solution. The bath was prepared with the required amount of water as per MLR of 1 : 20 and required amount of stock solution of chitosan. The pH of the bath was maintained between 4.5 and 5.5 for the entire duration of treatment. To increase the affinity and durability of chitosan for the substrate, 0.4% of a binder was also added to the treatment bath. The substrate was then immersed in the bath for 5 minutes at room temperature followed by one nip through the squeezing rollers to remove any excess liquor. The substrate was then cured at 149°C in the curing chamber for 5 minutes.

2.2.3. Laundering Procedure. The treated samples were laundered using AATCC Test Method 61, 3A with modifications to mimic harsher industrial laundry conditions. Launderings were carried out using MLR of 1 : 10 with 0.5% on weight of fabric of AATCC detergent and 100 steel balls at a temperature of 90°C for 30 minutes. Washing cycle was followed by rinsing in plain water at 40°C for 10 minutes followed by tumble drying. Since a single wash-dry cycle simulated 5 regular wash cycles, the laundering procedure was repeated 10 times to obtain samples after 50 wash cycles.

2.2.4. Evaluation Methods. Treated and washed samples were challenged with strains of *S. aureus* and *E. coli*. The antibacterial activity was evaluated using AATCC Test Method 147: Antibacterial Assessment of Textile Materials: Parallel Streak Method and AATCC Test Method 100: Assessment of Antibacterial Finishes on Textile Materials.

For AATCC Test Method 147, specimen sizes of 25 × 50 mm were used. A loop full of the respective culture was transferred to the surface of a sterile agar plate by making five streaks approximately 60 mm in length, spaced 10 mm apart, covering the central area of the Petri dish. The specimen was then gently pressed transversely across the five streaks to ensure intimate contact with the agar surface. The plates were incubated at 37°C for 18–24 hours and were examined for interruption of growth along the streaks of inoculum beneath the specimen and for a clear zone of inhibition beyond its edge. The average width of a zone of inhibition (ZOI) along a streak on either side of the test specimen was calculated using

$$W = \frac{(T - D)}{2}, \quad (1)$$

where W is the average width of clear zone of inhibition in mm, T is the total diameter of test specimen and clear zone in mm, and D is the diameter of the test specimen in mm.

For AATCC Test Method 100, three discs of treated and untreated samples measuring 48 mm were inoculated with 1 mL of inoculum (1-2 × 10^5 organisms) in two separate 100 mL jars. The jars were incubated for 24 hours at 37°C. The samples were then neutralized with 100 mL of Tween 80 solution (Sigma-Aldrich) and were vigorously shaken for 1 minute. The solution was then eluted and plated with appropriate dilution in order to obtain countable colonies. The percent reduction of bacteria was calculated using

$$R = \frac{(B - A) * 100}{B}, \quad (2)$$

where R is the percent reduction of bacteria, A is the number of bacteria recovered from the inoculated treated test samples in the jar incubated over the desired contact period, and B is the number of bacteria recovered from the inoculated untreated test samples in the jar incubated over the desired contact period.

2.2.5. Statistical Analysis. The antimicrobial properties and durability to laundering of the antimicrobial treated fabrics were compared using Statistical Analysis System (SAS version 9.2, Cary, NC). The null hypothesis was that there were no statistically significant differences between triclosan and chitosan on a textile substrate. A three-factor design was used to structure the experiment, wherein the independent factors were antimicrobial agents, wash cycles, and bacteria. The treatment procedure, laundering, and antimicrobial testing of samples were replicated three times. The statistical evaluation was done using analysis of variance (ANOVA) at a 95% confidence level.

3. Results and Discussion

The mean zone of inhibition (ZOI, mm) as calculated using AATCC Test Method 147 for the polyester/cotton treated with triclosan and chitosan is shown in Table 1. Actual illustrative photographs against *E. coli* are displayed in Figures 3 and 4.

For purposes of this study, a mean ZOI of 2 mm was considered an indication of effective antimicrobial activity. Triclosan was found to possess excellent antimicrobial action against both *S. aureus* and *E. coli* (Table 1 and Figure 3). The mean ZOI against *S. aureus* before laundering the treated substrate was 22.2 mm and after 50 laundry cycles there was a slight decrease to 21.1 mm conclusively proving

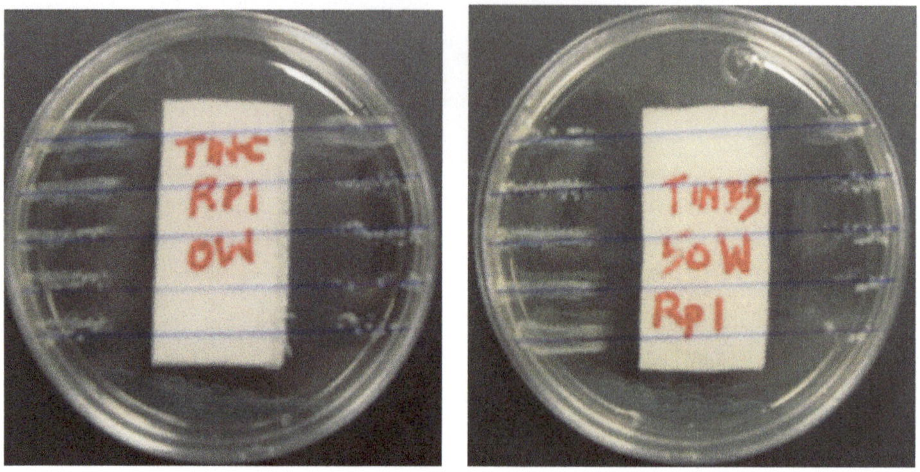

(a) (b)

FIGURE 3: Zone of inhibition of triclosan treated polyester/cotton blend against *E. coli*: after treatment (a) and after 50 laundry cycles (b).

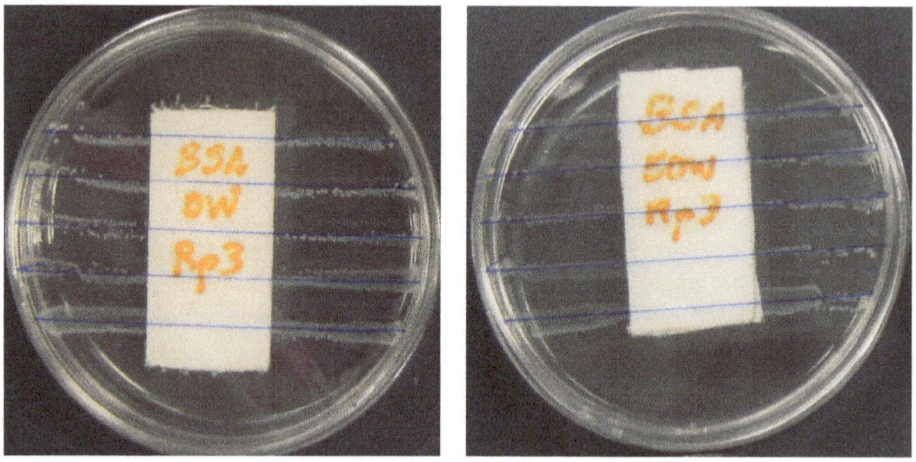

(a) (b)

FIGURE 4: Zone of inhibition of chitosan treated polyester/cotton blend against *E. coli*: after treatment (a) and after 50 laundry cycles (b).

TABLE 1: Mean zone of inhibition (mm) for treated polyester/cotton substrate against *S. aureus* and *E. coli* after treatment (0 laundry cycles) and after 50 laundry cycles.

Antimicrobial agent	*S. aureus* ZOI (mm)		*E. coli* ZOI (mm)	
	Laundry cycles			
	0	50	0	50
Triclosan	22.2	21.1	8.5	7.8
Chitosan	0.0	0.0	0.0	0.0

TABLE 2: Percentage reduction in bacteria of treated polyester/cotton substrate against *S. aureus* and *E. coli* after treatment (0 laundry cycles) and after 50 laundry cycles.

Antimicrobial agent	*S. aureus* (% reduction)		*E. coli* (% reduction)	
	Laundry cycles			
	0	50	0	50
Triclosan	100.0	100.0	100.0	100.0
Chitosan	100.0	92.0	99.6	51.0

the durability to laundering of the samples treated with triclosan. The values against *E. coli* were 8.5 mm before laundering and 7.8 mm after 50 laundry cycles again clearly underscoring the excellent and durable antimicrobial properties of triclosan based antimicrobials. The mean ZOI for chitosan treated substrates was 0 mm against both types of bacteria (Table 1 and Figure 4). This result is logical because AATCC Test Method 147 works best with *leaching* type of antimicrobials and as explained previously chitosan belongs to the bound or *nonleaching* type of antimicrobials. The non-leaching characteristic of chitosan was further emphasized by the absence of bacteria on the underside of the fabrics.

Table 2 reports the results obtained using AATCC Test Method 100. The data for triclosan are consistent with the results obtained using AATCC Test Method 147. Triclosan was extremely effective against both types of bacteria and

TABLE 3: Statistical comparison (P values) of polyester/cotton substrate against *S. aureus* and *E. coli* after treatment (0 laundry cycles) and after 50 laundry cycles.

Antimicrobial agent	*S. aureus*		*E. coli*	
	Laundry cycles			
	0	50	0	50
Triclosan (agent 1) / Chitosan (agent 2)	1	0.29	0.35	<0.0001

the efficacy did not diminish after 50 launderings. Antimicrobial activity of chitosan against *S. aureus* and *E. coli* and after treatment was excellent as well with 100% and 99.6% reduction of bacteria, respectively. After 50 laundering cycles, the activity against *S. aureus* remained high and only slightly decreased to 92%. However, the efficacy of chitosan against *E. coli* reduced significantly to 51% after 50 launderings. A probable explanation is that the stronger outer cell wall of Gram negative bacteria such as *E. coli* restricts the antimicrobial molecules from penetrating and killing the bacteria efficiently.

Statistical analysis (Table 3) showed that the efficacies of unlaundered samples ("0" wash cycles) of substrates treated with triclosan and chitosan against *E. coli* were not statistically different from each other (P value = 0.35). Similarly, against *S. aureus* both the agents exhibited 100% reduction of bacteria and therefore $P = 1$. After 50 launderings, the performance of chitosan against *E. coli* was reduced to 51% which was statistically significant compared to triclosan which exhibited 100% reduction (P value < 0.0001). Against *S. aureus*, chitosan at 92% bacterial reduction had less absolute efficacy than triclosan (100%) but this difference was not statistically significant (P value = 0.29).

4. Conclusions

The results of this study show that treatment of polyester/cotton substrate with the naturally occurring antimicrobial agent, chitosan, imparted antimicrobial activity against both *S. aureus* and *E. coli*. The treatment was very durable to laundering against the Gram positive bacteria but less durable against the Gram negative bacteria. The research thus suggests that chitosan based natural antimicrobial agent has the potential to be a viable alternative to chemically synthesized antimicrobial agents such as triclosan especially in light of the reported adverse impact of widespread use of chemically synthesized antimicrobial agents.

Conflict of Interests

The authors declare that there is no conflict of interests regarding the publication of this paper.

References

[1] Centers for Disease Control and Prevention, USA, 2014, http://www.cdc.gov/ncidod/dhqp/pdf/StateHAI_PlanningTemplate.pdf.

[2] Y. Gao and R. Cranston, "Recent advances in antimicrobial treatments of textiles," *Textile Research Journal*, vol. 78, no. 1, pp. 60–72, 2008.

[3] B. Simoncic and B. Tomsic, "Structures of novel antimicrobial agents for textiles: a review," *Textile Research Journal*, vol. 80, no. 16, pp. 1721–1737, 2010.

[4] R. Purwar and M. Joshi, "Recent developments in antimicrobial finishing of textiles—a review," *AATCC Review*, vol. 4, no. 3, pp. 22–26, 2004.

[5] M. Orhan, D. Kut, and C. Gunesoglu, "Use of triclosan as antibacterial agent in textiles," *Indian Journal of Fibre and Textile Research*, vol. 32, no. 1, pp. 114–118, 2007.

[6] U.S. Food and Drug Administration, 2014, http://www.fda.gov/forconsumers/consumerupdates/ucm205999.htm.

[7] M. Rinaudo, "Chitin and chitosan: properties and applications," *Progress in Polymer Science*, vol. 31, no. 7, pp. 603–632, 2006.

[8] S.-H. Lim and S. M. Hudson, "Review of chitosan and its derivatives as antimicrobial agents and their uses as textile chemicals," *Journal of Macromolecular Science*, vol. 43, no. 2, pp. 223–269, 2003.

[9] D. H. Young, H. Kohle, and H. Kauss, "Effect of chitosan on membrane permeability of suspension-cultured glycine max and *Phaseolus vulgaris* cells," *Plant Physiology*, vol. 70, no. 5, pp. 1449–1454, 1982.

[10] I. M. Helander, E.-L. Nurmiaho-Lassila, R. Ahvenainen, J. Rhoades, and S. Roller, "Chitosan disrupts the barrier properties of the outer membrane of Gram-negative bacteria," *International Journal of Food Microbiology*, vol. 71, no. 2-3, pp. 235–244, 2001.

14

A Computer-Based System for Evaluation of Slub Yarn Characteristics

R. Abd El-khalek, R. El-Bealy, and A. El-Deeb

Textile Engineering Department, Faculty of Engineering, Mansoura University, Mansoura 35516, Egypt

Correspondence should be addressed to R. Abd El-khalek; rehab_ali1976@yahoo.com

Academic Editor: Anindya Ghosh

This research aims to build a computer-based system for measuring slub yarn characteristics and evaluating the results using statistical methods. The measuring system was based on the electrical condenser of the Uster evenness tester as a measuring sensor. A digital storage oscilloscope was used to convert the analogue output signals into digital data to be recorded on a computer. A designed computer program was constructed using MatLab language for signal analysis based on a proposed method. Final results from this system contained graphical plots for the three measured slub yarn geometrical parameters slub length, slub distance, and slub thickness and a final report for all the measured slub yarn characteristics. The results of tested slub yarn samples showed that actual slub geometrical parameters differ from nominal values and the difference had two level: prevailing difference and outlier values. Outliers which are considered slub yarn defects were detected and counted in a final report using the box plot statistical method which could be an effective industrial tool for evaluation of both yarn quality and machine performance. The prevailing difference was assessed using histogram comparison which could help in adjusting the setting of slub yarn device to obtain the required appearance.

1. Introduction

Ground slub yarn is a simple fancy yarn whose slub appearance is gained by the variation of yarn linear density during spinning process and no additional yarn or process is required. It is used to produce an acceptable handle and fancy appearance mainly for plain conventional fabrics. Slub yarns could be produced by modifying ring spinning frame, such that the intermittent acceleration of the rollers will cause varying degrees of draft to be applied [1]. The programmed electronic control of two or more separate variable speed motors for each drafting roller enabled leading companies to produce such new drafting systems to obtain a wide range of slub yarn characteristics.

Evaluation of slub yarn characteristics which are different from conventional yarn is very important in the slub yarn manufacturing process as they determine the final appearance of the produced fabric. The traditional manual measuring method based on the black board and using standard measuring scales is time consuming. Different methods for measuring slub geometrical parameters, that is, slub length, slub distance, and slub thickness, were introduced in recent researches. Sudhakar [2] used and compared between different methods for measuring slub geometrical parameters. Slub length and slub thickness were measured using image analysis methods from slub yarn images captured using a flat scanner. Slub thickness was also measured using a yarn profile tester which is an imaging system that uses a charged coupled device (CCD) array to collect the yarn profile as the yarn moves through the imaging system. Slub length and slub distance were measured using the same imaging system of the yarn profiler tester and another software (Electronic inspection Board CTT EIB software). Pan et al. [3] used image analysis methods to measure slub length, slub distance, and slub thickness from slub yarn images captured using a flat scanner and used ID Fourier transform to identify the period of slubs for periodic slub yarns. Image analysis method is limited by size of captured image and parameters

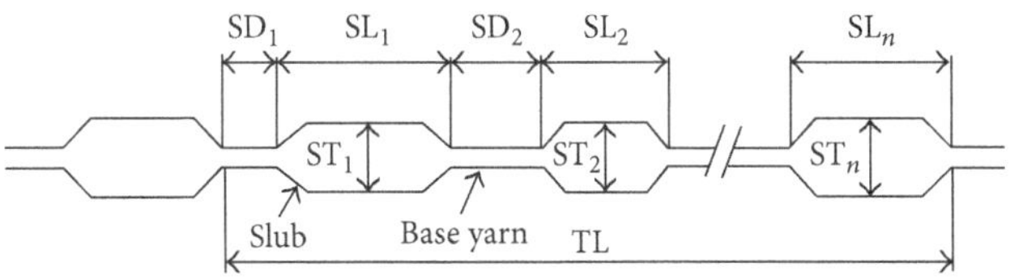

FIGURE 1: Slub yarn geometrical parameters.

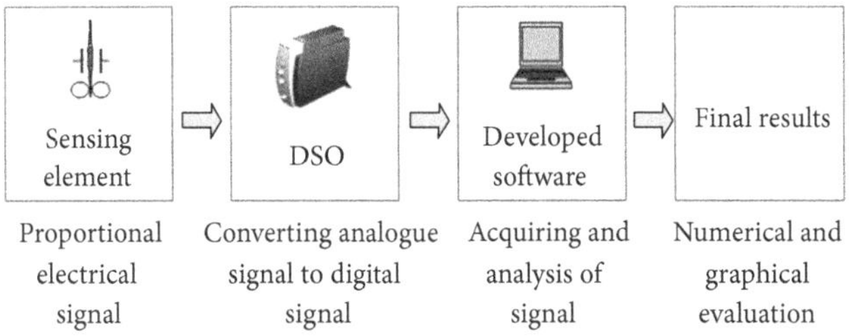

FIGURE 2: Block diagram for the measuring system.

that have length higher than the maximum image size of image capturing equipment cannot be measured. Thus, slub distance could not be measured by Sudhakar [2] or only slub yarn samples with short slub distance were used by Pan et al. [3]. Using image analysis method for measuring slub diameter which could be affected by the filling ratio and twist amount may not accurately represent slub thickness which is a relative mass change. Liu et al. [4] introduced an analytical method of determining two slub parameters: slub length and distance from voltage signals acquired from a capacitance type sensor. The method depends on eliminating the noise signal by low pass filtration, separating the slub part signal to build a pulse signal using the dynamic threshold, and confirming the parameters by observing the 2D image and pulse signal. Liu and Lu [5] introduced a method for recognition of the three slub yarn parameters based on FFT (fast Fourier transform) to filter the voltage signals from a capacitive-type sensor. Liu et al. [6] introduced a clustering analysis method for characterizing the two slub parameters: slub length and distance. Uster technologies AG introduced the slub yarn evaluation tool "Fancy yarn Profile" which is a software option to the Uster tester 5 [7] and pointed out that the appearance of the resulting slub does not necessary coincide with the setting at slub yarn device of the spinning machine, that is, nominal values which was also reported by Liu et al. [4–6].

This research aims to build a computer-based system using the electrical condenser of the Uster evenness tester as a sensor for measuring slub yarn characteristics and evaluating the results using statistical methods in order to detect and assess the difference between actual and nominal values.

2. Slub Yarn Characteristics

A ground slub yarn is formed of a single structure that has two parts: slub part and base yarn part. The basic geometrical parameters of slub yarn as shown in Figure 1 are slub length (SL), slub distance (SD) which represents the length of base yarn between two successive slubs, and slub thickness (ST) which represents the mass ratio between slub part and base yarn.

Other parameters that characterize a slub yarn produced with a repeat pattern containing n number of slubs could be calculated according to the following equations:

$$\begin{aligned}
&\text{Total slub length} = \sum_{i=1}^{n} \mathrm{SL}i = \mathrm{SL}_1 + \mathrm{SL}_2 + \cdots \mathrm{SL}_n \\
&\text{Total slub distance length} \\
&\quad = \sum_{i=1}^{n} \mathrm{SD}i = \mathrm{SD}_1 + \mathrm{SD}_2 + \cdots \mathrm{SD}_n, \\
&\text{Total pattern length (TL)} = \sum_{i=1}^{n} \mathrm{SL}i + \sum_{i=1}^{n} \mathrm{SD}i, \\
&\text{Slub per meter (S/m)} = \frac{n}{\mathrm{TL}}, \\
&\text{Sub length percentage (SL\%)} = \frac{\sum_{i=1}^{n} \mathrm{SL}i}{\sum_{i=1}^{n} \mathrm{SL}i + \sum_{i=1}^{n} \mathrm{SD}i} \cdot 100, \\
&\text{Slub distances percentage (SD\%)} \\
&\quad = \frac{\sum_{i=1}^{n} \mathrm{SD}i}{\sum_{i=1}^{n} \mathrm{SD}i + \sum_{i=1}^{n} \mathrm{SL}i} * 100, \\
&\text{Mean of slub length (SLm)} = \frac{\sum_{i=1}^{n} \mathrm{SL}i}{n}, \\
&\text{Mean of slub distance (SDm)} = \frac{\sum_{i=1}^{n} \mathrm{SD}i}{n}, \\
&\text{Mean of slub thickness (STm)} = \frac{\sum_{i=1}^{n} \mathrm{ST}}{n}.
\end{aligned} \tag{1}$$

3. Measuring System

Figure 2 shows a block diagram for the measuring system where the electric condenser of the USTER Evenness Tester I was used as the measuring sensor. The electric output signal was fed into a digital storage oscilloscope DSO (Velleman PcSU1000) to convert the analogue signal to digital data in order to be recorded on a computer using the installed software of the DSO and then analyzed by using a constructed software program.

3.1. Signal Recording. In order to obtain an accurate indication of the base yarn mean value, the measuring slot number was chosen according to the base yarn count. According to the application handbook for evenness testers by Uster [8], variation of more than 500% in the cross-section from the yarn mean value could be detected.

To process the analog signal in computers, it must be converted to "digital" form. While an analogue signal is continuous in both time and amplitude, a digital signal is discrete in both time and amplitude. To convert a signal from

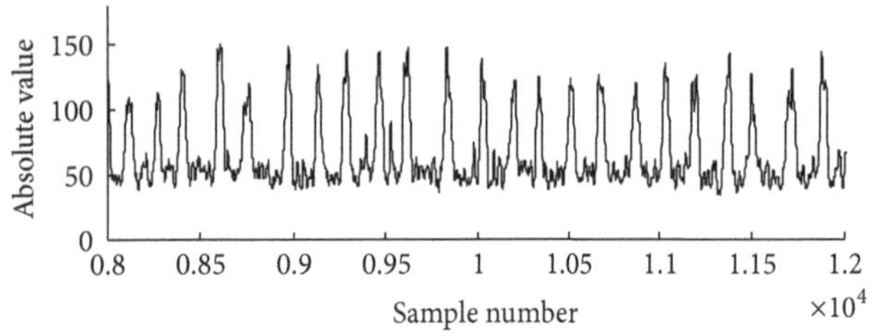

FIGURE 3: A part of a recorded signal for a slub yarn sample.

continues time to discrete time, a process called sampling is used. The value of the signal is measured at certain intervals in time. Each measurement is referred to as a sample. The sampling is done using a sampling rate or sampling frequency which is defined as the number of samples obtained in one second (samples per second). To keep the data organized, each sample is assigned a sample number or index. These are the numbers that appear along the horizontal axis of the digitized signal [9]. The number of samples recorded per unit length depends on both testing speed (length per unit time) and sample frequency (no. of samples per unit time). At higher speed and low sample frequency if the number of slubs per unit length increases this may results in losing of some slubs or recording low number of samples for each slub. In this study, signals were recorded at sample frequency of 100 sample/sec which is a specification of the DSO; thus, a speed of 8 m/min was convenient to ensure that every slub is represented by a quite sufficient number of samples.

Signal recording included two steps. Firstly, recording a signal for the tested slub yarn sample and secondly, recording the output value of measuring circuit when the base yarn is found between the measuring electrodes without drawing it through the measuring slot. Ten different values were recorded for ten different points along the base yarn. The aim of this step is to estimate an initial approximated value of base yarn mean.

3.2. Signal Analysis. Figure 3 shows a part of a recorded signal for a slub yarn sample in which the sample number on x-axis could be converted to its original value as time using the sample frequency and the absolute display value of the oscilloscope on the y-axis could be converted to its original value as volt using the voltage step (32 = 1 volt). A computer program was constructed using MatLab language (version 7.01) for signal analysis and evaluation of slub yarn characteristic. Figure 4 shows the main stages in the proposed signal analysis method.

After acquiring the digital data of signals the x-axis and y-axis were represented as distance and mass ratio, respectively, as follows.

(i) Separation of the base yarn signal from slub yarn signal: the aim of this step is to determine an accurate mean value of the base yarn. This process is done by thresholding slub yarn signal using the initial approximated value of base yarn mean as a threshold value. Figure 5 shows the separated base yarn signal from the part of recorded signal in Figure 3.

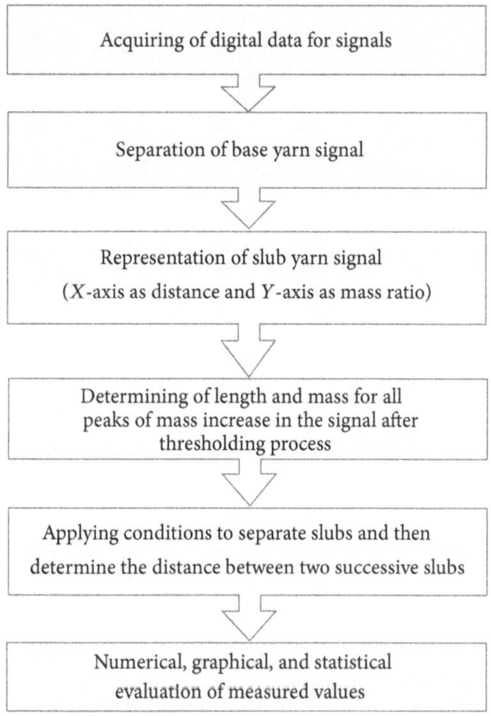

FIGURE 4: Main stages in the proposed method of signal analysis.

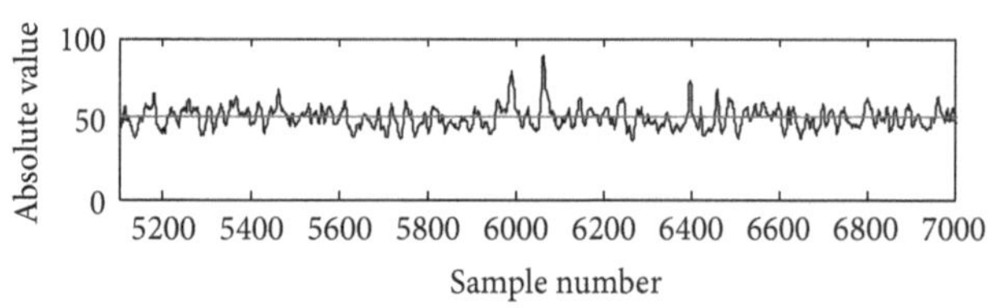

FIGURE 5: A part of separated base yarn signal from the recorded slub yarn signal in Figure 3.

(ii) The y-axis was converted from volt to mass ratio by dividing every sample value in the slub yarn signal by the determined base yarn mean value.

(iii) The x-axis was converted from time to length using a conversion constant (cm/sample) which is determined using test speed and sample frequency.

Figure 6 shows a zoomed part of a slub yarn signal after representing x and y-axis. Slub length was measured between the starts point of the slub where the yarn thickness start to increase and the end point of the slub where the yarn thickness reaches its original thickness. Slub distance was measured between the start of the slub to the end of the previous slub. As normal mass variation along slub length appears along the flat part of slub length therefore slub relative thickness was measured as the distance between the average peak value, that is, [maximum peak (A) + minimum trough (B)]/2 and the base yarn line. Slubs were then identified after thresholding and using of separation conditions to differentiate between slubs and other peaks of normal mass increase in the base yarn. The threshold value (1.15 $*$ base yarn mean value) was chosen based on the comparison between

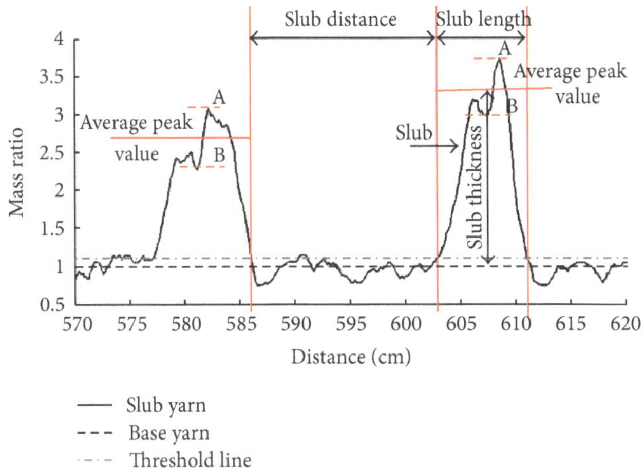

FIGURE 6: Feature extractions of slub yarn geometrical parameters from the recorded signal.

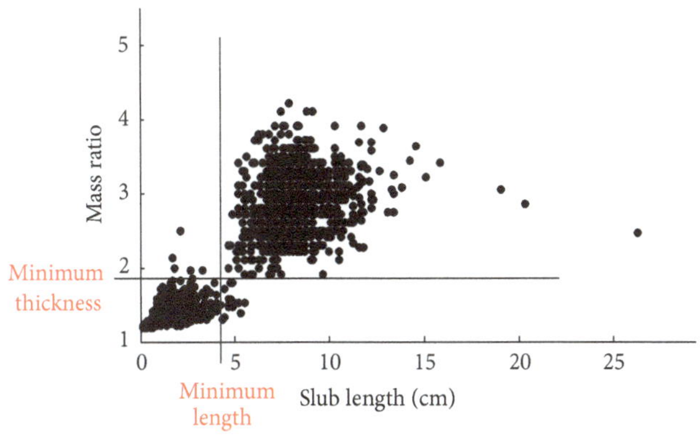

FIGURE 7: Identifying slubs after measuring the length and mass ratio for every peak exceeds the threshold value.

results from traditional measuring method and the proposed analysis method to obtain a high correlation. The length and thickness were measured for every peak that exceeds the threshold value. Exceeding the minimum limits of slub length and slub thickness is the separation condition as shown in Figure 7. After identifying slubs, distances between every two successive slubs were measured.

3.3. Reliability of the Measuring Method. A signal for a short sample of slub yarn marked by two knots at its both ends was recorded as shown in Figure 8. Slub lengths and distances between knots were measured using the conventional measuring with a ruler under magnifier and then compared with results from the previous proposed method for the same part of yarn. As shown from Figure 9, there is a very high correspondence between measurements from both methods.

By using the capacitor sensor the length of the shortest mass variations which can be recorded is mainly dependent on the length of the measuring field which is formed by the electrode plates. This length of measuring field results in the fact that the mean value of mass which lies between the measuring electrodes is determined and indicated. According to Uster-Manual of evenness testing variations which are

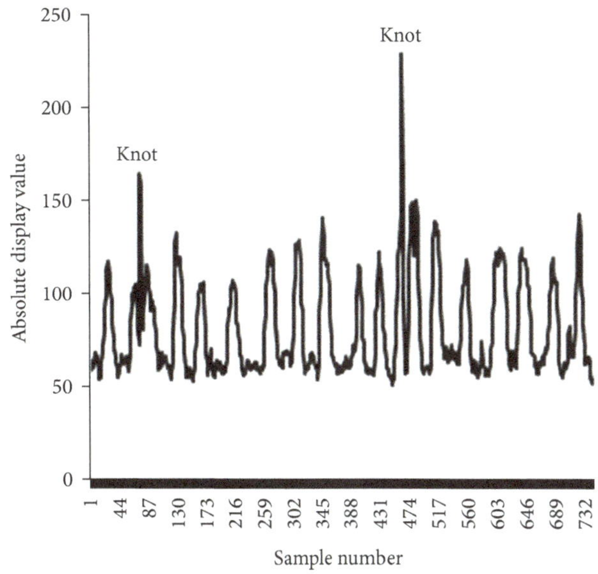

FIGURE 8: A recorded signal for a short length of slub yarn between two knots.

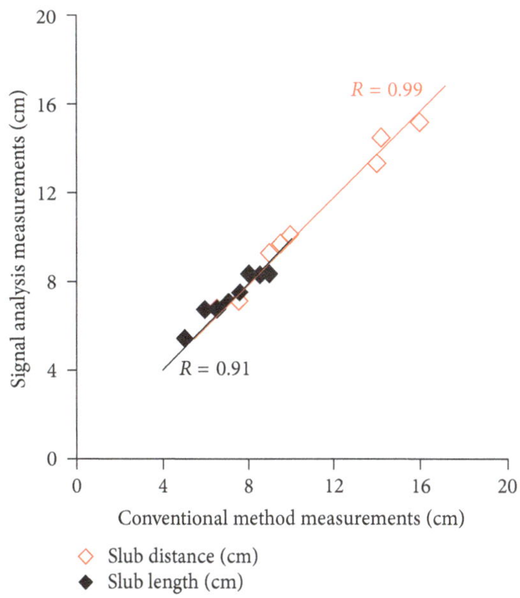

FIGURE 9: Measurements of signal analysis method plotted against measurements of conventional method.

shorter than the measuring field length cannot be indicated according to their full amount but appear as a reduced value according to a certain relationship. The resulting decrease of the indication at the 8 mm electrode (used in yarn testing) is negligibly small for fiber length larger than approximately 25 mm as prooved by Breny [10].

3.4. Statistical Evaluation of the Measured Slub Yarn Geometrical Parameters. In order to identify slub yarn samples with outlier characteristics box plot statistical method was used in which two levels of outliers could be identified [11]. If the measured value is at least 1.5 interquartile ranges (IQR)

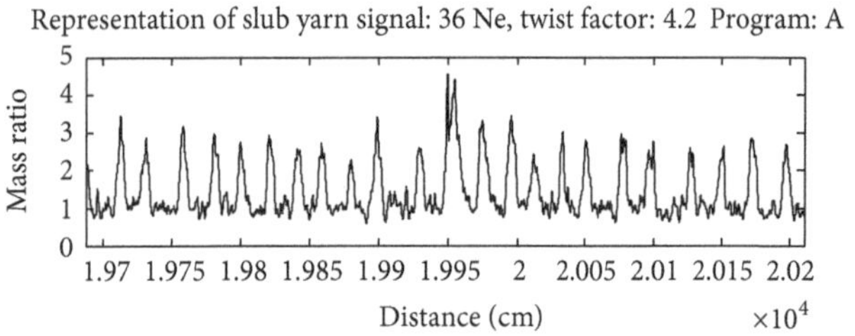

FIGURE 10: A part of the recorded signal for slub yarn sample number 1.

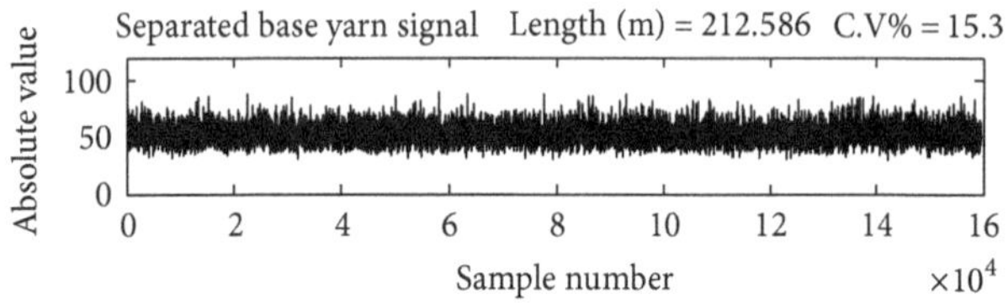

FIGURE 11: The separated base yarn signal for slub yarn sample number 1.

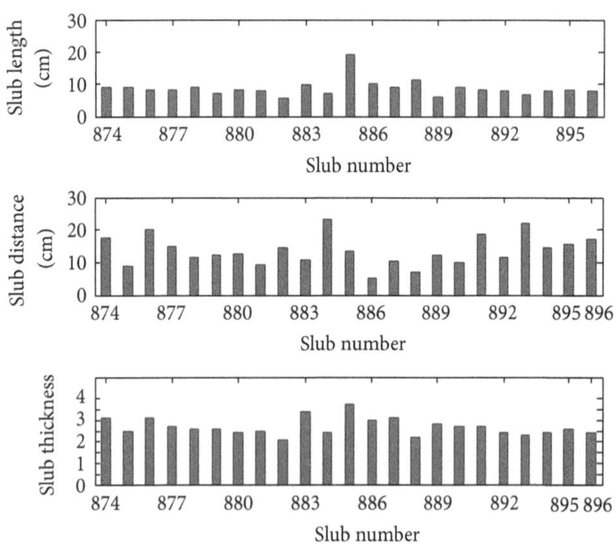

FIGURE 12: A part of the bar charts for the measured slub parameters for slub yarn sample number 1.

below the first quartile or at least 1.5 interquartile ranges (IQR) above the third quartile it is detected as a close outlier value, for example, long or low slub length, long or low slub distance, and high or low slub thickness. If the value is at least 3 interquartile ranges (IQR) below the first quartile or at least 3 interquartile ranges (IQR) above the third quartile it is detected as a very far outlier or extreme for example, very long or very low slub length, very long or very low slub distance, very high or very low slub thickness. Therefore, a final report which includes 12 different items of outliers was obtained. In order to assess the deviation between the nominal and the actual values of slub geometrical parameters, the system is prepared to compare between the two datasets using histograms.

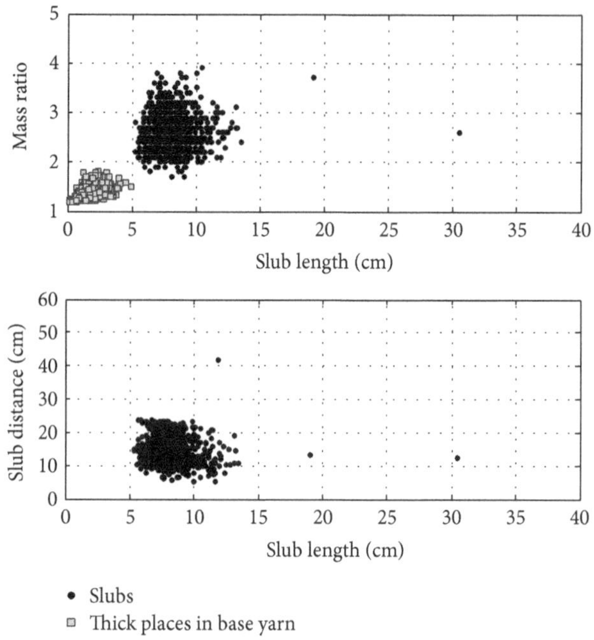

FIGURE 13: Measured values of slub length against slub thickness and slub distance for sample number 1.

Slub yarn: 36 Ne, twist factor: 4.2 Program: A

Total length of tested yarn (m) = 331.67
Number of slubs = 1473

Slub length

Min = 5.3
Max = 30.5
Average = 8.1

Slub distance

Min = 5.1
Max = 41.7
Average = 14.4

Slub thickness

Min = 1.7
Max = 3.9
Average = 2.6

Slub per meter = 4.44

Slub length (%) = 35.9

Slub distance (%) = 64.1

Yarn C.V (%) = 45.4

FIGURE 14: Final report for all the measured slub yarn characteristics in sample number 1.

4. Experimental Work

In this study, slub yarn samples with different count and twist factor produced by CAIPO slub device [12] were tested. The different patterns of slub yarn were designed using a software program installed on a computer. Each pattern of slub yarn consists of a number of steps equal to the number of slubs

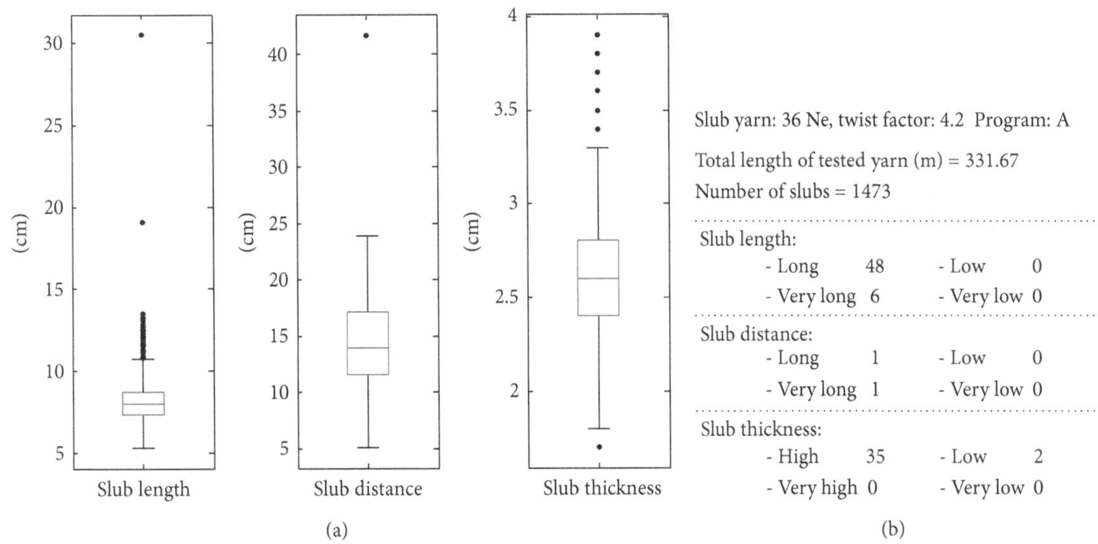

FIGURE 15: (a) Box plots for measured slub geometrical parameters. (b) Report of the number of detected outliers in slub yarn sample number 1.

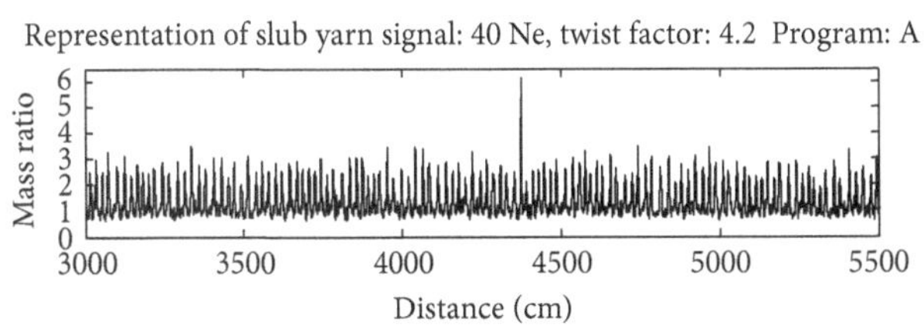

FIGURE 16: A part of the recorded signal for slub yarn sample number 2.

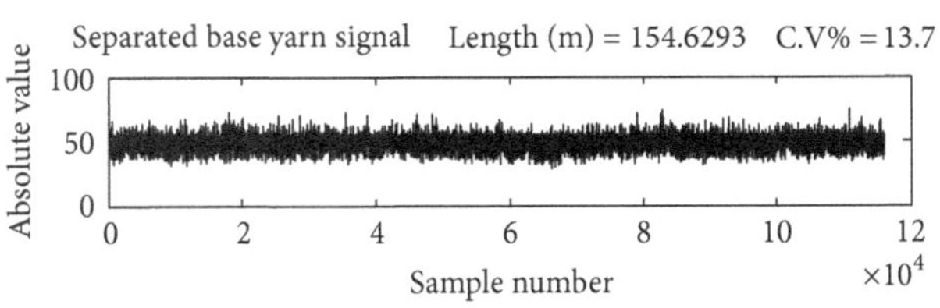

FIGURE 17: The separated base yarn signal for slub yarn sample number 2.

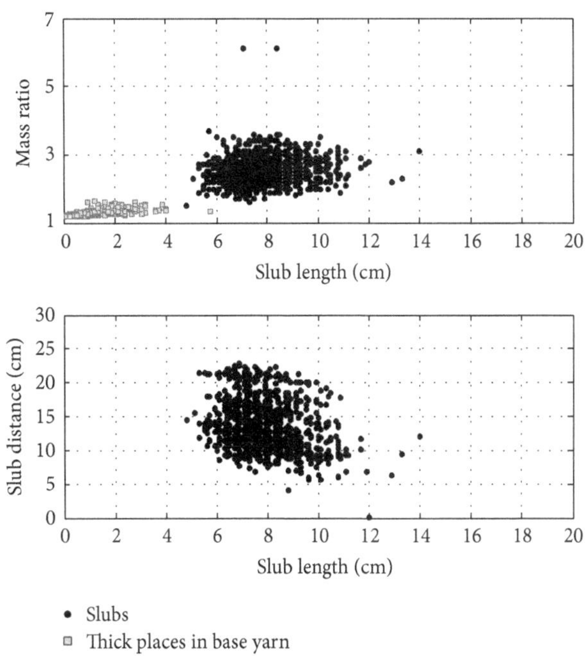

FIGURE 18: Measured values of slub length against slub thickness and slub distance for sample number 2.

per repeat. Each step holds the following information: slub distance (cm), slub length (cm), and slub thickness. This information is created randomly between a minimum and a maximum value for each slub parameter. Table 1 shows setting of each slub yarn pattern and slub yarn characteristics which were calculated from the detailed information of each pattern using equations in Section 2.

5. Results and Discussion

Final results from the signal analysis program for different slub yarn samples were as follows.

Sample Number 1. Figure 10 shows a part of a recorded signal for slub yarn sample of 36 Ne produced from carded Egyptian cotton with 4.2αe using slub yarn pattern A. Figure 11 shows the separated base yarn signal where the measured C.V of the 212.59 m was 15.3%. Figure 12 shows a part of the measured values of slub length, slub distance, and slub thickness for the 23 successive slubs shown in Figure 10. Figure 13 shows all the measured values of slub length plotted against slub thickness and slub distance. Figure 14 shows the final report of all the measured characteristics where the tested length was 331.67 m with 1473 slubs. Figure 15(a) shows box plots of the measured parameters: slub length, slub distance, and slub thickness and Figure 15(b) shows the final report of the number of detected outliers in which, for example, 6 slubs with extreme length (very long slub) and one slub with extreme slub distance (very long slub distance) were detected.

Table 1: Nominal characteristics of slub yarn samples.

Slub Yarn Pattern code	Setting of slub yarn device						Calculated slub yarn parameters							
	Slub length (SL) cm		Slub distance (SD) cm		Slub thickness (ST) ratio		No. of steps	SLm (cm)	SDm (cm)	STm (ratio)	S/m	TL (m)	SL%	SD%
	Min	Max	Min	Max	Min	Max								
Pattern A	4	6	12	25	2.3	2.8	99	5.2	17.4	2.55	4.424	22.4	23.1	76.9
Pattern B	3	6	30	70	3	4	99	4.4	50.5	3.49	1.822	54.3	7.1	92.7

TABLE 2: Difference between actual and nominal values.

Tested samples		Actual value – nominal value			
		SLm (cm)	SDm (cm)	STm (ratio)	TL (m)
No. 3	36 Ne, 3.6αe, pattern B	3.2	−4.8	0.18	−0.6
No. 4	36 Ne, 3.6αe, pattern A	2.6	−3	0.22	−0.5
No. 5	24 Ne, 3.6αe, pattern A	2.2	−2.8	0.25	−0.6

Slub yarn: 40 Ne, twist factor: 4.2 Program: A

Total length of tested yarn (m) = 244.18
Number of slubs = 1138

Slub length

Min = 4.8
Max = 14
Average = 7.9

Slub distance

Min = 0.1
Max = 22.8
Average = 13.6

Slub thickness

Min = 1.5
Max = 6.1
Average = 2.5

Slub per meter = 4.66

Slub length (%) = 36.7

Slub distance (%) = 63.3

Yarn C.V (%) = 44.5

FIGURE 19: Final report for all the measured slub yarn characteristics in sample number 2.

Sample Number 2. Figure 16 shows a part of a recorded signal for a slub yarn sample of 40 Ne produced from combed Egyptian cotton with 4.2αe using slub yarn pattern A. Figure 17 shows the separated base yarn signal where the measured C.V of the 154.63 m was 13.7%. Figure 18 shows all the measured values of slub length plotted against slub thickness and slub distance. Figure 19 shows the final report of all the measured characteristics in a tested length of 244.18 m with 1138 slubs. As shown in Figures 20(a) and 20(b) two extreme values of slub thickness (very high) and one extreme value of slub length (very long) for example were detected.

Results in the previous two samples showed that the difference between actual and nominal slub yarn geometrical parameters had two levels: the first is the prevailing difference and the second is the outlier values. The detected outliers in the measured slub geometrical parameters are probably caused by electronic signal malfunctioning. Very far outliers (extremes) such as very long slubs as marked by the arrow in Figures 21(a) and 21(b) or very long slub distance as marked by the arrow in Figure 21(c) will have a more noticeable adverse effect on fabric appearance than close outlier values such as low distance as marked by the arrow in Figure 22(a) or long slub distance as marked by the arrow in Figure 22(b). The decrease in slub thickness to the extent that it is classified as a mass increase in the base yarn after applying the separation condition to identify slubs resulted in detecting long slub distance as shown in Figure 22(c).

As shown in the final reports of the measured slub yarn characteristics in Figures 14 and 19 for sample number 1 and 2, respectively, actual mean of slub length (SLm) increased more than the nominal value by a value of 2.9 and 2.7 cm, actual mean of slub distance (SDm) decreased, than the nominal value by a value of 3 and 3.8 cm, and actual mean of slub thickness varied by a slight value of +0.05 and −0.05 in samples numbers 1 and 2, respectively. In order to assess this prevailing difference histogram comparison for three different slub yarn samples (samples numbers 3, 4, and 5) with no extreme characteristics were obtained. Slub yarn characteristics were measured in three different recorded signals for each sample. Each recorded signal represents one pattern repeat (99 successive slubs) with no extreme characteristics. Figures 23 and 24 show results of one of the three recorded signals for slub yarn sample number 3 of 36 Ne produced from carded Egyptian cotton yarn with 3.6αe using slub yarn pattern B.

Figures 23(a), 23(b), and 23(c) show histograms for nominal values in the above sections and actual values in the lower sections for slub length, slub distance, and thickness, respectively. The distribution of actual slub length extends through a wider range with a higher mean, median, and mode values. The distribution of actual slub distance extends through approximately the same range but with lower mean, median, and mode values. The distribution of actual slub thickness extends through a wider range with slight higher mean, median, and mode values. As shown in Figure 24 the increase in slub length resulted in an increase of percentage of slub length in the yarn (SL%) from 7.3 to 14.2% and the decrease of slub distance resulted in a decrease of percent of slub distance in the yarn (DL%) from 92.7 to 85.8%. The slight decrease in the total pattern length from 54.3 to 53.8 m leads to a very slight increase in number of slubs per meter from 1.822 to 1.84. Table 2 summarizes results for slub yarn samples numbers 3, 4, and 5 in which the average difference between actual and nominal values in the three recorded signals were determined for the parameters STm, SDm, STm, and TL in each tested sample.

The increase of slub length may be due to the time lag between the electronic signals of speed change and the mechanical response caused by inertia of rotating masses. This time lag could be different in the two cases of accelerating

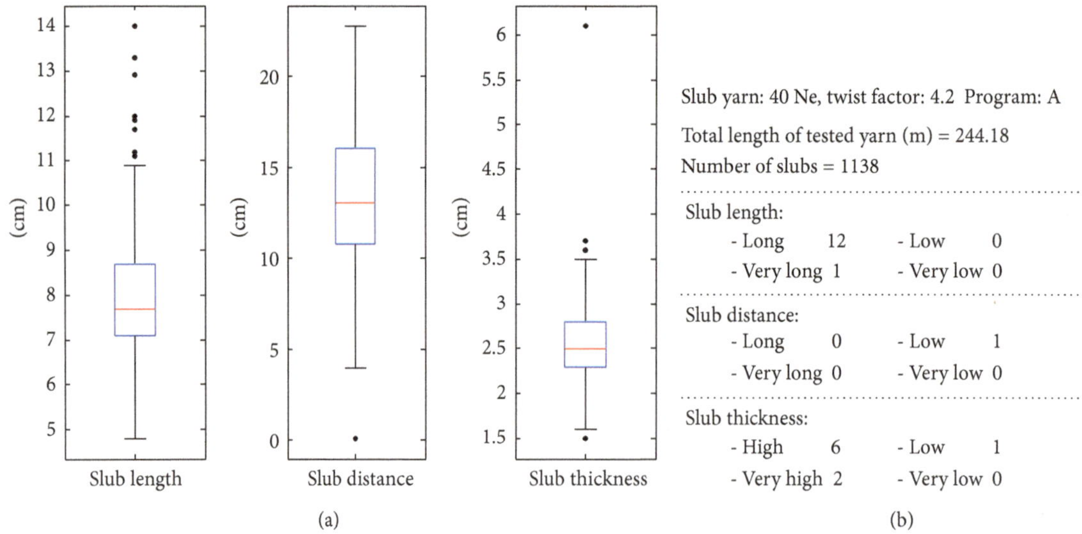

FIGURE 20: (a) Box plots for measured slub geometrical parameters. (b) Report of the number of detected outliers in slub yarn sample number 2.

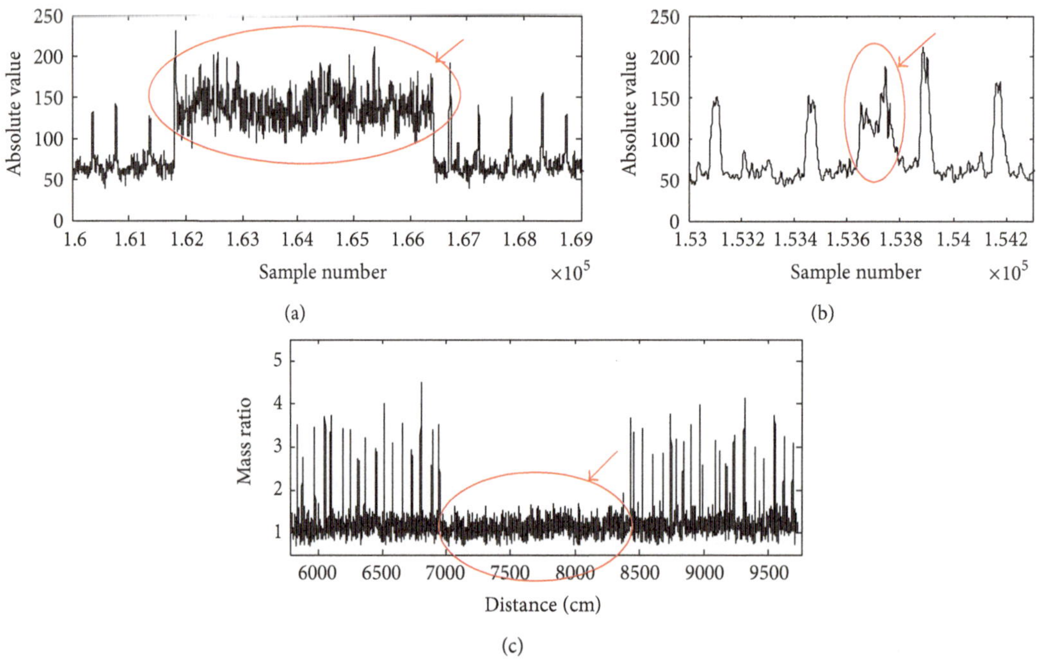

FIGURE 21: Examples of detected slubs with extreme characteristic.

and decelerating, that is, the time needed to decelerate a rotating roller from higher to lower speed is more than the time needed to accelerate it from lower to higher speed. As the decelerating signal is to end the slub part and in the same time to start the slub distance part, thus if the slub part extends this will decrease the length of the slub distance. Results in Table 2 show that the difference between actual and nominal mean slub length increased from 2.2 to 2.6 cm when the count increased from 24 Ne to 36 Ne in slub yarn samples produced using pattern A. Also, the difference between actual and nominal slub length increased from 2.6 to 3.2 cm in slub yarn samples of 36 Ne ($3.6\alpha e$) when the pattern changed from A to B which may be interpreted by the increase of nominal slub thickness from 2.55 in pattern A to 3.49 in pattern B, that is, the higher the speed the more time required to decelerate the roller to its original speed thus slub will extends more if slub thickness increased. Assessing the prevailing difference between actual and nominal values under different processing parameters needs more experimental study.

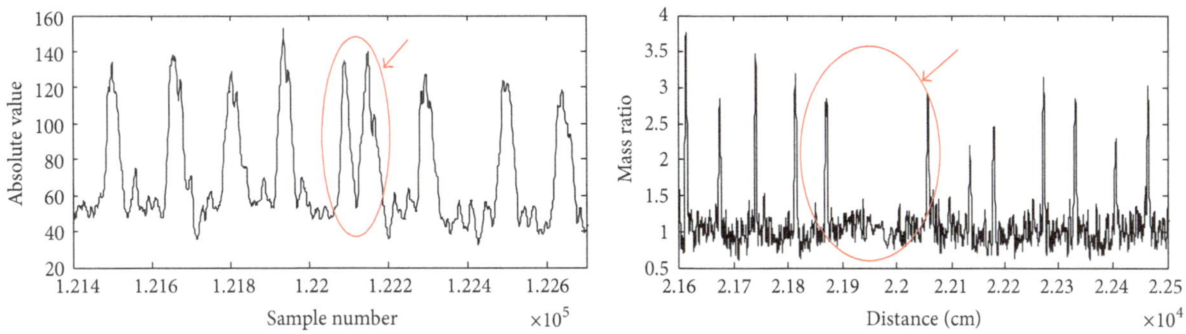

(a) Low slub distance

(b) Long slub distance due to missed slub over a short length

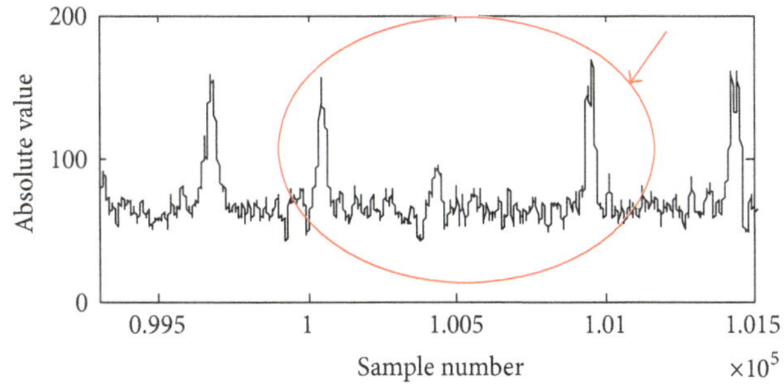

(c) Long slub distance due to very low slub thickness

FIGURE 22: Examples of detected low and long slub distance.

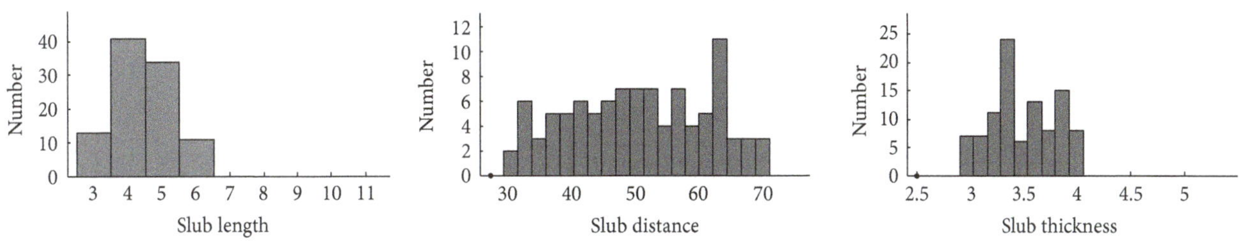

(A) Nominal: mean = 4.43, std. = 0.86, median = 4.00, approx. mode = 4.00

(A) Nominal: mean = 50.45, std. = 10.70, median = 51.00, approx. mode = 62.99

(A) Nominal: mean = 3.49, std. = 0.30, median = 3.50, approx. mode = 3.34

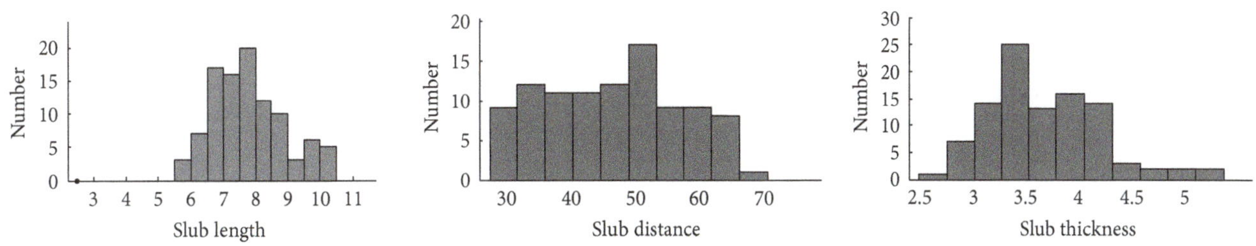

(B) Measured: mean = 7.71, std. = 1.10, median = 7.60, approx. mode = 7.75

(a)

(B) Measured: mean = 46.61, std. = 10.55, median = 46.90, approx. mode = 51.20

(b)

(B) Measured: mean = 3.66, std. = 0.53, median = 3.60, approx. mode = 3.41

(c)

FIGURE 23: Histogram comparison between actual and nominal slub geometrical parameters.

6. Conclusion

(i) A reliable system for evaluating different slub yarn characteristics based on the electrical condenser of the Uster evenness tester as a measuring sensor has been developed. The electrical output signal was fed into a digital storage oscilloscope and then analyzed using a proposed analysis method. Results from the proposed analysis method showed a very good correlation with the traditional manual measuring method.

Slub yarn: 36 Ne, twist factor: 3.6 Program: B

Total length of tested yarn (m) = 53.78
Number of slubs = 99

Slub length
- - - - - - - - -
Min = 5.7
Max = 10.4
Average = 7.7
- - - - - - - - -
Slub distance
- - - - - - - - -
Min = 27.6
Max = 66.4
Average = 46.6
- - - - - - - - -
Slub thickness
- - - - - - - - -
Min = 2.5
Max = 5.1
Average = 3.7
- - - - - - - - - - -
Slub per meter = 1.84
- - - - - - - - - - -
Slub length (%) = 14.2
- - - - - - - - - - -
Slub distance (%) = 85.8
- - - - - - - - - - -
Yarn C.V (%) = 56.2

Figure 24: Final report for all the measured slub yarn characteristics in sample number 3.

(ii) By using a computer-designed program for signal analysis, the system is capable of measuring slub length, slub distance, slub thickness, number of slub per meter, percentage of slub length, percentage of base yarn length, and CV% of tested slub yarn and base yarn. Final results from the system are graphical plots and a final report of all the measured characteristics.

(iii) Using box plot statistical method the system also makes it possible to detect slubs with outlier parameters which are considered slub yarn defects as very long slub length, very long slub distance, and very high slub thickness. A final report for the number of detected outliers was obtained which could be an effective industrial tool for evaluation of both yarn quality and machine performance. It is recommended also to evaluate the number of defects per constant unit length, for example, 1000 meter.

(iv) Actual slub geometrical parameters were compared by nominal values using histograms to assess their deviation. As the geometrical parameters of slub yarn determine the final appearance of the produced fabric, therefore the prevailing difference between actual and nominal values needs more experimental study to decide the causes and effect.

Conflict of Interests

The authors declare that there is no conflict of interests regarding the publication of this paper.

References

[1] R. H. Gong and R. M. Wright, *Fancy Yarns: Their Manufacture and Application*, Woodhead, Cambridge, UK, 2002.

[2] J. Sudhakar, *Characterization methods and physical properties of Novelty yarn [M.S. thesis]*, Textile Management and Technology, College of Textiles, Raleigh, NC, USA, 2005.

[3] R. Pan, W. Gao, J. Liu, and H. Wang, "Recognition the parameters of slub-yarn based on image analysis," *Journal of Engineered Fibers and Fabrics*, vol. 6, no. 1, pp. 25–30, 2011.

[4] J. Liu, Z. Li, Y. Lu, and H. Jiang, "Visualization and determination of the geometrical parameters of slub yarn," *Fibres & Textiles in Eastern Europe*, vol. 18, no. 1 (78), pp. 31–35, 2010.

[5] J. Liu and Y. Lu, "Recognition method for the slub yarn's apparent parameters based on FFT," in *Proceedings of the 86th Textile Institute World Conference*, pp. 2567–2573, 2008.

[6] J. Liu, Z. Xie, W. Gao, and H. Jiang, "Automatic determination of slub yarn geometrical parameters based on an amended similarity-based clustering method," *Textile Research Journal*, vol. 80, no. 11, pp. 1075–1082, 2010.

[7] E. Sandra, "Uster tester 5-S8000 application report measurement of slub yarns partl—basics," 2007, http://www.uster.com/en/service/downloadcenter/U_T5_S800_Measurement_of_slub_yarns_Basic.pdf.

[8] A. G. Z. Uster, "Application Handbook for evenness testers of the type Uster evenness Tester (GGP B/C)," Uster Tester I B/C and Uster Tester II B/C, CH-8610 Uster/Switzerland.

[9] W. Steven Smith, *The Scientist and Engineer's Guide to Digital Signal Processing*, California Technical Publishing, San Diego, Calif, USA, 2nd edition.

[10] H. Breny, "The calculation of the variance length curve from the length distribution of fibers," *The Journal of Textile Institute*, vol. 44, no. 1, pp. 1–9, 1953.

[11] D. L. Massart, J. Smeyers-Verbeke, X. Capron, and K. Schlesier, "Visual presentation of data by means of box plots," *LC-GC Europe*, vol. 18, no. 4, pp. 215–218, 2005.

[12] CAIPO, Biella, Italy, http://www.caipo.com/.

15

Mathematical Modeling to Predict the Geometrical and Physical Properties of Bleached Cotton Plain Single Jersey Knitted Fabrics

A. Fouda, A. El-Hadidy, and A. El-Deeb

Textile Engineering Department, Faculty of Engineering, Mansoura University, Mansoura 35516, Egypt

Correspondence should be addressed to A. Fouda; abdo_faheem@mans.edu.eg

Academic Editor: Ricardo Molina

This paper presents a novel mathematical model of bleached cotton plain single jersey knitted fabrics. The mathematical model is used to deduce the fabric geometrical relationships that can be useful for forecasting the properties of the fabric before production. A practical verification is carried out at different cotton yarn counts and twist factors. The obtained mathematical and practical results are deeply studied and analyzed. The results showed a good agreement between the proposed mathematical model and the practical one. The finished fabric weight is predicted at different yarn counts and loop lengths, and a forecasting weight chart is deduced. This chart will undoubtedly help the producers to enhance the fabric productions. In addition, an actual yarn diameter in the fabric measuring method is carried out and the fabric thickness is estimated consequently. The obtained results proved that the plain single jersey thickness is proportional to three times of the actual measured yarn diameter.

1. Introduction

Knitted fabrics are distinguished from the comfort point of view as compared to woven fabrics. Therefore, it is widely used in sportswear, underwear, and other comfortable fabrics. Over all types of knitted fabrics, plain single jersey knitted fabric is considered as a wide spread type. So, many research studies were focused on this type of fabrics.

Generally, the research studies depend on mathematical models to forecast the fabric properties that help to save frequent production time and efforts and in addition to achieve production of fabrics with high quality.

The knitted fabric weight per unit area is predicted using appropriate software called proKNIT [1]. But the prediction of the fabric weight is dependent on the calculated dimensional parameters of K_c, K_w, K_s, and R, which have been entered into the system, but these constants are used widely with wool knitted fabrics.

The denominated deviation rate DR [2] was found to have an excellent statistical correlation with the aspect of the knitted fabrics. For fine gauge knitted fabrics, it is recommended to have DR values low to 25%, for classified products those like of excellent quality, which is not verified experimentally.

Considering the analytical and experimental procedures for estimating elastic properties of a plain weft-knitted fabric and of polymer composite materials reinforced by it [3] and the elastic moduli of the cotton yarn and knitted fabrics, having different load span and knitting directions, based on the leaf and glaskin model, a numerical (FEM) elastic properties averaging model was elaborated.

The feasibility of assessing yarns with the wool comfort meter (WCM) to predict the comfort properties of the corresponding single jersey-knitted fabrics is examined [4]; inclusion of knitting gauge and cover factor slightly improved predictions. This indicates that evaluation at the yarn stage would be a reliable predictor of knitted fabric comfort, and thus yarn testing would avoid the time and expense of fabric construction; such correlation is only limited to wool knitted fabrics.

Considering the applicability of finite element method to analyze the bagging behavior of plain single jersey weft knitted fabrics in terms of bagging resistance [5], the numerical modeling of the bagging resistance of the plain knitted fabric using solid elements and yarn transverse isotropic properties is in good agreement with experimental values. Finally, this paper demonstrates a good practical verification of the proposed mathematical study, but in the rare field for uses of knitted fabrics.

Obviously, most of these researches did not differentiate between the mathematical and practical results, except for a few researches that differentiate the mechanical properties only. Most of these researches did not focus on bleached cotton plain single jersey knitted fabrics.

The aims of this paper are construction of a mathematical model that represents the geometrical properties of finished (bleached) plain single jersey knitted fabric, practical verification of the proposed mathematical model, and finally deducing some relationships to study the geometrical properties of the fabric.

2. Experimental Work and Tests Methodology

The specifications of the used machine are circular single jersey knitting machine, model ALBI, gauge 28, diameter 17 inch, and number of feeders 34. The fabrics are produced at loop length of 2.87 mm. The material used is cotton 100% Giza 86 and eight yarn counts are used (18, 22, 26, 30, 36, 40, 45, and $60N_e$). All of the yarns are produced at three levels of yarn twist factor (2.8, 3.5, and $4.2\alpha_e$).

Although it is known that the use of yarns with fine counts with respect to machine gauge and high twist factor gives spirality in plain single jersey knitted fabrics, a wide range of twist factor and yarn count was used to study the compatibility of the practical results with the proposed mathematical model.

All fabrics are finished (full bleached) before tests on Thies Jet machine. The bleached solution contains the following agents with liquor ratio L.R (1 : 8):

(1) sequestering agent (0.9 : 1.25 g/lit),

(2) soap (1.25 g/lit),

(3) sodium hydroxide 50° Be (5 g/lit),

(4) peroxide stabilizer (1.6 : 1.9 g/lit),

(5) hydrogen peroxide 50% (10 g/lit).

The process was carried out at 100°C for 40 min. At 80°C the optical bleaching agent was used (0.375 g/lit), followed by rising the temperature to 100°C again for 20 min. Washing off was carried at 80°C for 10 min, followed by acid bath washing using acetic acid (1.25 g/lit) at 60°C for 20 min. Cold rinsing was followed by hot rinsing at 80°C for 10 min. Softening agent (5 g/lit) was used at 45°C for 30 min, followed by the final cold rinsing.

WPC, CPC, thickness, and weight tests are carried out and measured practically.

During wales density test, two needles are removed and there is a spread of 288 needles between them. The distance between the two removed needles is measured in the finished fabric and the wales are calculated b:

$$(\text{WPC}) = \frac{288}{\text{distance between the removed needles (cm)}}. \quad (1)$$

The courses density test is carried out by inserting a different color cone during the production and the length of ten repeats is measured. The course of finished fabric is then calculated by

$$(\text{CPC}) = \frac{\text{Number of feeders} \times 10}{\text{length of ten repeats (cm)}}. \quad (2)$$

The fabric thickness is measured according to ASTM D1777 standard.

The fabric weight is measured using a digital balance of two decimal digits accuracy.

3. Mathematical Knitting Model

In this section the mathematical model of the plain single jersey knitted fabric is deduced. The model is presented through two stages, namely, idealized general mathematical model and geometrical model at maximum set in only wales directions. Regarding the first model, a general model is assumed with general assumptions, and then mathematical equations are proved. Through the second model, new assumptions are introduced and the final equations are deduced to find out the suitable mathematical model.

3.1. Idealized General Mathematical Model of Plain Single Jersey Knitted Fabrics. In this model, the general assumptions are as follows.

(1) Yarns are circular cross section and can be touched but they cannot be compressed.

(2) Yarns are formed in straight lines and half circles.

According to these assumptions, the yarn appearance is drawn as shown in Figure 1.

To find out the mathematical equations that represent the plain single jersey knitted fabric, one repeat is taken into account and its boundaries are $UXYZ$ as shown in Figure 1. The dimensions of the loop under study are illustrated in Figure 2.

As shown in Figure 2, the loop head and the two legs can be drawn graphically by assuming an appropriate value for their radius of curvature (R). To avoid jamming, the radius of curvature should satisfy the following condition:

$$R < \left(\frac{W}{2} - \frac{d}{2}\right), \quad (3)$$

where W is the wale space and d is the yarn diameter.

According to the assumptions, the right stem can be graphically drawn by connecting the points ($n\ m$) and ($p\ q$) which can be simply determined as $n\ p = m\ q = d$. Similarly, the left stem can be drawn by the same technique.

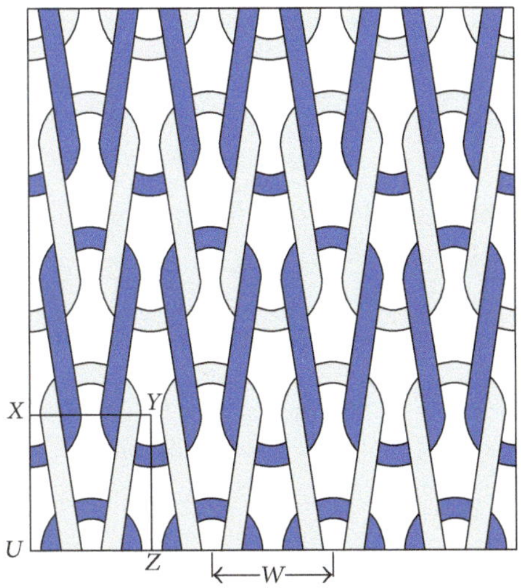

FIGURE 1: Simulated appearance of open plain single jersey knitted fabric.

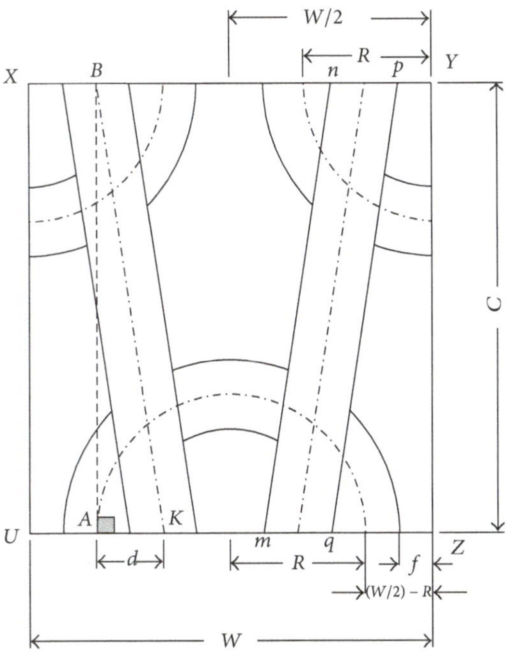

FIGURE 2: Graphical representation of plain single jersey general model.

Proof. According to Figure 2, the following relationships can be deduced:

$$f = \frac{W}{4} - d. \tag{4}$$

The value of R as a function of the dimensions can be determined as follows:

$$\frac{W}{2} = R + \frac{d}{2} + f,$$
$$\therefore R = \frac{W}{4} + \frac{d}{2}. \tag{5}$$

The loop length value (l) can be determined from the following equations:

$$l = 2\pi R + 2BK, \tag{6}$$

$$\therefore l = \frac{\pi W}{2} + \pi d + 2\sqrt{C^2 + d^2}, \tag{7}$$

where $BK = \sqrt{C^2 + d^2}$ and b is the course space.

The tightness factor (T.F) can be estimated according to the following relationship:

$$\text{T.F} = \frac{l \times d}{W \times C}. \tag{8}$$

Substituting by (8),

$$\therefore \text{T.F} = \frac{\pi W d/2 + \pi d^2 + 2d\sqrt{C^2 + d^2}}{W \times C}. \tag{9}$$

□

Regarding (7), it is found that the loop length (l) is a function of the yarn diameter (d). However, practically, it is found that there is no relationship between the loop length and the yarn diameter. Therefore, new assumptions should be introduced to find out a suitable model as discussed in the following section.

3.2. Geometrical Model at Maximum Set in Only Wales Directions. Due to the nature of knitted fabrics finishing processes in the form of rope, the knitted fabrics are exposed to tension in the length direction, which tends to close any spaces in between wales on the expense of increasing spaces in course direction.

Therefore, in this model, beside the general assumption, the following assumption is introduced.

(1) The loop length is fixed.

(2) The wale space at maximum set is equal to four times of the yarn diameter.

According the overall assumption, the wale space, course space, and tightness factor can be estimated as a function of yarn diameter.

Figure 3 shows the simulated appearance of the maximum set in wales direction for plain single jersey knitted fabric.

To derive the final mathematical equations for maximum set in wales direction of plain single jersey knitted fabric, the repeat $UXYZ$ is taken into account as shown in Figure 3. The dimensions of the loop under study are illustrated in Figure 4.

Proof. According to Figure 4, the following relationships can be deduced.

The value of wale space (W) can be determined as

$$W = 4d,$$

$$W = \frac{10}{\text{WPC}}\ (\text{mm}), \tag{10}$$

$$\therefore \text{Max .wales} = \frac{10}{W},$$

$$\because d = \frac{1}{28\sqrt{N_e}}\ \text{inch}. \tag{11}$$

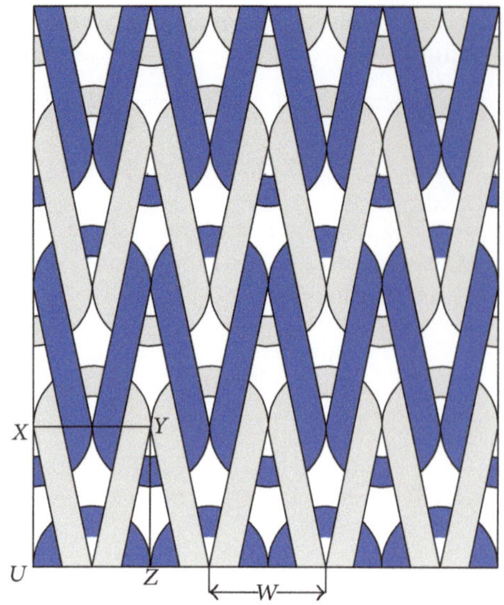

Figure 3: Simulated appearance of jersey knitted fabric touched in wales directions.

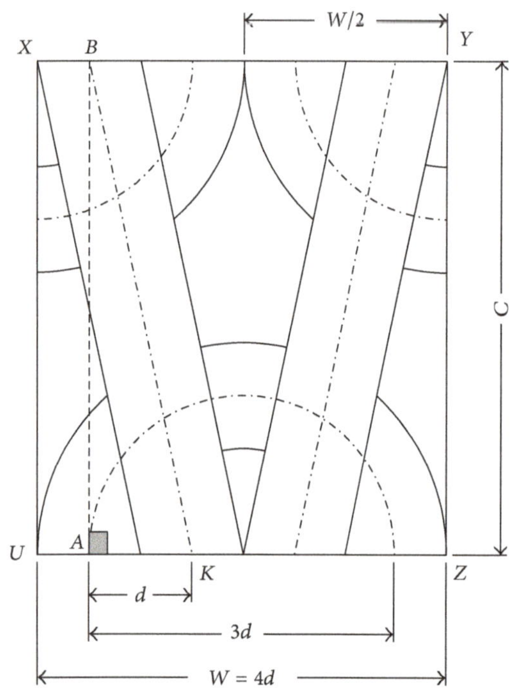

Figure 4: Graphical representation of touched in wales direction for plain single jersey model.

So, the maximum wales can be calculated as

$$\therefore \text{Max .wales} = 2.756\sqrt{N_e},$$
$$\therefore \text{Max .wales} = \frac{66.97}{\sqrt{\text{Tex}}}, \tag{12}$$

where N_e is English yarn count and Tex is Tex yarn count.

The value of loop length can be determined from the following equation:

$$l = 3\pi d + 2BK. \tag{13}$$

From the triangular ABK, the distance BK can be easily determined as

$$BK = \sqrt{d^2 + C^2}. \tag{14}$$

Substituted in (13) by the value of BK, the relationship of the course space (C) can be deduced as follows:

$$\therefore l = 3\pi d + 2\sqrt{d^2 + C^2},$$
$$\sqrt{d^2 + C^2} = \left(\frac{l - 3\pi d}{2}\right),$$
$$C^2 = \frac{(l - 3\pi d)^2}{4} - d^2,$$
$$\therefore C = \sqrt{\frac{(l - (3/4)\pi W)^2}{4} - \left(\frac{W}{4}\right)^2}, \tag{15}$$
$$\because C = \frac{10}{\text{CPC}}\,(\text{mm}),$$
$$\therefore \text{CPC} = \frac{10}{C}\,(\text{courses/cm}),$$
$$\therefore \text{Stitch density (SD)}$$
$$= \text{CPC} \times \text{WPC}\left(\text{S/cm}^2\right).$$

The T.F can be determined from the following equations:

$$\text{T.F} = \frac{\text{Covered area}}{\text{Total area}} = \frac{\text{loop length }(l) * d}{W * C},$$
$$\therefore \text{T.F} = \frac{\left((3/4)\pi W + 2\sqrt{(W/4)^2 + C^2}\right) \times d}{W \times C}, \tag{16}$$
$$\text{Fabric weight (F.W)} = \frac{l \times \text{SD} \times \text{Tex}}{100}\left(\text{g/m}^2\right),$$
$$\text{Fabric Bulkiness (F.B)} = \frac{\text{F.W}}{h \times 1000}\left(\text{g/cm}^3\right),$$

where SD is the stitch density (S/cm^2) and h is the fabric thickness (mm). □

The calculation of fabric thickness is discussed in details in Section 4.6.

Figure 5 shows the systematical steps that conclude the calculation procedures to get T.F and stitch density from the input known parameters which are yarn diameter and loop length.

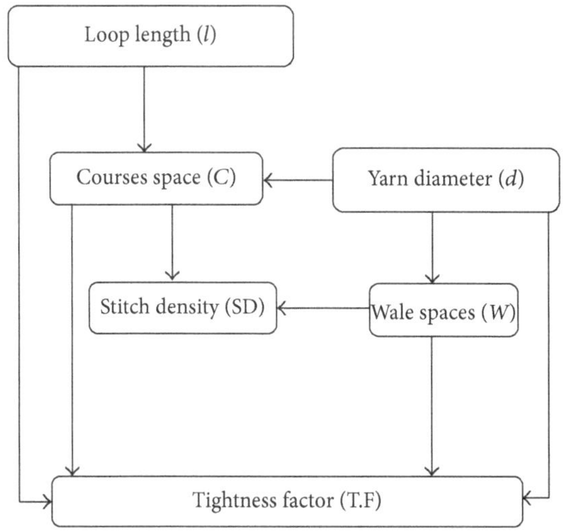

FIGURE 5: Output of the final mathematical model (touched model in wales direction).

4. Results and Discussions

In this section, a comparison between practical and theoretical results is presented. In addition, yarn count and twist factor effects on geometrical properties of plain single jersey knitted fabrics are studied and analyzed.

4.1. Wales Density (WPC). Figure 6 illustrates the actual effect of yarn count on the wales density at different levels of twist factors and a comparison with the results is obtained from the mathematical model.

Practically, it can be seen that. By increasing the yarn count from 9.8 to 32.8 Tex, the wales density decreased by 22 percent. On the other hand, the same trend is observed from the results obtained from the mathematical model, with different slope.

It can be justified that increasing the yarn diameter leads to increasing of the wale space; therefore, the wales density decreases.

Therefore, as compared to the mathematical model results, the yarn counts of 22.7 Tex can be considered as tight fabric because the yarns may be touched and compressed. On the other hand, the yarn counts less than 19.7 can be considered as open fabric.

Practically, it can be noted that the yarn twist factor has no significant effect.

4.2. Course Density (CPC). Figure 7 presents the practical and mathematical results of the yarn count effect and the yarn twist factor on the density courses.

From a practical point of view, it can be seen that, by increasing the yarn count from 9.8 to 32.8 Tex, the course space decreases and consequently the density courses increase by 33 percent. On the other hand, the same trend is observed from the results obtained from the mathematical model.

By a comparison with the mathematical model results, the fabric produced from the yarn that counts less than 19.7 Tex

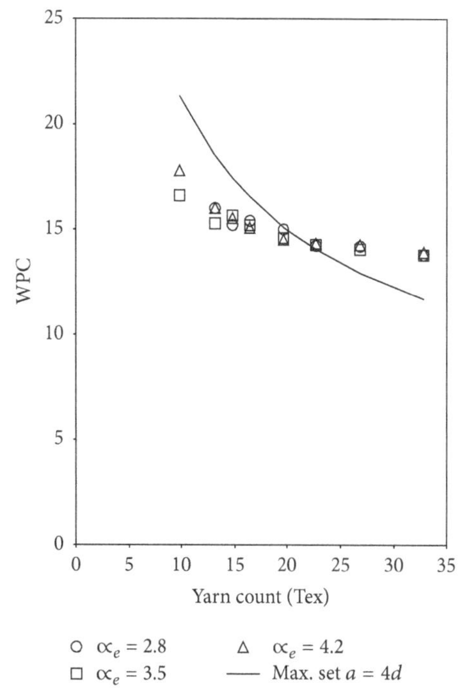

FIGURE 6: Effect of yarn count and twist factor on the wales density.

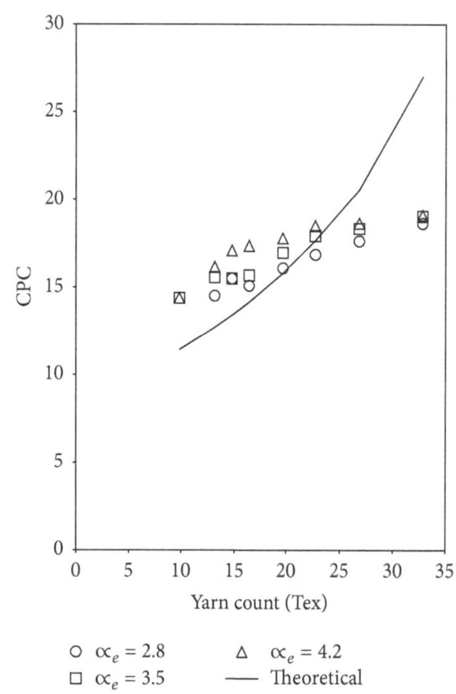

FIGURE 7: Effect of yarn count and twist factor on the course density.

could be tight in contrast with the results obtained for the wales density (see Figure 6). The nature of the knitted fabrics can be understood where the length increases as the width decreases and vice versa.

It should be noted that the yarn twist factor has a small effect at the fine counts. For instance, by increasing of the yarn

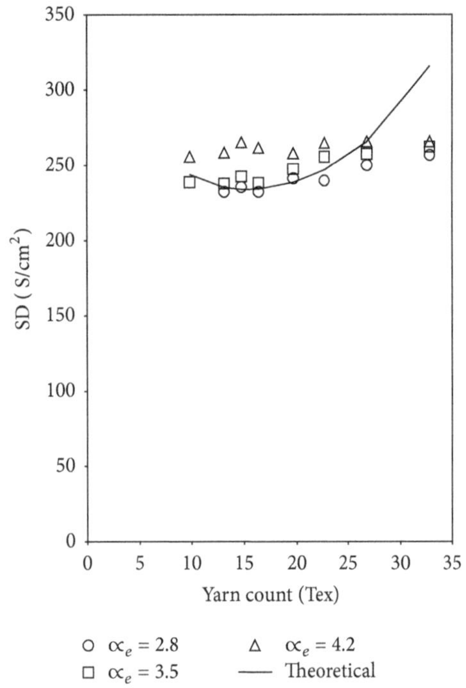

FIGURE 8: Effect of yarn count and twist factor on the stitch density.

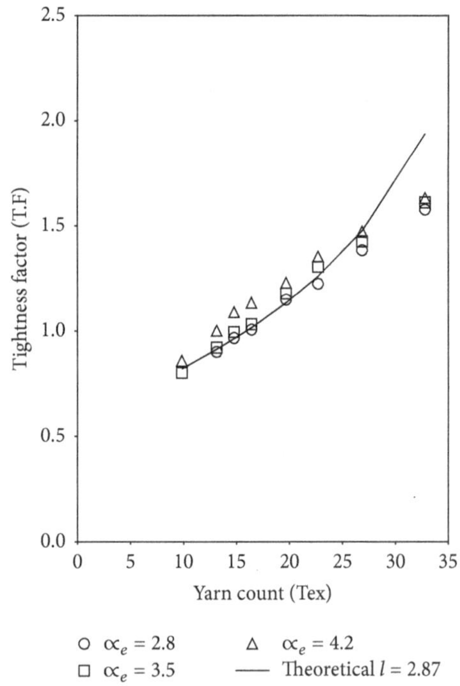

FIGURE 9: Effect of yarn count and twist factor on the fabric tightness.

twist factor from 2.8 to 4.2, the course density increases by 11%. It is due to the increasing of twist liveliness that leads to inclining of the knitting loop and in consequence the loop high decreases and as a result, the course density increases. Obviously, this is observed in the fine counts.

4.3. Stitch Density (SD). The effect of yarn count and yarn twist factor on the stitch density SD (S/cm^2) is shown in Figure 8.

It can be seen that the stitches are nearly constant over the range of yarn Tex tested, that is because the stitch density is the result of multiplying the course and wales densities, where the first factor is directly proportional and the other is inversely proportional to the yarn count.

On the other hand, the yarn twist factor has a small effect at the fine counts. For instance, by increasing of the yarn twist factor from 2.8 to 4.2, the stitch density increases by 12.6%.

It can be noted that there is an agreement between the practical and theoretical results except at the yarn count of 32.8 Tex; there is a difference of 18.9%.

4.4. Fabric Tightness. Figure 9 illustrates the effect of the yarn count and the yarn twist factor on tightness factor (T.F).

It should be noted that the tightness factor increased by 103.5% as the yarn Tex increased from 9.8 to 32.8 Tex. This happened due to increasing the yarn diameter and increasing the covered area because the loop length is constant.

Also, it can be seen that the yarn twist factor has no effect and there is a complete agreement between the practical and theoretical results.

From the obtained results, an equation is deduced to predict the tightness factor for finished plain single jersey knitted fabric. The equation is given by

$$\text{T.F} = 0.0014\ \text{Tex}^2 - 0.0113\ \text{Tex} + 0.8168, \qquad (17)$$

where T.F is the tightness factor and Tex is the yarn count. The correlation factor of the given equation is estimated by 0.99.

4.5. Fabric Weight. The effect of the yarn count and yarn twist factor on the weight (g/m^2) at loop length of 2.87 mm is presented in Figure 10.

It can be seen that, by increasing the yarn count from 9.8 to 32.8 Tex, the weight is increased significantly by three times (285%). It can be accounted that by increasing the yarn diameter the stitch weight increases and as a result, the overall weight increases.

Also, it should be noted that there is a total agreement between the practical and theoretical results and the effect of yarn twist factor is insignificant.

The weight trend changing at different loop lengths is obtained and shown in Figure 11.

As it is clear, the required weight can be obtained at different loop lengths and yarn counts. For instance, weight of 180 g/m^2 can be obtained at yarn count of 24 Tex or 29 Tex with loop length of 3.1 mm or 2.3 mm, respectively.

4.6. Fabric Thickness. Figure 12 shows the effect of yarn count and yarn twist factor on fabric thickness (mm). From a practical point of view, it can be seen that the thickness increased

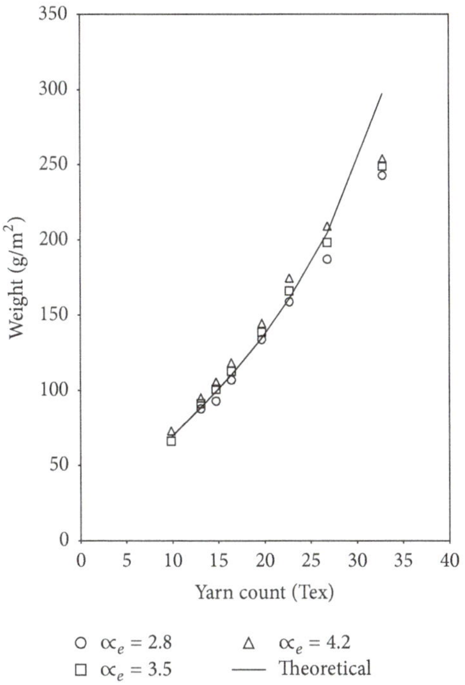

FIGURE 10: Effect of yarn count and twist factor on the fabric weight.

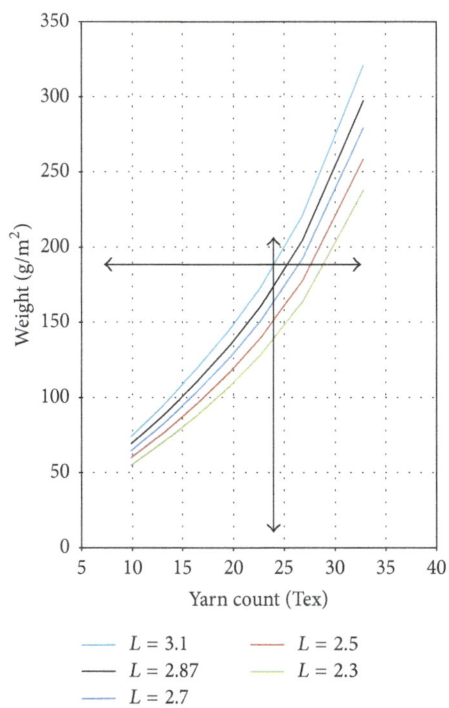

FIGURE 11: Empirical predicted weight of finished cotton plain single jersey knitted fabrics.

by a percentage of 48.15 as the yarn Tex increased from 9.8 to 32.8 Tex and it is logically acceptable because by increasing Tex the diameter increases.

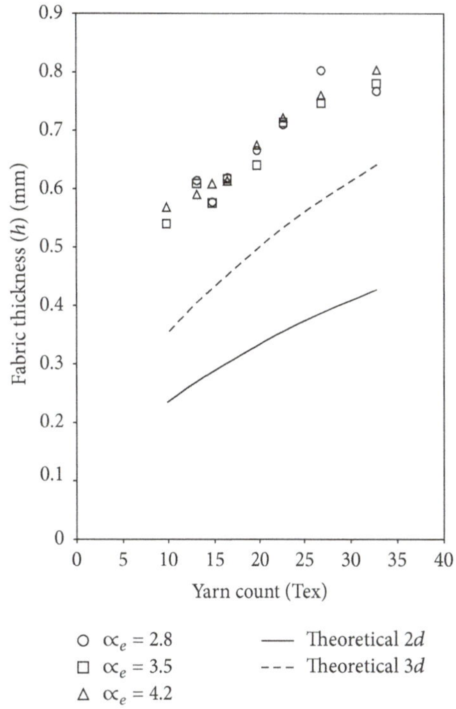

FIGURE 12: Effect of yarn count and twist factor on the fabric thickness.

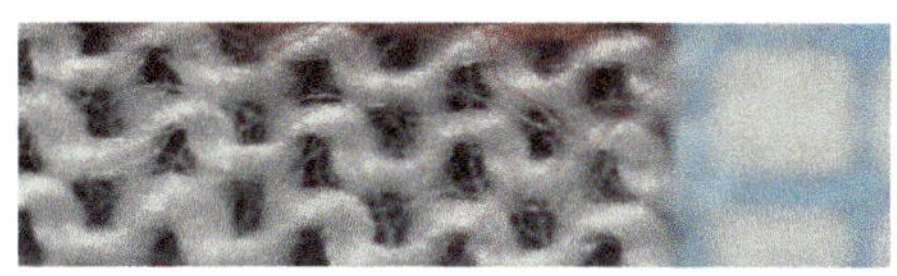

FIGURE 13: The optical pictures to represent the actual diameter measuring test.

Theoretically, the thickness of plain single jersey knitted fabric is estimated at two and three times of yarn diameter (d) and the results are plotted as solid and dashed curves as shown in Figure 12.

The yarn diameter can be calculated by (11).

It should be noted that neither $2d$ nor $3d$ is in agreement with the practical results and it dissents the well-known formulations published before which stated that the thickness of the plain single jersey knitted fabric is proportional to two times of the yarn diameter given by [6]

$$\text{Fabric thickness}\,(h) = 2 \times d\,(\text{mm})\,. \tag{18}$$

To find out the exact thickness at different yarn counts, an actual yarn diameter measuring test is carried out. Figure 13 presents the optical images for one plain single jersey knitted fabric sample of $26N_e$ yarn count with twist factor of $3.5\alpha_e$.

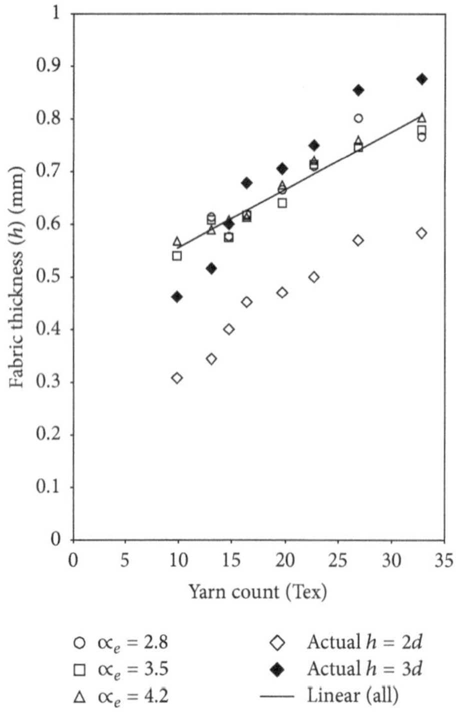

FIGURE 14: Verification of actual measured yarn diameter and fabric thickness.

TABLE 1: The actual measurements of yarn diameter.

N_e	18	22	26	30	36	40	45	60
d (mm)	0.292	0.285	0.25	0.235	0.226	0.2	0.186	0.177

For all samples, the actual yarn diameter in the fabric is measured by using scaled paper and 12.8 mega pixels camera; then the diameter of the yarn is measured at seven different positions for each sample, and then the average value is estimated. The results are recorded in Table 1.

Thickness of plain single jersey knitted fabric is estimated using the actual measured yarn diameter at different yarn counts. The thickness is estimated at two and three times of the yarn diameter and the results are plotted as shown in Figure 14.

It can be seen that there is an acceptable agreement between the practical results and the mathematical results obtained at three times of the actual measured yarn diameter especially at 14.8, 16.4, 19.7, and 22.7 yarn counts.

For clarification and justification purpose of the obtained results, a plain single jersey knitted fabric real sample, simulated appearance, and cross section are presented as shown in Figure 15.

It can be seen that, due to the incline of the loop as a result of the yarn twist liveliness existence, two legs of loops are overlapped one over the other as shown in Figure 15(a) and it is presented as shown in Figure 15(b). It should be noted that the wale space became three times of the yarn diameter instead of four times at the maximum set. Therefore, the area of interloping contains three yarn diameters as shown in Figure 15(c).

4.7. Fabric Bulkiness. The effect of the yarn count and yarn twist factor on the knitted fabric bulkiness (g/cm^3) at loop length of 2.87 mm is presented in Figure 16.

It can be seen that, by increasing the yarn count from 9.8 to 32.8 Tex, the fabric bulkiness is increased significantly by 167%. It can be accounted that the decreasing of wale space as a result of yarn diameter increasing leads to increasing of course space. Consequently, the overall fabric bulkiness increases.

Also, it should be noted that there is a complete agreement between the practical and theoretical results and the effect of yarn twist factor is also insignificant.

5. Conclusion

In this paper, a mathematical model of plain single jersey knitted fabric was presented and the results were verified and compared to practical results. Obviously, the results have shown a significant agreement between the mathematical and practical results.

In addition, a weight forecasting relationship was deduced and verified practically. An empirical chart was also presented at different loop lengths and yarn counts.

Furthermore, the yarn diameter was measured actually and the fabric thickness as a function of yarn diameter was deduced. The obtained thickness relationship showed that

(a) Simulated appearance

(b) Real image

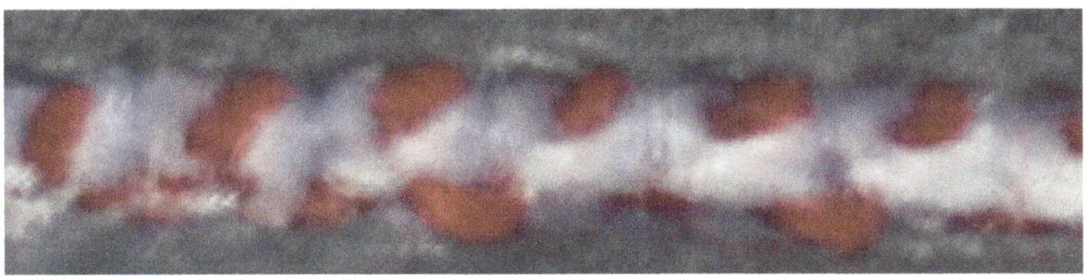

(c) Cross section

FIGURE 15: Plain single jersey vision by variety methods.

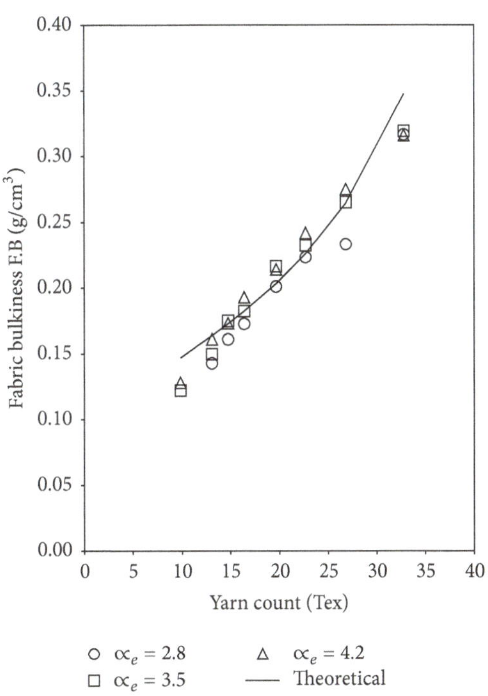

FIGURE 16: Effect of yarn count and twist factor on the fabric bulkiness.

the thickness is proportional to three times of the actual measured yarn diameter.

Geometrical Parameters

N_e:	English yarn count
Tex:	Direct yarn count
WPC:	Wales per cm
CPC:	Courses per cm
SD:	Stitches density (S/cm^2)
R:	Radius of loop head
W:	Wale space
C:	Course space
l:	Loop length (mm)
T.F:	Tightness factor
d:	Yarn diameter
F.W:	Fabric weight (g/m^2)
F.B:	Fabric bulkiness (g/cm^3)
h:	Fabric thickness (mm).

Conflict of Interests

The authors declare that there is no conflict of interests regarding the publication of this paper.

References

[1] E. Gravas, P. Kiekens, and L. van Langenhove, "Predicting fabric weight per unit area of single- and double-knitted structures using appropriate software," *AUTEX Research Journal*, vol. 6, no. 4, pp. 223–237, 2006.

[2] F. Marsal and P. Daniel, "Aspect prediction of the knitted fabrics from the yarn properties," in *Proceedings of the International Scientific Conference: Innovative Solutions for Sustainable Development of Textiles Industry*, Annals Of The Oradea University, Oradea, Romania, 2009.

[3] O. Kononova, A. Krasnikovs, K. Dzelzitis, G. Kharkova, A. Vagel, and M. Eiduks, "Modelling and experimental verification of mechanical properties of cotton knitted fabric composites," *Estonian Journal of Engineering*, vol. 17, no. 1, pp. 39–50, 2011.

[4] M. Naebe, V. Lutz, X. Wang, B. A. McGregor, and D. Tester, "Predicting comfort properties of knitted fabrics by assessing yarns with the Wool ComfortMeter," *Journal of the Textile Institute*, vol. 104, no. 6, pp. 628–633, 2013.

[5] M. Karimian and S. Ajeli, "Numerical modeling of bagging behavior of plain weft knitted fabric using finite element method," *Indian Journal of Fibers and Textile Research*, vol. 39, pp. 204–208, 2014.

[6] D. J. Spencer, *Knitting Technology a Comprehensive Handbook and Practical Guide*, Woodhead, Cambridge, UK, 3rd edition, 2003.

16

The Integration of Technology and Management in the Competitiveness of the United States Short Staple Yarn Manufacturing Industry

Suvalee T. Tangboonritruthai,[1] William Oxenham,[1] Nancy L. Cassill,[1] and Erin D. Parrish[2]

[1] *North Carolina State University, 1000 Main Campus Drive, Raleigh, NC 27695, USA*
[2] *East Carolina University, Rivers Building, East Fifth Street, Greenville, NC 27858, USA*

Correspondence should be addressed to Erin D. Parrish; parrisher@ecu.edu

Academic Editor: Joanne Yip

The purpose of this research was to investigate whether a balance of technology and management could be used in order to sustain a viable United States (US) short staple yarn production industry. The methodology used in this study consisted of primary research comprising two phases. The first phase consisted of interviews with industry executives in order to find out what US yarn manufacturers think are the important factors in maintaining a competitive yarn industry. The second phase was an online survey with industry customers in order to find out why weavers and knitters buy from US manufacturers. Results reveal that US manufacturers have competitive advantages over other countries because of skill level of workforce, variety of products offered, customer relationships, innovation and technology, and product quality. In order to keep these competitive advantages, "technology" and "management" strategies should be implemented. Results also reveal the areas in which the US yarn manufacturers should focus their resources in order to stay competitive or indeed survive.

1. Introduction

No one who has passed from the 20th to 21st century can deny that the world today has become significantly more competitive. With the advent of instantaneous communications and relatively fast transportation systems, there is truly a global manufacturing base where products can be sourced at competitive prices. It is essential that companies embrace these changes in the business environment to ensure their growth or, in some cases, survival.

Some industry experts have recommended that, for yarn manufacturers to stay competitive in a "global" environment, quick response in the hemisphere is very important. Others have suggested that not only is *quick response* important but *quality* in products that respond tocustomer needs is also important for the company to stay competitive [1].

In order to achieve these goals, it is necessary to optimize the use of available technologies. For example, it is obvious that in yarn manufacturing two immediate prerequisites to success are *high productivity* and *high quality*; however, these must also be accompanied by *correct pricing* and *customer service* (including *short delivery times* and *technical support*). There is, however, a paramount danger in focusing on just one of the above parameters. For example, since it is evident that if one only considers advances in technology which yield significant benefits in productivity, vortex spinning would be the only choice of spinning systems used, conversely, it is clear that there are many other spinning systems which are being successfully used because their deficiencies in productivity are offset by other attributes.

Overall, differing opinions are still debated on what is really important for the yarn industry to stay competitive in the market. The top two issues are "technology," which includes the high-technology machinery and highly innovative products, and "management," which includes "how," "when," "where," and "what" to sell, be it either product or service [1].

Based on this, the overall research question was as follows.

What are the relative significance of "technology" and "management" and the balance of these dimensions in the long-term sustainability of the US short staple yarn manufacturing industry?

Specific research questions developed from the overall research question were as follows.

(1) What are the US short staple yarn manufacturing competitive advantages in relation to other countries?

(2) Can technology and management strategies be optimally integrated to sustain a US short staple yarn manufacturing industry?

1.1. Significance of the Research. This study is significant for three reasons. First, while many researchers have conducted studies to find out the international competitive advantage of textile and apparel as a whole industry, this study specifically focuses on short staple yarn manufacturing. This is an interesting segment of the industry that still has a major manufacturing base in the US when most areas of the industry have moved offshore. Short staple yarn also has a significant export market which is vital to the competitiveness of the US textile industry. Second, this study identifies the major key drivers that influence the success of a stable US manufacturing industry and ascertains the relative importance of these drivers. Third, the results of this study explain how "technology" and "management" can be used together to create and sustain a competitive advantage.

2. Conceptual Framework

2.1. The Productivity Triad. "The productivity triad provides a framework for analyzing or X-raying an economy to identify its challenges and opportunity" [2].

The concept of productivity triad captures the systemic character of change at enterprise and regional levels. The three domains of production system, business model, and skill formation are not separable and additive components but interconnected subsystems. The focus of the attention that shapes a region's capability development is on three interactive domains: business model, production system, and skill formation (Figure 1).

2.1.1. Business Model. Chesbrough [3] explains that a business model performs two important functions: it creates value and captures a portion of that value. It creates value by defining a series of activities from raw materials through to the final consumer that will yield a new product or service with value being added throughout the various activities. The business model captures value by establishing a unique resource, asset, or position within that series of activities, where the firm enjoys a competitive advantage.

Other studies mentioned that the model comprises the combined elements of "who," "what," "when," "why," "where," "how," and "how much" involved in providing customers and end users with products and services [4]. The functions of business models were developed and defined as

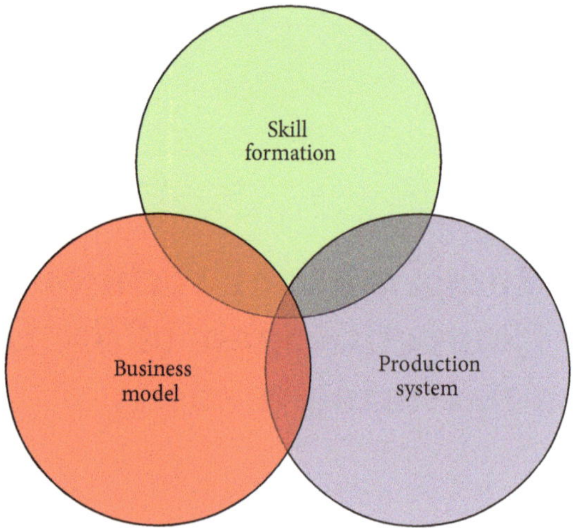

Figure 1: Diagram of the new competitive advantage. Source: Best [2].

(i) the structure of the value chain required by the firm to create the offering and determine the complementary assets needed to support the firm's position in this chain;

(ii) create the competitive strategy, and only the innovative firm will gain and hold advantage over rivals [3].

When business strategy is a long-term approach to implement a firm's business plans to achieve its business objectives, this strategy is used to facilitate gaining the competitive advantage in the market [5].

2.1.2. Production Systems. "Production capabilities are an expression of an underlying but unifying principle of production and organization. Competitive strategy and production systems are bound together" [2].

The fundamental principle of production is the development of production capabilities which includes labor productivity. Moreover, the integration of design and manufacturing is needed to compete and to keep up with rapid new product development.

2.1.3. Skill Formation. The growth process is limited by the number of scientific and engineering persons required to work for the growing companies. Best [2] stated that the condition for success depends on three conditions:

(1) characterization of the demand for specific technologic skills,

(2) investment in technical education,

(3) skill formation in the workplace.

Competitive advantage for a specified region can be built by establishing skill formation in anticipation of technology transitions. Moreover, in order to gain competitiveness with high-technology manufacturing, flexibility is needed. This flexibility depends on the ability of workforces that can learn how to use the complex new equipment quickly and to work well in groups.

2.2. The Interconnection. The interconnection of production system, business model, and skill formation has critical implications depending on governmental policymaking. For example, when firms and regions become immovable in the low productivity triad, the government may be the only institution that can coordinate organization change in all three domains of the productivity triad.

Based on the conceptual framework, the overall research question was as follows.

Can a balance of "technology" and "management" be used to sustain a globally competitive US short staple yarn production industry?

Before this can be answered, it is important to first understand the main factors impacting the competitiveness of the yarn manufacturing industry.

2.3. Supply Chain Structure. Logistics and transportation services have become a leading area of growth in the global market. The challenge for supply chain management is the ability to accomplish delivery of the right products at the right time, all for low costs [6].

According to Kurt Salmon Associates [7], the bigger retailers have a significant impact on supply chains, especially with their increased negotiation power when purchasing large volumes. In international markets, companies are required to pay attention and deliver satisfaction to consumers while finding and maintaining specialized, reliable, and highly competitive suppliers. Since the textile chain has changed from a supply chain to a demand chain, the key success factors for this chain in the future will be joint forecasting and planning and joint inventory management as a result of cooperation between all firms in the chain [8].

2.4. Technology and Innovation. Technological change is a form of learning about how to solve specific problems in a highly differentiated and volatile environment [9]. The technology is used to reduce costs, improve quality, and increase flexibility [10]. The impact of technology changes affects the local, regional, and worldwide industry. Economic theory, historical evidence, and growth models confirm that advancing technological knowledge is the most important single factor that contributes to long-term productivity and economic growth [11].

The spinning industry has a long history in terms of technological advances. Over the past few decades, this industry has had an almost continuous flow of new developments in terms of process methods in production. The three spinning technologies (ring, rotor, and air jet spinning), with their different focus on applications, cover the entire range concerning fineness and input materials. Machine speeds have risen, and automated operations have improved production capability and increased product quality demands [12]. One of the most important technological innovations in the spinning of yarn was the introduction of open-end spinning, which combined what were formerly three separate processes into a single process using rotors instead of spindles. Spinning speeds increased at least fourfold, and labor requirements were reduced by approximately 40 percent [9].

The innovations have led to further improvements, including the increased automation and increased speed of textile operations and the consequent reduction in the number of workers needed. The drive to introduce new technologies has been stimulated mainly by the need for the production in developed countries to be cost competitive in the face of the very low labor costs in developing countries [12]. Cost reduction is not the only benefit derived from the new technologies for developed countries to gain cost competitiveness over the low labor costs in developing countries. Time saving, as the result of the automated manufacturing, also provides greater benefits for the manufacturers to respond quickly to customer demand [9].

2.5. Quality. Quality has been defined as value [13], conformance to specifications [14], fitness for use [15], and meeting and/or exceeding customer's expectations [16]. In the textile industry, the importance of quality cannot be overstated. For example, the combed cotton yarn in Canada represents high quality products that can compete over low labor cost products. Better quality products and services on a consistent basis are needed for the industry to compete internationally [10]. With the competitiveness in low-wage countries, differentiated products with high quality have made the Italian manufacturer an example of a major exception in competing with low-price products. In general, the Italian producers have pursued a strategy of product specialization. They have focused on a fashion orientation with the aim of avoiding dependence upon those types of goods most strongly affected by low-cost competition.

2.6. The Proposed Framework. After the literature review, the proposed framework (Figure 2) was developed to try to visualize the problem and was used throughout this study. The goal of this "framework" was to organize the various components into the broad categories of "T = technology," "M = management," and "S = skill formation." For this particular model, "technology" is defined as spinning machinery, "management" is defined as the decisions relating to what, where, and how to sell yarn, and "skill formation" is defined as the invested amount of the resources, for example, time and money.

It was believed that this approach would help to conceptualize the problem and aid in the subsequent analysis, since it should be easier with the drivers defined in broad groups. The size of the circle represents the level of importance of the components, and the distances (L1, L2) show the amount of resources invested towards "technology" and "management."

To better explain or visualize the framework dynamic, the following paragraph depicts how this framework would work in different situations (Figure 3).

In the past, the ideal model shows the balance of US yarn manufacturing with an emphasis on the importance of technology due to investment in new, high speed machinery, because the producers needed to produce yarns faster. The size of circle in "T = technology" is bigger than "M = management," and "S = skill formation" moves closer to "technology" to create balance of this model (Figure 3(a)).

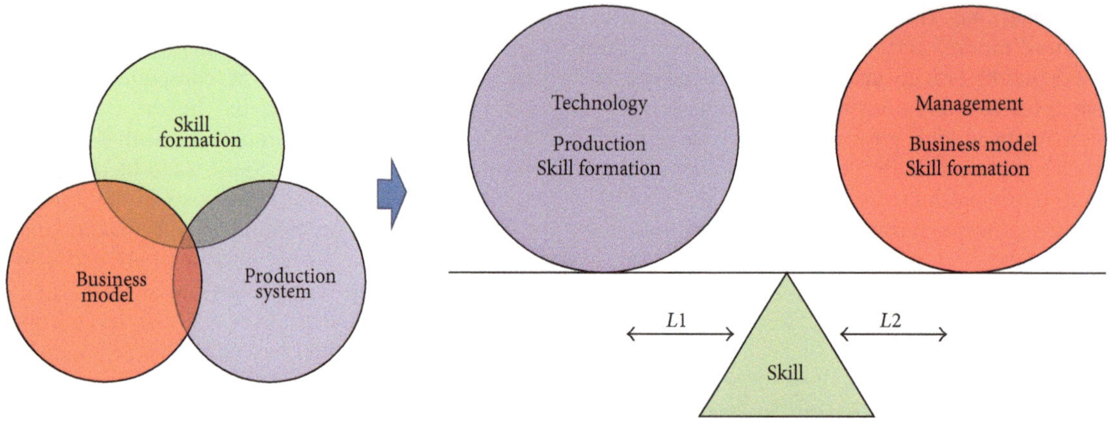

FIGURE 2: The proposed framework or "the ideal model."

When the distance L1 is shorter than the distance L2, the company is investing more towards technology resources, such as training programs for employees to have skills to work with high technology.

Another scenario shows less investment towards machinery in the United States. The ideal model (Figure 3(a)) is changed, with the importance laying more towards management and less towards technology. This situation creates an imbalance of the model shown in Figure 3(b). In order to maintain the balance, the company needs to adjust the resources invested towards management, thus restoring balance to the model (Figure 3(c)).

3. Methodology

In order to answer whether a balance of "technology" and "management" can be used to sustain a globally competitive yarn production industry, research was conducted in two phases: inductive and deductive. The first phase used an inductive approach and included semistructured interviews with industry executives in order to find out what US yarn manufacturers think are the important factors in maintaining a competitive yarn industry. The second phase used a deductive approach and included an online survey with industry customers to find out why weavers and knitters buy yarn from US manufacturers.

3.1. Phase I: US Yarn Manufacturers

3.1.1. Sample Selection. The sample of this study was US yarn spinning mills; however, the study focused only on the short staple yarn manufacturers due to its significance to the overall US textile industry. The following steps were performed in order to develop the sample list for Phase I.

Step 1. It includes identifying the company names from ReferenceUSA.com in the chosen NAICS code 313111 which represents "yarn spinning mills in the US" (population n = 384).

Step 2. Determining the company location: this study selected the companies located in Alabama, North Carolina, South Carolina, Virginia, and Georgia. It narrowed the group of companies which account for 93% of all spun yarn production in the United States (narrowed sample n = 203).

Step 3. It includes narrowing the identified companies by using product/business description related to product. The product must consist of "cotton yarn or yarn from cotton blends." Then, more information about the company was gathered from many databases and websites such as Manta, Davison's Textile Blue Book, Thomas Register, Lexis@Acadamic, Hoover's Company records, and http://www.cottonsourcingusa.com/.

Moreover, this step deleted duplicate company names.

Step 4. It includes reviewing the identified companies from step 3 by experts in the yarn industry. Then, additional companies were added in the final sample list (sample = 20).

3.1.2. Instrument Development. The interview questionnaire for Phase I included both qualitative and quantitative items and was developed by the researcher based on the research objectives and the literature review. Moreover, questions from other studies [17] were modified for use in the questionnaire. It consisted of the following three sections.

Company Information. The first section of the survey for US short staple manufacturers was used to document the company's background. The responses gave an overview of respondents' area of work, size of the plant, product offered by the plant, number of spinning systems, and the age of the machinery.

Globalization. The second section of the interview questionnaire focused on globalization including rating countries in terms of the importance as customers. Respondents were also asked what, in a spinner perspective, the key drivers that are important to their customers and the choice of business strategy for the company are and to rate the importance of component in yarn costs. For this section, questions used a quantitative Likert scale rating so that results from this phase could be directly compared to the results from Phase II. Additional questions were used to find out how globalization has impacted the company.

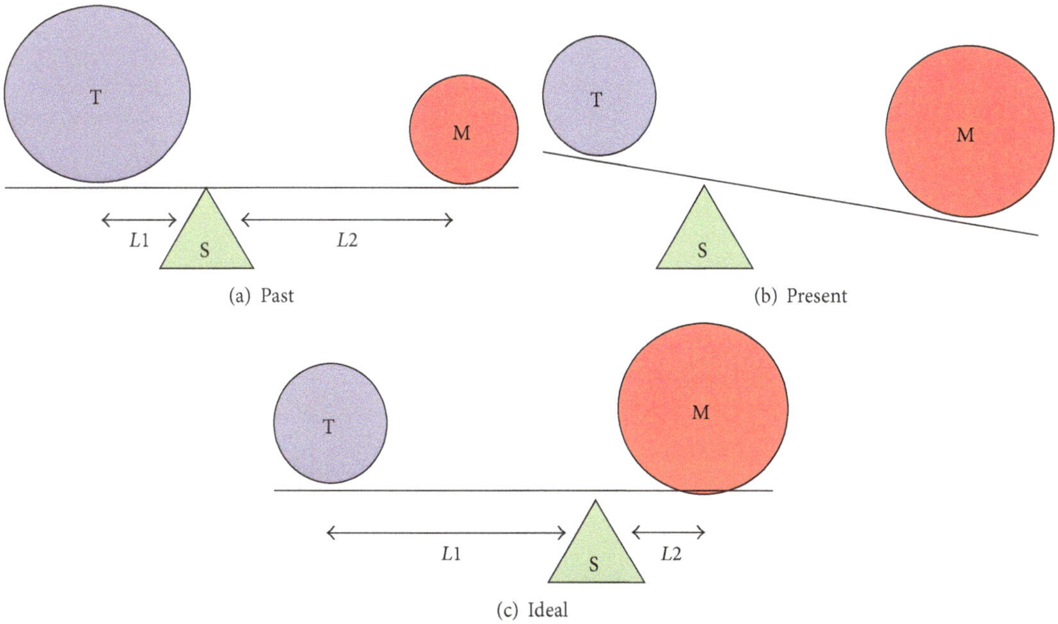

FIGURE 3: The visual graph of proposed framework.

Technology and Management. The third section of the questionnaire focused on integration of technology and management and asked the respondent's agreement regarding the impact of "technology"; the impact of "optimization in management"; and the percentage of transaction between subunits in their company. Additional questions asked the respondents to give their opinion on the integration of technology and management. This section consisted of open-ended questions.

After the questionnaire was field-tested with two industry experts to establish validity, the internal review board (IRB) approved the questionnaire.

3.1.3. Data Collection and Analysis. The selected US yarn manufacturers were contacted to participate in this study via email. The initial cover letter emails were sent out to the selected companies explaining the purpose of the study and requesting the company's participation in the research. Two weeks after sending out the initial cover letter, a follow-up email or phone call was made in an attempt to set up a visit or conference call. Once accepted, the semistructured interviews were conducted either in person or by telephone, and the questionnaire was used to guide the interviews, collecting both quantitative and qualitative data. The interview was used to reveal industry opinions and perceptions. However, respondents were free to add information where they deemed appropriate.

Once all the interviews were conducted, the results were analyzed based on the procedure of Yin [18] and Parrish [19]. The results were first transcribed into paragraph form. For further comparison, the information from the interviews was divided into tables. The comparison was made to determine similarities and differences regarding the integration of technology and management for the creation of long-term sustainability. Finally, the quantitative data was collected and grouped together. Mean scores were calculated so that the results from Phase I could be directly compared to the results from Phase II.

3.2. Phase II: US Yarn Manufacturing Customers

3.2.1. Sample Selection. Respondents in Phase I were asked to identify purchasers of short staple yarn and provided the initial sample for Phase II. Additional companies were added, but because the data for companies in other countries was limited, this study focused on companies located in Honduras, El Salvador, Columbia, Guatemala, and Dominican Republic. Furthermore, these countries represented more than 62.2 percent share of the total US export of cotton yarns in 2006 by value ($US).

These companies' contact information was acquired from respondents in Phase I and through apparel and textiles exporters' associations' company profiles, resulting in an initial sample size of 58. Fourteen usable responses were received, resulting in a 24 percent response rate.

3.2.2. Instrument Development. The survey for Phase II was quantitative and developed by the researcher based on the research objectives and the literature review. The first section of the survey for customers of US short staple yarn manufacturers was used to document company information. This section included the questions regarding size, origin of capital investment, product information, location, and whether or not the company buys yarn from US manufacturers. This section collected nominal data. In the second section, respondents, using a Likert scale, were also asked to rate the importance of key drivers that affected their decision to buy yarn from chosen manufacturers.

TABLE 1: Phase I: sample description.

Product category	Company	Interviewee	Number of employees
Fiber/yarn	A	Director of business development	2,000
Fiber/yarn	B	Senior vice president in operations, America & Strategic Procurement	3,200
Fiber/yarn	C	Vice president in operations	750
Fiber/yarn	D	New business director	n/a
Fiber/yarn	E	CEO/innovators	>300
Fiber/yarn	F	Senior vice president in manufacturing, manufacturing group vice president, Global Marketing manager, technical textiles division	1,600
Fiber/yarn	G	Manager of Six Sigma	n/a
Fiber/yarn	H	Executive vice president	1,700
Fiber/yarn	I	Director of fiber processing research and technical services	n/a

n/a: not available from the respondents.

After the questionnaire was field-tested with two industry experts to establish validity, the internal review board (IRB) approved the questionnaire.

3.2.3. Data Collection and Analysis. The sample of US yarn manufacturing customers was contacted by email. The survey invitation email included a link to the online survey.

The quantitative data was collected and grouped together. Mean scores of what the important key drivers that affect their decision to buy yarn from a chosen manufacturer (Phase II) are were calculated and directly compared to the mean scores of what, from a spinner's perspective, the key drivers that are important to their customers (Phase II) are.

4. Results

4.1. Phase I: Yarn Manufacturers' Sample. The goal of this phase of the research was to find out what US yarn manufacturers think are the important factors in maintaining a competitive yarn industry. The research sample consisted of 12 respondents from nine US textile companies of the fiber and yarn mills and auxiliary sectors. Table 1 provides information about the sample in Phase I, including product category, company identification, interviewee job positions, and size based on the number of employees. In general, the interviews followed the outline of the questionnaire. However, there were some deviations from the standard questions, which proved to be insightful and supplied additional factors that could be significant for consideration.

4.2. Phase I Results. In general, *short lead time* is thought by the spinners to be the most important driver influencing the customers' purchasing decision. Surprisingly, *lower cost* is thought by the spinners to be the least important driver. One executive commented that *market proximity* will become more important to customers because the freight cost tends to increase due to recent rises in oil prices. As long as there is the demand in the market, one respondent believes that he can sell yarn at a competitive price. Companies, which rated product quality as somewhat important, were asked further why product quality is not very important. The comment was "product quality is important, but it is not the most important." Overall, most of the respondents commented on the business strategy that "pull" or capability and willingness to produce products that can react to market is *more important* than "push" for the current market situation.

Most of the respondents strongly agreed that "optimization in management" improves quality, management decisions, productivity, profit, and opportunity of business. One of the respondents stated that it is not important to always have new technology to be able to produce the product that responds to customer needs. Sometimes old machinery can produce a product with certain desired properties. It was remarkable to see the disagreement on R&D issues from the interviews. One respondent indicated that while owning R&D is very important, using R&D as a major part in business strategy is not important for them.

In addition to the key drivers specifically asked about during the interviews, four additional drivers were discovered that need more attention in order to sustain the competitiveness of the US yarn manufacturing industry.

(i) Energy costs: US is competitive in the cost of electricity as long as the relationship between basic energy sources stays the same. When considering the cost to build a new plant in US and other countries, the cost of capital will be the same, for example, steel and cements for the building, but the difference depends on energy sources and standards to build the plant in different countries.

(ii) The limitation in sourcing fibers overseas: it is becoming harder to reliably source the fiber that customer wants, which is limited by trade compliance. An example given was the difficulty in sourcing black fiber for inclusion in T-shirt yarns. This is also important to the potential purchase of cotton fiber from sources other than the US, and this will impact the following concern.

TABLE 2: Phase II: sample description.

Company code	Area of work	Number of employees	The origin of capital investments	Where company buys yarn from: US	Other countries (excluding US)	Both US and other countries
A1	Purchasing	150 or more	US capital			X
B1	Production	0–49	Salvadoran capital			X
C1	Production	100–149	Guatemalan capital			X
D1	Other sales' administration	150 or more	Salvadoran capital			X
E1	Other full packages: costing, purchasing	150 or more	Salvadoran capital			X
F1	Production	150 or more	US Capital			X
G1	Purchasing, sourcing, and production	150 or more	Salvadoran capital			X
H1	Production	50–99	Salvadoran capital	X		
I1	Other planning area	150 or more	Dominican capital	X		
J1	Production	No response	Dominican capital	X		
K1	Purchasing, other administration	0–49	Costa Rican capital		X	
L1	Production	150 or more	Bolivian capital		X	
M1	Sourcing, production	0–49	US and Korean capital		X	
N1	Production	150 or more	Korean capital		X	

(iii) Long-term availability of the right quality of cotton: particular concern was expressed in obtaining long-term commitments from cotton merchants over the availability of the fiber into future years. Because of the competition and potentially higher prices paid for cotton by international competitors, US spinners may have difficulty in having sufficient fiber supplies to meet customers' needs.

(iv) Farm bill: this was viewed as a great benefit, even though there was some unhappiness expressed over the fact that some of the benefits would be taxed. The finance from the farm bill could provide useful resources to maintain the technological advantages of the US spinners, since the funds must be used for equipment purchases.

4.3. Phase II: Yarn Customers' Sample. The goal of this phase of the research was to find out why weavers and knitters buy yarn from US manufacturers. The research sample consisted of 14 respondents from 14 companies in the textile and apparel industry, and these were located in DR-CAFTA region. The results represented the opinions of the yarn customers on the main drivers that influence their decision to purchase yarn from either the US or from other countries.

Table 2 provides information on the sample used in Phase II, including company codes, area of work of the respondents, number of employees, the origin of capital investments for each company, and whether the company buys yarn from the United States, other countries, or both. The sample respondents included people who work in purchasing, production, sourcing, administration, and planning from various countries in Central American areas. The sample represents seven companies that buy yarn from both US and other countries, three companies that buy yarn from US (excluding other countries), and four companies that buy yarn from other countries.

4.4. Phase II Results. In the case of purchasers of yarn from US manufacturers, *product quality* was rated as the most important driver in purchasing decision followed by *tariff and trade benefits* and *short lead times* in second and third places, respectively. *Skill level of workforce* was considered as the least important driver, which influenced yarn purchasing decision from US manufacturers.

In the case of purchasers of yarn from other countries (excluding US manufacturers), *lower cost* and *short lead times* were *the most important drivers. Skill level of workforce* was considered as the *least important driver* in the purchasing decision.

4.5. Results Summary: A Comparison of Phase I and Phase II. Further analysis was conducted regarding key drivers including perceived drivers versus actual drivers that influenced yarn purchasing decisions from US manufacturers. A radar chart was developed to compare the quantitative responses from both Phase I and Phase II. The chart graphically shows the size of the gaps among the nine drivers that influenced customers that purchased yarn compared to what US yarn manufacturers believe are the main reasons for purchasing yarn. This radar chart (Figure 4) illustrates that the largest gap exists in *market proximity* where the spinners think this driver is very important, whereas in reality the customer rates this driver as less important than other drivers.

This radar chart shows that five out of nine drivers (*market proximity, short lead times, skill levels of workforce, customer relationship,* and *variety of product offers*) from yarn manufacturers' judgments exceed the customers' expectation; *product quality* and *tariff and trade benefits* almost reach customers' expectation.

Lower cost is what yarn manufacturers have to work on to reach the customers' expectation. The spinners' views indicated that *lower cost* is *the least important driver*, since the price is a given and customers pay for quality. On the other hand, customers' views did not agree on these. Customers

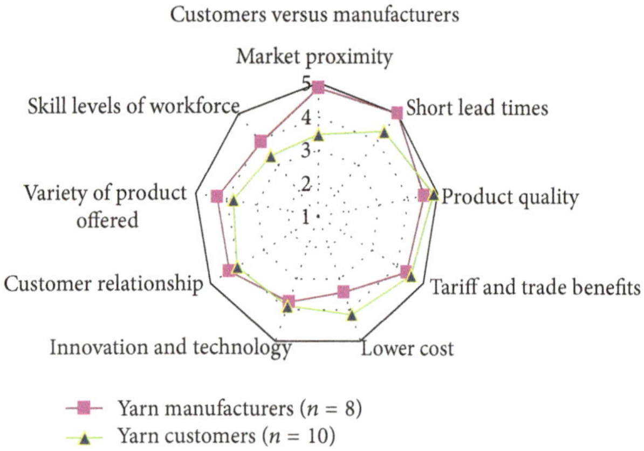

Figure 4: Radar chart shows a comparison of key drivers from what the spinners think to be important for customer *versus* what the customers actually think.

think that *lower cost* is a *very important* driver influencing yarn purchasing decisions.

4.6. Rethinking the Proposed Framework. After the data analysis of both the interviews and survey, it was clear that the simple visual (Figure 1) could not be used to illustrate the present dynamics in the short staple spinning industry. In summary, results lead to a conclusion that, instead of looking at each component individually, that is, "technology" or "management," it makes more sense to look at it as an integrated combination of both.

While "technology management" integration (Figure 2) may be purely a visual, it seems to better reflect the situation in the spinning industry as evidenced by the present research. From the interviews, it was gleaned that "technology" is now available to anyone who has sufficient finances. Thus, while in the past the US was regarded as being much more technologically advanced compared to China and India, this is no longer the truth. Respondents stated that these "developing countries" are actually investing significantly more in newer spinning technology than the United States.

The major advantage which the US yarn manufacturers claimed is having the ability to "optimize the use of new technology." This is essentially how to use "technology management" approach. This process starts with the following:

(i) selecting fiber for particular "high tech" machines which is based on meeting the ultimate requirements of the yarn purchasers,

(ii) optimizing the processing of fiber to not only produce a high quality yarn but also do this economically,

(iii) ensuring that there is an adequate current and future market for the yarn,

(iv) servicing the customers by creating customer relations (knowing what they want and how they want it).

It should therefore be evident that, in the present spinning enterprise, technology and management are so interwoven and interdependent; it is extremely difficult to absolutely define many drivers as either technology or management. Therefore, the integrated, diffuse visual (Figure 5) is considered more appropriate than the initial proposed visual (Figure 2).

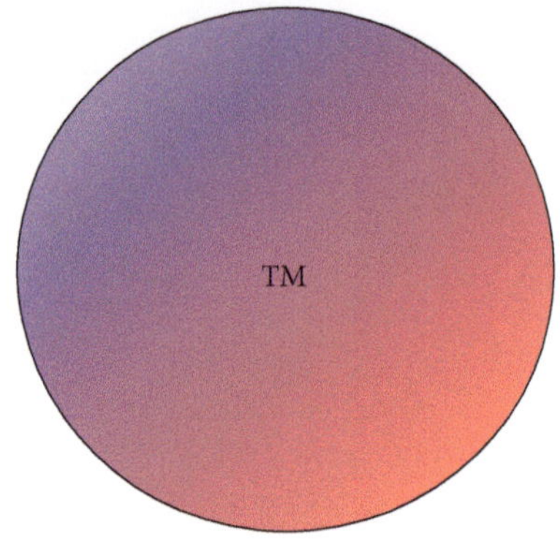

Figure 5: New "technology management" integrated system.

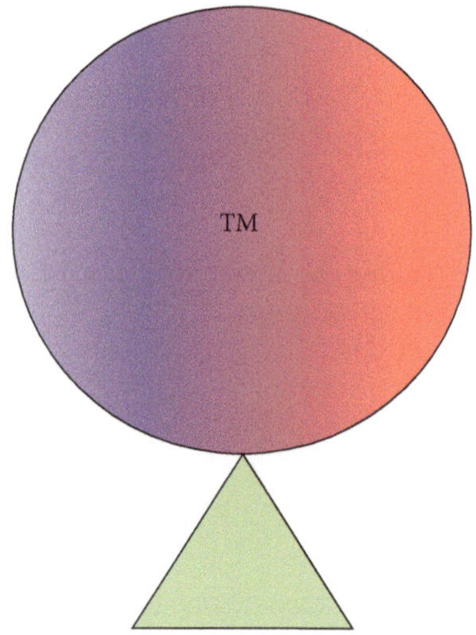

Figure 6: The balance of new "technology management" integrated system.

However, in the current climate of rapid changes in most aspects of the global economics, it is important to consider the new "technology management" integrated system as a carefully balanced sphere sitting at top of a pivot, representing both internal and external pressures that have a direct impact on the competitiveness of the US yarn manufacturing industry (Figure 6). Examples include energy costs, US trade policy, technology, and raw material costs. This technology management fusion has a combination of product, market, and service components. Any small changes caused by internal or external pressures could destabilize the system. In other words, there is much more volatility to sustaining a company than ten years ago.

4.7. Limitations of the Study. For this particular study, only "cotton yarn" was included which is a combination of 300 and 301 series in the US textile and apparel category. Even though this study only included one specific industry, the study's methodology can be replicated with different industries efficiently. Due to the limited sample size, both the survey and interview results can only be generalized to those companies that participated in the study and not the entire short staple yarn manufacturing industry.

5. Conclusions

The conclusions are based on the original research questions and while there are some interdependence of the responses and results, it was believed to be appropriate to discuss these as separate items.

5.1. What Are the US Short Staple Yarn Manufacturing Competitive Advantages in relation to Other Countries through Selected Regions and Countries? In the yarn purchasers' opinion, US manufacturers have competitive advantages over other countries because of these key drivers: *skill level of workforce, variety of products offered, customer relationships, innovation and technology*, and *product quality*. In order to keep these competitive advantages, "technology" and "management" strategies should be implemented. For example, *innovation and technology* can be created by using the "niche" approach in product differentiation.

US manufacturers should be aware and focus on the perceived key drivers which influence yarn purchasing decisions from yarn customers. This is demonstrated by the fact that yarn manufactures perceived *market proximity* as the most important driver and *low cost* as the least important driver for their customers. When in reality, the yarn customers perceived these drivers as opposite.

Government policy makers should be encouraged, since *tariff and trade benefits* are considered very important by both yarn spinners and yarn customers. Additional competitive advantages include US energy costs when compared to other countries, the limitation in sourcing fiber overseas, the long-term availability of the right quality of cotton, and the US farm bill.

5.2. Can Technology and Management Strategies Be Optimally Integrated to Sustain a US Short Staple Yarn Manufacturing Industry? Yes, the US yarn manufacturing industry can be sustained with the optimal integration of technology and management. The concept of "technology management" refuted the conceptual model in the literature review, which separates "technology" and "management." From Phase I discussions, it is evident that most US spinners think "technology" is less important than "management" in order to sustain a US yarn manufacturing industry in global competition. Their reasons are that technology is available to anyone who has the money; however, "managing the technology" and "managing the enterprise" are what is needed to make the technology successful. As time goes on, separating management and technology is becoming more difficult. For example, the use of cotton data (EFS) was considered "management," while the use of software to control the subunit transactions in the company was considered "technology blended with management."

In summary, there was universal agreement that while both technology and management are important, it is the new "technology management" integrated system that is the key to staying competitive. Thus, the skill of optimally using the latest equipment, matched with fiber knowledge, in order to yield the best yarn (quality and cost) demonstrates a critical blend of management and technology know-how.

Conflict of Interests

The authors declare that there is no conflict of interests regarding the publication of this paper.

References

[1] J. Phillips, "Consumer product recalls prompt product visibility questions," Textile World, 2007, http://www.textileworld.com/News.htm?CD=3847&ID=13599.

[2] M. H. Best, *The New Competitive Advantage: The Renewal of American Industry*, Oxford University Press, New York, NY, USA, 2001.

[3] H. Chesbrough, *Business Models: How to Thrive in the New Innovation Landscape*, Harvard Business School Press, Boston, Mass, USA, 2006.

[4] D. Mitchell and C. Cole, "The ultimate competitive advantage of continuing business model innovation," *Journal of Business Strategy*, vol. 24, no. 5, pp. 15–21, 2003.

[5] BNET Business Network, "BNET business dictionary," 2008, http://www.webcitation.org/getfile?fileid=5be050e5d3888039-e5d99c9aa2c2f40209390955.

[6] Plunkett's Research, "Apparel and textiles industry trends," 2004, http://www.plunkettresearch.com/.

[7] Kurt Salmon Associates (KSA), *Vision and Perspective of Consumer Behavior and Trends in Clothing*, Deutscher Fachverlag GmnH, Frankfurt, Germany, 2004.

[8] KSA, "Trends in consumer's behavior and the global clothing market," *Textile Outlook International*, vol. 118, pp. 120–150, 2005.

[9] P. Dicken, *Global Shift: Reshaping the Global Economic Map in the 21st Century*, The Guilford Press, New York, NY, USA, 4th edition, 2003.

[10] O. Sabanci, "Meeting the challenge of global competition," *International Fiber Journal*, vol. 20, no. 4, pp. 4–6, 2005.

[11] C. Freeman, "The "national system of innovation" in historical perspective: technology, globalization and economic performance," *Cambridge Journal of Economics*, vol. 19, no. 1, pp. 5–24, 1995.

[12] P. H. Dinsdale, "Developing investments in textiles in emerging markets the search for quality projects," in *Proceedings of the Textile Institute 83rd World Conference (TIWC '04)*, vol. 3, pp. 1193–1199, Shanghai China, May 2004.

[13] L. Abbott, *Quality and Competition*, Columbia University Press, New York, NY, USA, 1955.

[14] H. L. Gilmore, "Product conformance cost," *Quality Progress*, vol. 7, no. 15, pp. 16–19, 1974.

[15] J. M. Juran and F. M. Gryna, *Juran's Quality Control Handbook*, McGraw-Hill, Hightstown, NJ, USA, 4th edition, 1988.

[16] A. Parasuraman, V. A. Zeithaml, and L. L. Berry, "A conceptual model of service quality and its implications for future research," *Journal of Retailing*, vol. 69, no. 1, pp. 140–147, 1993.

[17] S. A. Zahra and J. G. Covin, "Business strategy, technology policy and firm performance," *Strategic Management Journal*, vol. 14, no. 6, pp. 451–478, 2003.

[18] R. K. Yin, *Case Study Research: Design and Methods*, Sage, Thousand Oaks, Calif, USA, 1994.

[19] E. D. Parrish, *Niche market opportunities in the global marketplace [Ph.D. dissertation]*, North Carolina State University, Raleigh, NC, USA, 2003.

17

Optimization of Bromelain Treatment pH with Wool for Antifelting and Reduced Pilling Behaviour: Objective Assessment Approach

Amandeep Kaur and J. N. Chakraborty

National Institute of Technology, Jalandhar 144 011, India

Correspondence should be addressed to J. N. Chakraborty; chakrabortyjn@gmail.com

Academic Editor: Phillip W. Gibson

Wool fabric possesses unique properties like resiliency, warmth, wide aesthetic qualities, and comfort. However, felting-shrinkage and pilling tendency of wool after wear and repeated launderings are its inherent limitations. Proteolytic enzymes are extensively used as biocatalysts for eco-friendly wool antifelting treatments. However, uncontrolled wool protein hydrolysis by protease may lead to excessive weight loss leading to weakening of fiber and reduction in its tensile value, which further have detrimental effect on the pilling propensity of protease treated wool during use of woolen textiles. Optimum selection of protease treatment parameters like concentration, pH, temperature, and so forth can help in controlled reaction to achieve the desired effect, which generally involves cumbersome sampling and analysis. Optimization for pH of bromelain (proteolytic enzyme having activity in acidic conditions) treatment with wool has been done while aiming at the desired areawise shrinkage value, minimum weight loss, and reduced pilling by using standard methods and objective fast Fourier transformation technique for pilling evaluation specifically.

1. Introduction

Wool constitutes a minor yet important segment of total textile fibers produced globally in the commodity as well as luxury apparel range [1]. The remarkable properties like resiliency, cool in summer, warm in winter, and wide range of available weight and thickness makes wool suitable for traditional as well as contemporary apparels. Chemically, wool is composed mainly of proteins and lipids within heterogeneous morphological parts: the cuticle and cortex structure [2]. Cuticles are composed of overlapping cells (scales) and surround the cortex. With numerous advantages, felting arising due to scaly structure of wool is the limiting behaviour of woolen textiles that affects the consumer acceptability after repetitive wear and launderings. Many commercial treatments like chlorination, resin application, and combined methods for example, chlorine-Hercosett and so forth, have been developed to reduce wool felting as well as subsequent pilling [3]. Conventional antifelting methods have serious environment hazards, for example, release of absorbable organic halides (AOX), excess water consumption, harm to aquatic life, and so forth [4]. Use of proteolytic enzymes as biocatalysts for protein hydrolysis is a promising alternative in view of eco-friendly wool antifelting treatments. Antifelting property, improvement in fabric handle with reduction in fiber bending stiffness through structural protein degradation, reduced pilling, and improved dyeability are major claimed developments of protease treatments [5–11]. However, uncontrolled protein hydrolysis at the interior of fiber cortex by protease leads to excessive weight loss of wool leading to weakening of fiber and reduction in its tensile value. It has been observed that subsequent wear and launderings lead to additional abrasion or wear, and wool fabric becomes even thinner; previously formed pills break off and eventually lead to the formation of holes [12]. Hence weight loss due to proteolytic treatment may have detrimental effect on the pilling propensity of protease treated wool.

Felting tendency of wool fabric is expressed in terms of area-shrinkage % (Woolmark specification AW-1: flat woven, pile woven, and pressed felt apparel products). According to this, after standard washing methods if wool fabric possesses maximum total shrinkage as 3% each in both warp and weft directions (area shrinkage comes out to ~6%), then wool may be termed as machine washable. For pilling measurement, basically subjective, objective, and descriptive rating tests methods are applied depending on the severity of pilling after going through standard abrasion technique which may not correlate with the intended end-use [13]. Moreover balance between initial rate of pill formation and rate of loss of pills by breaking-off affects the density of pills [14]. For wool particularly, because of large number of possible variations like diet, breed, health of the sheep, climate, and so forth, physical properties like diameter, length, crimp, and their chemical composition vary. These factors in turn lead to even higher uncertainty in pilling grades obtained by replicate fabric specimen with respect to actual grades in controlled laboratory tests. Moreover, characterization of pill size from small to large and corresponding grade rating are not comparable [15]. Limitation of lesser validity of subjective pilling evaluation due to wide grade variability by judges has been also mentioned [16]. Other methods suggesting searing off pills and calculating mean pill mass, number, and total weight indicated a good correlation between weight of pills and ratings given by subjective visual assessment but had limitations like laborious, time consuming, and skill requirement for searing off pills delicately [17, 18]. Many image processing based pilling evaluation methods are based on segmenting pills from background while considering density and spatial domains [19, 20].

Objective measurement of pilling evaluation has been topic of many research publications. Laser triangulation method [21], image synthesis based on the Karhunen-Loeve transform [22], digital image analysis [23], two-dimensional Gaussian fit theory to train pill template while using actual pill images for template matching [24], frequency domain analysis algorithm [25], light projection image analysis [26], noncontact three-dimensional measurement system using slit laser beam projector and CCD camera [27], wavelet reconstruction scheme using discrete wavelet transform [28], and fast Fourier transformation technique [13, 29] are important to mention in this regard.

Desired antifelting effect and observed weight loss is affected by number of factors like protease concentration, activity, time, temperature, pH, rate of enzyme diffusion, and so forth during treatment. Accordingly, establishing correlated behaviour between antifelting effect, weight loss, and pilling tendency arising from weakening of wool is time consuming and cumbersome. The aim of the present study is to optimize the treatment pH by using objective fast Fourier transformation technique for pilling evaluation specifically and areawise shrinkage as well as weight loss of protease treated pilled wool fabric.

2. Materials and Methods

Woven bleached wool fabric (warp: 2/52 Nm, weft: 1/30 Nm, epi: 84, ppi: 54, gsm: 224) was used in the study. Proteolytic enzyme bromelain was procured from Excellent Biotechnologies, Bangalore, India. Nonionic surfactant Sandoclean PCJ (Clariant) and other analytical grade chemicals (SDFCL, India) were used. Wool samples were cut according to Martindale test specimen cutting template (38 mm diameter) and conditioned in standard atmosphere. Samples were used in triplicate for a specific process and were tested to calculate the standard mean and error.

2.1. Primary Optimization of Pilling Cycle Threshold. For selecting running cycle threshold, pilling test (Woolmark test method TM 196) of bleached wool samples was repeated on Martindle abrasion tester (Prolific Engineering) with respect to increasing number of cycles, that is, 30, 60, 90, 120, 150, and 180. Pilled images were taken with the help of USB Optical Microscope 800x (Micro-measure) and image analysis for average pill number and area was done by custom developed program in MATLAB software. For this, the coloured image of the samples was converted to gray and the corresponding frequency domain (by fast Fourier transformation), where spectral peaks located greater than a specified threshold, was replaced by zero (representing the high frequency repeat unit). Otsu's algorithm was used for deciding the global threshold value for every image which is then converted to binary image [30]. White part of this binary image representing the pill located on the fabric was counted in terms of "ON" pixel. Parameters related to pilling evaluation like number, area, and rating were also obtained [29].

2.2. Treatment of Wool with Bromelain at Different pH Values. Conditioned bleached and dried wool samples were treated with fixed amount of bromelain (1% owf) in presence of surfactant (0.5% owf) at 60°C for 45 minutes in BOD incubator (Globe Tex Industries) with rotation speed of 160 rpm while varying pH values from 4 to 9 (at fixed interval of 0.5) at 1 : 30 bath ratio. Half of treated samples were tested for areawise shrinkage (Woolmark TM31) and weight loss (%). Remaining treated samples were abraded for 30 (optimized) cycles on abrasion tester and both subjective (Woolmark TM 96) and objective pill analyses (MATLAB software) were done. The weight loss of bromelain treated-pilled wool was calculated by subtracting the initial weight of bromelain treated wool samples (before pilling) and final weight of the pilled fabric. Microscopic evaluation of fiber surface was done with SEM (Jeol-JSM-6610LV).

3. Results and Discussions

3.1. Primary Optimization of Pilling Cycle Threshold. Pilled and bleached wool samples, after passing through different cycles were assessed for average pill number and pixel area in MATLAB software. Obtained trend with standard error bars is plotted in Figure 1.

It was observed that with rise in cycle till 90, pill number increased slightly and then started to decrease with little reduction in pills. Generally with increase in abrasion, fabric becomes thinner and previously formed pills break off and may be attributed to observed reduction in pill number. However, the abraded area under the pills increased which

TABLE 1: Effect of pH on average areawise shrinkage (%), weight loss (%), pill numbers, and area of bromelain treated and bromelain treated-pilled wool.

pH	Bromelain treated: Area-wise shrinkage (%)	Bromelain treated: Average weight loss (%)	Bromelain treated-pilled: Subjective visual pill rating (P_V)	Objective pilling evaluation: Average pill rating (P_I)	Objective pilling evaluation: Average pill area	Objective pilling evaluation: Average pill numbers	Average weight loss (%)
4	31.6 (1.22)	0.12 (0.77)	1.32 (0.55)	1.47 (0.31)	167.33 (0.67)	38.33 (0.88)	0.77 (0.765)
4.5	25.6 (1.22)	0.53 (0.88)	1.5 (0.70)	1.52 (0.16)	156.00 (0.88)	33.00 (0.67)	1.53 (0.876)
5	22.3 (0.90)	1.341 (0.74)	1.82 (0.82)	2.01 (0.35)	121.67 (1.33)	21.00 (0.88)	2.45 (0.739)
5.5	5.67 (0.85)	4.710 (0.35)	4.33 (0.92)	4.23 (0.58)	63.00 (1.45)	7.67 (0.88)	6.88 (0.347)
6	5.1 (0.62)	5.156 (0.70)	3.92 (0.69)	3.88 (0.23)	69.67 (0.58)	7.33 (0.33)	6.94 (0.70)
6.5	5.4 (0.63)	5.713 (0.34)	3.56 (0.41)	3.72 (0.33)	71.00 (0.88)	8.67 (0.67)	7.74 (0.34)
7	5.9 (0.82)	7.767 (0.55)	3.01 (0.35)	3.23 (0.15)	81.67 (0.33)	12.33 (0.33)	11.12 (0.55)
7.5	7.9 (0.95)	9.610 (1.08)	2.01 (1.04)	1.78 (0.23)	99.33 (0.67)	15.67 (0.88)	15.24 (1.08)
8	8.8 (0.85)	11.810 (0.80)	2.78 (0.72)	2.9 (0.30)	100.67 (0.67)	20.67 (1.33)	17.23 (0.80)
8.5	18.7 (1.02)	13.63 (1.48)	2.08 (0.92)	1.9 (0.36)	133.67 (0.88)	23.67 (1.33)	19.83 (1.48)
9	21.9 (1.12)	14.810 (1.47)	2.23 (1.11)	1.98 (0.69)	122.33 (0.88)	21.67 (1.33)	22.87 (1.47)

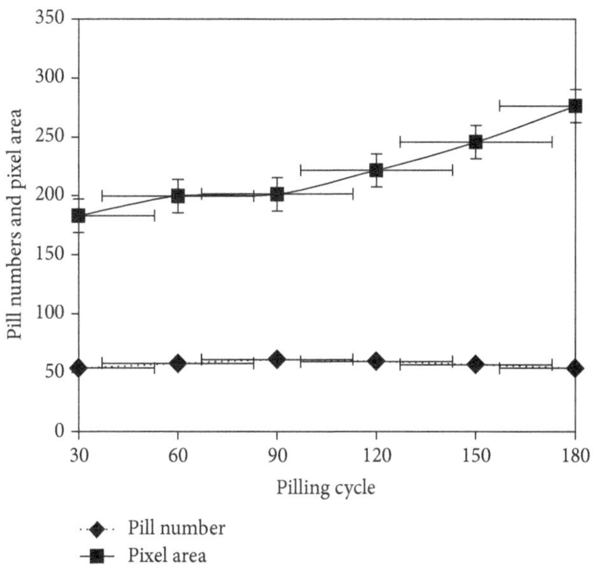

FIGURE 1: Average pill numbers and area for bleached wool with varying pilling cycle.

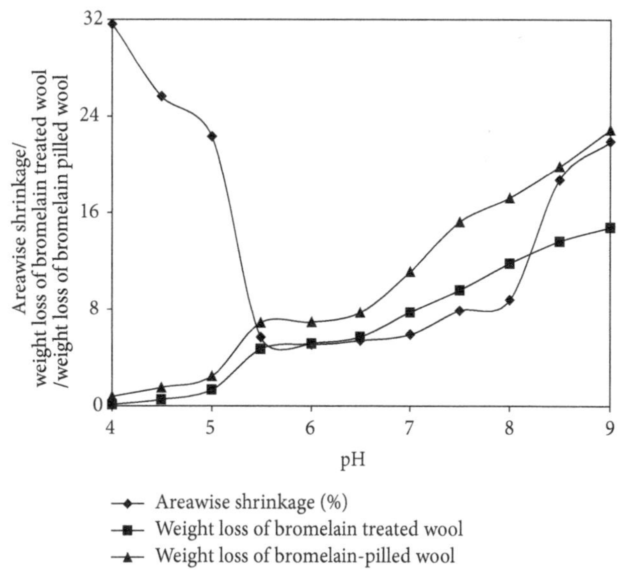

FIGURE 2: Variation in areawise shrinkage (%) and weight loss (%) with varying pH (standard errors of mean < 1.5).

may be due to multiple splitting with increase in pilling cycles. As no significant increase in pill number was observed after the first cycle count, hence 30 cycles were assumed to be primarily responsible for the pill generation and were selected for further pills analysis.

3.2. Treatment of Wool with Bromelain at Different pH Values. Results obtained with respect to only bromelain as well as bromelain treated-pilled wool for shrinkage, weight loss, and pill analyses with respective standard errors of mean (data in parenthesis in respective columns) are shown in Table 1.

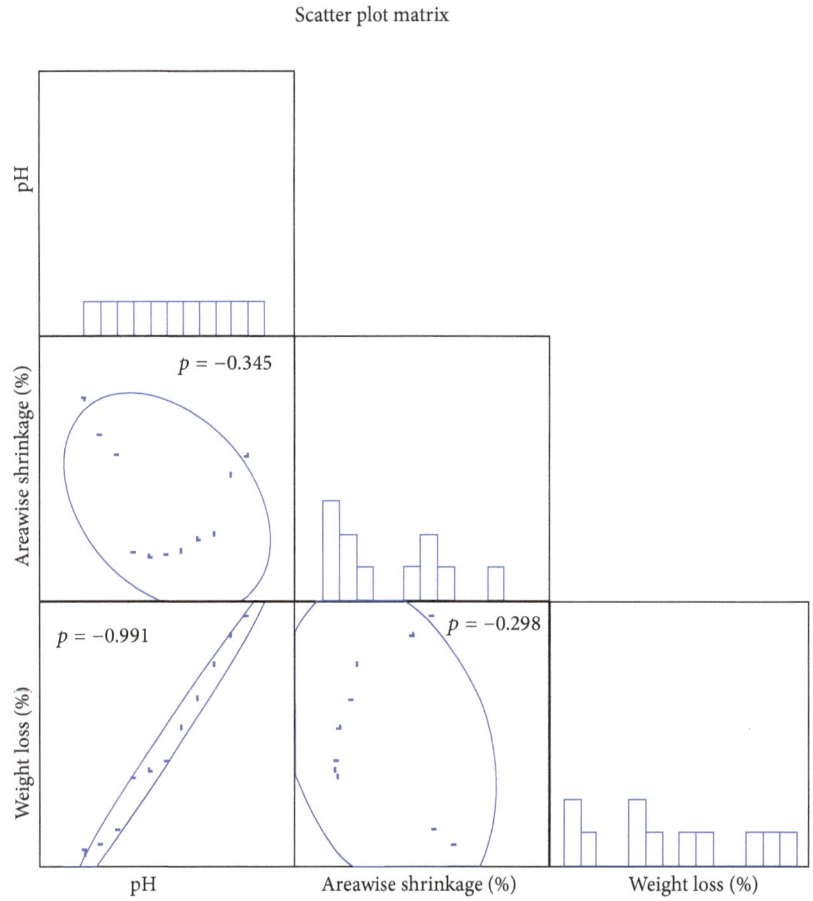

FIGURE 3: Scatter plot for pearson correlation between areawise shrinkage and weight loss with varying pH.

3.2.1. Areawise Shrinkage (%). To obtain desirable machine washable wool, pH of proteolytic treatment should provide less than 6% areawise shrinkage (Woolmark specification AW-1). However, among experimental range of pH, higher values have been obtained for pH ~ 4–5.5 (22–32%) as well as for pH ~ 7.5–9 (8–22%). Only for pH ranging from ~5.5 to 7.0, desirable values were obtained (Figure 2).

Negative Pearson correlation ($p = -0.345$) was observed for area shrinkage values with change in pH. Observed high area shrinkage value in corresponding range may appear due to negligible or very substantial action of bromelain on wool leading to either ineffective or damaging effect, respectively, while affecting the weight loss too. Hence it may be assumed that relative weight loss observed during bromelain treatment may indicate either the amount of complete fiber protein hydrolysis or focused surface effect only. However, while studying the correlated behavior, only moderate negative relationship (p; Pearson correlation = −0.298) was observed with in areawise shrinkage and weight loss (Figure 3). Very strong positive relationship ($p = 0.991$) was observed for weight loss with change in treatment pH. Scanning electron microscopy indicating the changes induced at the fiber surface was carried out to understand the above experiential trend with varying pH.

3.2.2. Weight Loss (%). Proposed assumption for observed bromelain action was ascertained by assessing the trend of relative weight loss after proteolytic treatment and compared after pilling too. As observed from Figure 4, higher weight loss values were obtained for pilled samples as expected due to abrasion on fabric surface, though obtained trend was highly correlated too with changing pH ($p = -0.996$).

Considering the effect within pH, almost similar trend has been obtained in case of bromelain treated and bromelain treated-pilled wool for pH ~ 4–6.5, and after that the gap between the observed values widened for pH ~ 7–9. Very less weight loss values have been obtained for pH ~ 4–4.5 for both wool samples, suggesting very less or almost negligible action of bromelain in this pH range.

While comparing SEM image, scales were almost present on pilled wool treated with bromelain at pH 4.5 (Figure 5(b)) as visible on bleached wool itself (Figure 5(a)). Only lifting of scale and wear due to pilling was found on the fiber surface. Similarly slight reduction in pill number and area was observed indicating almost similar pilling behavior in this pH.

From pH 5.5 to 6.5, rise in weight loss was observed that ascertained the rise in activity of bromelain in this range leading to protein hydrolysis and subsequent weight loss.

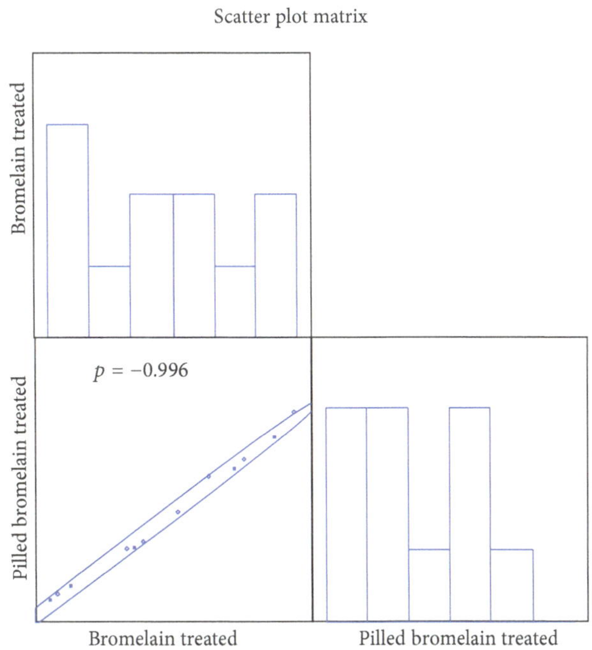

FIGURE 4: Scatter plot for Pearson correlation between weight loss of bromelain treated and bromelain treated-pilled wool with varying pH.

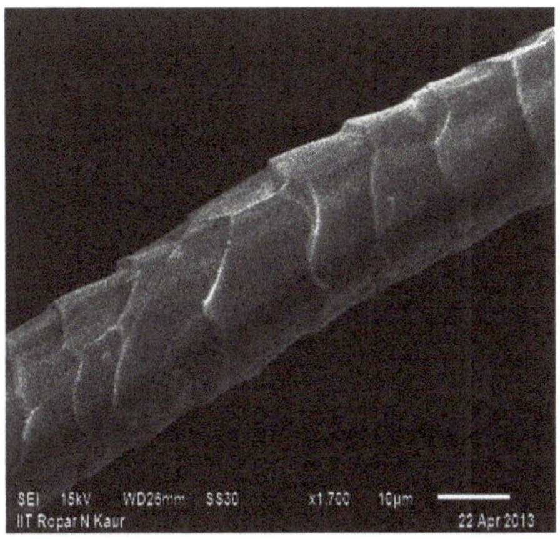

(a) Bleached wool

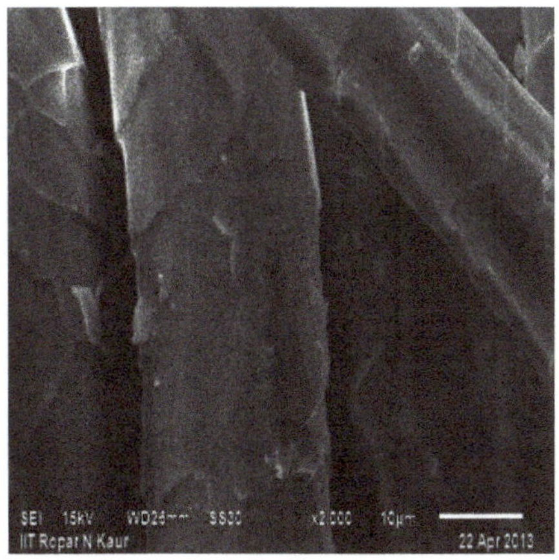

(b) Bleached wool treated with protease at pH 4.5 and pilled

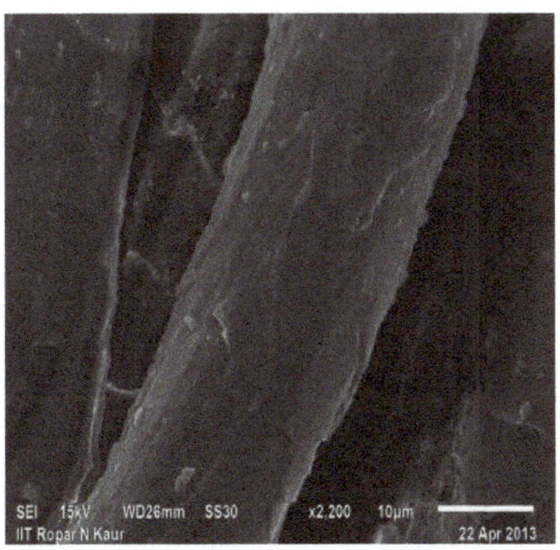

(c) Bleached wool treated with protease at pH 6.0 and pilled

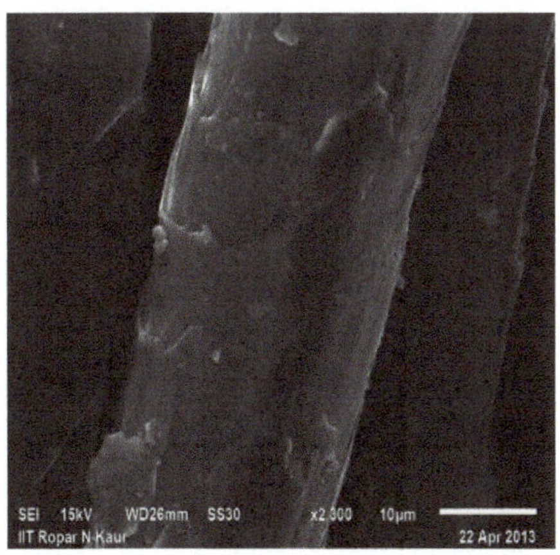

(d) Bleached wool treated with protease at pH 9.0 and pilled

FIGURE 5: SEM (a) bleached wool, (b) bromelain treated and pilled wool at pH ~ 4.5, (c) bromelain treated and pilled wool at pH ~ 6.0, and (d) bromelain treated and pilled wool at pH ~ 9.0.

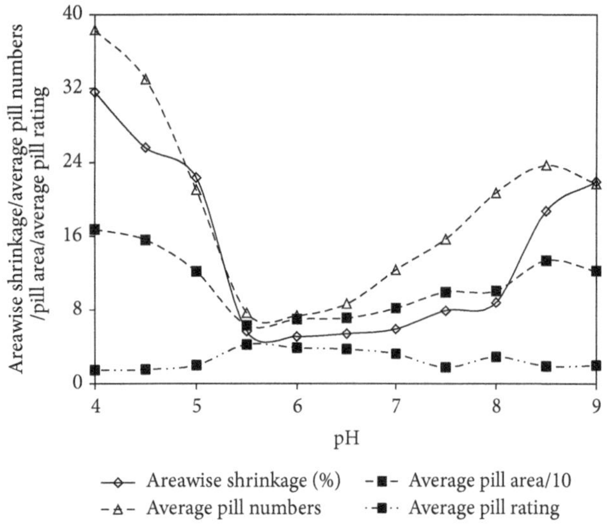

FIGURE 6: Objective pill analysis and areawise shrinkage (%) with varying pH (standard errors of mean < 1.5).

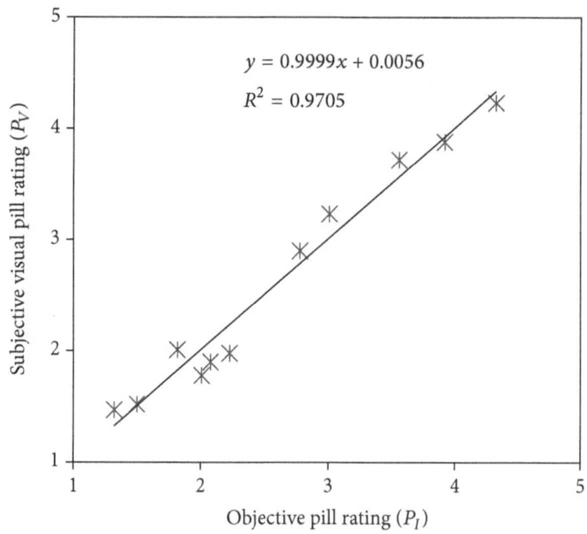

FIGURE 7: Regression trend between objective and subjective pill ratings.

Fiber surface abraded due to pilling, but significant removal of scales for wool treated at pH ~ 6.0 suggested desirable action of bromelain at this pH (Figure 5(c)).

High weight loss was observed for bromelain treated wool in pH ~ 7–9 suggesting intense action of bromelain at this pH leading to excessive protein hydrolysis and subsequent thinning of wool fabric. However, as observed in Figure 5(d), only partially removed scales lifted on the surface of wool were found with significant wear at pH ~ 9. Hence, it may be proposed that bromelain at this pH might have exhibited reduced proteolytic action; still alkaline treatment condition may have led to increased damage to wool. Inherent general stability of wool in acidic conditions as compared to alkaline also ascertained the observed trend [31]. Increase in the comparative difference between weight losses in this range also assisted the assumption as thinning or weakening of fabric may lead to higher abrasion as well as weight loss after pilling, according to observation.

In view of areawise shrinkage effect, pH ~ 5.5–7.0 has been obtained as optimum pH range. But apparent weight loss and probable damage to wool beyond pH 6.5, suggested narrowing down optimum range to ~ 5.5 to 6.5.

3.2.3. Pill Analysis. Pill number and area followed "U" trend while showing initial less values indicating almost similar pilling behavior at pH 4–5.5 (Figure 6). Optimum level of pill reduction has been obtained at pH ~ 5.5–6.5 suggesting removal of short protruding wool fibres during bromelain treatment resulting in smooth surface and generation of lesser number of pills.

Significant lifting of scales and wear was observed in all pilled samples in comparison to raw wool, though higher wear was observed for bromelain treated below as well as beyond optimum pH 5.5–6.5 (Figure 5). Observed behavior is in agreement with appearance of high number of pills and pill area for the respective pH range too.

For validation of objective evaluation method, regression analysis was performed comparing ratings given by the image analyser by MATLAB software with the average ratings given by expert according to standard subjective method and shown in Figure 7 (average root mean square difference = 4.62 and p; Regression coefficient = 0.970).

4. Conclusion

Optimized bromelain treatment with bleached wool at pH ~ 5.5–6.5 range resulted in desirable area shrinkage effect (≤6%), weight loss (4.7–5.7%), and optimum level of pill reduction (average pill number 7–9; pill rating 3.7–4.2). Proposed fast Fourier transform technique can be preferably applied for objective evaluation of pilling behavior of treated wool in terms of pill number, area, and rating. Strong positive correlation between standard subjective test method and objective test method ascertained the utility of proposed quicker and easier method.

Conflict of Interests

The authors declare that there is no conflict of interests regarding the publication of this paper.

References

[1] J. N. Sneddon, J. A. Lee, and G. N. Soutar, "Making sense of consumers' wool apparel preferences," *Journal of the Textile Institute*, vol. 103, no. 4, pp. 405–415, 2012.

[2] E. Heine and H. Höcker, "Enzyme treatments for wool and cotton," *Review of Progress in Coloration and Related Topics*, vol. 25, no. 1, pp. 57–70, 1995.

[3] Q. H. Chen, K. F. Au, C. W. M. Yuen, and K. W. Yeung, "Development of wool shrinkproofing, 1969–99," *Textile Asia*, vol. 31, no. 4, pp. 38–43, 2000.

[4] J. G. Dominguez, P. Erra, A. de la Maza, M. R. Julia, A. Barella, and T. Shaw, "The application of the Hercosett-anionic-surfactant process to impart shrink-resistance to wool," *Journal of Textile Institute*, vol. 71, pp. 165–168, 1980.

[5] S. V. Chikkodi, S. Khan, and R. D. Mehta, "Effects of biofinishing on cotton/wool blended fabrics," *Textile Research Journal*, vol. 65, no. 10, pp. 564–569, 1995.

[6] H. Nolte, D. P. Bishop, and H. Höcker, "Effects of proteolytic and lipolytic enzymes on untreated and shrink-resist-treated wool," *Journal of the Textile Institute*, vol. 87, no. 1, pp. 212–226, 1996.

[7] D. P. Bishop, J. Shen, E. Heine, and B. Hollfelder, "The use of proteolytic enzymes to reduce wool fibre stiffness and prickle," *Journal of the Textile Institute*, vol. 89, no. 3, pp. 546–553, 1998.

[8] A. Riva, I. Algaba, and R. Prieto, "Dyeing kinetics of wool fabrics pretreated with a protease," *Coloration Technology*, vol. 118, no. 2, pp. 59–63, 2002.

[9] M. Parvinzadeh, "Effect of proteolytic enzyme on dyeing of wool with madder," *Enzyme and Microbial Technology*, vol. 40, no. 7, pp. 1719–1722, 2007.

[10] N. A. Ibrahim, E. A. Allam, M. B. El-Hossamy, and W. M. El-Zairy, "Enzymatic modification of cotton/wool and viscose/wool blended fabrics," *Journal of Natural Fibers*, vol. 5, no. 2, pp. 154–169, 2008.

[11] R. Araújo, C. Silva, R. Machado et al., "Proteolytic enzyme engineering: a tool for wool," *Biomacromolecules*, vol. 10, no. 6, pp. 1655–1661, 2009.

[12] W. D. Cooke and M. Goksoy, "Problem of predicting pilling performance using laboratory test methods," *Melliand Textilberichte*, vol. 69, pp. 250–254, 1988.

[13] B. K. Behera and T. E. M. Mohan, "Objective measurement of pilling by image processing technique," *International Journal of Clothing Science and Technology*, vol. 17, no. 5, pp. 279–291, 2005.

[14] J. W. S. Hearle, B. Lomas, and W. D. Cooke, *Atlas of Fibre Fracture and Damage to the Textiles*, Woodhead Publishing, Cambridge, UK, 1998.

[15] C. H. Hsi, R. R. Bresee, and P. A. Annis, "Characterizing fabric pilling by using image-analysis techniques. Part I: pill detection and description," *Journal of the Textile Institute*, vol. 89, no. 1, pp. 80–95, 1998.

[16] D. G. Neilly, "Grades of pilling," *Textile Asia*, vol. 3, pp. 68–69, 1990.

[17] W. Conti and E. Tassinari, "II—the numerical distribution of fibres in pills," *Journal of the Textile Institute*, vol. 66, no. 2, pp. 73–79, 1975.

[18] V. A. Williams, "A kinetic model for pilling of wool knitwear," *Textile Research Journal*, vol. 55, pp. 312–320, 1985.

[19] H. C. Abril, M. S. Millán, Y. Torres, and R. Navarro, "Automatic method based on image analysis for pilling evaluation in fabrics," *Optical Engineering*, vol. 37, no. 11, pp. 2937–2947, 1998.

[20] R. B. Ramgulam, J. Amirbayat, and I. Porat, "Objective assessment of fabric piling. Part I: methodology," *Journal of the Textile Institute*, vol. 84, no. 2, pp. 221–226, 1993.

[21] J. Amirbayat and M. J. Alagha, "The objective assessment of fabric pilling. Part II: experimental work," *Journal of the Textile Institute*, vol. 85, no. 3, pp. 397–401, 1994.

[22] H. C. Abril, Y. M. Torres, and M. S. Millan, "Image synthesis of pilled textiles by karhunen-Loeve transform," in *3rd Iberoamerican Optics Meeting and 6th Latin American Meeting on Optics, Lasers, and Their Applications*, vol. 3572 of *Proceedings of SPIE*, pp. 254–258, July 1999.

[23] K. L. Jensen and J. M. Carstensen, "Fuzz and pills evaluated on knitted textiles by image analysis," *Textile Research Journal*, vol. 72, no. 1, pp. 34–38, 2002.

[24] B. Xin, J. Hu, and H. Yan, "Objective evaluation of fabric pilling using image analysis techniques," *Textile Research Journal*, vol. 72, no. 12, pp. 1057–1064, 2002.

[25] I. M. Dar, W. Mahmood, and G. Vachtsevanos, "Automated pilling detection and fuzzy classification of textile fabrics," in *5th Machine Vision Applications in Industrial Inspection*, vol. 3029 of *Proceedings of SPIE*, pp. 26–36, San Jose, Calif, USA, February 1997.

[26] X. Chen and X. B. Huang, "Evaluating fabric pilling with light-projected image analysis," *Textile Research Journal*, vol. 74, no. 11, pp. 977–981, 2004.

[27] T. J. Kang, D. H. Cho, and S. M. Kim, "Objective evaluation of fabric pilling using stereovision," *Textile Research Journal*, vol. 74, no. 11, pp. 1013–1017, 2004.

[28] S. C. Kim and T. J. Kang, "Image analysis of standard pilling photographs using wavelet reconstruction," *Textile Research Journal*, vol. 75, no. 12, pp. 801–811, 2005.

[29] A. Khandual, A. Luximon, and A. K. Pattanayak, "Optimization of acid cellulose enzyme concentration to reduce pilling of bamboo fabric: an objective assessment approach," *Fibers and Polymers*, vol. 12, no. 6, pp. 816–820, 2011.

[30] N. Otsu, "A threshold selection method from gray-level histograms," *IEEE Transaction of System, Man and Cybernetics*, vol. 9, no. 1, pp. 62–66, 1979.

[31] A. M. Sookne and M. Harris, "Electrophoretic studies of wool," *Textile Research Journal*, vol. 9, no. 12, pp. 437–443, 1939.

18

Exploratory Investigation of Chitosan as Mordant for Eco-Friendly Antibacterial Printing of Cotton with Natural Dyes

M. D. Teli, Javed Sheikh, and Pragati Shastrakar

Department of Fibers and Textile Processing Technology, Institute of Chemical Technology, Mumbai 400019, India

Correspondence should be addressed to M. D. Teli; textileudct@yahoo.com

Academic Editor: Gajanan S. Bhat

Although a lot of antibacterial finishes are available for cotton, the user safety and durability of the finish are always the important issues. Some of the natural dyes are known to possess antibacterial properties; however, most of them are nonsubstantive. Hence, they are used in conjunction with eco-friendly natural mordants. Although metallic mordants are effective, they are environmentally pollutants, hence not desirable. In the current work, the novel natural mordant like chitosan with different concentrations (10% and 20% of weight of fiber) was applied on cotton and fabric was printed with natural dyes such as catechu, turmeric, and marigold using screen printing method. All the three dyes were used in powder form in print paste with different concentrations (1%, 5%, and 10%) and guar gum was used as thickener. The printed samples using chitosan as a mordant showed the equivalent color values to that of prints obtained using alum as a mordant. The printed fabric showed excellent antibacterial activity against both gram-positive and gram-negative bacteria. The method may be considered suitable for eco-friendly printing and antibacterial finishing of textile materials.

1. Introduction

Textiles, especially those made of natural fibers, are an excellent medium for the growth of microorganisms when the basic requirements for their growth such as nutrients, moisture, oxygen, and appropriate temperature are present. The large surface area and ability to retain moisture of textiles also assist the growth of microorganisms on the fabric [1]. The growth of microorganisms on textiles inflicts a range of unwanted effects not only on the textile itself but also on the wearer. These effects include the generation of unpleasant odor, stains, discoloration in the fabric, a reduction in the tensile strength of the fabric, and an increased likelihood of contamination [2]. In the last few decades, with the increase in new antimicrobial fiber technologies and the growing awareness about cleaner surroundings and healthy lifestyle, a range of textile products based on synthetic antimicrobial agents, such as triclosan, metal and their salts, organometallics, phenols, and quaternary ammonium compounds, have been developed and quite a few are also available commercially [3]. Even though the excellent antimicrobials are available, their user ecology is always a question. Some of the natural dyes are reported to possess antimicrobial properties.

Synthetic dyes are here to remain as one can tailor make them depending upon the required hue, substantivity, and fiber. The development of synthetic dyes at the beginning of the twentieth century led to a more complete level of quality and more reproducible techniques of application. As a result, a distinct lowering in the dyestuff costs per kg of dyed goods was achieved [4]. Also, a full gamut of color is available for application on various types of fibers.

However, during the last few decades, the use of synthetic dyes is gradually receding due to an increased environmental awareness and harmful effects, of either toxic degraded products or their nonbiodegradable nature. In addition to previously mentioned, some serious health hazards like allergenicity and, carcinogenicity are associated with some of the synthetic dyes based on azo chromophores. As a result, a ban has been imposed all over the world including European Economic Community (EEC), Germany, USA, and India on the use of some synthetic dyes (e.g., azodyes) which finally triggered active research and development to revive world heritage and traditional wisdom of employing safer natural dyes [5]. Consumers nowadays are becoming more and more concerned about environmental issues and hence are demanding for natural product incorporating natural ingredient. Due to increasing awareness of environmental issues and pollution controls, natural dyes are gaining importance as they are obtained from renewable resources and they present no health hazards and some of them sometimes act as a health care product [6]. Hence, it is important to understand that, although natural colors cannot substitute use of synthetic dyes completely, there is definitely increasing market for such complete eco-friendly dyed or printed materials.

Natural dyes with a few expectations are nonsubstantive and hence must be used in conjunction with mordants. Mordant is a chemical, which can fix itself on the fiber and also combines with the dyestuff. The challenge before the natural dyers in application of natural color is the necessity to us metallic mordants which themselves are pollutant and harmful. Due to the environmental hazard caused by metallic mordant while dyeing of textile fabric, dyers are always looking for safe natural mordant for natural dyes. In this paper, this issue of screening a natural mordant for printing is explored.

The main objective of printing is to produce colored patterns with sharp boundaries on textile materials without any dye spreading beyond the boundaries of the motif of design [7]. Although a lot of research has been reported in the area of printing and natural dyeing of textiles, the research in the area of natural dye printing is reported to a limited extent. Also application of natural dye prints as antibacterial finish has also remained unexplored. The applications of chitosan for different applications in textiles are reported [8–24], but the application of such functional biopolymer as a mordant in natural dyeing has been quite rare in the literature. In the current work, chitosan extracted from waste shrimp shells [24] was utilized as a mordant for simultaneous natural dye printing and antibacterial finishing of cotton in comparison with commonly used metal mordants. The efficacy of chitosan as ecofriendly mordant and antibacterial finish has been studied. The antibacterial activity against both gram-positive and gram-negative bacteria has also been screened.

2. Materials and Methods

2.1. Materials. Chitosan (mol. wt. 124999.8, degree of deacetylation, 89.06%, nitrogen content, 7.1%) was extracted from waste shrimp shells [24] and used for the study. Ready for dyeing (RFD) Cotton fabric (Plain weave, EPI-69, PPI-88, GSM-122.95) was purchased from the market. Catechu, turmeric, and marigold were purchased from the market. All other chemicals used were of laboratory grade.

2.2. Methods

2.2.1. Preparation of Mordant. Stock solutions of alum and $CuSO_4$ were made by dissolving in double distilled water. Chitosan was dissolved in aqueous acetic acid (20 grams per liter, gpl). Chitosan powder was added to acetic acid solution followed by warming at 60°C for 2 h with continuous stirring using mechanical stirrer.

2.2.2. Preparation of Dye for Printing. All the dyes used were first converted to powder form. For this purpose, catechu, turmeric, and marigold were first dried in an oven at 50° for 24 h and later grinded in mixer grinder. The powder so obtained was filtered through the 60 mesh nylon fabric. The fine powder of natural dye was used for printing.

2.2.3. Mordanting of Cotton Fabric. Mordanting of cotton was carried out with different mordants at two different concentrations on the weight of fabric (alum; 10% and 20%, copper sulphate; 1% and 2%, and chitosan; 10% and 20%) at 90°C for 60 mins with material to liquor ratio as 1 : 30. The mordanted fabric was squeezed and dried in open width form at 100°C for 5 mins and then used for printing.

2.2.4. Printing. The stock paste of guar gum was prepared using 2% of guar gum powder. The guar gum powder was sprinkled slowly in water with continuous stirring in order to prevent lump formation. The paste was stirred continuously at 90°C for 1 h.

The print paste was prepared in different dye concentrations (1%, 5%, and 10%) on the basis of print paste. For preparation of 1% of print paste, 1 gm of dye powder was first pasted with small quantity of water followed by addition of 99 gm of guar gum stock paste. The print paste was continuously stirred for 30 mins. The mordanted cotton fabrics were then printed with two strokes of squeeze, steamed at 102°C for 10 mins. The samples were washed with water and then dried in air.

2.3. Analysis of Printed Fabrics

2.3.1. Color Value by Reflectance Method. The printed samples were evaluated for the depth of color by reflectance method using 10-degree observer. The absorbance of the dyed samples was measured on Spectraflash SF 300 (Datacolor International, USA) equipped with reflectance accessories. The K/S values were determined using expression

$$\frac{K}{S} = \frac{(1-R)^2}{2R}, \quad (1)$$

where R is the reflectance at complete opacity, K is the absorption coefficient, S is the Scattering coefficient

Dyed fabrics were simultaneously evaluated in terms of CIELAB color space (L^*, a^* and b^*) values using the Rayscan Spectrascan 5100+. In general, the higher the K/S value, the higher the depth of the color on the fabric. L^* corresponding to the brightness (100 = white, 0 = black), a^* to the red-green coordinate (+ve = red, −ve = green), and b^* to the yellow-blue coordinate (+ve = yellow, −ve = blue). As a whole, a combination of all these parameters enables one to understand the tonal variations.

2.3.2. Washing Fastness. Evaluation of color fastness to washing was carried out using ISO II methods [25]. A solution containing 5 gpl soap solution was used as the washing liquor. The cotton samples were treated for 45 min at 50°C using material to liquor ratio of 1 : 50 in rota machine. After rinsing and drying, the change in color of the sample and staining on the undyed samples was evaluated on the respective standard

TABLE 1: Effect of mordants and catechu concentration on color strength of cotton.

Mordant (%)	Catechu (%)	Colour value K/S	CIE colour coordinates L^*	a^*	b^*	Washing fastness Change	Staining	Light fastness
Alum								
10	1	0.2514	81.626	4.566	8.739	5	5	6
10	5	0.6024	83.211	4.991	11.255	5	4-5	7
10	10	1.0746	83.757	5.434	12.167	4-5	4-5	7
20	1	0.299	82.104	3.562	8.851	4-5	4-5	6
20	5	0.5852	83.216	5.055	11.454	4-5	4	7
20	10	1.1067	83.842	5.752	12.448	4	4	7
Chitosan								
10	1	0.3363	79.477	4.235	9.042	5	5	6
10	5	0.5287	78.845	6.323	8.972	5	5	6
10	10	1.0201	79.248	7.689	10.13	4-5	4-5	7
20	1	0.4182	78.529	4.718	7.628	4-5	4-5	7
20	5	0.8044	79.117	7.286	9.349	4-5	4-5	7
20	10	1.3245	79.183	8.214	9.867	4-5	4	7
Copper sulphate								
1	1	0.7087	72.994	6.824	14.267	4-5	5	7
1	5	1.6571	71.891	9.586	13.176	4-5	4-5	7
1	10	3.0619	73.111	9.882	14.614	4-5	4	7
2	1	0.6359	72.575	6.913	13.488	4	4-5	7
2	5	3.7264	73.109	10.569	14.928	4	4	7
2	10	3.7948	72.734	10.114	14.604	4-5	4	8

TABLE 2: Effect of mordants and turmeric concentration on color strength of cotton.

Mordant (%)	Turmeric (%)	Colour value K/S	CIE colour coordinates L^*	a^*	b^*	Washing fastness Change	Staining	Light fastness
Alum								
10	1	0.8084	84.007	−4.01	33.849	4-5	4-5	3
10	5	4.3847	89.655	2.389	43.257	4-5	4	3
10	10	5.9102	90.404	1.295	44.146	4	4	3
20	1	0.937	84.934	4.716	35.625	4	3-4	3
20	5	3.7634	90.364	3.028	44.609	4-5	3-4	3
20	10	7.9375	91.327	3.129	45.076	4	4	4
Chitosan								
10	1	1.1319	83.174	5.348	39.866	5	4-5	3
10	5	2.4573	84.203	1.755	42.078	4-5	4-5	4
10	10	6.3261	86.902	1.418	45.928	4-5	4	4
20	1	1.1838	81.587	3.784	37.307	4	4	3
20	5	2.4426	84.3	0.586	42.285	4	3-4	3
20	10	6.6156	87.28	1.097	46.202	4	4	4
Copper sulphate								
1	1	0.6626	81.256	3.234	27.299	4-5	4	5
1	5	2.6011	87.844	3.495	37.35	4-5	4	5
1	10	5.0764	89.695	4.177	39.489	4-5	3-4	5
2	1	0.7542	82.015	3.598	28.505	4	3-4	5
2	5	4.0985	88.032	2.253	37.751	3-4	3-4	5
2	10	7.3138	89.973	−3.06	40.382	4	3-4	6

TABLE 3: Effect of mordants and marigold concentration on color strength of cotton.

Mordant (%)	Marigold (%)	Colour value	CIE colour coordinates			Washing fastness		Light fastness
		K/S	L^*	a^*	b^*	Change	Staining	
Alum								
10	1	1.1593	81.411	2.234	32.553	5	4-5	6
10	5	3.6119	85.186	−1.16	36.621	4-5	4	6
10	10	6.2023	86.014	1.086	37.852	4-5	3-4	7
20	1	1.1186	80.382	−1.66	31.881	4-5	3-4	6
20	5	4.3349	85.02	0.058	37.712	4	3-4	7
20	10	7.2649	86.29	0.494	38.388	4-5	4	7
Chitosan								
10	1	1.3023	72.346	1.266	21.601	5	5	7
10	5	2.2313	76.981	0.533	24.224	5	4-5	7
10	10	3.4716	78.828	2.247	28.088	4-5	4-5	7
20	1	1.5061	74.502	0.916	23.481	5	4-5	7
20	5	3.5484	78.269	2.107	26.801	4-5	4-5	7
20	10	4.561	78.963	3.354	28.089	4-5	4	7
Copper sulphate								
1	1	1.0403	74.96	2.073	22.475	5	5	7
1	5	3.0923	78.166	1.362	26.119	5	5	7
1	10	7.7622	81.004	0.148	29	4-5	4-5	7
2	1	1.4072	75.411	1.78	23.091	4	4	7
2	5	4.9077	79.504	1.514	28.381	4-5	4	7
2	10	12.4811	82.349	0.323	30.92	4	4	7

scales (rating 1 : 5; where 1: poor; 2: fair; 3: good; 4: very good and 5: excellent).

2.3.3. Light Fastness. Printed cotton fabric was tested for color fastness to light according to ISO 105/B02 [26]. The light fastness was determined using artificial illumination with Xenon arc light source, Q-Sun Xenon Testing Chamber with black standard temperature of 65°C with relative humidity of the air in the testing chamber as 40% and daylight filter, wavelength, $k = 420$ nm. The cotton samples were compared with the standard scale of blue wool reading (ratings, 1 : 8; where 1: poor; 2: fair; 3:moderate; 4: good; 5: better; 6: very good; 7: best and 8: excellent).The color fastness to light was measured using test method.

2.3.4. Determination of Antibacterial Activities of Printed Fabrics. The antibacterial activity of the printed fabrics was estimated by AATCC Test Method 100-2004 [27].

3. Results and Discussion

3.1. Color Values of Natural Dye Printed Cotton. The efficacy of chitosan as mordant was screened in comparison with metal mordants and the results are summarized in Tables 1, 2, and 3.

The initial attempt was to explore the role of chitosan as a mordant for the localised coloration (printing) with natural dyes in the study. For this purpose, the cotton fabric was printed with all the three dyes with and without mordants. The prints were very light in shade and the color bled heavily after washing where mordants were absent. However, in case of chitosan, like other metal mordants, the color values obtained were higher and the prints were washed fast and this trend was valid for all the three dyes studied. The observations confirmed the role of chitosan as a mordant in natural dye printing. The role of chitosan as a mordant may be attributed to the presence of amino groups which get protonated and offer the sites for attachment of the dye which is mainly anionic in nature. The mechanism of dye attachment to the fiber is similar to that in case of metal mordants.

The mordant concentrations used were 10 and 20% for alum and chitosan and 1% and 2% for $CuSO_4$. The results in Tables 1–3 indicate the dye specific trends of color values. In case of catechu as a dye, $CuSO_4$ showed the highest color values among the mordants. However, the color values for chitosan and alum mordants were comparable. In case of turmeric, alum mordant showed highest color values while in case of marigold $CuSO_4$ showed highest color values. The varying combinations of mordant and dye resulted in different shades and tones of dyed cotton fabric. In case of all the dyes and mordant combinations, keeping the dye concentration constant, the increasing mordant concentrations resulted in increasing color values. The same trend was valid when dye concentration was varied keeping mordant concentration constant.

The fastness properties of the printed cotton fabrics were studied and the results are presented in Tables 1–3 which indicate the comparative fastness properties in the case of all the mordants. The only dyed samples showed much inferior

TABLE 4: Effect of mordant and dye combinations on antibacterial properties.

Mordant	Mordant conc. (%)	Dye	Dye conc. (% of print paste)	Bacterial reduction$ (%) *S. aureus*	Bacterial reduction$ (%) *E. coli*
Alum	20	Catechu	10	98.25	97.50
Chitosan	20		10	99.75	99.80
$CuSO_4$	2		10	100	100
Alum	20	Turmeric	10	98.25	97.00
Chitosan	20		10	99.25	99.10
$CuSO_4$	2		10	100	100
Alum	20	Marigold	10	99.50	99.00
Chitosan	20		10	99.90	99.75
$CuSO_4$	2		10	100	100

$Represents average value of 3 determinations.

fastness properties which were improved with mordanting (irrespective of the type) which indicate the positive role of mordant in holding the dye on fabric. The improvements in fastness properties with mordant concentration clearly indicate the positive role of mordants in case of natural dyes. The washing fastnesses obtained varied in the range of good to excellent grades. Light fastness was found to be improving with improving K/S values. $CuSO_4$ showing very high light fastness as well as high degree of antibacterial activity which may be attributed to its positive contribution in improving the stability of dye-metal complex to light as well as its known toxic nature.

3.2. Antibacterial Activity of Printed Cotton Fabric. The quantitative antibacterial assessment was made using AATCC-100(2004) test method and the results are presented in Table 4.

The results for mordanted and printed fabrics indicate the excellent antibacterial activity for all the mordants. The cotton fabric showed no antibacterial activity against both *S. aureus* and *E. coli.*

As far as antibacterial activity is concerned, the chitosan mordanted samples showed comparative antibacterial activity with that of $CuSO_4$ and higher activity than that of alum in case of all the dyes studied. Chitosan is known for antibacterial activity against broad spectrum bacteria and mechanism of antibacterial activity was explained by researchers [12, 28]. The chitosan hence can be claimed as eco-friendly mordant cum antibacterial finishing agent for cotton.

4. Conclusion

The concept of printing of cotton fabric with natural dyes in fine powder form was successfully carried out using chitosan as an eco-friendly mordant. The color values varied with the dye mordant combination; however, the chitosan mordant showed good color values. The excellent antibacterial activity was displayed by the printed samples. The printed cotton also displayed excellent fastness properties. The role of chitosan as a mordant in natural dye printing has been confirmed through this study. The proof of demonstration of concept sample can open door for further research to develop antibacterial printed cotton using natural dyes.

Conflict of Interests

The authors do not have any stake in Data color international or AATCC and their names appear in the paper purely in coincidental and academic nature. Their reference is made to indicate a standard equipment and test method were made use of in this work and in no way it advocates these brands.

References

[1] W. Su, S. S. Wei, S. Q. Hu, and J. X. Tang, "Antimicrobial finishing of cotton textile with nanosized silver colloids synthesized using polyethylene glycol," *Journal of the Textile Institute*, vol. 102, no. 2, pp. 150–156, 2011.

[2] Y. Gao and R. Cranston, "Recent advances in antimicrobial treatments of textiles," *Textile Research Journal*, vol. 78, no. 1, pp. 60–72, 2008.

[3] M. Joshi, S. W. Ali, R. Purwar, and S. Rajendran, "Ecofriendly antimicrobial finishing of textiles using bioactive agents based on natural products," *Indian Journal of Fibre and Textile Research*, vol. 34, no. 3, pp. 295–304, 2009.

[4] T. Bechtold, A. Turcanu, E. Ganglberger, and S. Geissler, "Natural dyes in modern textile dyehouses—how to combine experiences of two centuries to meet the demands of the future?" *Journal of Cleaner Production*, vol. 11, no. 5, pp. 499–509, 2003.

[5] J. K. Kumary and A. K. Sinha, "Resurgence of natural colourants: a holistic view," *Natural Product Letters*, vol. 18, no. 1, pp. 59–84, 2004.

[6] K. H. Prabhu, M. D. Teli, and N. G. Waghmare, "Eco-friendly dyeing using natural mordant extracted from Emblica officinalis G. Fruit on cotton and silk fabrics with antibacterial activity," *Fibers and Polymers*, vol. 12, no. 6, pp. 753–759, 2011.

[7] V. A. Shenai, *Technology of Printing*, Sevak Publications, Mumbai, India, 1985.

[8] J. A. Rippon, "Improving the dye coverage of immature cotton fibres by treatment with chitosan," *Journal of the Society of Dyers and Colourists*, vol. 100, no. 10, pp. 298–303, 1984.

[9] S.-H. Lim and S. M. Hudson, "Application of a fibre-reactive chitosan derivative to cotton fabric as a zero-salt dyeing auxiliary," *Coloration Technology*, vol. 120, no. 3, pp. 108–113, 2004.

[10] S.-H. Lim and S. M. Hudson, "Application of a fiber-reactive chitosan derivative to cotton fabric as an antimicrobial textile finish," *Carbohydrate Polymers*, vol. 56, no. 2, pp. 227–234, 2004.

[11] S. K. Tiwari and M. M. Gharia, "Characterization of chitosan pastes and their application in textile printing," *AATCC Review*, vol. 3, p. 25, 2003.

[12] Z. Zhang, L. Chen, J. Ji, Y. Huang, and D. Chen, "Antibacterial properties of cotton fabrics treated with Chitosan," *Textile Research Journal*, vol. 73, no. 12, pp. 1103–1106, 2003.

[13] S. Sharaf, K. Opwis, D. Knittel, and J. S. Gutmann, "Comparative investigations on the efficiency of different anchoring chemicals for the permanent finishing of cotton with chitosan," *Autex Research Journal*, vol. 11, no. 2, pp. 71–77, 2011.

[14] K. El-Tahlawy, "Chitosan phosphate: a new way for production of eco-friendly flame-retardant cotton textiles," *Journal of the Textile Institute*, vol. 99, no. 3, pp. 185–191, 2008.

[15] K. F. El-Tahlawy, M. A. El-Bendary, A. G. Elhendawy, and S. M. Hudson, "The antimicrobial activity of cotton fabrics treated with different crosslinking agents and chitosan," *Carbohydrate Polymers*, vol. 60, no. 4, pp. 421–430, 2005.

[16] T. Oktem, "Surface treatment of cotton fabrics with chitosan," *Coloration Technology*, vol. 119, no. 4, pp. 241–246, 2003.

[17] A. Abou-Okeil, A. El-Shafie, and A. Hebeish, "Chitosan phosphate induced better thermal characteristics to cotton fabric," *Journal of Applied Polymer Science*, vol. 103, no. 3, pp. 2021–2026, 2007.

[18] B. N. Bandyopadhyay, G. N. Sheth, and M. M. Moni, "Chitosan can cut salt use in reactive dyeing," *International Dyer*, vol. 183, no. 11, pp. 39–42, 1998.

[19] R. S. Davidson and Y. Xue, "Improving the dyeability of wool by treatment with chitosan," *Journal of the Society of Dyers and Colourists*, vol. 110, no. 1, pp. 24–29, 1994.

[20] S. Rattanaphani, M. Chairat, J. B. Bremner, and V. Rattanaphani, "An adsorption and thermodynamic study of lac dyeing on cotton pretreated with chitosan," *Dyes and Pigments*, vol. 72, no. 1, pp. 88–96, 2007.

[21] M. Jassal, R. B. Chavan, R. Yadav, and P. Singh, "Chitosan as thickner for printing of cotton with pigment colors," in *Chitin and Chitosan: Opportunities & Challenges*, P. K. Dutta, Ed., vol. 94, SSM International Publication, Contai, India, 2005.

[22] S. A. Bahmani, G. C. East, and I. Holme, "The application of chitosan in pigment printing," *Journal of the Society of Dyers and Colourists*, vol. 116, no. 3, pp. 94–99, 2000.

[23] M. D. Teli and J. Sheikh, "Simultaneous pigment dyeing and antibacterial finishing of denim fabric using Chitosan as a binder," *International Dyer*, vol. 197, no. 4, p. 28, 2012.

[24] M. D. Teli and J. Sheikh, "Extraction of chitosan from shrimp shells waste and application in antibacterial finishing of bamboo rayon," *International Journal of Biological Macromolecules*, vol. 50, no. 5, pp. 1195–1200, 2012.

[25] E. R. Trotmann, *Dyeing and Chemical Technology of Textile Fibers*, Charles Griffin and Company, London, UK, 1984.

[26] ISO technical manual, Geneva, Switzerland, 2006.

[27] "American Association of Textile Chemists and Colorists," AATCC Technical Manual 76, AATCC, Research Triangle Park, NC, USA, 2007.

[28] S.-H. Lim and S. M. Hudson, "Review of chitosan and its derivatives as antimicrobial agents and their uses as textile chemicals," *Journal of Macromolecular Science*, vol. 43, no. 2, pp. 223–269, 2003.

19

Evaluation of Bed Cover Properties Produced from Double Fabric Based on Honeycomb

A. A. Salama, A. S. El-Deeb, and I. M. El-shahat

Textile Engineering Department, Mansoura University, Mansoura 35516, Egypt

Correspondence should be addressed to A. A. Salama; eng_amanysalama@yahoo.com

Academic Editor: Sambandam Anandan

This research aims to innovate a new fabric structure, which could be used as a bed cover based on double honeycomb fabric with self-stitching. The honeycomb air pockets were aimed at facing each other to form closed small air chambers which work to sequester the air. The double fabric increases fabric thickness. Thus, the opportunity to improve thermal comfort could be achieved. A number of samples were produced with different densities and counts of weft yarn. Thermal insulation and water vapour permeability were measured and compared with bed covers produced from reversible weft backed structure. Geometrical properties, abrasion resistance, and air permeability were also measured. The results showed that the innovated structure had higher values of thermal insulation than reversible weft backed structure at certain weft counts and densities.

1. Introduction

Clothing is considered as a second skin of the human body as it works to protect the human body from hazardous substances [1]. It also controls the thermal equilibrium of man in his environment and preserves the human body from losing its temperature, since fabric organizes the heat transfer and gas exchange between the human body and the ambient atmosphere in different environmental conditions.

Comparing awaking state to sleep state, we can find that the energy emitted by the human body drastically reduces from 1500 watts as a maximum limit to 100 watts as a minimum limit during sleep [2]. Therefore, extra textiles are needed to achieve thermal equilibrium. These textiles are called bed covers, which support clothing in organizing heat and gas exchange. The main required properties of bedcovers are thermal insulation and water vapour permeability in addition to a reasonable weight.

These types of textiles vary according to the above-mentioned properties from light bed sheets to heavy blankets and medium weight coverlet which is considered to be the most suitable bed covers in mid seasons.

The main traditional fabric structures of Egyptian coverlets are reversible weft backed, single honeycomb, and wadded double plain weave with interchangeable layers.

This research is an attempt to innovate and evaluate a fabric structure that could be used to produce a coverlet and compare its thermal properties to reversible weft backed coverlet made from the same warp and weft yarns.

2. Review of Literature

Wilson et al. described variables affecting thermal resistance of infant bedding materials such as different types of bedding, different sleep positions, methods of tucking, and two types of mattress. It was found that thermal resistance of bedding increased as thickness of bedding increased [3]. Fukazawa et al. measured thermal and water-vapour resistances of three kinds of cold-protective clothing ensembles in a cold environment. Sweating rate affected skin wittedness [4]. Qian and Fan showed that the surface thermal insulation is little affected by moisture transfer by using the sweating fabric manikin-Walter under varying climatic conditions and walking speeds [5]. Lin and Deng measured the total insulation values for a wide range of bedding systems. The insulation values were influenced by bedding and its percentage coverage over a human body and sleepwear [6]. Matusiak and Sikorski produced 19 variants of cotton woven fabrics of different weaves, linear densities of the weft, and different

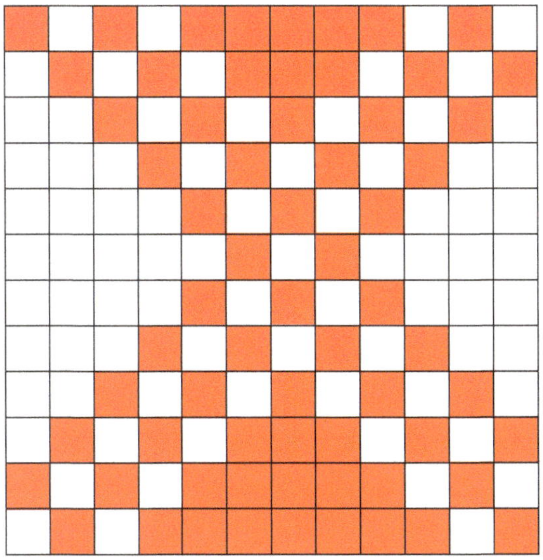

(a) Face structure

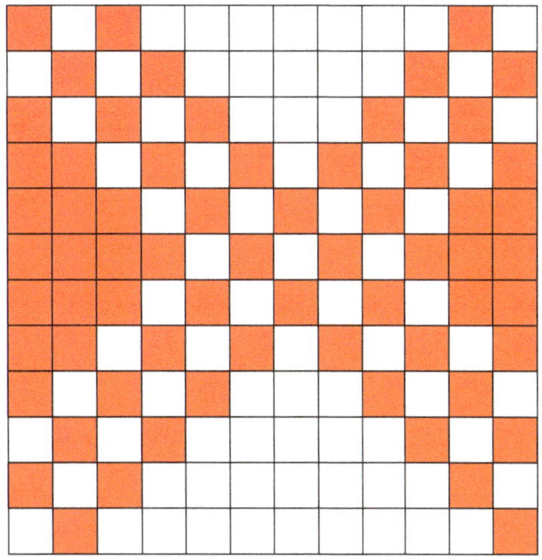

(b) Back structure

FIGURE 1: Face and back structure.

weft densities. On the basis of the results, the weave of woven fabrics, the linear density of weft yarn, and fabric thickness influence their thermal insulation properties significantly. The mass per square meter of fabrics affected their thermal conductivity. Closed air inside the fabric structure plays the role in thermal insulation. A strong and statistically significant correlation exists between the thermal insulation properties of woven fabrics and their structural parameters [7].

The previous researches showed that each of the fiber type, yarn structure, yarn count, yarn density, fabric structure, and type of finishing affects fabric thermal properties. Also the increase of fabric thickness which works to lock up the air inside the weave plays an important role in fabric thermal insulation especially in bed covers [3].

The innovated structure which this research introduces is based on honeycomb to benefit from the fabric thickness and air pockets and producing double honeycomb so that air pockets can face each other to form air chambers. The effect of weave structure at different densities and counts of weft yarns on the thermal insulation of the double honeycomb was studied and compared to reversible weft backed weave based on twill 3/1 which is commonly used in the Egyptian environment.

FIGURE 2: Self-stitched double honeycomb structure.

3. Design of the Innovated Structure

Double honeycomb structure was designed with self-stitching where the face weave was honeycomb repeated on (12 × 12) as shown in Figure 1(a). The back weave was the same structure but it was shifted half repeat warp direction and half repeat weft direction as shown in Figure 1(b).

The arrangement of warp and weft threads for face and back was (1 : 1). Self-stitching was carried out diagonally on both sides of fabric by dropping a face end under a back pick and lifting a back end over a face pick. By this arrangement the air pockets in the face fabric will face the air pockets in the back fabric hopefully to form closed air chambers. Figure 2 shows one complete repeat (24 × 24) based on face and back weaves of Figure 1.

4. Experimental Work

Ten samples of the innovated double honeycomb weave were produced with two levels of weft density (16 and 22 p.p.cm) and each level of weft density was produced with five weft yarn counts (14, 12, 10, 8, and 6 Ne). For the sake

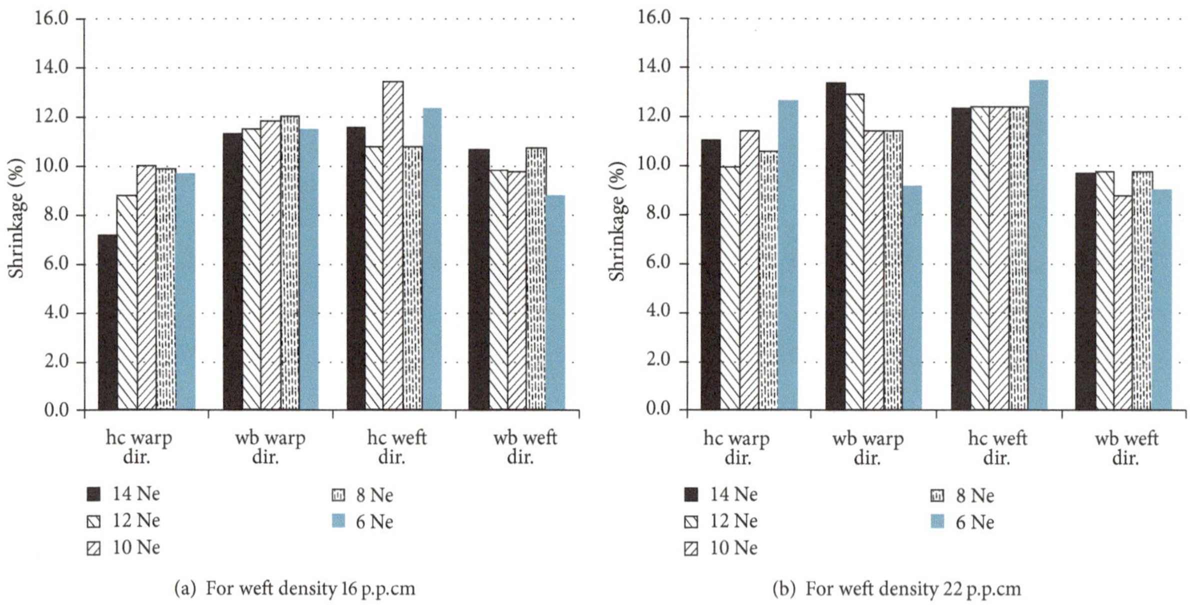

FIGURE 3: Shrinkage percent for double honeycomb and reversible weft backed weave in warp and weft direction at two levels of pick density.

of comparison, other ten samples of reversible weft backed fabric were produced with the same previous densities and counts of weft yarn. All samples were produced from the same warp (24/2 Ne) and (24 ends per cm) on sulzertextil TPS 600 weaving machine. The two structures were designed with almost equal average float.

All samples were washed three cycles by using normal cotton program at a temperature of 90°C. Geometrical and physical properties were measured after complete fabric relaxation. Thickness was measured according to ASTM D1777. Heat insulation was measured according to ASTM D 1518. Air permeability was measured according to ASTM D 737. Fabric stiffness was measured according to ASTM D4032. Abrasion resistance was measured according to ASTM D 3885. Water vapor permeability was measured according to ASTM E 96. Then, relative water vapor permeability (RWVP) was determined from the following equation:

$$\text{RWVP}(\%) = \frac{G_S * 100}{G_o}, \quad (1)$$

where G_S = weight change with sample; G_o = weight change without sample.

The crimp percent, shrinkage percent, and the fabric weight per unit area were measured, and bulk density was calculated from the following formula:

$$\text{Bulk density}\left(\text{kg/m}^3\right) = \frac{W\left(\text{g/m}^2\right)}{t\,(\text{mm})}, \quad (2)$$

where W = fabric weight per unit area; t = fabric weight.

5. Experimental Results and Discussions

Statistical analysis of all test results of physical and geometrical properties mentioned above was carried out by using Multiway ANOVA by using SPSS program to examine the significance of the structure, density, and count of weft yarns on these properties.

5.1. Shrinkage. Figures 3(a) and 3(b) show the shrinkage percent of double honeycomb and reversible weft backed weave for 16 and 22 p.p.cm, respectively, for both warp and weft direction.

At the same count and density, shrinkage percent of double honeycomb in warp direction (hc warp dir.) is less than reversible weft backed weave (wb warp dir.) and vice versa concerning the weft direction since the shrinkage percent of double honeycomb (hc weft dir.) is higher than the reversible weft backed weave (wb weft dir.). Statistical analysis showed a significant effect of weave structure on shrinkage percent in each of the warp and weft directions, but there is no significant influence of each of the count and density of the weft thread.

5.2. Fabric Thickness. Figure 4 presents the effect of weft count on the thickness of both double honeycomb (hc) and reversible weft backed (wb) samples at weft density of 16 and 22 p.p.cm. Generally, the thickness of double honeycomb fabric is higher than the reversible weft backed with a maximum of 77% at 22 p.p.cm and weft count of 6 Ne because of the surface nature of each layer of stitched double honeycomb fabric. The thickness also decreases when weft count increases for both structures because of the inversed relation between yarn count (Ne) and its diameter. Statistical analysis showed a significant effect of fabric structure and weft count and insignificant effect of weft density on fabric thickness.

5.3. Crimp Percent. Figure 5 shows the effect of weft count at the two levels of weft densities on warp crimp percent for

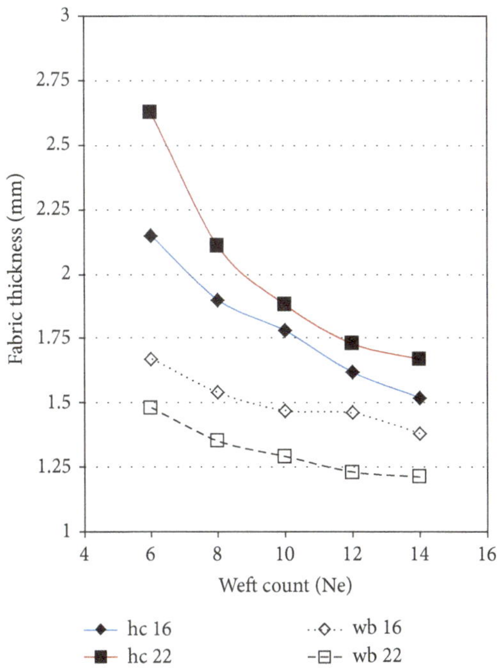

FIGURE 4: The effect of weft count on the thickness of both double honeycomb and reversible weft backed fabric at two weft densities.

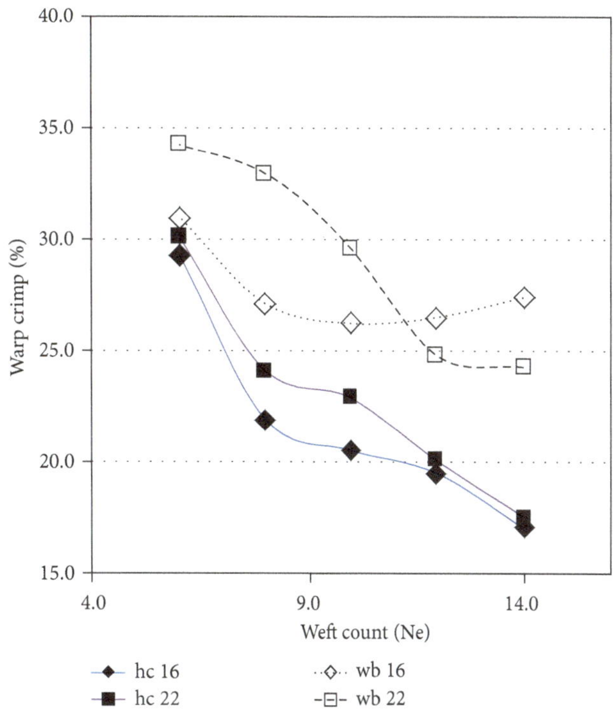

FIGURE 5: Warp crimp percent for double honeycomb and reversible weft backed samples.

both double honeycomb (hc) and reversible weft backed (wb) samples. The warp crimp percent decreased by weft count (Ne) increasing for both structures by different rates as crimp percent is greatly affected by densities and counts of crosswise threads. The crimp percent of wb structure is higher than the crimp percent of hc structure probably because warp threads in case of wb structure interlaced between two levels of weft while warp threads of each layer in hc structure interlaced within the same layer except at stitching point. Statistical analysis showed that the effect of weft count and structure on warp crimp is significant.

Figure 6 shows a slight effect of weft count on weft crimp percent for both double honeycomb (hc) and reversible weft backed (wb) samples at the two levels of weft densities as the warp count and density are the same in all samples. The crimp percent of hc sample is higher than the wb sample crimp percent. Statistical analysis showed a significant effect of fabric structure and insignificant effect of weft count on the weft crimp.

The warp crimp increases at the expense of weft crimp in the case of the reversible weft backed sample, while the weft crimp increases at the expense of warp crimp in the case of the double honeycomb sample. This is due to the crimp interchange; that is, the crimp of one direction increases at the expense of the other direction.

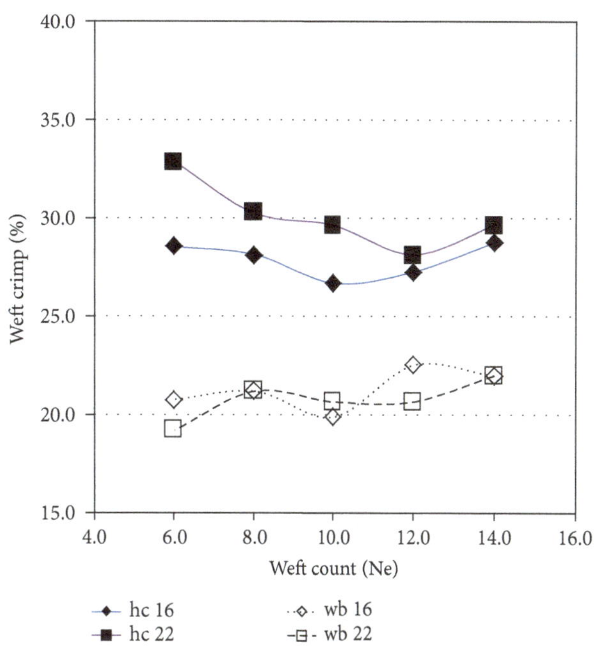

FIGURE 6: Weft crimp percent for double honeycomb and reversible weft backed samples.

5.4. Fabric Density. Figure 7 presents the effect of weft count on fabric weight on both double honeycomb (hc) and reversible weft backed (wb) samples at two weft densities.

At the same weft count and density, the weight per unit area in both structures is almost the same which means that the production cost of each structure is the same. Statistical analysis showed that the effect of structure on fabric weight is insignificant.

5.5. Fabric Bulk Density. Figure 8 shows the effect of weft count on the bulk density for double honeycomb (hc) and reversible weft backed (wb) samples at two weft densities. The

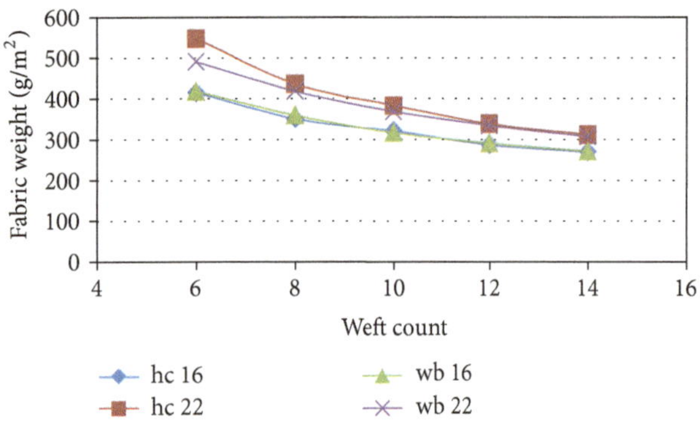

Figure 7: The effect of weft count on fabric weight of double honeycomb and reversible weft backed samples at two weft densities.

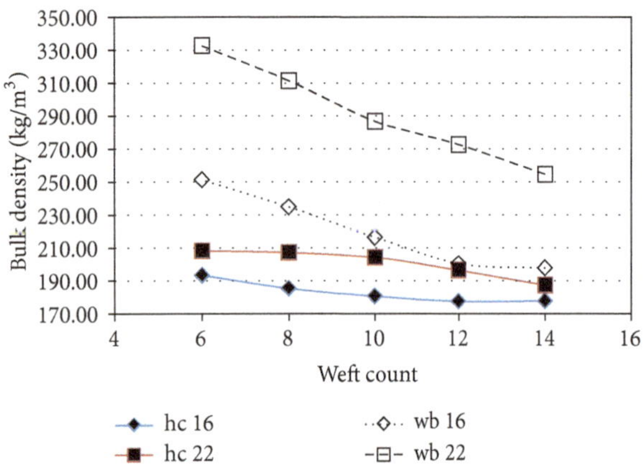

Figure 8: The effect of weft count on the bulk density of double honeycomb and reversible weft backed samples at two weft densities.

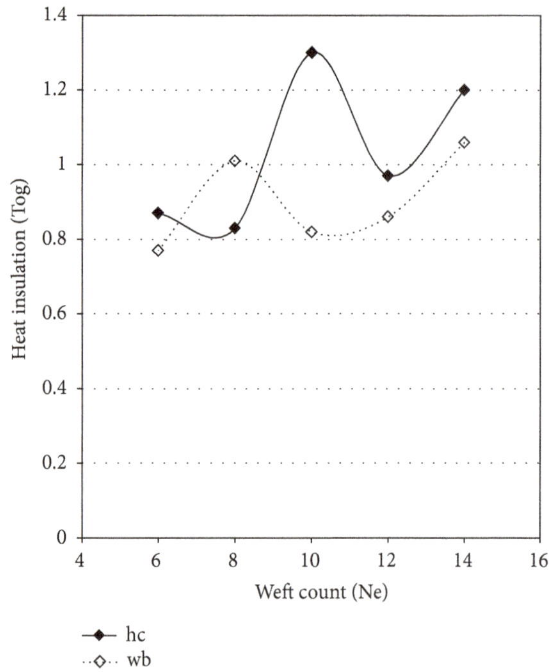

Figure 9: Effect of weft count on thermal insulation of double honeycomb and reversible weft backed samples at 16 p.p.cm.

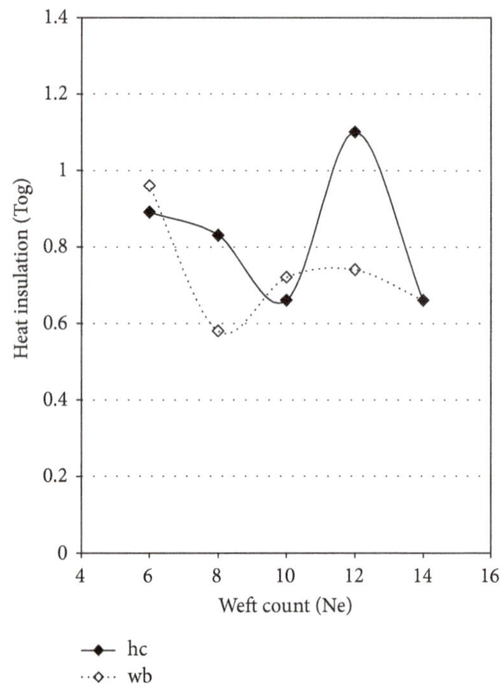

Figure 10: Effect of weft count on thermal insulation of double honeycomb and reversible weft backed samples at 22 p.p.cm.

bulk density of the weft backed cloth is higher than the double honeycomb at the same weft count and density. Therefore, the amount of air trapped inside the double honeycomb fabric is higher than weft backed cloth and this may lead to an expected increase of thermal insulation.

5.6. Heat Insulation. Figure 9 shows a fluctuation effect of weft count on thermal insulation for both double honeycomb (hc) and reversible weft backed (wb) samples at 16 p.p.cm. Thermal insulation of hc fabric increases by 59% compared to weft backed fabric at count 10 Ne.

Figure 10 shows also a fluctuation effect of weft count on thermal insulation for both double honeycomb (hc) and reversible weft backed (wb) samples at 22 p.p.cm. Thermal insulation of hc fabric is higher than wb fabric at count 12 Ne by 49% and at count 8 Ne by up to 43%.

The thermal insulation decreases by an average of 19% in both structures with increasing of weft density, as the bulk density (kg/m^3) for the produced samples from density 16 p.p.cm is less than the produced samples from density 22 p.p.cm by an average of 9% in honeycomb and by an average of 24% in weft backed sample. Therefore the produced samples with lower density (16 p.p.cm) contain more air than the produced samples with the highest density (22 p.p.cm); therefore the opportunity to produce higher insulation at lower cost could be achieved. Statistical analysis shows a significant effect of structure, weft count, and weft density on the thermal insulation. So, there are certain values of weft density and count of double honeycomb fabric to obtain

(a) Warp cross section

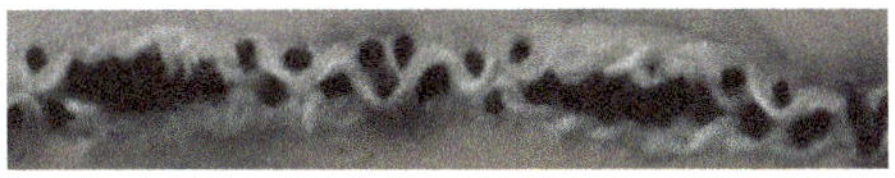

(b) Weft cross section

FIGURE 11: Cross section in double honeycomb fabric.

higher thermal insulation values than reversible weft backed fabric.

Figures 11(a) and 11(b) show warp and weft cross sections for two successive repeats of self-stitched double honeycomb at weft density 16 p.p.cm and weft count 10 Ne, respectively, since maximum heat insulation was achieved. From both figures it is clear that the fabric is not of uniform thickness, that is, fabric of less thickness at stitching point and higher thickness at maximum float. It is also clear that both warp and weft spacing are not uniform, that is, widely spacing at stitching point and closer spacing at max float. Therefore, air pockets did not appear as expected at fabric zones of higher thickness and closer spacing which might reflect on thermal properties.

5.7. Relative Water Vapour Permeability. Figures 12(a) and 12(b) present the effect of fabric structure on the relative water vapor permeability at 16 and 22 p.p.cm, respectively. In spite of the presence of two layers of the double honeycomb fabric, the water vapor permeability was almost equal to the backed cloth. Statistical analysis showed that the effect of structure, count, and density is insignificant on the relative water vapor permeability. The increase of thermal insulation by 59% with maintaining water vapour permeability means better thermal comfort.

The most insulated double honeycomb sample (at 16 p.p.cm and weft yarn count of 10 Ne) was tested for abrasion resistance, fabric stiffness, crease resistance, and air permeability. The test results were compared with reversible weft backed sample at the same weft yarn count and density. The abrasion resistance of double honeycomb fabric was higher than reversible weft backed fabric by 48.6% which could be interpreted by the honeycomb (hc) fabric that is not a uniform thickness (two levels at stitching points and at maximum float) while the thickness of weft backed (wb) fabric is a uniform thickness. Therefore, the area exposed to the abrasion in hc fabric is less than the area in wb fabric; so the weight loss due to abrasion for double fabric is less than the weight loss for backed fabric. Flexural rigidity of hc fabric was less than wb fabric by 52% in the weft direction while it increased by 40.5% in the warp direction. The crease resistance of hc fabric was less than wb fabric by 9.6% in the weft direction, while in the warp direction it decreased by 3.4%. Air permeability for both structures was almost the same.

Therefore, from all samples and experimental tests, weft count of 10 Ne and weft density of 16 p.p.cm were

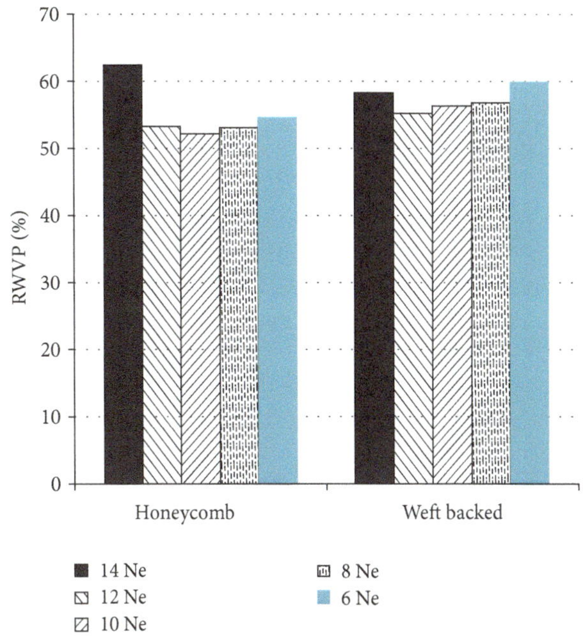

(a) For weft density 16 p.p.cm

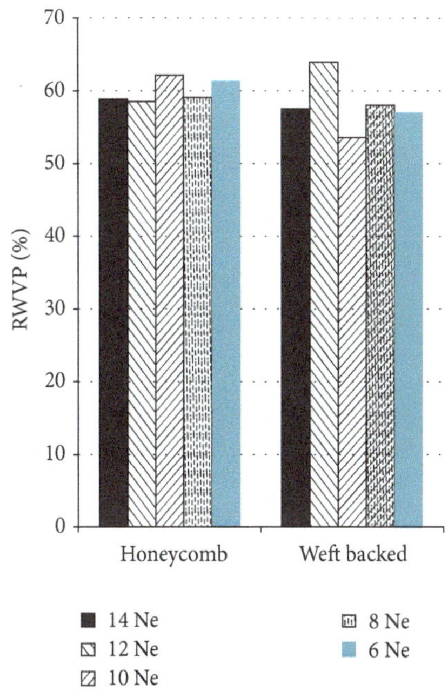

(b) For weft density 22 p.p.cm

FIGURE 12: Influence of double honeycomb and reversible weft backed weave on relative water vapor permeability at two levels of pick density.

recommended to produce a good bed cover from double honeycomb compared with reversible weft backed fabric made from the same weft yarn.

6. Conclusion

(i) An attempt to innovate a fabric structure that could be used to produce a coverlet based on self-stitched double honeycomb was made.

(ii) The innovated structure was compared to reversible weft backed structure made from the same weft counts and densities.

(iii) Double honeycomb fabric has higher values of thermal insulation than reversible weft backed fabric at certain weft counts and densities.

(iv) In spite of the presence of two layers of double honeycomb, its water vapor permeability is almost equal to the reversible weft backed fabric.

(v) The thermal comfort of the innovated structure could be improved by using bulky filament yarns as wadded weft threads to pry the fabric layers.

Conflict of Interests

The authors declare that there is no conflict of interests regarding the publication of this paper.

References

[1] A. K. Haghi, S. Thomas, and L. A. Pothan, *Heat and Mass Transfer in Textiles: Theory and Applications*, WSEAS Press, 2007.

[2] B. P. Saville, *Physical Testing of Textiles*, North America by CRC Press LLC 2000 Corporate Blvd, NW, Boca Raton, Fla, USA, 2000.

[3] C. A. Wilson, R. M. Laing, and B. E. Niven, "Estimating dry thermal resistance of multiple-layer bedding materials—re-examining the problem," *Journal of the Human-Environment System*, vol. 2, no. 1, pp. 69–85, 1999.

[4] T. Fukazawa, G. Lee, T. Matsuoka, K. Kano, and Y. Tochihara, "Heat and water vapour transfer of protective clothing systems in a cold environment, measured with a newly developed sweating thermal manikin," *European Journal of Applied Physiology*, vol. 92, no. 6, pp. 645–648, 2004.

[5] X. Qian and J. Fan, "Interactions of the surface heat and moisture transfer from the human body under varying climatic conditions and walking speeds," *Applied Ergonomics*, vol. 37, no. 6, pp. 685–693, 2006.

[6] Z. Lin and S. Deng, "A study on the thermal comfort in sleeping environments in the subtropics—measuring the total insulation values for the bedding systems commonly used in the subtropics," *Building and Environment*, vol. 43, no. 5, pp. 905–916, 2008.

[7] M. Matusiak and K. Sikorski, "Influence of the structure of woven fabrics on their thermal insulation properties," *Fibres & Textiles in Eastern Europe*, vol. 19, no. 5 (88), pp. 46–53, 2011.

Effect of Fabric Layers on the Relationship between Fabric Constructional Parameters and Percentage Reflectance Values of Polyester Fabrics

Mine Akgun, Behcet Becerir, and Halil Rifat Alpay

Textile Engineering Department, Faculty of Engineering, Uludag University, Gorukle, 16059 Bursa, Turkey

Correspondence should be addressed to Behcet Becerir; becerir@uludag.edu.tr

Academic Editor: Phillip W. Gibson

This paper focused on the assessment of the relation between constructional properties and percentage reflectance values of fabrics woven from polyester yarns through fabric layer numbers. Reflectance measurements were conducted on pretreated but undyed fabric samples at five different fabric layers. Twelve polyester fabrics having different constructional parameters were used and fabrics differed from each other by their weft yarn densities, weave patterns, and weft yarn filament fineness. Warp yarn properties (type, count, and density) were the same at all the fabrics. Percentage reflectance values of the fabrics changed according to yarn density, weave pattern, and filament fineness in accordance with fabric layer numbers during reflectance measurement. Percentage reflectance values gradually increased as fabric layer numbers increased. The highest reflectance values were obtained at 16 layers of fabric. The effects of single constructional parameters on reflectance values disappeared as fabric layer numbers increased. Percentage reflectance values were analyzed according to ANOVA (Analysis of Variance) and statistical results revealed the cross relations obtained. Light-trap phenomenon was discussed according to reflectance characteristics of woven fabrics.

1. Introduction

Color measurements on opaque surfaces (textile fabrics) are made by measuring the percentage reflectance of surfaces and the measured reflectance values are transferred to color computing software for the determination of color coordinates and other color related properties, that is, whiteness. Whiteness of textiles is an important aspect which is considered in daily life. White textiles are usually preferred by customers especially in summer because of comfort reasons. Also white, being an achromatic color, has a physiological effect on human sensation.

Fibers which are used in textile fabric production for domestic purposes have intrinsic colors of their own mainly given at the production stage. Polyester fibers, being synthetic ones, are usually produced in white color from matt to bright by the usage of delustering agents such as titanium dioxide. The white color of fabrics produced from polyester yarns could be changed to a whiter shade by wet treatments using textile auxiliary agents or by applying fluorescent whitening agents (FWAs) to fabrics. Depending on the domestic usage, polyester fabrics could be used without any application of FWAs.

When the constructional properties of fabrics are considered in relation with their color, it is recognized that the visual color could change according to the effect of constructional properties on appearance of fabric samples. Under these circum-stances appearance properties of fabric structures become important.

Appearance is a variable of the perceptual world detected by the visual system. Characterization of appearance, new approaches to appearance definition, past considerations, and appearance related subjects were well documented and referenced by Lozano and Caivano [1, 2]. According to Lozano and Caivano, appearance can be divided into three components: color, cesia, and spatial variation or spatiality. Also color is divided into three parts: lightness, chroma, and hue. The last two variables form what we call chromaticity. Lightness is related to cesia when it is converted to luminosity, also giving place luminous transmittance or reflectance. Color will include whiteness and yellowing and is related to optical power spectral properties of the stimulus detected by observers [1].

In most cases color is an important factor in the production of the material and is often vital to the commercial success of the product. The color of an object depends on many factors, such as lighting (illumination), size of sample, and background and surrounding colors. Much more importantly, color is a subjective phenomenon and depends on the observer [3, 4].

When a beam of light encounters the surface of an object, a small portion of it is reflected and does not penetrate. The actual amount of reflected light is dependent upon the smoothness of the surface, the refractive index of the material, and the angle at which the beam strikes the surface. The light striking an object will be affected by interaction with the object in a number of different ways. The resulting light distributions give us our expressions of what that object looks like. The diffuse reflection caused by the scattering of light within a material, where no selective absorption by dyes and pigments is involved, causes a white appearance. In the majority of objects specular (shiny) reflection, diffuse reflection by scattering, and absorption operate which result in the color and appearance of the objects [5].

Steen and Dupont [6] researched the reflectance measurements of the highly structured white textiles. White can be defined as an aspect of color corresponding to a high luminosity and an absence of the hue. The impression of whiteness of a surface implies two conditions: a diffuse reflectance which is as flat as possible (achromatic) and the highest possible reflectance values. A white textile can rarely be compared to a sheet of white paper or a flat surface which is painted white. A textile surface is often highly structured, permeable to light, and composed of different materials. Highly structured surfaces (lace and embroidery in the lingerie sector) can mean that difficulties arise when measuring the color or whiteness of such a material.

There is also another phenomenon linked to structured white textiles, known as the light-trap rule. A so-called "flat textile" is structured with more or less regularly spaced hollows and bumps. This regularity allows good colorimetric measurements to be made in certain conditions (i.e., maximum aperture, sample layered until it is opaque, and multiple measurements). These simple rules mean that it is possible to limit the differences between two measurements of the same flat textile article. Nevertheless, each hollow between the yarns or fibers is a light trap which darkens the color of the fabric [6].

It was concluded by Steen and Dupont that the lighter a textile was, the greater the influence of its physical structure on the color would be and the less tolerant they would be in accepting it. With regard to whites, they observed light-trap phenomenon for a large number of different structures. The magnitude of the phenomena was linked to the physical structure of the textiles. Steen and Dupont showed that the distribution of the measurements of a white textile sample gets more dispersed as the sample gets more structured.

Relations between the physical structure of textile surfaces and light reflectance through fiber fineness were researched by Kobsa et al. and Rubin et al. [7, 8] in detail. The greater amount of the reflected light which appeared from textile surfaces containing microfiber (microdenier) yarns was attributed to the peculiar properties of these fibers during production, to their physical forms, and to their physical existence and settlement in textile structures. The light-trap phenomenon was discussed from another aspect related with microfiber properties.

The relation between fabric cover factors and light reflectance of polyester fabrics was researched by Akgun et al. [9]. A novel formula was proposed and its results were tested in accordance with reflectance measurements.

Many research works [1, 2, 6–12] showed that physical structure of fabric samples strongly affects reflectance measurements and color results obtained. When reporting about color of a fabric surface, the physical structure of fabric surface surely plays an important role in obtaining the results. The complex phenomena of color when altered with physical structure implied that color could be considered as a subgroup of appearance. More strict rules might be needed in color measurement to avoid the results to be dependent more on physical structure. Also the term of "appearance measurement" could be proposed to measure the reflectance properties of the highly structured textiles.

This paper focused on the assessment of the relation among constructional properties and percentage reflectance values of fabrics woven from polyester yarns through fabric layer numbers. The relation phenomenon between physical structure and percentage reflectance values was considered from the view point of fabric layers of woven fabrics. Percentage reflectance values of fabrics changed according to the yarn density, weave pattern, and filament fineness in accordance with fabric layer numbers during reflectance measurement. Cross relations among fabric constructional properties and fabric reflectance values were obtained.

2. Materials and Methods

2.1. Materials. Fabrics consist of weft yarns having the same yarn count but different filament numbers (filament counts), and yarn densities were used in the experimental part. Also different weave patterns were considered to assess the relations among fabric percentage reflectance values and properties of fibers, yarns, and fabrics through fabric layer numbers.

Twelve polyester woven fabrics with different constructional parameters were used. Fabrics were woven under controlled mill conditions in order to obtain exact constructional properties. After weaving, woven fabrics were pretreated (washed with 2 g/L nonionic agent at 60°C for 30 minutes and later stentered without tension at 180°C for 60 seconds) under mill conditions and prepared for reflectance measurement in the laboratory. Constructional parameters of the fabrics were given in Table 1. Twill and sateen fabrics were weft faced ones. Fibers of the yarns had round cross-sectional shapes and they were semidull fibers (commercially available from a spinning mill in Bursa). Warp yarn properties and warp densities were the same for all the fabrics under consideration. Weft yarn properties differed from each other in terms of weft yarn densities and weft yarn filament fineness.

TABLE 1: Constructional parameters of the fabrics.

Fabric weave	Warp count [denier/filament]	Warp density [thread/cm]	Weft count [denier/filament]	Weft density [thread/cm]
Plain	70/36	60	100/36	15
				20
Plain	70/36	60	100/144	15
				20
1/2 twill	70/36	60	100/36	15
				20
1/2 twill	70/36	60	100/144	15
				20
1/5 sateen	70/36	60	100/36	15
				20
1/5 sateen	70/36	60	100/144	15
				20

Reflectance measurements were made on ready-to-dye white polyester fabric samples. Two different weft yarns with 100-denier yarn count were used composed of 36 and 144 filaments, respectively. 100/36 (yarn denier/filament number) yarn could be considered as polyester yarn with conventional fiber fineness (~2.7 denier per filament) and 100/144 (yarn denier/filament number) yarn could be considered as polyester yarn with microfiber (microdenier) fiber fineness (~0.60 dpf). The fabrics were produced at two weft densities, namely, 15 and 20 threads/cm. These weft densities were recalled in "Results and Discussions" part as 15- and 20-thread weft yarn density.

2.2. Methods. Percentage reflectance values of the samples were measured on a Macbeth Reflectance Spectrophotometer (MS 2020+) between 400–700 nm (at 20 nm intervals) under a D65/10° illuminant and in SCI (Specular Component Included) mode. Percentage reflectance values were recorded according to AATCC Evaluation Procedure 6 [13].

Samples were prepared as four replicas for reflectance measurement. Four reflectance measurements were made on each replica with a sample rotation of 90° and percentage reflectance of a sample was calculated by using total sixteen measurements. Reflectance measurements of the replicas were made at different fabric layers (from the single layer up to 2, 4, 8 and 16 layers) on the achromatic black background (black background color coordinates; average %R: 4.88; average K/S: 9.27; L^*: 26.3; a^*: 0.04; b^*: −0.29; C^*: 0.29; h°: 277.69).

3. Results and Discussions

Percentage reflectance values of fabric samples according to increasing fabric layer numbers were presented in Figures 1–16 for the evaluation of effects of yarn densities, weave patterns, and filament fineness on reflectance properties. Fabric samples were laid on an achromatic black background during reflectance measurements in order to absorb all illumination which passed through the samples. The usage of calibration tile (as a white background) disabled the reflectance results to be clearly obtained.

A general consideration of all the figures showed that percentage reflectance values obtained at fabrics with single fabric layer gradually decreased as the wavelength increased. The reason could be the single layer which could not reflect all the light which fell upon it during illumination because of the black background. The insufficient cover of single layer fabrics could not reflect the illuminating light and some part of the light was primarily absorbed by the black background. The D65/10° illumination is high at the blue region but it is considerably low in the red region of the spectrum. The illumination distribution during measurement could have let the reflectance distribution be obtained in such a characteristic manner in single layers. Also the comparatively higher reflectance values of 20-thread weft yarn density at single fabric layer supported the results. The intrinsic light reflection properties of the single layer fabrics were obtained.

A general similarity of all the figures was that percentage reflectance values gradually increased as fabric layer numbers and wavelengths increased. Percentage reflectance—wavelength distribution of the one and more layered white fabrics showed the same characteristics of white surfaces as given by McDonald [3]. Percentage reflectance values of 16 layers of fabrics were a little higher than those obtained at 8-layer fabrics. Distinct differences in percentage reflectance values were obtained among single and up to 16-layer fabrics.

As pointed in AATCC [13] technical manual most textile samples are nonopaque to some extent and it is advisable to layer the material until light no longer penetrates through layers. The required thickness of any given specimen can be determined by measuring it with increasing layers until no further change in the reflectance data is observed. The results of experimental part showed that minimum number of the layers in reflectance measurement of white polyester fabric samples must be 8 or 16 fabric layers. However, cross relations were obtained because of the light-trap phenomenon of the fabric surfaces. The results and discussions given below indicated the reflectance behavior of 16 layers of fabric otherwise pointed.

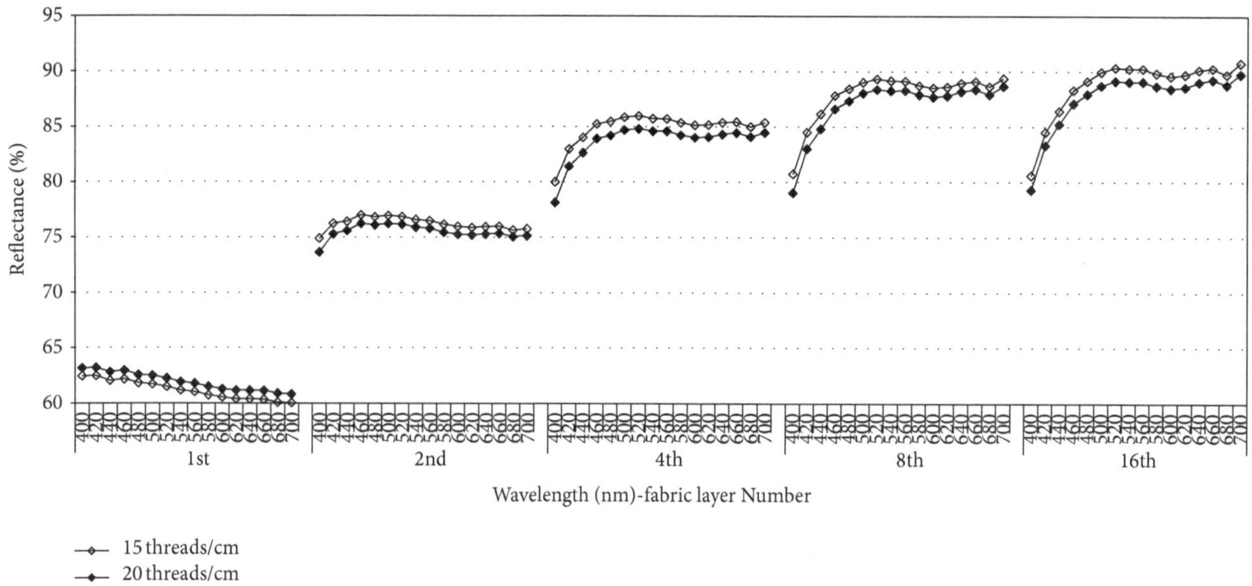

FIGURE 1: Percentage reflectances according to weft density (yarn count: 100/36 and weave pattern: plain).

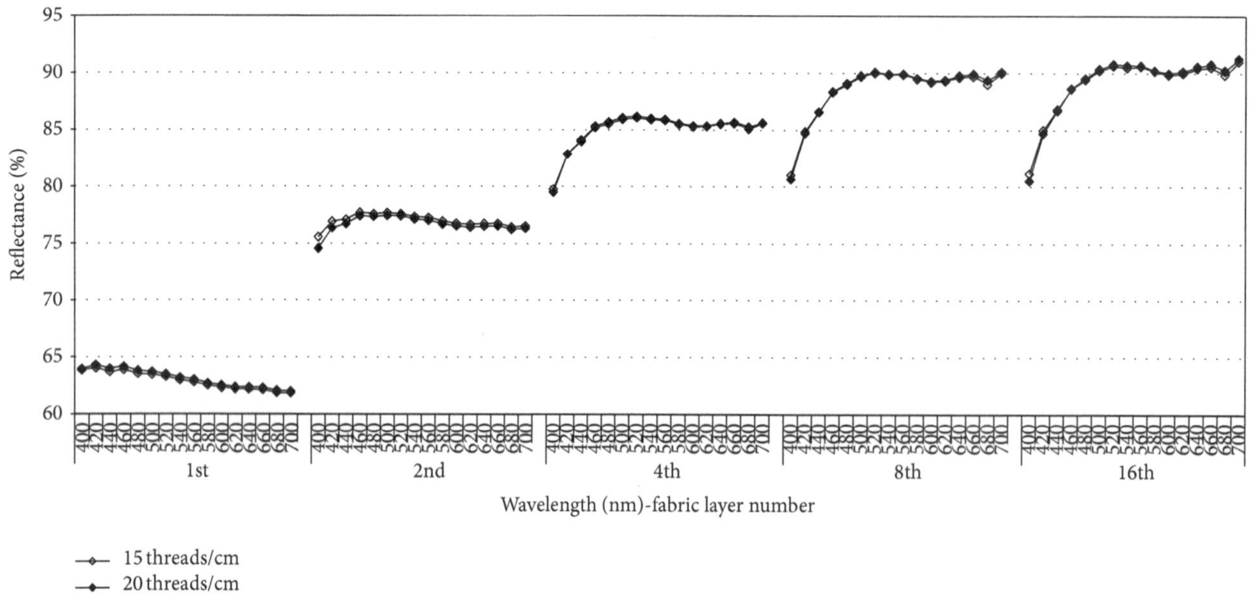

FIGURE 2: Percentage reflectances according to weft density (yarn count: 100/36 and weave pattern: 1/2 twill).

3.1. Effect of Yarn Densities on Percentage Reflectance. Effects of yarn densities on percentage reflectance values of plain, twill, and sateen fabrics were presented in Figures 1–3 for 100/36 yarns and in Figures 4–6 for 100/144 yarns. In Figures 1–3, reflectance values of samples woven with conventional fibers were presented. Percentage reflectance values of sateen weave were slightly higher than those of plain and twill weaves at single, 2, and 4 fabric layers. Also percentage reflectance values of plain and sateen weaves woven with 15-thread weft yarn density were higher than that of samples woven with 20-thread weft yarn density as the fabric layer numbers increased. The difference between percentage reflectance values increased as fabric layers increased. Percentage reflectance values of twill fabrics presented in Figure 2 were the same for both weft yarn densities (15- and 20-thread weft yarn density).

In Figures 4–6, reflectance values of samples woven with microfibers were presented. In Figure 4, percentage reflectance values of samples woven in plain weave were almost the same for the two weft yarn densities. However, percentage reflectance values of samples woven in twill and sateen weaves were much different from each other especially

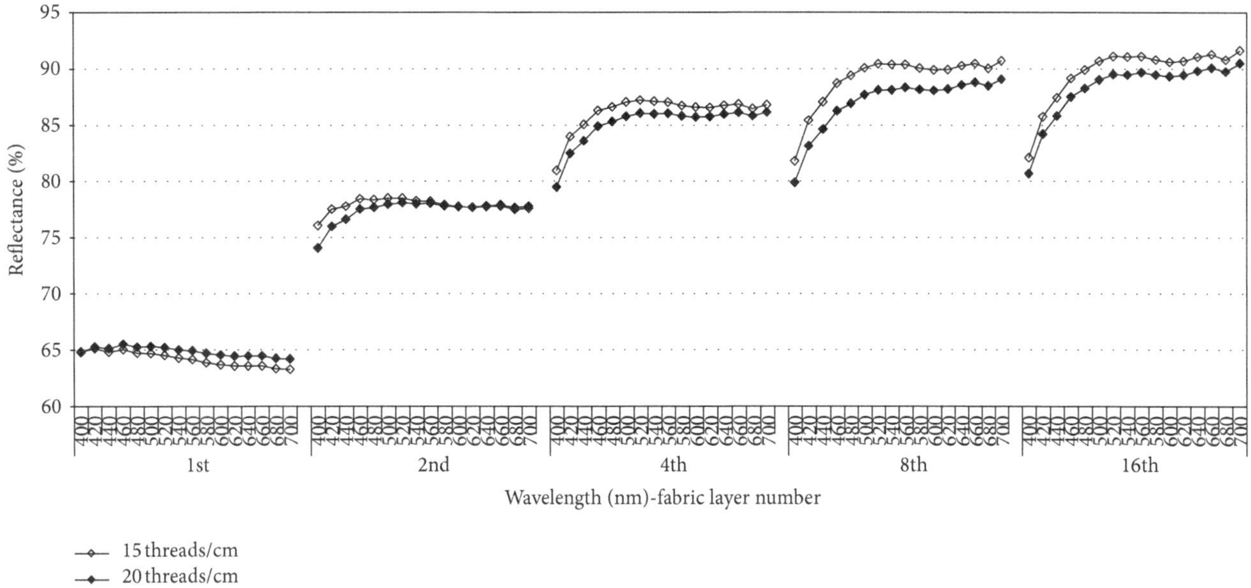

FIGURE 3: Percentage reflectances according to weft density (yarn count: 100/36 and weave pattern: 1/5 sateen).

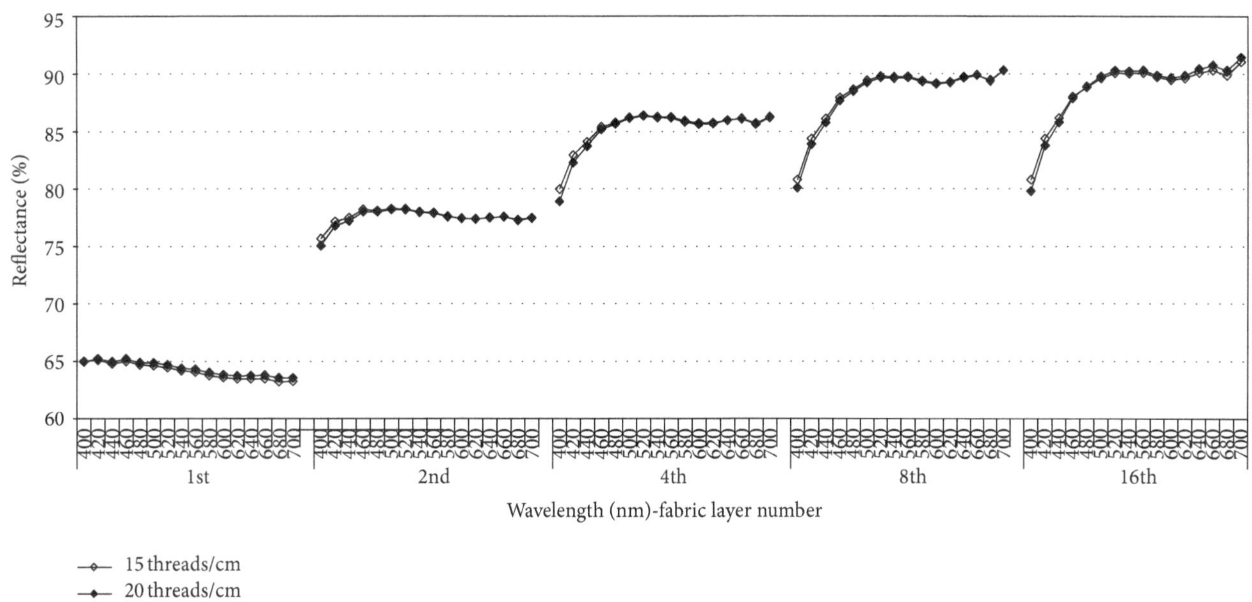

FIGURE 4: Percentage reflectances according to weft density (yarn count: 100/144 and weave pattern: plain).

at single fabric layer. In Figure 6, the difference in reflectances decreased as fabric layer numbers increased.

The higher percentage reflectance values of 15-thread weft yarn density fabrics obtained in Figures 1 and 3 at 2 to 16 fabric layers implied that more regular surfaces were obtained in plain and sateen fabrics. This regularity could be a unique result of the combination of yarn density and weave pattern. High yarn densities would be expected to cause more remarkable light reflection than low densities because of the closer settlement of yarns in fabric structure. But this was not observed in Figures 1, 3, and 5. The increase of fabric layers in reflectance measurement maximized the effect of yarn densities on fabric percentage reflectance values because distinct differences were obtained between percentage reflectances as the fabric layers increased from 1 to 16. This could be discussed as the presence of a cross relation between the constructional parameters and percentage reflectances, not a direct one. Percentage reflectance values obtained in Figures 1–6 showed cross relations among weave pattern, yarn density, and filament fineness on light reflection properties of fabric samples.

3.2. Effect of Weave Patterns on Percentage Reflectance. Effects of weave patterns (plain, twill, and sateen) on percentage

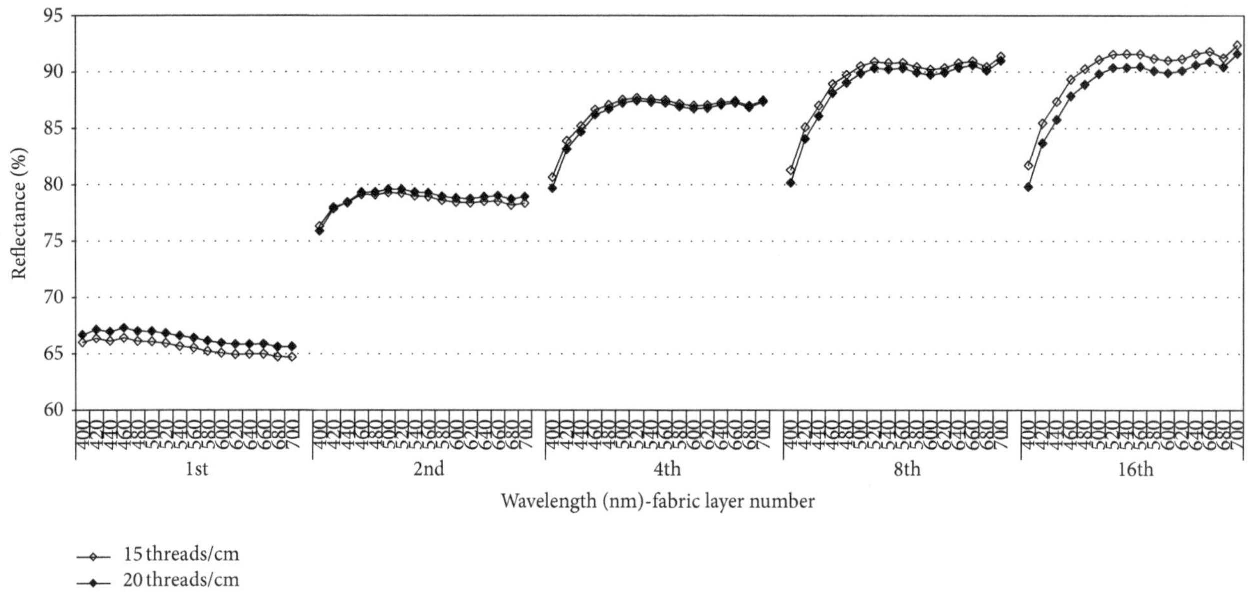

Figure 5: Percentage reflectances according to weft density (yarn count: 100/144 and weave pattern: 1/2 twill).

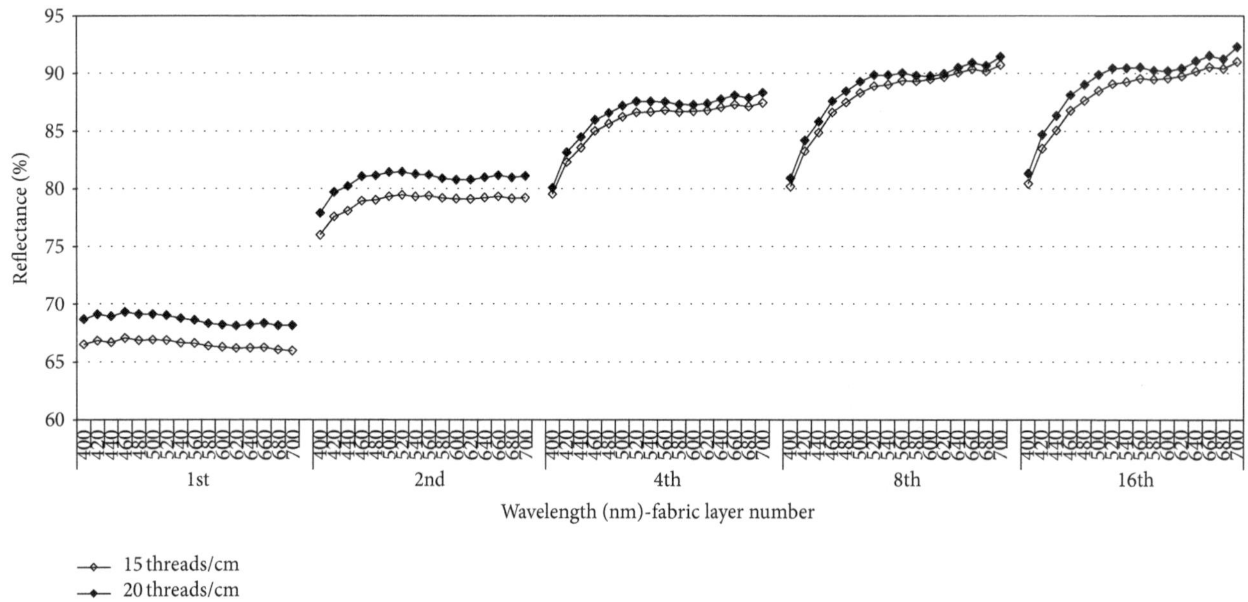

Figure 6: Percentage reflectances according to weft density (yarn count: 100/144 and weave pattern: 1/5 sateen).

reflectance values of fabric samples were presented in Figures 7 and 8 for 100/36 yarns and in Figures 9 and 10 for 100/144 yarns.

Percentage reflectance values of the fabrics produced from 100/144 yarns (Figures 9 and 10) were considerably higher than percentage reflectance values of the fabrics produced from 100/36 yarns (Figures 7 and 8) especially at the single, 2, and 4 fabric layers. The reason could be explained considering the different filament fineness of the two yarns. These results were in close relation with the former researches [7, 8]. Microfibers reflect higher proportion of light which falls upon them than conventional fibers do. Also a backscattering effect is obtained in light reflection properties of microfibers. Sateen weave fabrics could be expected to reflect more light especially when woven at high yarn densities. A cross relation was obtained which could be the result of the interaction of different constructional properties because the unique properties of sateen fabrics could not be obtained.

Effect of weave patterns on percentage reflectance values of fabrics was more apparent at single, 2, and 4 fabric layers in Figures 7-8 and in Figures 9-10. The differences which could have been caused from the weave patterns were maximized especially at single fabric layers.

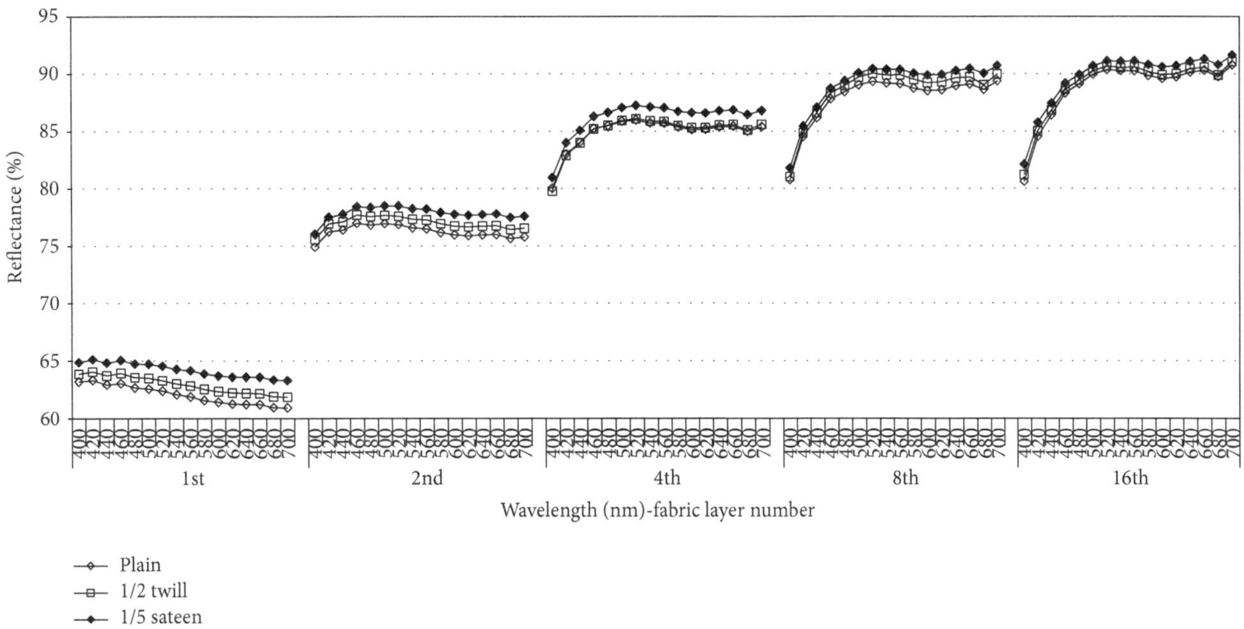

FIGURE 7: Percentage reflectances according to weave pattern (yarn count: 100/36 and weft density: 15 threads/cm).

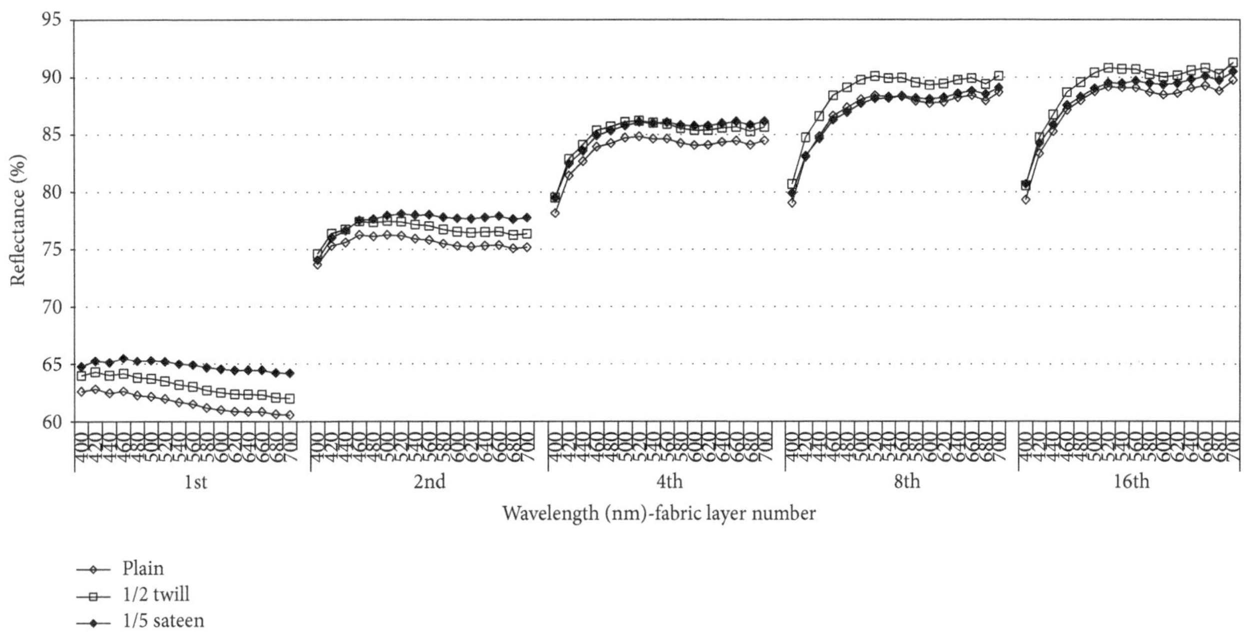

FIGURE 8: Percentage reflectances according to weave pattern (yarn count: 100/36 and weft density: 20 threads/cm).

Intrinsic reflectances of the fabric samples obtained at single, 2, and 4 fabric layers in Figures 7–10 revealed the well-known and established reflectance behaviors of the plain, twill, and sateen fabrics and microfiber yarns.

Effect of weave patterns on percentage reflectance values was found to be dependent on the filament fineness and yarn density especially at single, 2, and 4 fabric layers. The fabrics produced from conventional fibers and microfibers began to be similar to each other as the fabric layers increased in terms of fabric percentage reflectance values. An overall consideration of Figures 7–10 showed that the effect of weave patterns on percentage reflectance values changed mainly with the fabric layer numbers.

3.3. Effect of Filament Fineness on Percentage Reflectance. Effects of filament fineness on percentage reflectance values

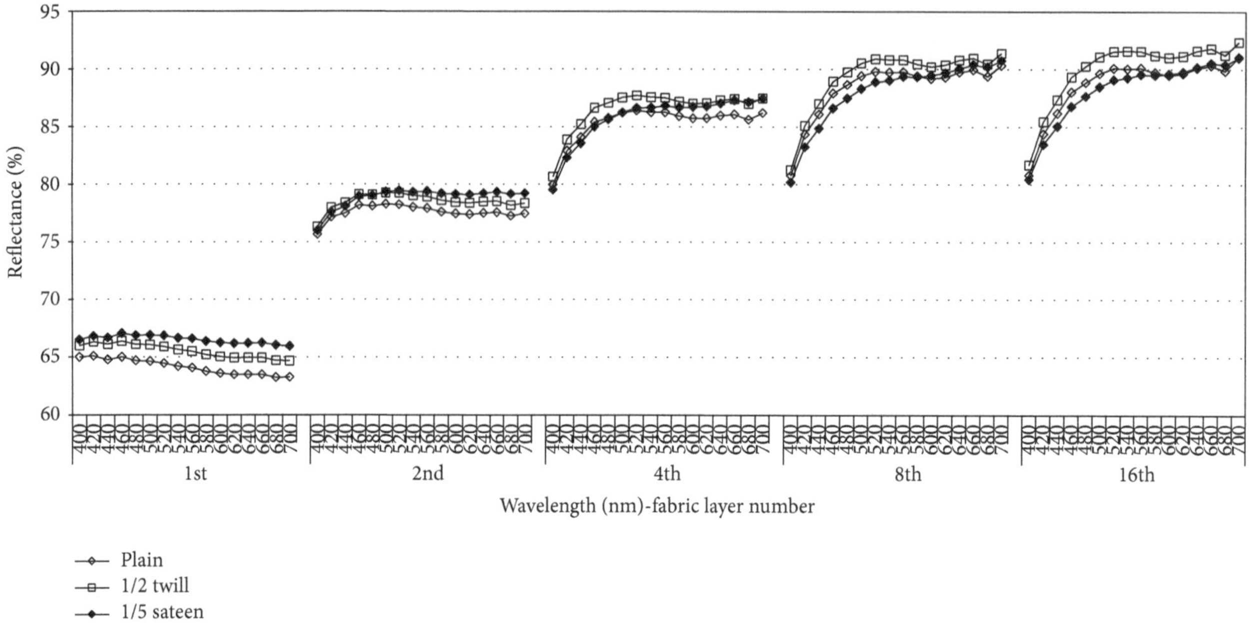

FIGURE 9: Percentage reflectances according to weave pattern (yarn count: 100/144 and weft density: 15 threads/cm).

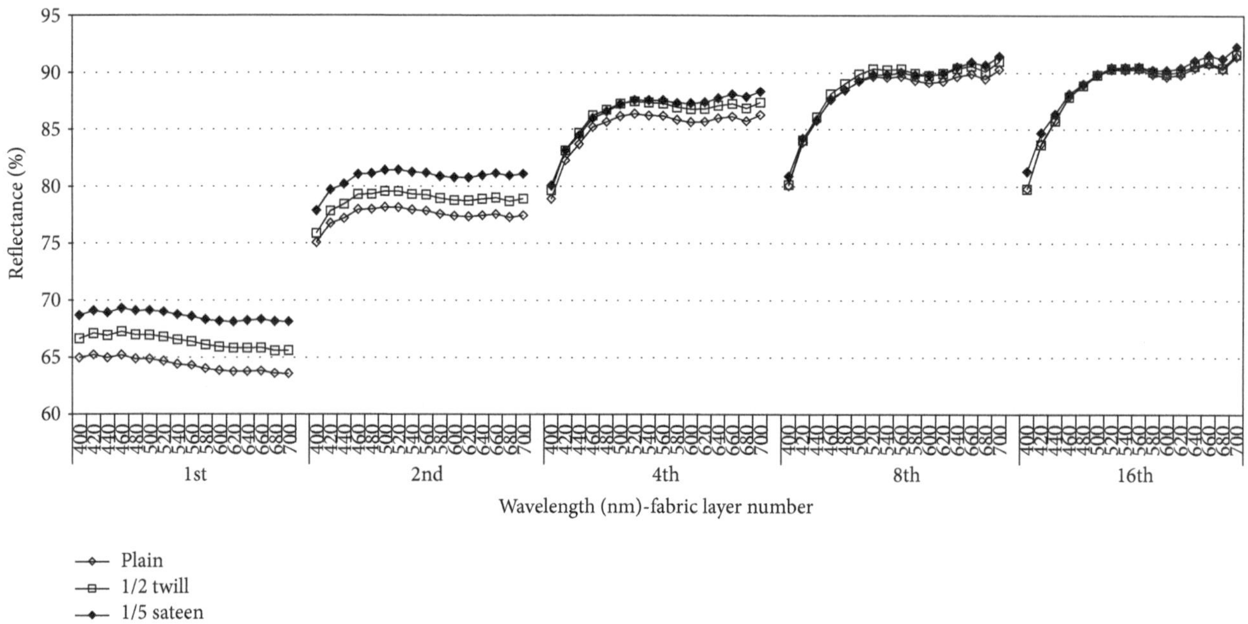

FIGURE 10: Percentage reflectances according to weave pattern (yarn count: 100/144 and weft density: 20 threads/cm).

of plain, twill, and sateen patterned fabrics were presented in Figures 11–16 for the two different weft yarn densities (15 and 20 threads/cm).

Percentage reflectance values presented in Figures 11–16 differed according to filament fineness when weave patterns were considered. Effect of filament fineness on reflectance values disappeared at plain fabrics as fabric layers increased (Figures 11 and 12). However, percentage reflectance values presented exact differences at twill and sateen weaves as fabric layers increased. Percentage reflectance values obtained at samples woven with 100/144 yarns were always higher than those of fabrics woven with 100/36 yarns in Figures 13, 15, and 16, except for 16 layers in Figure 14. In Figures 11 and 12, the increase of fabric layers hindered the effect of filament fineness on percentage reflectance values in plain fabrics. Percentage reflectance values became close to each other as

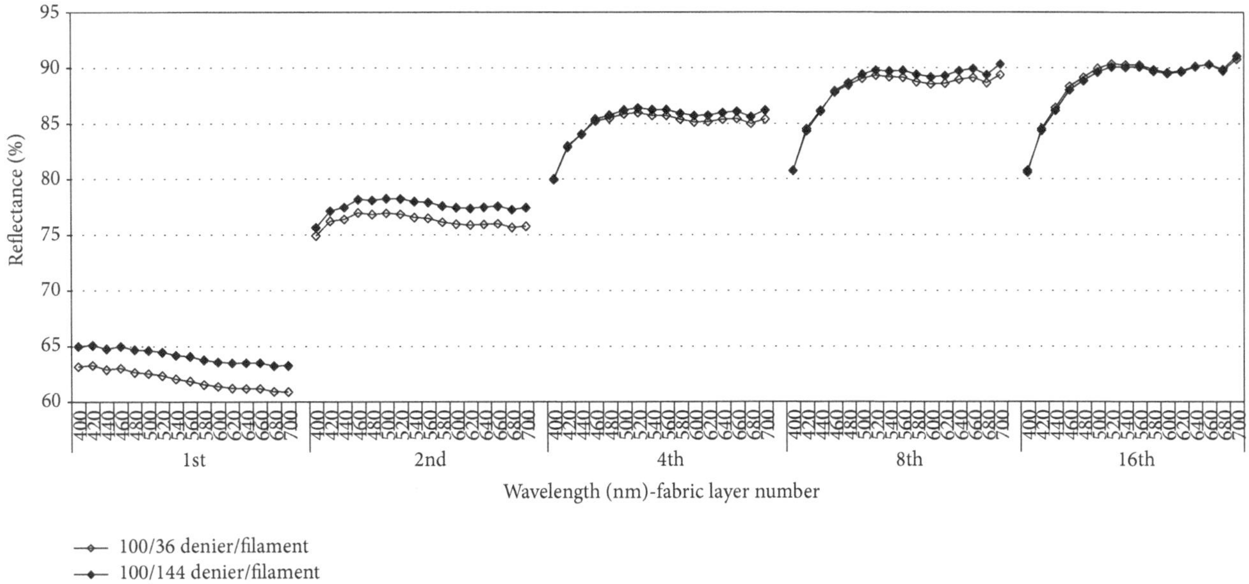

FIGURE 11: Percentage reflectances of fabrics according to filament numbers (weft density: 15 threads/cm and weave pattern: plain).

FIGURE 12: Percentage reflectances of fabrics according to filament numbers (weft density: 20 threads/cm and weave pattern: plain).

fabric layer numbers increased. Almost the same percentage reflectance values were obtained at the 16 times layered fabrics in Figures 11 and 12.

Percentage reflectance values obtained at single fabric layer in Figures 15 and 16 were considerably higher than the ones in Figures 11–14. However, percentage reflectance values obtained at 15-thread weft yarn density (Figure 15) were slightly higher than the ones obtained at 20-thread weft yarn density (Figure 16). Although both figures presented the results obtained with sateen weaves, there were considerable cross results within the figures.

When all the figures of percentage reflectance values were considered, it could be stated that direct relations could not be obtained among constructional parameters of fabrics and percentage reflectances, but cross relations.

3.4. Statistical Analysis. Statistical assessments and corresponding ANOVA (Analysis of Variance) (COSTAT Software) results were presented in Tables 2 and 3. Percentage reflectance values of fabric samples according to layer numbers in connection with the factors of weft density, filament number, and weave pattern were presented in Table 2.

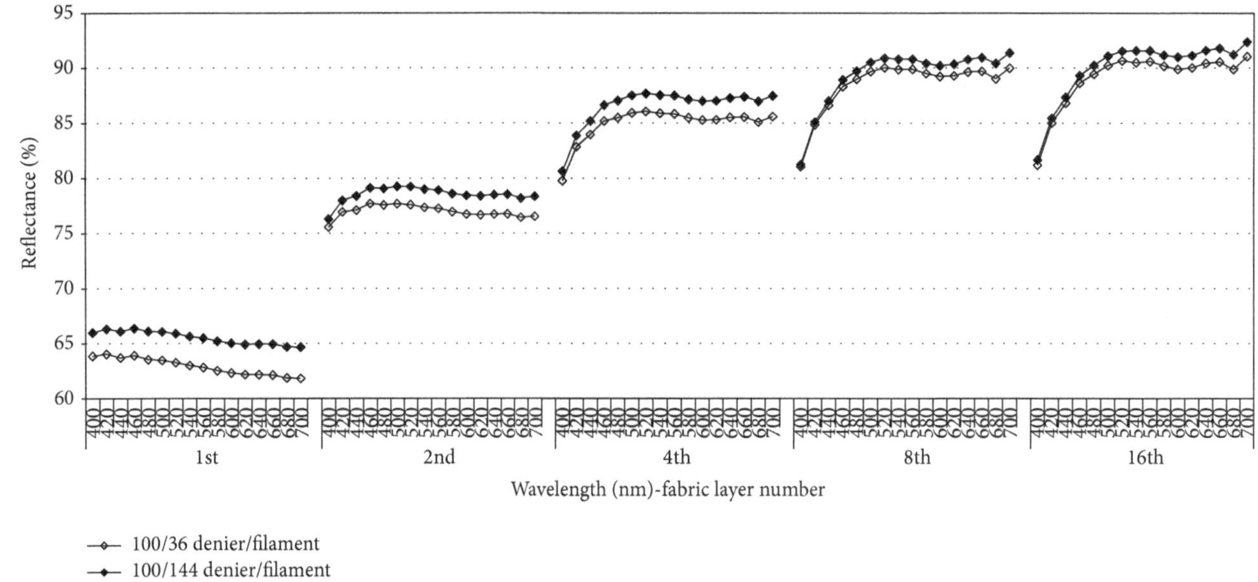

Figure 13: Percentage reflectances of fabrics according to filament numbers (weft density: 15 threads/cm and weave pattern 1/2 twill).

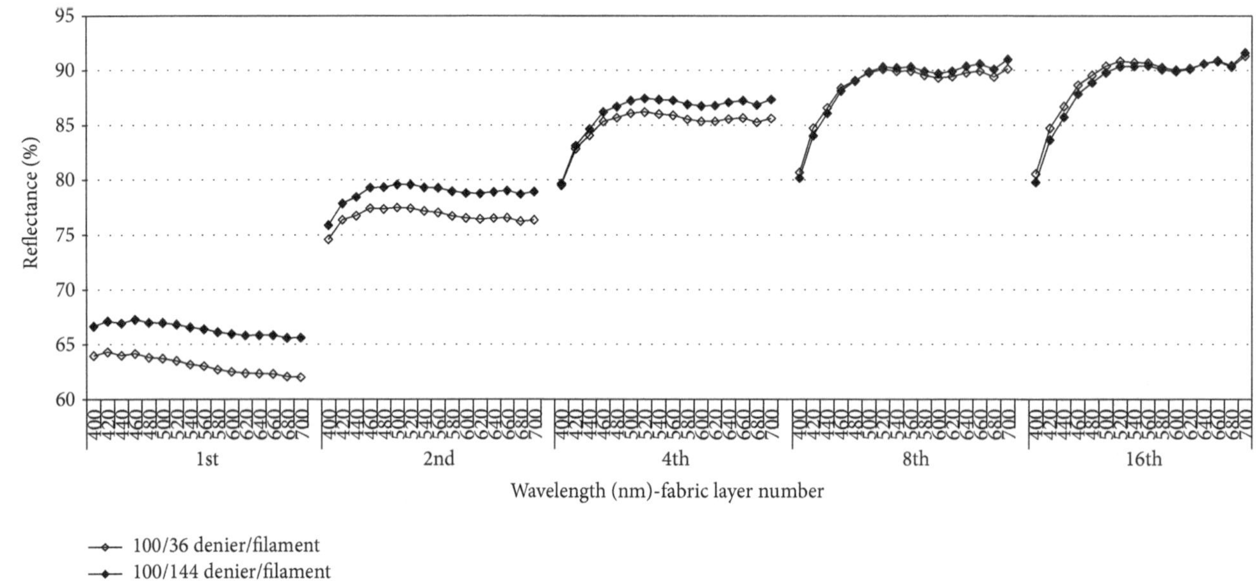

Figure 14: Percentage reflectances of fabrics according to filament numbers (weft density: 20 threads/cm and weave pattern: 1/2 twill).

The ranks were presented from the highest reflectance values to the lowest ones. Fabric samples which had the highest percentage reflectance values changed in connection with the fabric layer numbers. Table 2 revealed the cross relations among the fabric constructional properties and percentage reflectance values from the viewpoint of statistical analysis.

According to ANOVA analysis presented in Table 2, fabrics woven with microfibers (100/144 yarns) had higher percentage reflectance values than the ones woven with conventional fibers (100/36 yarns). This was an expected result because microfibers reflect incident light better than conventional fibers as a result of their dimensional characteristics [7, 8]. The backscattering effect of microfibers also plays an important role in reflectance behavior of fabrics including these fibers. Except single fabric layer number, percentage reflectance values obtained with 15-thread weft yarn density were higher than the ones obtained with 20-thread weft yarn density. This was an unexpected result. The low reflectance results of 20-thread weft yarn density could be ascribed to the light-trap phenomenon. The high reflectance results of 15-thread weft yarn density implied that a more regular fabric surface was formed. Except 8 and 16 fabric layers, sateen fabrics had the highest percentage reflectance values. In 8 and 16 fabric layers, twill fabrics had the highest percentage

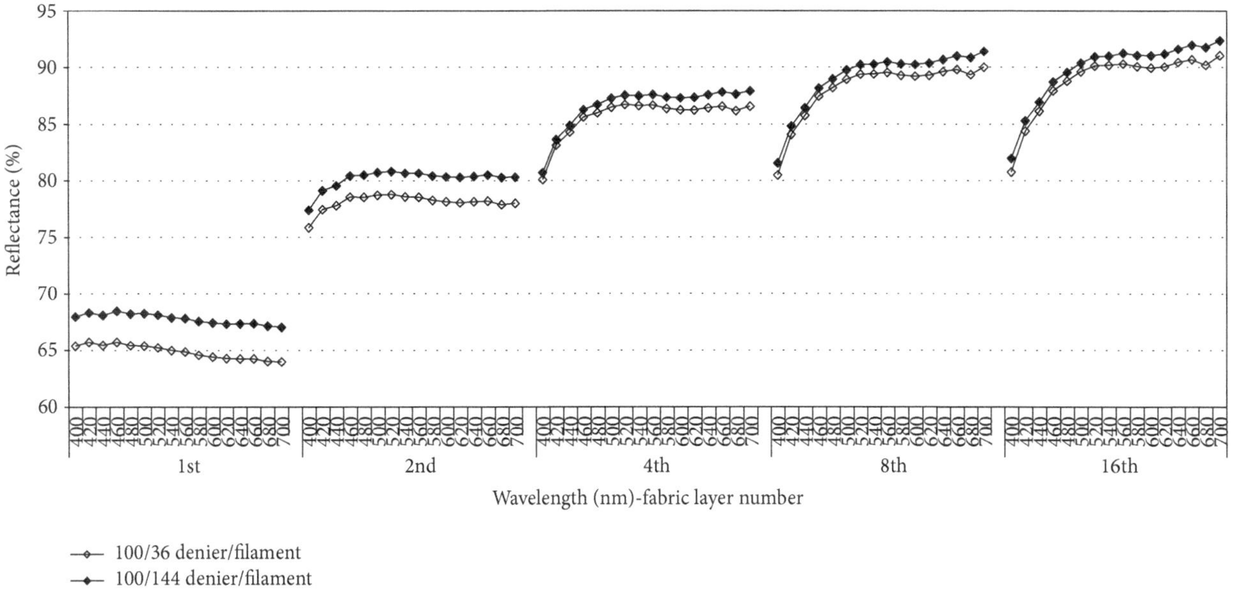

FIGURE 15: Percentage reflectances of fabrics according to filament numbers (weft density: 15 threads/cm and weave pattern: 1/5 sateen).

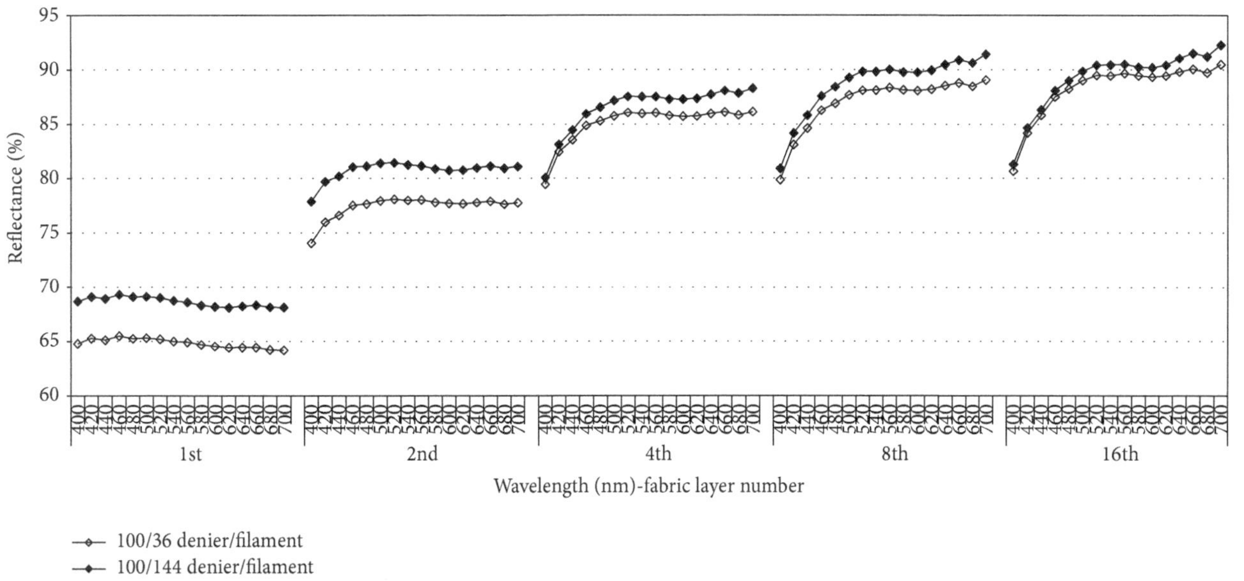

FIGURE 16: Percentage reflectances of fabrics according to filament numbers (weft density: 20 threads/cm and weave pattern: 1/5 sateen).

reflectance values. The higher percentage reflectance values of twill fabrics were another unexpected result because sateen fabrics are well known with their glossy appearance.

The higher percentage reflectance values of twill weaves accompanied with the higher percentage reflectance values of 15-thread weft yarn density implied that there were cross relations between fabric constructional properties and percentage reflectance values.

ANOVA results of percentage reflectance values according to fabric layer numbers were presented in Table 3. The highest reflectance values were obtained at the measurement with 16 fabric layers. It was obvious that percentage reflectance values increased as number of fabric layers increased under illumination. Number of fabric layers for reflectance measurement of white textiles must be at least 8 layers. But depending on filament fineness, yarn density, and weave pattern, number of layers must be 16 when very fine yarns, for example, microfibers, were considered. Also the percentage reflectance curves in the figures showed that there were significant differences between the values obtained up to eight fabric layers while the differences were much less between the values obtained at eight and sixteen fabric layers.

Yarn density played an important role on percentage reflectance values when the weave patterns changed from

Table 2: ANOVA results according to fabric constructional parameters.

Fabric layer number	Factors	Treatment	Fabric reflectance *Rank
1th	Weft density (1/cm)	1 (15)	2
		2 (20)	1
	Filament number (denier/filament number)	1 (100/36)	2
		2 (100/144)	1
	Weave pattern	1 (plain)	3
		2 (twill 1/2)	2
		3 (sateen 1/5)	1
2nd	Weft density (1/cm)	1 (15)	1
		2 (20)	2
	Filament number (denier/filament number)	1 (100/36)	2
		2 (100/144)	1
	Weave pattern	1 (plain)	3
		2 (twill 1/2)	2
		3 (sateen 1/5)	1
4th	Weft density (1/cm)	1 (15)	1
		2 (20)	2
	Filament number (denier/filament number)	1 (100/36)	2
		2 (100/144)	1
	Weave pattern	1 (plain)	3
		2 (twill 1/2)	2
		3 (sateen 1/5)	1
8th	Weft density (1/cm)	1 (15)	1
		2 (20)	2
	Filament number (denier/filament number)	1 (100/36)	2
		2 (100/144)	1
	Weave pattern	1 (plain)	2
		2 (twill 1/2)	3
		3 (sateen 1/5)	1
16th	Weft density (1/cm)	1 (15)	1
		2 (20)	2
	Filament number (denier/filament number)	1 (100/36)	2
		2 (100/144)	1
	Weave pattern	1 (plain)	2
		2 (twill 1/2)	3
		3 (sateen 1/5)	1

*The rank is presented from the highest percentage reflectance to the lowest percentage reflectance.

plain to sateen. Filament fineness played an important role on percentage reflectance values when number of fabric layers and weave patterns changed.

Reflectance behavior of different fabrics became similar to each other as the number of layers increased. Also, the same discussion could be made for the fabrics with plain weave, for the fabrics with low weft density, and for the fabric woven with conventional fibers. The light-trap phenomenon was also observed in relation with the fabric layers [6]. Loose structures with small layer numbers (1, 2, and 4 layers) caused most of the incident light to be captured by the achromatic black background and by interfabric holes. Also, the measured reflectance values decreased in fabrics with high yarn density as the layer numbers increased. This could be also attributed to light-trap phenomenon.

4. Conclusions

Effects of yarn densities, weave patterns, and filament fineness on percentage reflectance values of white polyester fabric

Table 3: ANOVA results according to fabric layer number.

Factors	Treatment	Fabric reflectance *Rank
Fabric layer number	1 (1th)	5
	2 (2nd)	4
	3 (4th)	3
	4 (8th)	2
	5 (16th)	1

*The rank is presented from the highest reflectance to the lowest reflectance.

samples were reported in accordance with fabric layer numbers. Percentage reflectance results showed that the inherited properties of white fabric samples were lost when they were layered more than eight times for reflectance measurement.

Constructional parameters affected the measured percentage reflectance values at single, 2, and 4 fabric layers at most. The differences in constructional parameters lost their effects on percentage reflectance values when fabric layer numbers increased up to 16 layers. Fabric constructional parameters showed cross relations with percentage reflectance values. Percentage reflectance values were directed in different manners by constructional parameters and it became impossible to make an exact conclusion on the single effects of these parameters on reflectance. The cross relations were also revealed by statistical results. Change of percentage reflectance values with constructional properties implied that the intrinsic reflectance properties of fabric structures must be measured at single fabric layer in order to avoid the cross relations. Considering the results which were presented in this paper and in the former papers [1, 2, 6–9, 14], the usage of the term of "appearance measurement" could be proposed instead of "color or reflectance measurement" to consider the reflectance measurements of the white and patterned fabric structures.

Textile fabrics have constructional characteristics changing from noncompact structures to compact ones. For that reason, reflectance behaviors of fabric samples depended on constructional factors (yarn density, yarn filament numbers, weave pattern, etc.) which directed the compactness of structures. The optimum layer number suitable for each fabric could be determined in the beginning of measurements. The light-trap phenomenon was observed both in fabrics with loose structures during measurement at low fabric layers and in fabrics with compact structures during measurement at high fabric layers. The determination of suitable fabric layer for reflectance measurement should be made prior to the measurement cycles to perform a better application regarding the fabric compactness.

Conflict of Interests

The authors do not have any stake in Macbeth Kollmorgen (former) or AATCC and their names appear in the paper purely in coincidental and academic nature. Their reference is made to indicate standard equipment and test method made use of in this work and in no way it advocates these brands.

References

[1] R. D. Lozano, "A new approach to appearance characterization," *Color Research & Application*, vol. 31, no. 3, pp. 164–167, 2006.

[2] J. L. Caivano, "Cesia : a system of v isual signs complementing color," *Color Research and Application*, vol. 16, no. 4, pp. 258–268, 1991.

[3] B. Rigg, "Colorimetry and the CIE system," in *Colour Physics for Industry*, R. McDonald, Ed., Society of Dyers and Colourists, Bradford, UK, 1987.

[4] R. Berns, *Billmeyer and Saltzmann's Principles of Color Technology*, John Wiley & Sons, New York, NY, USA, 2000.

[5] R. S. Hunter, *The Measurement of Appearance*, John Wiley & Sons, New York, NY, USA, 1975.

[6] D. Steen and D. Dupont, "Control of structured white textiles," *Coloration Technology*, vol. 119, no. 4, pp. 205–211, 2003.

[7] H. Kobsa, B. Rubin, S. M. Shearer, and E. M. Schulz, "Using optical ray tracing to explain the reduced dye yield of microdenier yarns," *Textile Research Journal*, vol. 63, no. 8, pp. 475–479, 1993.

[8] B. Rubin, H. Kobsa, and S. M. Shearer, "Modelling the dependence of fabric reflectance on denier per filament," *Textile Research Journal*, vol. 64, no. 11, pp. 685–689, 1994.

[9] M. Akgun, B. Becerir, and H. R. Alpay, "Assessing the relationship among fabric constructional parameters, fractional reflectances and cover factors of polyester fabrics by experimental and mathematical methods," *Fibers and Polymers*, vol. 11, no. 2, pp. 291–302, 2010.

[10] B. Becerir and M. A. Iskender, "Comparison of some process components of polyester microfibre dyeing by using mathematical and experimental methods," *Fibers and Polymers*, vol. 5, no. 4, pp. 309–315, 2004.

[11] B. Becerir, "A novel approach for estimating the relation between K/S value and dye uptake in reactive dyeing of cotton fabrics," *Fibers and Polymers*, vol. 6, no. 3, pp. 224–228, 2005.

[12] B. Becerir, "Assessment of the results of different color difference formulae under different illuminants by wash fastness tests," *Fibers and Polymers*, vol. 12, no. 7, pp. 946–956, 2011.

[13] American Association of Textile Chemists and Colorists, *AATCC Technical Manual*, vol. 74, AATCC, Research Triangle Park, NC, USA, 2005.

[14] M. Akgun, B. Becerir, and H. R. Alpay, "Abrasion of polyester fabrics containing staple weft yarns: color strength and color difference values," *AATCC Review*, vol. 6, no. 3, pp. 40–43, 2006.

21

Absolute and Relative Activity of Microencapsulated Natural Essential Oils against the Larvae of Carpet Beetle *Anthrenus flavipies* (LeConte)

Jayant Udakhe,[1] Neeraj Shrivastava,[1] Smita Honade,[1] Dhanashree Banait,[2] and Namita Sonawane[2]

[1] *Textile Chemical and Colour Department, Wool Research Association, Thane 400607, India*
[2] *Department of Fibres and Textile Processing Technology, I.C.T., Mumbai 400019, India*

Correspondence should be addressed to Jayant Udakhe; jsudakhe@wraindia.com

Academic Editor: Phillip W. Gibson

This study focuses on finding natural ecofriendly alternatives to the existing commercial *Anthrenus flavipies* resist chemicals. Eucalyptus, lavender, and citronella microcapsules were explored as natural alternatives. Chemical contents of microcapsules and fragrance releasing property were tested using gas chromatography. Absolute (proofing) and relative (repellent) activities of microcapsule treated fabrics were tested against the larvae of carpet beetle *Anthrenus flavipies* (LeConte). Proofing activity test results revealed that natural essential oils act as a deterrent for *Anthrenus flavipies*, but give lesser protection compared to commercial chemical permethrin. Repellency test results also affirmed these findings and it was observed that *Anthrenus flavipies* prefers to eat untreated fabric compared to its treated counterpart.

1. Introduction

The damage to the woollen textile products, hair, feather, and fur by wool moth (insect that feeds on woolen textiles) has been a continuing subject of investigation over the years. The damage to the woollen textiles by moth larvae throughout the world is estimated to cost millions of dollars every year. Clark [1] and Hartley et al. [2] have estimated the damage caused by a single larva to be 92.5 pounds in one year and 100 pounds in 280 days, respectively. Thus, moth proofing is very essential to wool for its long life as it is a costly commodity. Without effective moth proofing this damage would increase further and will cause permanent damage to wool which is considered as high quality natural bicomponent textile fibre. The insects usually remain undetected until the damage has occurred. Fabrics, of course, can be damaged by textile pests within a few weeks of initial infestation.

Different kinds of commercial finishing treatments are given to the wool product. Some of the colourless preparations available are Dieldrin [3], Mitin LP [4], Eulan WA [5], and synthetic pyrethroid permethrin [6, 7]. These moth proof chemicals are generally applied in a single bath during the dyeing of wool. Continuous discharge of dyeing effluents containing these chemicals provide significant toxic load to each of the affected effluent treatment systems and associated downstream river area. Permethrin has been shown to be highly toxic for aquatic arthropods, fish, and honey bees because they have lower levels of carboxyl esterase activity than mammals [8].

For any effective insecticidal (moth proof) finish to be used sustainably in the near future, it should have biodegradability as well as low mammalian and aquatic toxicity. There are natural insecticidal alternatives available such as natural pyrethrins. The natural pyrethrins which are insecticidal esters from *Chrysanthemum cinerariaefolium* (Compositae) are superior to many other synthesized insecticides in their quick-killing power, low mammalian toxicity, biodegradability, and relatively low resistance development in insects [9–11]. Many plants contain compounds that act in preventing attack from phytophagous (plant eating) insects. These chemicals fall into several categories, including repellents, feeding deterrents, toxins, and growth regulators. Most of them can be grouped into five major chemical categories, that is, nitrogen compounds (primarily alkaloids), terpenoids, phenolics, proteinase inhibitors, and growth regulators. The repellency of plant materials to certain insects has well been exploited for thousands of years by humans, most simply by hanging bruised plants in houses, a practice that is still in wide use throughout the developing countries. Plants like eucalyptus, lavender, and citronella are well reported to have insect repellent properties especially against mosquito [12–16].

The aim of this work was to study and explore natural essential oils for their activity against the larvae of carpet beetle *Anthrenus flavipies* (LeConte). Activity of these essential oils in terms of proofing and repellency is rarely reported in the literature. Microencapsulated natural essential oils for example, citronella, lavender, and eucalyptus oils were characterized for their chemical compositions using gas chromatography-mass spectroscopy (GC-MS). Wool fabrics were treated with these microcapsules and analyzed for their proofing and repellent activity along with the release of fragrance. The durability of these treatments to washing and dry-cleaning was also reported.

2. Materials and Methods

2.1. Materials. Microencapsulated citronella, lavender, and eucalyptus natural essential oils were supplied by Tanishka Products, Mumbai (India). Wool fabric was procured from Bhutti weaver's co-operative society Ltd., H.P. (India). Merino wool fabric (22.5 μm wool, warp, and weft count of 2/47 Nm, 2/2 twill weave, 52/40 EPI/PPI) was used in the study. Imerol PCJ liquid (wetting agent) was supplied by Clariant India Limited, Mumbai (India). Eulan SPA was supplied by Tanatex Chemicals India Pvt. Ltd., Thane (India). All other chemicals used were LR grade.

2.2. Fabric Scouring. In order to remove lubricants and antistatics, fabric was scoured using 0.5% (on the weight of fabric) Imerol PCJ liquid (wetting agent) at 60°C for 30 min at material to liquor ratio of 1 : 30, hydro extracted, dried at room temperature, and used for further treatments.

2.3. Application of Permethrin (Eulan SPA). Eulan SPA (commercial insects resist agent) was applied on to the wool fabric by exhaustion process. A concentration of 0.125% v/w Eulan SPA (recommended by International Wool Secretariat) was used in this study. The applied concentration was of level 3, which imparts appropriate protection to woollens to be used in warmer, drier climates against the carpet beetle *Anthrenus flavipies* (LeConte) [17]. Along with Eulan SPA, 2.5 gpl sodium acetate, 2.5 gpl acetic acid, and 5 gpl lyogen WSN (levelling agent) were used. Material to liquor ratio was kept at 1 : 20. The bath was heated up to 100°C with a rate of rise of 3°C/min and held for 60 min for complete exhaustion. Then, the sample was washed with cold water and dried at room temperaure.

2.4. Application of Microcapsules. The formulations were prepared using 0.5 gpl Imerol PCJ liquid and 20 gpl microencapsulated citronella, lavender, and eucalyptus oils. The pH of the solutions was maintained at 4.5. The wool fabric was soaked in the solution for 15 min. Afterwards the fabric was padded with a wet pick up of 80%, dried at 100°C, and cured at 130°C for 2 min.

2.5. Durability to Washing and Dry-Cleaning. For testing durability of finished fabrics, all samples were washed three times using Electrolux Wascator (Model FOM 71 CLS) laboratory washing machine with 7A washing programme as per standard test method TM 31 and fabric was also separately dry-cleaned three times as per standard test method ISO 3175-2 (1998).

3. Testing and Characterization

3.1. Particle Size Analysis. The particle size analysis was carried out using Malvern Mastersizer 2000 model with Laser light scattering principle. Microcapsules of eucalyptus, citronella, and lavender were tested in emulsion state using Hydro 2000S accessory.

3.2. Solid Capsule Contents. The microcapsules slurry (10 gm) was dried at 60°C for 6 hrs and final weight of dried capsule was taken. Solid content was determined from initial and final weight of the capsule slurry.

3.3. Extraction of Essential Oils from Capsules and Fabric Samples. The dried capsules were crushed using Retsch MM 400 mixer mill machine in the presence of hexane using 10 mm stainless steel balls. The frequency of oscillation was kept at 20 cycles/second for 15 min; similar method was used for extraction of essential oil from unwashed, washed, and dry-cleaned fabrics. The extracts were then analyzed using GC-MS to determine the essential oil contents in capsules as well as in fabric samples.

3.4. Gas Chromatography Mass Spectrometry (GC-MS) Analysis. The chemicals present in eucalyptus, citronella, and lavender essential oils were analyzed using GC-MS system. Thermo Scientific Focus GC-DSQ II system was used in the study. Helium was used as a carrier gas, with an injector temperature of 230°C. The initial oven temperature was kept at 80°C for 2 min which was further increased to 280°C at 10°C/min and held for 10 min. Spit ratio used was 1 : 20. The analysis was done using MSD detector. Component identification was carried out by comparing the obtained MS data with those reported in Library. The amount of essential oils present in the capsules, unwashed, washed, and dry-cleaned fabric was estimated against standard essential oil samples.

3.5. High Performance Liquid Chromatography (HPLC) Analysis. The sensitivity of permethrin in GC-MS was observed to be less. Hence, the analysis of permethrin content in unwashed, washed, and dry-cleaned fabric was carried out using HPLC. The permethrin treated samples were refluxed in methanol for 3 hrs and the extracts were analyzed using HPLC system.

3.6. Scanning Electron Microscopy (SEM). Wool fabric samples were coated with gold using JEOL JEC-550 twin coater before carrying out scanning electron microscopy. JEOL JSM-5400 scanning electron microscope was used for sample analysis. The surface morphologies of untreated, citronella unwashed, washed, and dry-cleaned samples were studied.

3.7. Intensity of Fragrance

3.7.1. Qualitative Evaluation. The relative intensity of fragrance was evaluated by a panel of ten evaluators. All evaluators were asked to take two-three whiffs of unwashed, washed, and dry-cleaned fabric samples. Samples were rated according to intensity of fragrance as very strong, strong, moderate, faint, and very faint. The evaluators were also given a smell of coffee beans in between to neutralize the fragrance of previous samples.

3.7.2. Quantitative Evaluation. The release rate of fragrance was tested over a period of 15 days at an interval of every five days. The eucalyptus, citronella, and lavender treated unwashed samples were kept in incubator at 37 ± 2°C to simulate human body temperatures. Qualitative analyses were also performed for these samples. The total essential oil content was determined using GC-MS system. The release rate was then calculated using the following formula:

$$\text{Release }\% = \frac{\text{Initial concentration} - \text{Final concentration}}{\text{Initial concentration}} \times 100. \quad (1)$$

3.8. Proofing Testing against the Larvae of Anthrenus flavipies. The biological assays were done with the larvae of *Anthrenus flavipies* (LeConte) according to visual observation of the extent of damage, weight loss of the test sample, and larval condition of the test larvae as per the procedure described by ISO 3998-1977 (E). The untreated and treated fabric samples were cut into discs of 40 mm diameter and conditioned at 65 ± 2% relative humidity at 27 ± 2°C temperature for 24 hrs and initial weight was noted down. The larvae of *Anthrenus flavipies* (LeConte) available at Wool Research Association were used for evaluation. Eleven-week-old larvae weighing about 0.8 to 1.2 mg were used for the test. The fabric samples of known weights were placed in contact with 15 larvae in glass Petri dishes for 14 days in controlled atmosphere of 27 ± 2°C temperature and 65 ± 2% relative humidity in a dark cabinet. Samples were weighted after completion of test. The loss in mass (Δm) of each test specimen and voracity control, due to feeding, was determined as follows:

$$\Delta m = \frac{m_0 \times m_3}{m_2} - m_1, \quad (2)$$

where m_0 is the mass of the test specimen or voracity control before exposure to larvae; m_1 is the mass of the test specimen or voracity control after exposure to larvae; m_2 is the mean initial mass of appropriate moisture regain controls; m_3 is the mean final mass of appropriate moisture regain controls.

All the samples were assessed for damage by visual observation in terms of extent of cropping and holes formation. The larvae condition was also accessed in terms of live, dead, or pupating.

3.9. Repellency Testing against the Larvae of Anthrenus flavipies. To study the repellent activity of the finished fabric, we have devised a test method based on ISO 3998-1977 (E). In this test two fabric samples were kept in single dish in which one sample was untreated and the other one was finished. This test was designed to study the preference of moth for a particular fabric. All other testing parameters of ISO 3998-1977 (E) were unchanged.

4. Results and Discussion

Detailed chemical composition of the citronella, lavender, and eucalyptus essential oils is shown in (Table 1). It was found that eucalyptol is major chemical constituent of the eucalyptus oil and comprises about 82% of total volume. Citronellal (45%) is major chemical content present in the citronella oil. Isononyl ester is major component of Lavender oil and contributes 26% of the total volume. These major chemical components impart characteristic fragrance to the respective essential oils.

Average particle size (Table 2) of microencapsulated eucalyptus, lavender, and citronella essential oils in terms of volume weighted mean was found to be in the range of 3–5 μm with mean specific surface area of 1-2 m^2/g.

Solid polymer content (Table 3) ranged from 34 to 37% for microcapsules used in this research. Loading of essential oil inside the microcapsules was found to be in the range of 24–41% on the basis of microcapsule slurry.

Lavender essential oil content was found to be higher on the unwashed finished fabric compared to other two essential oils (Table 4). Capsules were not found to be durable for washing process as essential oil content reduces up to trace level. Dry-cleaning treatment also reduced the essential oil content on the fabric. Loss of finish due to washing was greater than that of dry-cleaning, as the washing treatment is far more severe than that of dry-cleaning. Permethrin content was also found to reduce after washing and dry-cleaning but being a small molecule permethrin penetrates into the fibre structure and remains inside, thus showing better durability to washing and dry-cleaning processes.

SEM images (Figure 1) of citronella capsules treated fabric have shown uniform distribution of capsules on the fibre surface. After washing, most of the capsules get removed and few capsules can be seen on the fabric surface. Dry-cleaning also reduced the capsule content and presences of chunks were observed on the fibre surface. This indicates that the capsules content reduces drastically after washing and dry-cleaning treatments.

Table 1: Chemical composition of natural essential oils.

	Chemical compounds	Molecular formula	Molecular weight	Composition (%)
Eucalyptus oil	Oxirane, 2,2-dimethyl-3-propyl	$C_7H_{14}O$	114	2.32
	α-Pinene	$C_{10}H_{16}$	136	11.20
	Benzene, 1-methyl-2-(1-methylethyl)	$C_{10}H_{14}$	134	4.03
	Eucalyptol	$C_{10}H_{18}O$	154	82.46
	Pentan-2-ol, 4-allyloxy-2-methyl-	$C_9H_{18}O_2$	158	13.10
	3,4-Hexanediol, 2,5-dimethyl	$C_8H_{18}O_2$	146	6.83
Citronella oil	6-Octen-1-ol, 3,7-Dimethyl	$C_{10}H_{18}O$	154	17.35
	6-Octenal, 3,7-Dimethyl	$C_{10}H_{20}O$	156	7.26
	2,6-Octadien-1-ol, 3,7-Dimethyl-(E) (Citronellal)	$C_{10}H_{18}O$	154	44.95
	2,6-Octadienal, 3,7-dimethyl	$C_{10}H_{16}O$	152	6.24
	2,6-Octadien-1-ol, 3,7-dimethyl-acetate	$C_{12}H_{20}O_2$	196	4.25
	3-Ethyl-2,2-Dimethyl-Oxirane	$C_6H_{12}O$	100	3.90
Lavender oil	3,4-Hexanediol, 2,5-Dimethyl	$C_8H_{18}O_2$	146	2.11
	1R-α-Pinene	$C_{10}H_{16}$	136	14.04
	α-Pinene	$C_{10}H_{16}$	136	2.20
	1,6-Octadien-3-ol, 3,7-Dimethyl	$C_{10}H_{18}O$	154	14.11
	Bicyclo [2.2.1]heptan-2-one, 1,7,7-trimethyl, (1S)	$C_{10}H_{16}O$	152	3.6
	Isononyl ester	$C_{11}H_{22}O_2$	186	25.93
	(1s,2r,4r)-beta-Fenchol	$C_{10}H_{18}O$	154	8.29
	1,6-Octadien-3-ol, 3,7-Dimethyl-, 2-Aminobenzoate	$C_{17}H_{23}NO_2$	273	10.94
	2-Norpinene-2-ethanol, 6,6-Dimethyl-, acetate	$C_{13}H_{20}O_2$	208	14.88

Table 2: Particle size analysis.

Microcapsules	Surface weighted mean (μm)	Volume weighted mean (μm)	Specific surface area m^2/g
Eucalyptus	3.453	5.556	1.74
Lavender	2.646	3.560	2.27
Citronella	2.984	3.762	2.01

Table 3: Loading of essential oils in microcapsules.

Microcapsules	Solid polymer content in formulation %	Essential oil content in slurry %
Eucalyptus	34.45	24.50
Citronella	37.39	41.05
Lavender	36.95	36.00

The microcapsule treated unwashed fabric sample (Table 5) shows good fragrance durability and slow release of fragrance. The fabric shown in Figure 1 released about 17–25% of the essential oil content over a prolonged period of fifteen days. The fragrance release ratings were also found to be on very strong over a prolonged period of time. All the evaluators have rated these samples to be very strong.

The *Anthrenus flavipies* proofing properties of treated and untreated fabric samples is shown in (Table 6). The weight loss for untreated samples was 49.73 mg (32%) and several large holes can be seen on the wool fabric. The surface characteristics of the treated and untreated samples are shown in (Figure 2). Untreated fabric sample was heavily cropped and several large holes were formed, showing an inadequate resistance towards the test larvae, whereas the weight loss for unwashed eucalyptus, lavender, and citronella microcapsule treated samples were 20%, 12%, and 10%, respectively. The surface damage was also found to be less compared to the untreated sample. The activity against the test larvae was found to reduce drastically after washing and dry-cleaning of the samples. This can be attributed to the removal of microcapsules from the fabric surface as evidenced from the SEM images (Figure 1) and subsequent reduction in the essential oils present on the fabric samples as shown in (Table 4). The commercial *Anthrenus flavipies* proofing chemical permethrin gives satisfactory protection to wool fabric. The weight loss for permethrin finish unwashed, washed, and dry-cleaned fabric samples was found to be 2%, 4%, and 7%, respectively.

FIGURE 1: SEM images of (a) untreated, (b) citronella unwashed, (c) washed, and (d) dry-cleaned fabric samples.

TABLE 4: Chemical content, fragrance ratings, and durability of finish to washing and dry-cleaning.

Samples	Chemical content (%)	Fragrance ratings	Chemical release (%)
Eucalyptus unwashed	0.1944	Very strong	—
Washed	Traces	Very faint	100
Dry-cleaned	0.0211	Moderate	89
Citronella unwashed	0.3659	Very strong	—
Washed	Traces	Very faint	100
Dry-cleaned	0.0489	Moderate	87
Lavender unwashed	0.4683	Very strong	—
Washed	Traces	Very faint	100
Dry-cleaned	0.0329	Moderate	93
Permethrin Unwashed	0.0187	—	—
Washed	0.0104	—	44
Dry-cleaned	0.0080	—	57

Anthrenus flavipies repellent properties of the finished fabric are shown in (Table 7). In this study, larvae were provided with a choice of eating an untreated fabric kept alongside the treated one. *Anthrenus flavipies* were observed to migrate towards untreated samples and they preferred eating untreated sample compared to its treated counterpart irrespective of treatment type given to the samples. This indicates that eucalyptus, lavender, or citronella possess repellent activity against the larvae of *Anthrenus flavipies*. Lavender and citronella show more repellence compared to eucalyptus finished unwashed fabric sample. Two untreated samples placed side by side showed weight loss of 30–40% whereas weight loss of eucalyptus, lavender, and citronella treated unwashed samples were 30%, 9%, and 5%, respectively, with weight loss of accompanied untreated sample was in the range of 50–65%. The washing and dry-cleaning process were found to reduce the repellency activity. The weight loss for permethrin treated unwashed, washed, and dry-cleaned samples were found to be negligible compared to their untreated counterparts.

Sample photographs (Figure 3) depict the extent of surface damage after 14 days of testing. Representative untreated sample shows heavy surface cropping compared to the treated counterpart indicating their preference to the untreated one.

Table 5: Chemical content and fragrance ratings for a prolonged period.

Samples	Chemical content %	After 5 days Fragrance ratings	Chemical release %	Chemical content %	After 10 days Fragrance ratings	Chemical release %	Chemical content %	After 15 days Fragrance ratings	Chemical release %
Eucalyptus unwashed	0.1868	Very strong	4	0.1730	Very strong	11	0.1615	Very strong	17
Citronella unwashed	0.3443	Very strong	6	0.3041	Very strong	17	0.2781	Very strong	24
Lavender unwashed	0.4487	Very strong	4	0.4441	Very strong	5	0.3854	Very strong	18

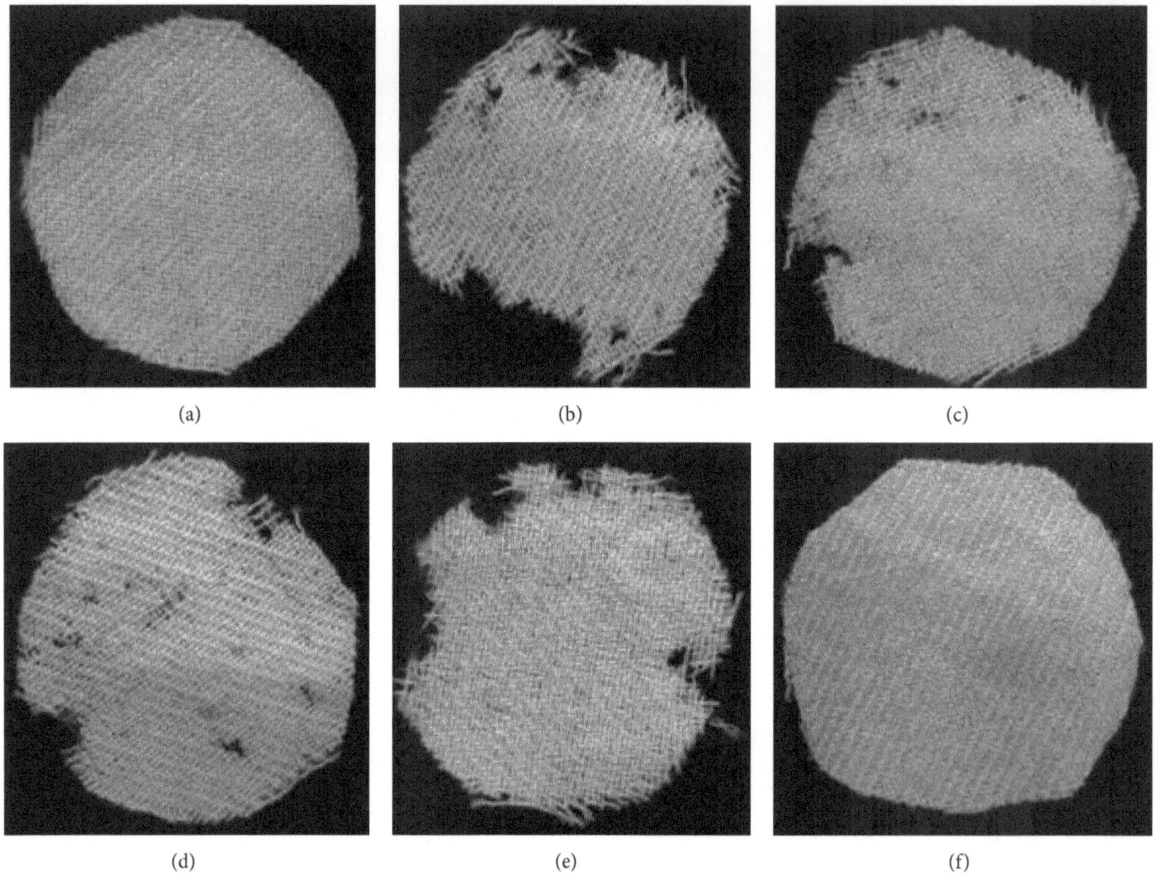

Figure 2: Samples after 14 days of exposure to the larvae of *Anthrenus flavipies* (proof testing): (a) control, (b) untreated, (c) citronella, (d) lavender, (e) eucalyptus, and (f) permethrin treated unwashed fabric samples.

Table 6: *Anthrenus flavipies* proof properties of fabrics.

Sample placed in Petri dish	Weight loss (mg)	Visual observations			Rating/Remark
		Surface cropping	Surface holes	Larval condition	
Untreated	49.73	4. very heavy cropping	D. several large holes	All alive	4D/inadequately resistant
Eucalyptus unwashed	35.27	4. very heavy cropping	D. several large holes	All alive	4D/inadequately resistant
Washed	42.02	4. very heavy cropping	D. several large holes	All alive	4D/inadequately resistant
Dry-cleaned	38.33	4. very heavy cropping	D. several large holes	All alive	4D/inadequately resistant
Lavender unwashed	26.11	4. very heavy cropping	D. several large holes	Two dead	4D/inadequately resistant
Washed	43.56	4. very heavy cropping	D. several large holes	All alive	4D/inadequately resistant
Dry-cleaned	32.36	4. very heavy cropping	D. several large holes	All alive	4D/inadequately resistant
Citronella unwashed	23.60	4. very heavy cropping	D. several large holes	All alive	4D/inadequately resistant
Washed	36.41	4. very heavy cropping	D. several large holes	All alive	4D/inadequately resistant
Dry-cleaned	25.52	4. very heavy cropping	D. several large holes	All alive	4D/inadequately resistant
Permethrin unwashed	3.75	1. no detectable damage	A. no detectable damage	Two dead	1A/satisfactorily resistant
Washed	11.33	4. very heavy cropping	D. several large holes	Two dead	1A/satisfactorily resistant
Dry-cleaned	13.45	1. no detectable damage	A. no detectable damage	One dead	1A/satisfactorily resistant

5. Conclusion

The chemical composition study using GC-MS shows eucalyptol (82%), citronellal (45%), and isononyl ester (26%) as the major constituents of eucalyptus, citronella, and lavender essential oils. The loading of essential oils in microcapsules were found to be in the range of 24–41% with mean particle size of 3–5 μm. The SEM images confirm uniform distribution of microcapsules over the fabric surface. The microcapsule treated unwashed fabric samples have shown

TABLE 7: *Anthrenus flavipies* repellent properties of fabrics.

Samples placed in Petri dish	Weight loss (mg)	Visual observations		Larval condition	Rating/Remark
		Surface cropping	Surface holes		
Untreated-1	29.21	4. very heavy cropping	D. several large holes	All alive	4D/inadequately resistant
Untreated-2	44.83	4. very heavy cropping	D. several large holes		4D/inadequately resistant
Untreated	88.24	4. very heavy cropping	D. several large holes	All alive	4D/inadequately resistant
Eucalyptus unwashed	49.70	4. very heavy cropping	D. several large holes		4D/inadequately resistant
Untreated	64.49	4. very heavy cropping	D. several large holes	All alive	4D/inadequately resistant
Washed	50.58	4. very heavy cropping	D. several large holes		4D/inadequately resistant
Untreated	81.49	4. very heavy cropping	D. several large holes	All alive	4D/inadequately resistant
Dry-cleaned	39.43	4. very heavy cropping	D. several large holes		4D/inadequately resistant
Untreated	76.37	4. very heavy cropping	D. several large holes	All alive	4D/inadequately resistant
Lavender unwashed	11.84	1. no detectable damage	A. no detectable damage		1A/satisfactorily resistant
Untreated	76.37	4. very heavy cropping	D. several large holes	All alive	4D/inadequately resistant
Washed	39.58	4. very heavy cropping	D. several large		4D/inadequately resistant
Untreated	55.82	4. very heavy cropping	D. several large holes	All alive	4D/inadequately resistant
Dry-cleaned	40.25	4. very heavy cropping	D. several large holes		4D/inadequately resistant
Untreated	43.88	4. very heavy cropping	D. several large holes	All alive	4D/inadequately resistant
Citronella unwashed	6.95	1. no detectable damage	A. no detectable damage		1A/satisfactorily resistant
Untreated	35.32	4. very heavy cropping	D. several large holes	All alive	4D/inadequately resistant
Washed	21.22	4. very heavy cropping	D. several large holes		4D/inadequately resistant
Untreated	29.52	4. very heavy cropping	D. several large holes	All alive	4D/inadequately resistant
Dry-cleaned	29.52	4. very heavy cropping	D. several large holes		4D/inadequately resistant
Untreated	119.79	4. very heavy cropping	D. several large holes	Two dead	4D/inadequately resistant
Permethrin unwashed	0.063	1. no detectable damage	A. no detectable damage		1A/satisfactorily resistant
Untreated	89.37	4. very heavy cropping	D. several large holes	One dead	4D/inadequately resistant
Washed	1.76	4. very heavy cropping	A. no detectable damage		1A/satisfactorily resistant
Untreated	90.34	4. very heavy cropping	D. several large holes	Three dead	4D/inadequately resistant
Dry-cleaned	5.98	1. no detectable damage	A. no detectable damage		1A/satisfactorily resistant

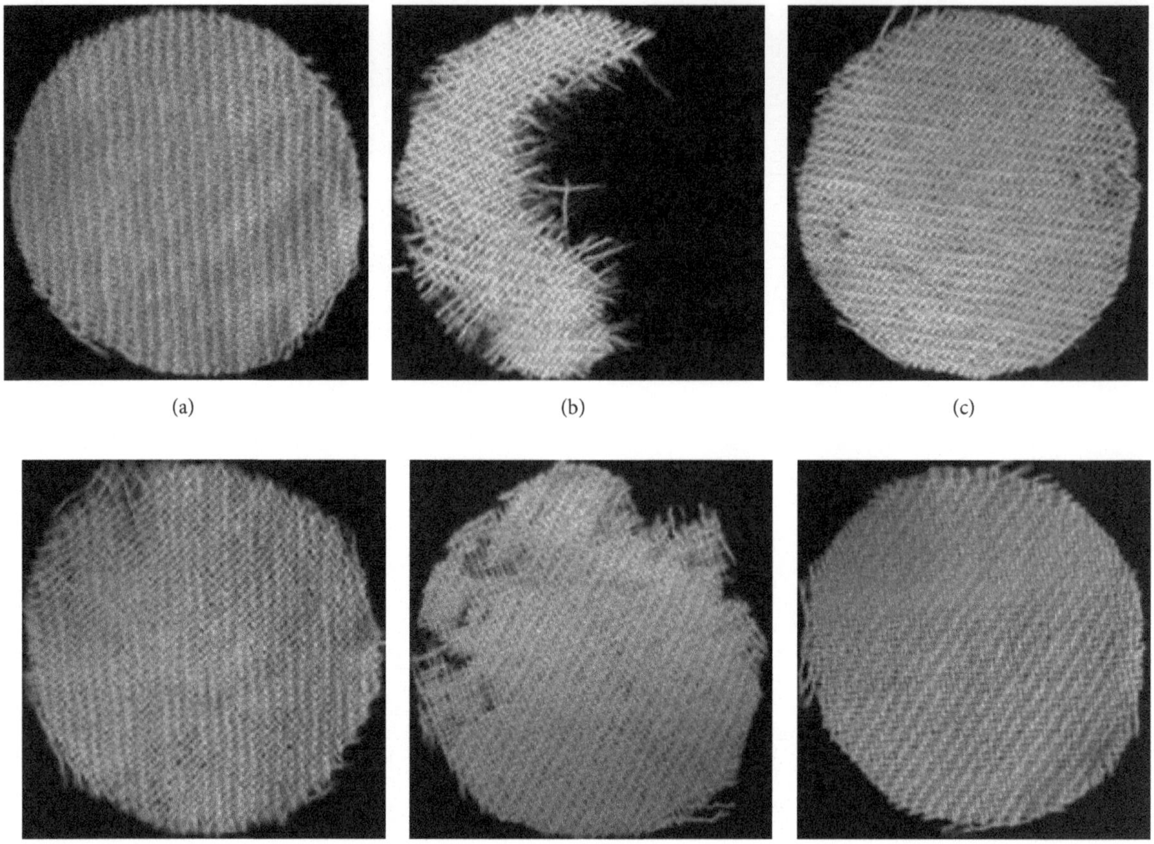

(a) (b) (c)

(d) (e) (f)

FIGURE 3: Samples after 14 days of exposure to the larvae of *Anthrenus flavipies* (repellency testing): (a) control, (b) untreated, (c) citronella, (d) lavender, (e) eucalyptus, and (f) permethrin treated unwashed fabric samples.

good fragrance durability and slow release of fragrance. Wool fabrics finished with microcapsules of citronella, lavender, and eucalyptus essential oils provides good resistance against the Larvae of carpet beetle *Anthrenus flavipies* (LeConte); however, the activity of the microencapsulated essential oils is less compared to commercial insect resist agent permethrin. The effectiveness of microcapsule finish reduces after washing and dry-cleaning treatments. The microcapsule treated samples performs better when tested along with the untreated samples. It was observed that *Anthrenus flavipies* (LeConte) prefer eating untreated samples than their treated counterpart. The performance of essential oils as a natural alternative repellent against the test larvae was in the order citronella > lavender > eucalyptus. As permethrin is highly toxic to the aquatic life, natural essential oils provide a natural, biodegradable, and less toxic alternative with an added advantage of pleasant fragrance over a long period of time.

Conflict of Interests

The authors declare that there is no conflict of interests regarding the publication of this paper.

References

[1] C. O. Clark, "The protection of animal fibres against clothes moths and dermestid beetles," *Journal of Textile Institute*, vol. 19, pp. 295–320, 1928.

[2] R. S. Hartley, F. F. Elsworth, and J. Barrit, "The mothproofing of wool," *Journal of the Society of Dyers and Colourists*, vol. 59, no. 12, pp. 266–271, 1943.

[3] M. Lipson and J. R. McPhee, "The mothproofing of wool with dieldrin," *Textile Research Journal*, vol. 28, no. 8, pp. 679–686, 1958.

[4] R. J. Mayfield, "Mothproofing," *Textile Progress*, vol. 11, no. 4, pp. 1-11, 1982.

[5] D. E. Wells and S. J. Johnstone, "High performance liquid chromatography of polychloro-2-(chloromethyl sulphonamido) diphenyl ethers and their impurities in the mothproofing agent, Eulan WA New, and in water," *Journal of Chromatographic Science*, vol. 19, no. 3, pp. 137–143, 1981.

[6] P. A. Duffield, "Mothproofing with permethrin," *Pesticide Science*, vol. 8, no. 3, pp. 279–283, 1977.

[7] S. W. Carter and P. A. Duffield, "Preliminary evaluation of the pyrethroid NRDC 143 as an industrial mothproofer," *The Journal of The Textile Institute*, vol. 67, no. 3, pp. 77–81, 1976.

[8] D. Wang, H. Naito, and T. Nakajima, "The toxicity of fenitrothion and permethrin," in *Insecticides—Pest Engineering*, F. Perveen, Ed., pp. 85–98, InTech, Shanghai, China, 2012.

[9] R. D. O'Brien, "Insecticides and environmental health," in *Insecticides, Action and Metabolism*, R. D. O'Brien, Ed., pp. 291–304, Academic Press, New York, NY, USA, 1967.

[10] I. Yamamoto, "Problems in mode of action of pyrethroids," in *Biochemical Toxicology of Insecticides*, R. D. O'Brien and I. Yamamoto, Eds., pp. 193–200, Academic Press, New York, NY, USA, 1970.

[11] W. F. Barthel, "Toxicity of pyrethrum and its constituents to mammals," in *Pyrethrum, The Natural Insecticides*, J. E. Casida, Ed., pp. 123–142, Academic Press, New York, NY, USA, 1973.

[12] M. F. Maia and S. J. Moore, "Plant-based insect repellents: a review of their efficacy, development and testing," *Malaria Journal*, vol. 10, no. 1, article S11, 2011.

[13] E. K. Patel, A. Gupta, and R. J. Oswal, "A review on: mosquito repellent methods," *International Journal of Pharmaceutical, Chemical and Biological Sciences*, vol. 2, no. 3, pp. 310–317, 2012.

[14] N. Rani, A. Wany, A. S. Vidyarthi, and D. M. Pandey, "Study of Citronella leaf based herbal mosquito repellents using natural binders," *Current Research in Microbiology and Biotechnology*, vol. 1, no. 3, pp. 98–103, 2013.

[15] D. R. Barnard and R. Xue, "Laboratory evaluation of mosquito repellents against *Aedes albopictus*, *Culex nigripalpus*, and *Ochlerotatus triseriatus* (Diptera: Culicidae)," *Journal of Medical Entomology*, vol. 41, no. 4, pp. 726–730, 2004.

[16] M. Arancibia, A. Rabossi, P. A. Bochicchio et al., "Biodegradable films containing clove or citronella essential oils against the mediterranean fruit fly *Ceratitis capitata* (Diptera: Tephritidae)," *Journal of Agriculture and Food Technology*, vol. 3, no. 3, pp. 1–7, 2013.

[17] International Wool Secretariat, "ETB-32 Initial concentration levels (ICLs) for insects resist agents on wool carpets, carpet yarns, rug yarns and upholstery," *Technical Information Bulletin*, no. 1, 1994.

A Study on Coloring Properties of *Rheum emodi* on Jute Union Fabrics

Lopamudra Nayak

Orissa University of Agriculture & Technology, Bhubaneswar 751003, India

Correspondence should be addressed to Lopamudra Nayak; licha.lopa@gmail.com

Academic Editor: Anindya Ghosh

Jute-cotton and jute-wool union fabrics have been printed with colorant extracted from *Rheum emodi* with guar gum and albumin as thickening agents and aluminium sulphate, copper sulphate, and ferrous sulphate as mordants. Printing with *Rheum emodi* colorant with different mordants resulted in different shades ranging from yellowish brown, deep brown, reddish brown, and gray shades to olive black. The *K/S* value showed that mordants, namely, $CuSO_4$ and $FeSO_4$, exhibited high colour absorption regardless of the nature of material used. All the printed samples have good to excellent wash, rubbing and light fastness properties regardless of the nature of materials, mordants, and thickeners used. Guar gum is closely at par with albumin as a thickening agent and considering the cost and availability of raw material, guar gum is the better choice for the textile industry.

1. Introduction

Renewable, biodegradable, and sustainable natural fibers and natural colorants could be potential substitutes for energy-intensive and petrochemical-based synthetic fibers and colorants in many applications and serve as an effective way of reducing the impact textiles have on the environment. The most classically known fibers such as jute, cotton, and wool are environmentally friendly, healthy, and comfortable. In addition, use of natural colorants on these fibers has become very much popular throughout the world due to its eco-friendliness, and aesthetic approach.

Jute is a lingo-cellulosic composite natural bast fiber; its major constituent components being cellulose, hemicellulose, and lignin. It has high tensile strength with low extensibility, which helps to make the best quality industrial yarn and fabric. The inherent drawback of jute such as harshness, brittleness, and low extensibility pose a hindrance for its use as furnishings and apparels [1, 2]. To overcome these problems jute fabrics could be produced with blend or union with other natural and manmade fibers. Furthermore, these fabrics are subjected to coloration to improve its appearance.

Though renewable dye resources provide important alternatives for fossil-based colorants, from the sustainability point of view, it is desirable to use natural dyes to a greater extent. *Rheum emodi* commonly known as Himalayan rhubarb is one such natural colorant found in the sub-alpine and alpine Himalayas at an altitude of 3000–5000 meters. *Rheum emodi* is a medicinal herb and belongs to the family Polygonaceae. The Indian rhubarb (*Rheum emodi*) commonly known as "Dolu" (Sanskrit: Amlaparni, Revatchini; Hindi: Revand chini) are reported to be used for curing of a variety of diseases, its main effect being a positive and balancing effect upon the whole digestive system. The rhizomes and roots of *Rheum emodi* yield a yellow coloring matter which could be used for textile coloration. The plant contains 3–5% anthracene derivatives. In the plant the anthraquinoid components are present in reduced form as anthraglycosides, which hydrolyze and oxidize to the respective anthraquinone. The extracted dye contains a complex mixture of anthraquinones; relevant compounds are Emodin, Chrysophanol, Aloeemodin, Rhein, and Physcion and the chief colouring component is chrysophanic acid [3–6].

The alkaline extract of root gives a brown colored powdered dye. The structures of the molecules are shown in Figure 1.

The formation of complexes due to the presence of metal salt from mordanting considerably improves dye uptake and all round fastness properties. Mordant brings about change

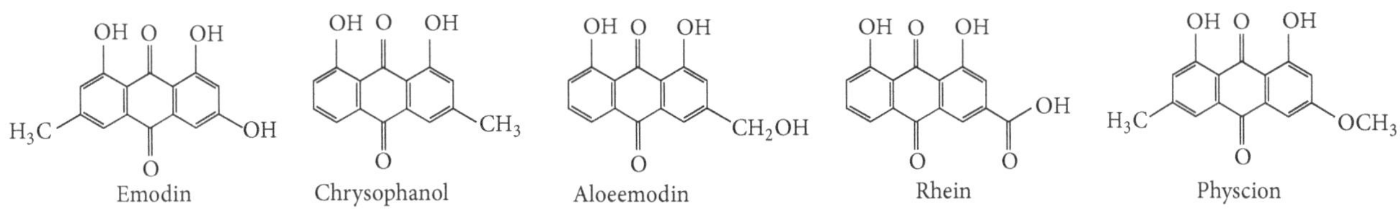

FIGURE 1: Some of the anthraquinoid dye structure of *Rheum emodi*: Emodin, Chrysophanol, Aloeemodin, Rhein, and Physcion.

of the hue of the natural colorants, which can be utilized with advantage to produce different shades on protein and cellulosic substrates. Keeping environment related issues and adoption of cleaner technology in view, inorganic salts such as, aluminium sulphate, potash alum, and ferrous sulphate can be employed as mordanting agent in view of their good complex forming ability and environmental safe character. Myrobalan is used as a source of tannin either alone or in combination with inorganic salts as mordanting agents especially for cotton [5, 6].

Thickeners are needed to help control the dye, to give it some bulk when printing with the dye. Guar gum can be used for most types of printing as it is very stable in all pH levels. Guar is a native to the Indian subcontinent and offers a cheaper alternative to Gum Arabic and Gum Senegal. Albumin can be used both as a thickening and a fixing agent in textile printing. They are widely distributed in plant and animal tissues, for example, legumelin of peas, leucosin of wheat, ovalbumin of egg, serum albumin of bovine, and lactalbumin of milk and so forth [7, 8].

The present study examines the coloring properties of natural colorant *Rheum emodi* on jute/cotton and jute/wool union fabrics and compares the performances of albumin and guar gum as thickening agents along with aluminium sulphate, copper sulphate, and ferrous sulphate as mordants.

2. Materials and Methods

2.1. Materials. Mill-scoured and bleached jute/cotton (60 : 40) union fabric with cotton yarn in the warp and jute/cotton yarns in the weft direction (plain weave, thickness 1.95 mm, EPI:17; PPI:17) and jute/wool (60 : 40) union fabric with wool yarn in the warp and jute/wool yarns in the weft direction (plain weave, thickness 1.97 mm, EPI:17; PPI:14) were purchased from the market. Jute yarns used in the fabrics were cream colored (not bleached white).

Natural dye powder, *Rheum emodi* (alkaline extract; brown colored powder), and myrobalan (*Terminalia chebula*) were obtained from M/s ALPS Industries Ltd., Ghaziabad, India. Thickeners, namely, guar gum and albumin, and mordants such as alum, copper sulphate ($CuSO_4$), ferrous sulphate ($FeSO_4$) and all other chemicals such as acetic acid and glycerin used were of laboratory grade.

2.2. Methods

2.2.1. Scouring. Jute/cotton and jute/wool union fabrics were washed with nonionic detergent (2 g/L) and washing soda (0.5 g/L) for half an hour at a temperature of 50°C–60°C, keeping the material to liquor ratio at 1 : 30 to facilitate subsequent mordanting and printing operation. The scoured material was thoroughly washed with tap water and dried at room temperature.

2.2.2. Treatment with Myrobalan. Half of the fabric samples were treated with myrobalan (20 g/L) for 15 minutes at a temperature of 40°C–60°C, keeping the material to liquor ratio at 1 : 20. The materials were taken out from bath and squeezed evenly and dried. Precaution is to be taken, not to dry the treated cloth in direct sun light as the surface exposed to sun becomes darker resulting uneven shades in printing.

Myrobalan is an important tannin-based mordant for cellulosic fiber, namely, cotton. Treatment with tannin introduces additional hydroxyl and carboxyl groups on the fiber matrix. A subsequent treatment of tannin treated cellulose with metal salts introduces the respective cation in the fiber that leads to the formation of metal complexes.

2.2.3. Preparation of Thickening Agents

(1) Guar Gum. The guar gum powder was sprinkled slowly in lukewarm water with continuous stirring in order to prevent lump formation. The mixture was cooked for half an hour and allowed to be cooled.

(2) Albumin. Albumin was dissolved in cold water and kept for overnight to dissolve properly. For most purposes one part of albumin is mixed with one part of gum.

2.2.4. Preparation of Stock Paste. Mordants were first dissolved in water and then added to cold thickening agents (see Table 1).

TABLE 1: Quantity of mordant and thickener.

Mordant name	Quantity of mordant (gm)	Quantity of thickener (gm)	Total quantity (gm)
Alum	6	94	100
Copper sulphate	4	96	100
Ferrous sulphate	4	96	100

Table 2: Printed sample coding.

Fabric	Sample code	Mordant + thickener	Sample code	Mordant + thickener
Untreated jute/cotton	UJC1	No mordant + guar gum	UJC1′	No mordant + albumin
	UJC2	Alum + guar gum	UJC2′	Alum + albumin
	UJC3	$CuSO_4$ + guar gum	UJC3′	$CuSO_4$ + albumin
	UJC4	$FeSO_4$ + guar gum	UJC4′	$FeSO_4$ + albumin
Myrobalan treated jute/cotton	MJC1	Alum + guar gum	MJC1′	Alum + albumin
	MJC2	$CuSO_4$ + guar gum	MJC2′	$CuSO_4$ + albumin
	MJC3	$FeSO_4$ + guar gum	MJC3′	$FeSO_4$ + albumin
Untreated jute/wool	UJW1	No mordant + guar gum	UJW1′	No mordant + albumin
	UJW2	Alum + guar gum	UJW2′	Alum + albumin
	UJW3	$CuSO_4$ + guar gum	UJW3′	$CuSO_4$ + albumin
	UJW4	$FeSO_4$ + guar gum	UJW4′	$FeSO_4$ + albumin
Myrobalan treated jute/wool	MJW1	Alum + guar gum	MJW1′	Alum + albumin
	MJW2	$CuSO_4$ + guar gum	MJW2′	$CuSO_4$ + albumin
	MJW3	$FeSO_4$ + guar gum	MJW3′	$FeSO_4$ + albumin

2.2.5. Print Paste

(1) Jute/Cotton Union Fabric. The printing pastes for jute/cotton union fabric were prepared using the following recipe:

Rheum emodi dye powder: 10 gm,

glycerin: 2.5 gm,

stock paste: 87.5 gm,

total weight of the paste: 100 gm.

(2) Jute/Wool Union Fabric. The printing pastes for jute/wool union fabric were prepared using the following recipe:

Rheum emodi dye powder: 10 gm;

glycerin: 2.5 gm;

acetic acid: 5 gm;

stock paste: 82.5 gm;

total weight of the paste: 100 gm.

The print paste was continuously stirred by a mechanical stirrer.

2.2.6. Printing Procedure. Printing was carried out on both tannin pretreated fabrics and untreated fabrics using the flat screen technique. Printed samples were then dried. The samples were steamed in a saturated atmosphere for 30 minutes in a star steamer. Printed samples were washed with water and then dried in air (see Table 2).

2.3. Measurement of Color Strength. The printed samples were evaluated for the depth of the prints by light reflectance measurements using a JAYPAK 4802 colour matching system at D65 illuminate/10 Deg Observer. The color strength (K/S value) was assessed using the Kubelka-Munk equation:

$$\frac{K}{S} = \frac{(1-R)^2}{2R}, \quad (1)$$

where R is the reflectance at specific wavelengths, K is the absorption coefficient, and S is the scattering coefficient.

In general, higher the K/S values the higher the depth of the colour on the fabric.

The colour of printed fabric samples was also measured in terms of CIELAB colour space, namely, color parameters L^* (+ lightness; − darkness), a^* (+ red; − green), and b^* (+ yellow; − blue), by using Computer Colour Matching System [9, 10]. The total colour deference ΔE^* was also evaluated using the following equation:

$$\Delta E^*_{ab} = \left[(\Delta L^*)^2 + (\Delta a^*)^2 + (\Delta b^*)^2\right]^{1/2}, \quad (2)$$

where

$$L^* = 116\left(\frac{Y}{Y_n}\right)^{1/3} - 16,$$
$$a^* = 500\left[\left(\frac{X}{X_n}\right)^{1/3} - \left(\frac{Y}{Y_n}\right)^{1/3}\right], \quad (3)$$
$$b^* = 200\left[\left(\frac{Y}{Y_n}\right)^{1/3} - \left(\frac{Z}{Z_n}\right)^{1/3}\right].$$

2.4. Tests for Fastness Properties. The fastness properties of the printed samples towards washing, rubbing, and light were assessed using standard methods.

Evaluation of color fastness to washing was determined as per IS: 764–1984 method using a Sasmira launder-O-meter and following IS-3 wash fastness method. The wash fastness rating was assessed using grey scale as per ISO-05-A02 (loss of shade depth) and ISO-105-AO3 (extent of staining). Colour fastness to rubbing (dry and wet) was assessed as per IS: 766–1984 method using a manually operated Paramount crock meter. Colour fastness to light was determined as per IS: 2454–1984 method. The sample was exposed to xenon light in a Paramount digiLIGHT fade-O-meter along with the eight blue wool standards (BS 1006: BOI: 1978). The fading

Table 3: Yielding of colours with different mordants.

Sample code	Color obtained
UJC1	Yellow ochre
UJC2	Light brown
UJC3	Brown
UJC4	Gray (light grayish ash)
MJC1	Brown (cinnamon-copper shade)
MJC2	Brown (dark cinnamon shade)
MJC3	Gray-black
UJW1	Yellowish brown
UJW2	Deep brown
UJW3	Deep brown (russet)
UJW4	Gray
MJW1	Brownish yellow
MJW2	Deep brown (russet)
MJW3	Deep brown
UJC1′	Earth yellow
UJC2′	Deep brown
UJC3′	Deep brown (brown-nose)
UJC4′	Deep brown (seal brown)
MJC1′	Golden-brown
MJC2′	Brown
MJC3′	Yellowish brown
UJW1′	Yellowish brown
UJW2′	Dark goldenrod
UJW3′	Dark goldenrod
UJW4′	Coffee brown
MJW1′	Brownish yellow
MJW2′	Grayish brown
MJW3′	Olive black

of each sample was observed against the fading of blue wool standards (1–8).

3. Results and Discussion

The K/S value (Figures 2 and 3) indicates that mordants, namely, $CuSO_4$ and $FeSO_4$ exhibited high colour absorption (K/S) regardless of the nature of material used, which may be attributed to their ability to form strong coordination complexes with the dye molecules. Furthermore, the high colour absorption and colour difference (Tables 3 and 4) to a certain extent may be due to the changes in scattering because of the chemical interactions between fibers and tannin and/or metallic salts along with the additional inherent colour input of the corresponding mordants.

Jute-wool union printed fabrics pretreated with myrobalan showed comparatively higher K/S values than the untreated fabrics; furthermore, the K/S values in case of jute-wool union fabrics were in most cases higher than their corresponding samples of jute-cotton union fabrics exception being jute-cotton untreated fabric particularly UJC2′ and UJC3′ printed using albumin as thickening agent which yielded darker and shining shades. This may be due to

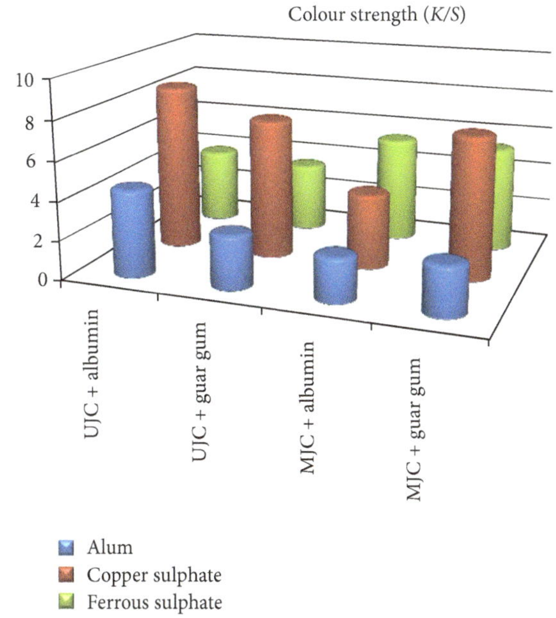

Figure 2: Effect of mordants and thickeners on color strength of jute/cotton union fabrics.

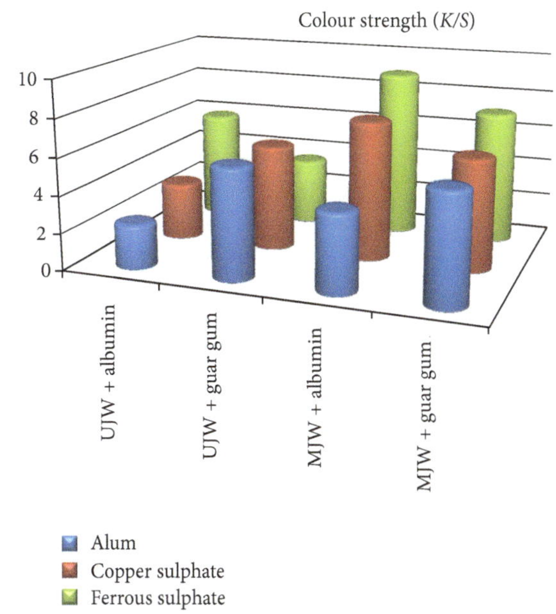

Figure 3: Effect of mordants and thickeners on color strength of jute/wool union fabrics.

the fact that myrobalan contains natural colour as well as natural tannin and in addition to that in case of protein fiber such as wool, three types of bonds, namely, hydrogen bond, ionic bond, and covalent bond, are formed. However, in case of cellulosic fabrics the tannin could form only two types of bonds, that is, hydrogen bond and covalent bond. Moreover, the penetration of mordant and colour in wool is easier than that of cellulose fibers, as the fiber is porous, softer, and more elastic; hence higher colour absorption occurs in case

Table 4: CIELAB lab measurements.

Sample code	Mordant + thickener	Delta CIELAB			
		ΔE^*	ΔL^*	Δa^*	Δb^*
UJC1	No mordant + guar gum	—	—	—	—
UJC2	Alum + guar gum	8.36	7.45	−2.30	−3.03
UJC3	$CuSO_4$ + guar gum	25.13	−23.98	7.17	−2.36
UJC4	$FeSO_4$ + guar gum	12.91	−7.06	1.63	−10.69
UJC1′	No mordant + albumin	8.42	−2.21	0.50	8.11
UJC2′	Alum + albumin	15.40	−10.18	0.53	11.55
UJC3′	$CuSO_4$ + albumin	29.15	−28.32	6.76	−1.59
UJC4′	$FeSO_4$ + albumin	14.05	−13.95	1.15	−1.29
MJC1	Alum + guar gum	6.12	4.24	0.46	−4.39
MJC2	$CuSO_4$ + guar gum	25.85	−24.31	7.21	5.12
MJC3	$FeSO_4$ + guar gum	20.64	−20.23	−0.11	4.11
MJC1′	Alum + albumin	2.98	−2.82	0.99	0.04
MJC2′	$CuSO_4$ + albumin	13.17	−11.47	1.36	6.34
MJC3′	$FeSO_4$ + albumin	20.67	−20.43	1.35	−2.86
UJW1	No mordant + guar gum	7.66	−2.95	5.72	4.16
UJW2	Alum + guar gum	22.89	−22.24	3.56	4.11
UJW3	$CuSO_4$ + guar gum	20.81	−19.81	6.18	−1.59
UJW4	$FeSO_4$ + guar gum	10.26	−10.15	1.02	−1.17
UJW1′	No mordant + albumin	11.01	0.25	−0.37	11.01
UJW2′	Alum + albumin	3.60	3.48	−0.82	0.4
UJW3′	$CuSO_4$ + albumin	9.19	−6.40	−0.01	−6.60
UJW4′	$FeSO_4$ + albumin	22.39	−20.95	6.72	−4.16
MJW1	Alum + guar gum	23.38	−22.46	2.27	6.11
MJW2	$CuSO_4$ + guar gum	22.87	−22.30	0.12	5.11
MJW3	$FeSO_4$ + guar gum	25.08	−23.75	8.02	−0.94
MJW1′	Alum + albumin	15.04	−14.95	1.22	−1.14
MJW2′	$CuSO_4$ + albumin	26.18	−25.61	2.71	4.75
MJW3′	$FeSO_4$ + albumin	30.25	−29.02	7.76	−3.59

Table 5: Fastness properties of the printed jute/cotton union fabrics towards washing, rubbing, and light.

Sample code	Mordant + thickener	Washing fastness			Rubbing fastness		Light fastness
		CC	CS		Dry	Wet	
			C	W			
UJC1	No mordant + guar gum	3	3-4	3	3-4	3	4
UJC2	Alum + guar gum	4-5	5	4-5	5	4-5	6
UJC3	$CuSO_4$ + guar gum	5	5	5	5	5	7
UJC4	$FeSO_4$ + guar gum	4	4	4-5	4-5	4	6
UJC1′	No mordant + albumin	4-5	4-5	4-5	4-5	4-5	6
UJC2′	Alum + albumin	4-5	4-5	4-5	5	4-5	6-7
UJC3′	$CuSO_4$ + albumin	5	5	5	5	5	7
UJC4′	$FeSO_4$ + albumin	5	5	4-5	4-5	4-5	7
MJC1	Alum + guar gum	4-5	4-5	4-5	5	4-5	6-7
MJC2	$CuSO_4$ + guar gum	4-5	5	4-5	5	5	7
MJC3	$FeSO_4$ + guar gum	4-5	4-5	4-5	4-5	4-5	6
MJC1′	Alum + albumin	4-5	4-5	5	5	4-5	6-7
MJC2′	$CuSO_4$ + albumin	5	5	5	5	5	7
MJC3′	$FeSO_4$ + albumin	5	5	5	5	4-5	7

CC: color change; CS: color staining; C: cotton; W: wool.

TABLE 6: Fastness properties of the printed jute/wool union fabrics towards washing, rubbing, and light.

Fabric	Mordant + thickener	Washing fastness			Rubbing fastness		Light fastness
		CC	CS		Dry	Wet	
			C	W			
UJW1	No mordant + guar gum	3	4	3-4	3-4	3-4	4
UJW2	Alum + guar gum	4-5	4-5	4-5	4-5	5	6-7
UJW3	$CuSO_4$ + guar gum	5	5	5	5	5	6-7
UJW4	$FeSO_4$ + guar gum	5	4-5	5	5	4-5	6
UJW1′	No mordant + albumin	4-5	4-5	4-5	5	4-5	6-7
UJW2′	Alum + albumin	5	5	4-5	5	5	7
UJW3′	$CuSO_4$ + albumin	5	5	5	5	5	7
UJW4′	$FeSO_4$ + albumin	5	4-5	5	5	5	7
MJW1	Alum + guar gum	4-5	4-5	4-5	4-5	4-5	6-7
MJW2	$CuSO_4$ + guar gum	4-5	4-5	5	5	5	6-7
MJW3	$FeSO_4$ + guar gum	5	4-5	5	5	4-5	6-7
MJW1′	Alum + albumin	4-5	5	4-5	5	4-5	7
MJW2′	$CuSO_4$ + albumin	5	5	4-5	5	5	7
MJW3′	$FeSO_4$ + albumin	5	5	5	5	5	7

CC: color change; CS: color staining; C: cotton; W: wool.

of myrobalan treated jute-wool union printed fabrics. Regarding jute-cotton untreated fabric yielding darker shades with albumin as thickening agent, it may be due to the fact that albumin being a heat-coagulating protein acts as a mordant thus increasing the colour absorption for a number of colors, despite that guar gum as a thickening agent is also nearly as efficient as albumin. Furthermore, considering the cost, availability and ethical issue, guar gum is the best choice as guar is a native plant abundantly available, the most economical, and can be used for most types of printing owing to its stability in all pH levels and, therefore, suitable print-paste thickener unlike albumin.

Colour analysis in terms of lightness (ΔL^*) and the Cartesian coordinates Δa^* and Δb^* showed that the overall principal colours yielded on jute-cotton and jute-wool union fabrics were ranged from yellow ochre, yellowish brown, deep brown, grey, and coffee to olive black and affected by the type of mordant used. The addition of a mordant to an appropriate colorant results in sudden change in color owing to the incorporation of the metal atom which has low energy levels, into the delocalized electron system of the dye, subsequently lowering of the overall energy. The absorbance of the hue and thus its color is related to this phenomenon.

On the L^* plane, the measurement difference of values was positive in case of a few samples printed with alum mordant (UJC2, MJC1, UJW1′, and UJW2′) corresponding to lighter shades, whereas the ΔL^* values were found to be negative in case of all other samples mordanted with alum, copper sulphate, and ferrous sulphate corresponding to darker shades. It is evident that the type of shades produced and the level of absorption of colorant were affected by using the different mordants and pretreatment with myrobalan whose chemical reaction with the natural colorant depended on the unique structures of the colour components and on the strength of the tannin and/or metal-colorant coordination complex formed during the printing process.

Tables 5 and 6 revealed the fastness properties of the printed samples. Both untreated and myrobalan treated jute union fabrics printed with copper sulphate and ferrous sulphate mordants showed excellent colorfastness properties. Furthermore, the overall fastness properties such as washing, rubbing and light for all mordanted samples were very satisfactory (rating lies between good to excellent) regardless of the nature of materials, mordants, and thickeners used. Thus, it is evident that the complex mixture of anthraquinone structures of *Rheum emodi* is capable of forming stable complexes with the metal ions, resulting in improved colourfastness to washing, rubbing, and lightfastness. Besides, presence of anthraquinone structure as is found in *Rheum emodi* shows in general an increased resistance to light fading [11]. Moreover, albumin being a heat-coagulating protein acts as fixing agent and, hence, further enhances the fastness properties of jute union fabrics.

4. Conclusions

(i) The choice of mordants, thickeners, and fibers types has significant influence on colour values. Printing with *Rheum emodi* (Himalayan rhubarb) colorant with different mordants resulted in different shades ranging from yellowish brown, deep brown, reddish brown, and gray shades to olive black. The K/S value showed that mordants, namely, $CuSO_4$ and $FeSO_4$ exhibited high colour absorption (K/S) regardless of the nature of material used.

(ii) All the printed samples have good to excellent wash, rubbing, and light fastness properties regardless of the nature of materials, mordants, and thickeners used. Both untreated and myrobalan treated jute/cotton fabrics printed with copper sulphate and ferrous sulphate mordants showed excellent all round colorfastness properties.

(iii) Guar gum is closely at par with albumin as a thickening agent and considering the cost, availability, and ethical issue, guar gum is the best choice as guar is a native plant abundantly available and cheap unlike albumin.

It can be concluded from the present investigation that the natural colorant *Rheum emodi* can be successfully employed to print jute-cotton and jute-wool union fabrics.

Conflict of Interests

The author declares that there is no conflict of interests regarding the publication of this paper.

Acknowledgment

The author expresses Sincere thanks to Ms. Ankita Saxena for her encourage and help to complete this minor project.

References

[1] S. P. Mishra, *Text Book of Fiber Science and Technology*, Newage International, 2005.

[2] B. N. Bandyopadhyay and N. R. Salaskar, "Prospects of jute," *Indian Textile Journal*, vol. 12, no. 5, pp. 12–25, 1997.

[3] T. Bechtold and R. Mussak, Eds., *Natural Colorants-Quinoid, Naphthoquinoid and Anthraquinoid Dyes*, John Wiley and Sons, 2009.

[4] M. S. Akhtar, M. Amin, M. Ahmad, and Alamgeer, "Hepatoprotective effect of *Rheum emodi* roots (*Revand chini*) and Akseer-e-Jigar against paracetamol-induced hepatotoxicity in rats," *Ethnobotanical Leaflets*, vol. 13, pp. 310–315, 2009.

[5] D. Das, S. R. Maulik, and S. C. Bhattacharya, "Colouration of wool and silk with *Rheum emodi*," *Indian Journal of Fibre and Textile Research*, vol. 33, no. 2, pp. 163–170, 2008.

[6] A. K. Samanta and A. Konar, "Dyeing of Textiles with Natural Dyes," in *Natural Dyes*, P. A. Kumbasar, Ed., pp. 30–56, InTech, 2011.

[7] M. Dressler, P. Fischer, and E. J. Windhab, "Rheological characterization and modeling of aqueous guar gum solutions," in *Proceedings of the 3rd International Symposium on Food Rheology and Structure*, pp. 249–253, 2002.

[8] "The Columbia Electronic Encyclopedia," Columbia University Press, 2013, http://cup.columbia.edu/.

[9] "Applications Notes: insight on Color, CIE $L^*a^*b^*$ Color Scale," *HunterLab*, vol. 8, no. 7, pp. 1–4, 2008.

[10] P. B. Tayade and R. V. Adivarekar, "Dyeing of silk fabric with *Cuminum Cyminum* L as a source of natural dye," *International Journal of ChemTech Research*, vol. 5, no. 2, pp. 699–706, 2013.

[11] N. S. Allen, "Photofading mechanisms of dyes in solution and polymer media," *Review of Progress in Coloration and Related Topics*, vol. 17, no. 1, pp. 61–71, 1987.

23

Zirconium Oxychloride as a Novel Mordant for Natural Dyeing of Wool Yarns

M. Forouharshad,[1] M. Montazer,[2] and B. Yadollah Roudbari[1]

[1] *Young Researchers Club, Islamic Azad University, South Tehran Branch, Tehran 1981916791, Iran*
[2] *Textile Department, Center of Excellence, Amirkabir University of Technology, Tehran, Iran*

Correspondence should be addressed to M. Forouharshad; forouharmahdi@gmail.com

Academic Editor: Joanne Yip

Zirconium oxychloride, a well-known flame retardant, is introduced as a novel mordant for wool dyeing. This has been indicated by mordanting of wool yarns with zirconium oxychloride in different conditions and then dyeing with madder as a famous natural dye. The tenacity, color coordinates, washing, and light fastnesses of the dyed wool confirmed the positive influence of zirconium oxychloride as a useful mordant for wool dyeing with a low impact on the color coordinates. Further, an appropriate washing and light fastness were obtained for the zirconium-oxychloride-mordanted wool yarns. The central composite design (CCD) was used to design the experiments with four variables on the results of tensile strength. Statistical analysis confirmed the optimum conditions obtained through the experimental results.

1. Introduction

Wool with high water absorption, good dye ability, high elasticity, reasonable bulkiness, and fire resistance has been used in clothing and floor coverings [1]. The dyeing of wool also poses a challenge for textile chemists due to its complexity of chemical and physical structure [2]. Wool yarns were initially dyed with natural dyes which then the dyeing process improved and transferred from one generation to the other. Although the consumption of natural dyes is high, it has been decreasing by the invention of synthetic dyes. Natural dyes were replaced by synthetic dyes due to their ability to easily match a desired color, increasing variety, high purity, cheap price, and easy processing. However, through production and application of synthetic dyes, producers and consumers have observed several ecological and biological problems [3].

Recently, an interest of using natural dyes in textile coloration has been growing due to the environmental regulation appearing in response to the toxic and allergic reactions associated with synthetic dyes. However, natural dyes are friendly to the environment and exhibit better biodegradability with higher environment compatibility [4].

Madder is a main source of a natural dye producing a variety of anthraquinone dyes in its roots and rhizomes. The main components are di- and tri-hydroxy-anthraquinones, alizarin and purpurin and their derivatives; ruberythric acid (alizarin-primeveroside), pseudopurpurin, and lucidin-primeveroside. Rubiadin, munjisti, quinizarin (1,4 dihydroxyanthraquinone), lucidin, nordamnacanthal, xanthopurpurin, and 1,8-dihydroxyanthraquinone are also identified from plant tissues [5–7]. Farizadeh et al. [8] determined the amount of alizarin, purpurin, and quinizarin in the Iranian madder. It is an old and famous dye for dyeing wool, silk, and cotton fibers [7] which has been widely distributed in southern and southeastern Europe, the Mediterranean area, and in Asia [9]. The chemical structure of more important coloring compounds is presented in Figure 1.

Finding new natural dye and obtaining easy application process are two considerable facts which are interesting for chemists. These are including dye content in the natural source, combination usage, other unknown properties, effect of mordants and auxiliaries on color, light fastness, improvements in production methods, and the development of alternative sources [2]. Gulrajani et al. studied the colorimetric properties of a range of natural yellow dyes,

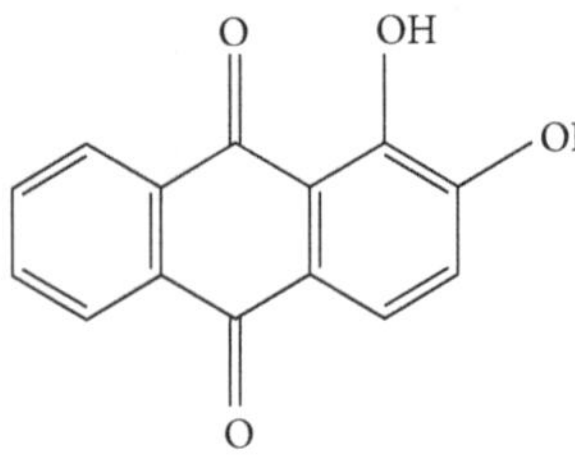

Alizarin

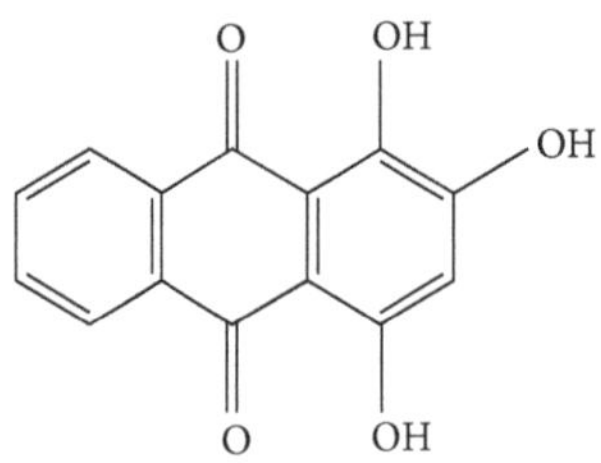

Purpurin

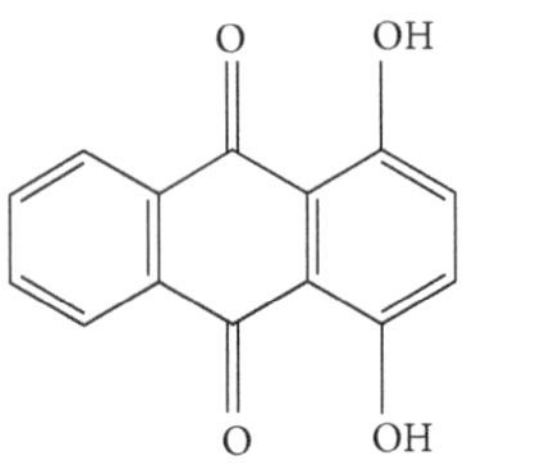

1,4-Dihydroxyanthraquinone

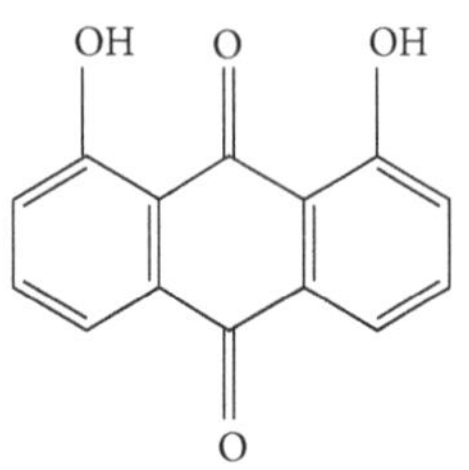

1,8-Dihydroxyanthraquinone

FIGURE 1: Chemical structure of the main compounds of madder.

TABLE 1: Central composite design for mordanting of wool with HCl.

Run number	A: zirconium oxychloride (wt%)	B: temperature (°C)	C: citric acid (o.w.w.%)	D: HCl (37%) (o.w.w.%)	Tenacity (g/tex)
1	10.60	95.00	12.80	5.65	8.62
2	5.60	77.00	12.80	5.65	8.57
3	8.10	70.86	9.55	8.00	8.86
4	8.10	86.00	9.55	8.00	9.21
5	8.10	86.00	9.55	4.05	8.76
6	10.60	77.00	6.30	10.35	8.86
7	8.10	86.00	9.55	8.00	9.17
8	8.10	86.00	9.55	8.00	8.95
9	8.10	101.14	9.55	8.00	7.65
10	8.10	86.00	9.55	8.00	9.03
11	8.10	86.00	9.55	8.00	9.56
12	12.30	86.00	9.55	8.00	8.57
13	5.60	95.00	6.30	10.35	9.12
14	5.60	95.00	12.80	10.35	9.13
15	3.90	86.00	9.55	8.00	8.95
16	8.10	86.00	15.02	8.00	8.34
17	8.10	86.00	4.08	8.00	8.9
18	10.60	77.00	12.80	10.35	9.5
19	5.60	77.00	6.30	5.65	9.52
20	10.60	95.00	6.30	5.65	9.19
21	8.10	86.00	9.55	11.95	8.38

including henna, dolu, kamala, onion skins, and turmeric, with various mordants.Horrocks investigated the color and fastness properties of natural saffron [2]. There are relatively few papers studying the effect of mordant on color [2].

Zirconium compounds have been used on cotton to improve dye fastness and functional properties and on wool to improve its thermal behavior [10–13]. Zirconium has no known biological role, and its compounds are of low toxicity. The human body contains only 1 mg zirconium, and daily intake is approximately 50 μg per day. Zirconium content in human blood is as low as 10 ppb. 70% of plants have no detectable zirconium content or as little as 5 ppb [14].

Forouharshod et al. investigated the effect of zirconium compounds on the flame-retardant wool indicating higher heat resistance on zirconium-salt-treated wool [11–13]. The optimum conditions of wool flame retardant obtained contain 9.74% $ZrOCl_2$, 9.6% citric acid and 6.3% formic acid, at 95°C and 8.96% $ZrOCl_2$, 9.85% citric acid, and 9.01% hydrochloric acid at 92°C [11, 12].

To the best of our knowledge there is no report on mordanting of wool before dyeing; however, there are some reports on flame-retardant wool with zirconium salts. Here, the bi-functionality of zirconium salt on wool as mordant and flame retardant has been introduced. The flame retardancy properties of zirconium salts have already been reported in our previous papers. In this paper the functionality of zirconium salt as a mordant on wool dyeing was focused on. Also the influence of zirconium oxychloride on the physical properties of the wool yarns at different temperatures and acid concentrations during wool mordanting were investigated. Further the colorimetric properties, light fastness, and washing fastness properties of the dyed wool with madder were reported and discussed.

2. Experimental

2.1. Materials. The coarse wool yarn with 432/2 tex and 144 tpm from Iran, a commercial nonionic detergent for wool scouring from local market (Iran), zirconium oxychloride (35% ZrO_2) from Shanghai Yancui Co., China, hydrochloric acid, formic acid, and citric acid from Merck, Germany, and madder roots from local market (Yazd, Iran) in the powder form were used.

2.2. Scouring. The wool yarns were first scoured with 0.5% nonionic detergent at 50°C for 30 min (L : G = 40 : 1), then washed with tap water, and finally dried at room temperature.

2.3. Mordanting. The mordanting bath was prepared with different concentrations of hydrochloric acid, formic acid, citric acid, and $ZrOCl_2$ with L : G = 20 : 1 according to Tables 1 and 2.

Hydrochloric acid or formic acid was added to the bath solution to maintain a pH = 2 and 3 during mordanting, respectively. The mordanting was started at 40°C, continued for 20 min, and then heated to reach the specified temperature for 30 min followed at the same temperature for 45 min.

2.4. Dyeing. Dye solutions were prepared 24 h prior to dyeing by adding madder powder to water (50% o.w.f., L : G = 50 : 1). The dyeing process was started at 40°C, followed by heating at 85°C within 20 min, and remained at the same temperature for 1 h. The pH was adjusted at 5 by using acetic acid solution.

2.5. Tensile Strength. The tensile strength of the mordanted and unmordanted wool yarns was obtained by using an Instron TE-500 from Farayab (Iran) with gauge of 20 cm and a cross-head speed of 25 cm/min. The samples were conditioned according to ASTM D2256 before testing.

2.6. Reflectance. The reflectance of the dyed samples was recorded by using a GretagMacbeth Coloreye 7000A spectrophotometer integrated with an IBM personal computer. CIELAB color coordinates (L^*, a^*, b^*, C^*, and h) was calculated from the reflectance data for 10° angle observer and illuminant D65.

2.7. Color Fastness. The washing fastness properties and degree of staining on the adjacent yarns of the samples were measured according to ISO 105-C01 using a gray scale with rating from 1 to 5.

The color hue changes of yarn and were measured after exposure to the daylight for 2 and 7 days according to ISO 105-B01. The color fastness of various dyed wool samples was assessed by using a blue scale rating from 1 to 8.

2.8. Scanning Electron Microscopy (SEM-EDXS). Morphology of the samples was observed by scanning electron microscopy (VEGA TESCAN, Czech Republic) as the samples were fixed to SEM holders and coated with a thin layer of gold prior to SEM observation.

2.9. Experimental Design. Central composite design (CCD) was used to obtain experimental plan with four variables according to Table 3. Four variables were zirconium oxychloride, formic acid or hydrochloric acid/citric acid, and temperature with the indicated ranges as indicated in Table 3. Also the influence of the variable on the results of tensile strength (C.L.) based on g/tex was fitted in the following second order polynomial function:

$$\text{C.L.} = b_0 + \sum b_i x_i + \sum b_{ij} x_i x_j + \sum c_i x_i^2, \quad i \geq j, \quad i = 1, 2, 3, 4. \tag{1}$$

In this equation, b_0 is an independent term according to the mean value of the experimental plan, b_i is regression coefficient explaining the influence of the variables in their linear form, b_{ij} is regression coefficient of the interaction terms between variables, and c_i is the coefficient of quadratic form of variables.

3. Results and Discussion

3.1. Tensile Strength. The initial modulus, tenacity, and extension at break of the untreated and zirconium-oxychloride treated wool yarns are presented in Table 4. The extension at break for the treated samples is higher than the untreated one due to that the influence of mordanting with zirconium salt led to producing more elastic wool fibers. Also the sample treated with HCl shows a small decrease in tenacity comparing to the sample treated with formic acid thereby resulting in higher adsorption of zirconium salt. Further, the initial modulus of the treated wool with zirconium salt decreases that is lower for the fabric treated in very low pH (HCl) that can be due to the acid impact on the wool fabric.

TABLE 2: Central composite design for mordanting of wool with formic acid.

Run number	A: zirconium oxychloride (wt%)	B: temperature (°C)	C: citric acid (o.w.w.%)	D: formic acid (o.w.w.%)	Tenacity (g/tex)
1	10.30	95.00	12.80	5.65	8.97
2	10.30	77.00	12.80	10.35	8.27
3	4.00	86.00	9.55	8.00	8.92
4	5.60	95.00	12.80	10.35	8.36
5	10.30	95.00	6.30	5.65	9.20
6	7.95	101.14	9.55	8.00	8.57
7	5.60	95.00	6.30	10.35	9.16
8	11.90	86.00	9.55	8.00	8.50
9	7.95	86.00	15.02	8.00	9.50
10	5.60	77.00	12.80	5.65	9.23
11	7.95	86.00	9.55	8.00	9.65
12	7.95	86.00	9.55	8.00	9.71
13	7.95	86.00	9.55	8.00	8.95
14	7.95	86.00	4.08	8.00	8.58
15	5.60	77.00	6.30	5.65	9.94
16	7.95	86.00	9.55	8.00	8.37
17	7.95	70.86	9.55	8.00	8.64
18	7.95	86.00	9.55	11.95	8.85
19	10.30	77.00	6.30	10.35	8.68
20	7.95	86.00	9.55	4.05	8.40
21	7.95	86.00	9.55	8.00	8.58

TABLE 3: Range of variables.

Variable	Lower limited	Upper limited
Temperature (°C)	77	95
Zirconium oxychloride (wt%)	5.60	10.60
Citric acid (o.w.w%)	6.30	12.80
Formic acid (o.w.w%)	5.65	10.35
Hydrochloric acid (37%) (o.w.w%)	5.65	10.35

TABLE 4: Tensile properties of untreated and zirconium-oxychloride treated wool yarns (SD. was less than 5%).

Samples	Initial modulus (g/tex)	Extension at break (%)	Tenacity (g/tex)
Raw wool	49.93 (0.3)	16.20 (4.4)	10.52 (0.9)
Scoured wool	42.47 (0.6)	15.75 (2)	8.70 (0.7)
$ZrOCl_2$-formic acid[a]	33.93 (0.5)	20.33 (2)	8.97 (0.4)
$ZrOCl_2$-hydrochloric acid[b]	27.67 (0.6)	24.75 (4.4)	8.90 (0.9)

[a]Wool treated with 9.74% $ZrOCl_2$, 9.6% citric acid, and 6.3% formic acid at 95°C.

[b]Wool treated with 8.96% $ZrOCl_2$, 9.85% citric acid, and 9.01% hydrochloric acid at 92°C.

3.2. Statistical Analysis. The variance analysis has indicated that all of the expression models are meaningful (Tables 5 and 6). Also, the suitable P value test, which did not pass 0.050 for samples treated with formic acid and HCl, showed the suitability of the models.

Table 5: ANOVA for response surface quadratic model for samples treated with formic acid.

Source	Sum of squares	df	Mean square	F value	P value Prob > F
Model	0.000	0			
Residual	4.75	20	0.24		
Lack of fit	3.31	16	0.21	0.58	0.8081
Pure error	1.43	4	0.36		
Cor total	4.75	20			

Table 6: ANOVA for response surface quadratic model for samples treated with HCl.

Source	Sum of squares	df	Mean square	F value	P value Prob > F
Model	0.000	0			
Residual	4.02	20	0.20		
Lack of fit	3.80	16	0.24	4.31	0.0839
Pure error	0.22	4	0.055		
Cor total	4.02	20			

According to the ANOVA, suitable models for each treatment are

(i) tenacity = 8.98 (for treated wool with formic acid),

(ii) tenacity = 8.91 (for treated wool with hydrochloric acid).

3.3. Color. The color coordinates of zirconium-treated wool yarns dyed with madder are shown in Table 7. A small decrease is indicated in a*, b*, C*, and L* for the treated wool with HCl in comparison with the treated samples with formic acid in the two different processes (filtered dye solution and unfiltered dye solution). The treated wool yarn with zirconium oxychloride in the presence of hydrochloric acid showed a higher adsorption of zirconium on the wool and higher metallic complex formation with wool [11, 12] led to the lower L* comparing to the treated wool with formic acid.

These results are in accordance with the results reported by Montazer and Parvinzaden on dyeing of wool premordanted with alum [3]. Therefore, zirconium oxychloride can be used as a commercial mordant in dyeing of wool with multifunctional purposes.

Also the filtration has a good influence on the lightness of the dye as the filtered dye indicated a higher lightness on the wool yarns. The samples dyed with unfiltered dye showed half lightness comparing to the filtered dye samples due to the higher dye uptake. The color changes in Table 7 indicated the same results.

3.4. Color Fastness. The results of washing and light fastness tests are reported in Table 8. The appropriate washing fastness was obtained on the samples treated with zirconium oxychloride with two diverse methods. Also the samples treated with formic and hydrochloric acids showed appropriate light fastness comparing to untreated wool after 2 days of the

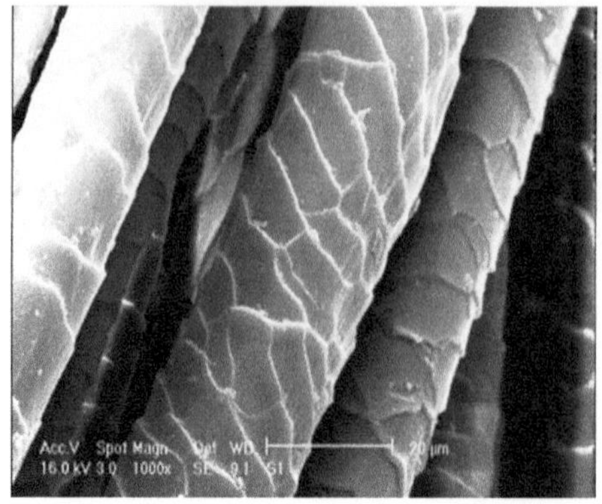

(a)

(b)

Figure 2: SEM of (a) wool treated with 9.74% ZrOCl2, 9.6% citric acid, and 6.3% formic acid at 95°C and (b) wool treated with 8.96% $ZrOCl_2$, 9.85% citric acid, and 9.01% hydrochloric acid at 92°C.

daylight exposure. Zirconium oxychloride and hydrochloric acid led to increased dye penetration into the wool fiber and produced darker shades. Hence, they are less vulnerable to the light [15, 16]. Further, no more fading has been observed on the samples exposed to the daylight for 7 days. Thus, the color fading of the samples was limited to a certain period of the daylight exposure. The results are similar to the results of light fastness obtained on the dyed wool with madder pretreated with alum reported by Montazer and Parvinzaden. Overall, zirconium oxychloride premordanting helps to obtain higher fastness properties [3]. Further, filtration of the madder dye led to higher fastness properties as some impurities removed from the dye assisted better fastness features.

3.5. SEM. The SEM was used to show the effect of mordant on the surface of the fabric. Surface morphology of the treated wool samples is shown in Figure 2. There are no significant changes on the fiber surface of the mordanted wool; however, presence of some particles on the fiber surface may be related to the loaded zirconium mordant. It has been already shown that the adsorption of zirconium is higher with hydrochloric acid in comparison with formic acid [12].

Presence of zirconium and other elements on the wool-treated surfaces is investigated by EDXS analysis and results are reported in Table 9 following the analysis of one particle observed in Figure 2. The treated wool with hydrochloric

TABLE 7: Color coordinates of zirconium treated madder dyed wool with filtered and unfiltered solution dye.

Sample	a*	b*	C*	h	L*	ΔE
$ZrOCl_2$-formic acid[a]	29.80 (1.2)	15.64 (1.0)	33.66 (1.1)	27.69 (1.8)	24.68 (1.0)	20.8
$ZrOCl_2$-formic acid[b]	28.22 (0.8)	17.98 (1.0)	33.46 (0.4)	32.51 (0.5)	45.29 (0.7)	
$ZrOCl_2$-hydrochloric acid[c]	26.39 (0.7)	12.02 (0.4)	29.00 (1.5)	24.48 (0.9)	21.00 (0.7)	24
$ZrOCl_2$-hydrochloric acid[d]	27.69 (0.4)	16.56 (0.1)	32.27 (0.8)	30.88 (0.8)	44.56 (0.4)	

[a] Wool treated with 9.74% $ZrOCl_2$, 9.6% citric acid, and 6.3% formic acid at 95°C and dyed with unfiltered madder dye solution.
[b] Wool treated with 9.74% $ZrOCl_2$, 9.6% citric acid, and 6.3% formic acid at 95°C and dyed with filtered madder dye solution.
[c] Wool treated with 8.96% $ZrOCl_2$, 9.85% citric acid, and 9.01% hydrochloric acid at 92°C and dyed with unfiltered madder dye solution.
[d] Wool treated with 8.96% $ZrOCl_2$, 9.85% citric acid, and 9.01% hydrochloric acid at 92°C and dyed with filtered madder dye solution.

TABLE 8: Wash fastness and light fastness properties of zirconium-treated madder dyed wool yarns with filtered and unfiltered solution dye.

Sample	Wash fastness			Light fastness	
	Wash fastness	Staining on wool	Staining on cotton	After 2 days	After 7 days
$ZrOCl_2$-formic acid[a]	4-5	4-5	4-5	5	4-5
$ZrOCl_2$-hydrochloric acid[b]	4-5	4-5	4-5	5	5
$ZrOCl_2$-formic acid[c]	5	5	5	5	4-5
$ZrOCl_2$-hydrochloric acid[d]	5	5	5	5	5
Untreated wool	3-4	3-4	3-4	3-4	3-4

[a] Wool treated with 9.74% $ZrOCl_2$, 9.6% citric acid, and 6.3% formic acid at 95°C and dyed with unfiltered madder dye solution.
[b] Wool treated with 8.96% $ZrOCl_2$, 9.85% citric acid, and 9.01% hydrochloric acid at 92°C and dyed with unfiltered madder dye solution.
[c] Wool treated with 9.74% $ZrOCl_2$, 9.6% citric acid, and 6.3% formic acid at 95°C and dyed with filtered madder dye solution.
[d] Wool treated with 8.96% $ZrOCl_2$, 9.85% citric acid, and 9.01% hydrochloric acid at 92°C and dyed with filtered madder dye solution.

TABLE 9: Elemental analysis of treated wool (normalized) (wt%).

Sample	Zr	S	Cl	Au	Total
$ZrOCl_2$-formic acid[a]	4.97	2.94	0.27	91.82	100
$ZrOCl_2$-hydrochloric acid[b]	6.74	2.13	0.46	90.68	100

[a] Sample was obtained with 9.74% $ZrOCl_2$, 9.6% citric acid, and 6.3% formic acid at 95°C.
[b] Sample was obtained with 8.96% $ZrOCl_2$, 9.85% citric acid, and 9.01% hydrochloric acid at 92°C.

acid is capable to absorb more salts comparing to the wool treated with formic acid as the more zirconium measured on hydrochloric-treated wool.

4. Conclusion

Zirconium oxychloride, a well-known flame-retardant compound for wool, is introduced as a new mordant for wool dyeing with madder with reasonable properties. This is a colorless mordant for natural dyeing of wool with dye adsorption enhancement properties and light/washing fastness characteristics improvement. Further, using hydrochloric acid in dyeing process produced a darker shade property comparing to formic acid as reported elsewhere. Finally, the optimum mordanting conditions were obtained through using statistical models.

Conflict of Interests

The authors do not have a direct financial relation with the commercial identity mentioned in their paper.

References

[1] M. Montazer, M. Parvinzadeh, and A. Kiumarsi, "Colorimetric properties of wool dyed with natural dyes after treatment with ammonia," *Coloration Technology*, vol. 120, no. 4, pp. 161–166, 2004.

[2] A. R. Horrocks, "Fire retardant materials," in *Textiles*, A. R. Horrocks and D. Price, Eds., pp. 128–181, Woodhead, New York, NY, USA, 2001.

[3] M. Montazer and M. Parvinzadeh, "Effect of ammonia on madder-dyed natural protein fiber," *Journal of Applied Polymer Science*, vol. 93, no. 6, pp. 2704–2710, 2004.

[4] F. A. Nagia and R. S. R. EL-Mohamedy, "Dyeing of wool with natural anthraquinone dyes from Fusarium oxysporum," *Dyes and Pigments*, vol. 75, no. 3, pp. 550–555, 2007.

[5] P. Banyai, I. N. Kuzovklina, L. Kursinszki, and E. Szoke, "HPLC analysis of alizarin and purpurin produced by *Rubia tinctorum* L. Hairy root cultures," *Chromatographia Supplement*, vol. 63, pp. 66–72, 2006.

[6] G. C. H. Derksen, M. Naayer, T. A. V. Beek et al., "Chemical and enzymatic hydrolysis of anthraquinone glycosides from Madder roots," *Phytochemical Analysis*, vol. 14, no. 3, pp. 137–144, 2003.

[7] G. C. H. Derksen, G. P. Lelyveld, T. A. V. Beek, A. Capelle, and E. D. Groot, "Two validated HPLC methods for the quantification of alizarin and other anthraquinones in *Rubia tinctorum* cultivars," *Phytochemical Analysis*, vol. 15, no. 6, pp. 397–406, 2004.

[8] K. Farizadeh, M. E. Yazdanshenas, M. Montazer, R. M. A. Malek, and A. Rashidi, "Kinetic studies of adsorption of madder on wool using various models," *Textile Research Journal*, vol. 80, no. 9, pp. 847–855, 2010.

[9] F. Nakanishi, Y. Nagasawa, Y. Kabaya, H. Sekimoto, and K. Shimomura, "Characterization of lucidin formation in *Rubia*

tinctorum L.," *Plant Physiology and Biochemistry*, vol. 43, no. 10-11, pp. 921–928, 2005.

[10] N. A. Ibrahim, A. R. El-Gamal, M. Gouda, and F. Mahrous, "A new approach for natural dyeing and functional finishing of cotton cellulose," *Carbohydrate Polymers*, vol. 82, no. 4, pp. 1205–1211, 2010.

[11] M. Forouharshad, M. Montazer, M. Bameni Moghadam, and O. Saligheh, "Flame retardancy of wool fabric with zirconium oxychloride optimized by central composite design," *Journal of Fire Sciences*, vol. 28, no. 6, pp. 561–572, 2010.

[12] M. Forouharshad, M. Montazer, M. B. Moghadam, and O. Saligheh, "Flame retardant wool using zirconium oxychloride in various acidic media optimized by RSM," *Thermochimica Acta*, vol. 516, no. 1-2, pp. 29–34, 2011.

[13] M. Forouharshad, M. Montazer, M. B. Moghadam, and O. Saligheh, "Preparation of flame retardant wool using zirconium acetate optimized by CCD," *Thermochimica Acta*, vol. 520, no. 1-2, pp. 134–138, 2011.

[14] J. Emsley, *Nature's Building Blocks*, Oxford University Press, Oxford, UK, 2001.

[15] M. R. Wendelin and C. C. Patricia, "Influence of shade depth on the effectiveness of selected ultraviolet absorbers in reducing fading," *Textile Research Journal*, vol. 63, no. 4, pp. 231–238, 1993.

[16] J. Oakes, "Photofading of textile dyes," *Review of Progress in Coloration and Related Topics*, vol. 32, pp. 21–28, 2001.

24

Knitting Force Measurement on Flat Knitting Machines

A. Fouda, A. El-Hadidy, and A. El-Deeb

Textile Engineering Department, Faculty of Engineering, Mansoura University, Mansoura 35516, Egypt

Correspondence should be addressed to A. Fouda; abdo_faheem@mans.edu.eg

Academic Editor: Nabil Ibrahim

Knittability can be defined as the ability of yarns to run on knitting machines without problems. Knittability can be achieved when less stress is applied on the knitting machine parts by the knitting yarns. This paper presents a novel measuring system for the knitting force needed to perform knitting yarns on flat knitting machine based on data acquisition system (DAS). The proposed system is used to measure the knitting force at different machine settings and different properties of the knitting yarns to determine the optimal production conditions. For this reason, three types of knitted fabric structures (single jersey, Rib 1 × 1, and full cardigan) with three different loop lengths and five different twists of ply yarn were produced. The obtained results showed the optimal yarn ply twist factor (α_e) which gave minimum knitting force (less stress on needles or knitting yarns) at different loop lengths for each structure.

1. Introduction

In textile industry, there is mutually stress between materials and machine parts, such as traveler and spinning yarn on spinning machines, heald eyes and warp threads on weaving machines, and sewing needle and sewing thread, sewing needle and fabrics on sewing machines. In the knitting industry there is a stress between the needle hook and knitting yarn during withdrawal of the new loop and between stem, latch of the needle, and old loop during knock-over process.

Consequently, the study of these stresses helps to design and manufacture each knitting needle and knitting machines parts. Also, it helps to determine knitting yarns specifications in order to preserve the parts of knitting machinery from damage for a longer period and keep the knitting yarns from breakages to avoid all of the defects and low production.

The previous searches aimed to study knitting force theoretically and practically at different knitting variables such as the knitting machine settings (machine speed, type of needles, cam setting, input, and take-down tension), knitted fabric structures, and knitted yarn properties.

Wray and Burns [1, 2] developed a transducer system to measure the dynamic forces between a latch needle and the stitch and guard cams during the formation of a knitted loop.

Dönmez and Marmarali [3] recorded the number of machine stops, yarn breaks, and holes, for Rib 1 × 1 produced with thirty different yarns. They proved that two equations depending on yarn characteristics can be used to predict yarn-needle and yarn-yarn friction.

Hu and Zhu [4] proposed a quantitative method used for assessing the degree of glass filament breakage during the weft knitting, and also they studied and analyzed the effects of different factors such as cam setting, knitted structures, and yarn parameters.

Liu et al. [5] studied the effect of yarn parameters on the knittability of glass ply yarn. They found that a finer fiber diameter, higher twist, and optimized ply structure can make the yarn stronger and more durable during the knitting process.

Kowalski et al. [6] illustrated the characteristics of dynamic thread loads in the knitting zone of weft knitting machines based on a probabilistic model. They found that the forces increase if the knocking-off depth of needles in the knitting zone is increased.

However, all these studies can be considered as an indirect method to demonstrate the knittability because they measured indirect parameters such as number of holes in fabrics and number of thread breakage. Also, most of these studies did not present the optimal settings during the production

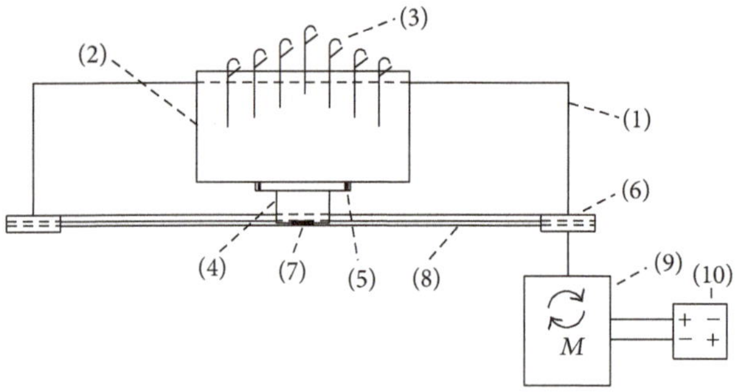

(1) Needle bed
(2) Machine carriage
(3) Latch needles
(4) Connecting piece
(5) Fixed screw
(6) Pulley
(7) Link
(8) Belt
(9) DC motor
(10) Electric current

FIGURE 1: Diagrammatic sketch for flat knitting machine.

of knitted fabric structures which means the production with less stresses on needles or knitting yarns.

Therefore, the first aim of this research is to attempt to measure these stresses dynamically on knitting machine during production under actual operating condition with multiple settings of the machine and disparate yarns specifications. The second aim is to determine the optimal production setting for different knitted fabric structures.

2. Test Methods

To achieve the first aim, a measuring system is implemented practically to collect the data with the help of DAS. The principle of this measuring system is based on the change in a measured voltage as an indication to the change in knitting force. The change in voltage is resulting from a change in electrical resistance of a sensing element (strain gauge) which is affected by the knitting force. For the purpose of data analysis and to achieve the second aim, a subroutine program was written in Matlab environment.

2.1. Sensing Element. The diagrammatic sketch of the flat knitting machine is shown in Figure 1.

The sensing element is fixed on the connecting piece (4) which transfers the reciprocated movement from the DC motor (9) to the machine carriage (2). The reciprocated movement is transferred from the DC motor to the connecting piece by the belt (8) and the pulley (6). The connected piece is a metallic part of stainless steel on the form of double angle leaver 2 mm thickness as shown in Figure 2(a). Two strain gauges each of 120 Ω foil resistance are fixed on the connecting piece as shown in Figure 2(b). The best position of the strain gauge resistances is determined by ANSYS Program.

Due to the up and down of the needles fed with different knitting yarns, stresses are applied on the connecting piece and in consequence cause a tension/compression displacement in the strain gauges, as we see Figure 3.

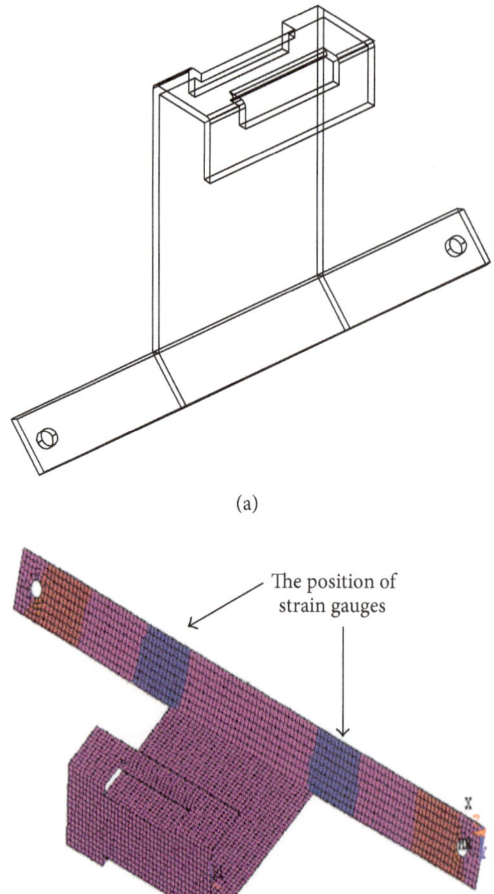

(a)

(b)

FIGURE 2: The connecting piece: (a) isometric shape and (b) ANSYS picture.

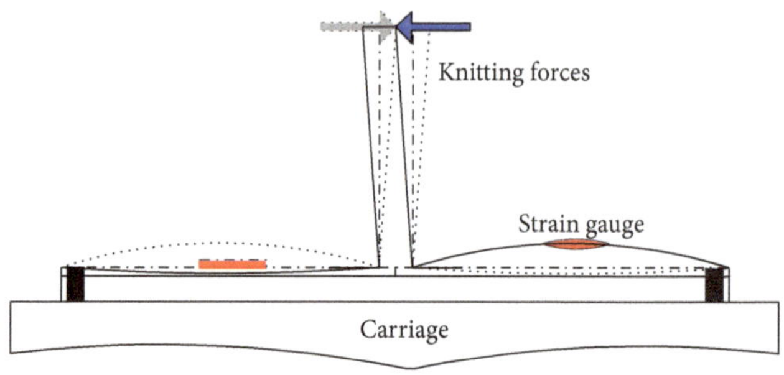

FIGURE 3: The stresses on the sensing element due to knitting force.

According to the change of the strain gauge dimension, the output voltage is changed as a representation of the knitting force.

It can be noted that the change of the output voltage of the strain gauge is small due to its structure technique, so a voltage amplification circuit is needed to allow recording of the output voltage.

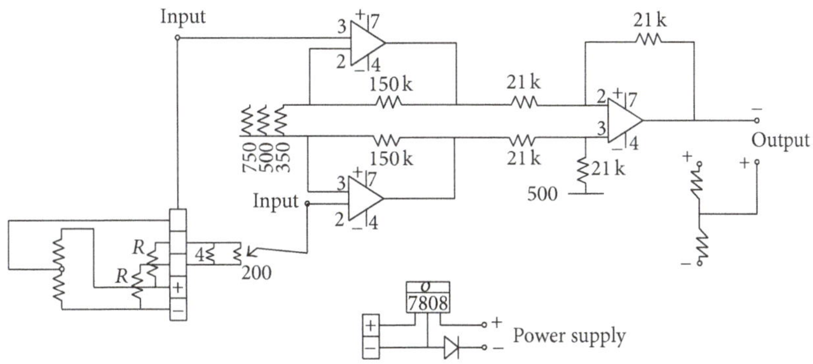

FIGURE 4: Measuring circuit.

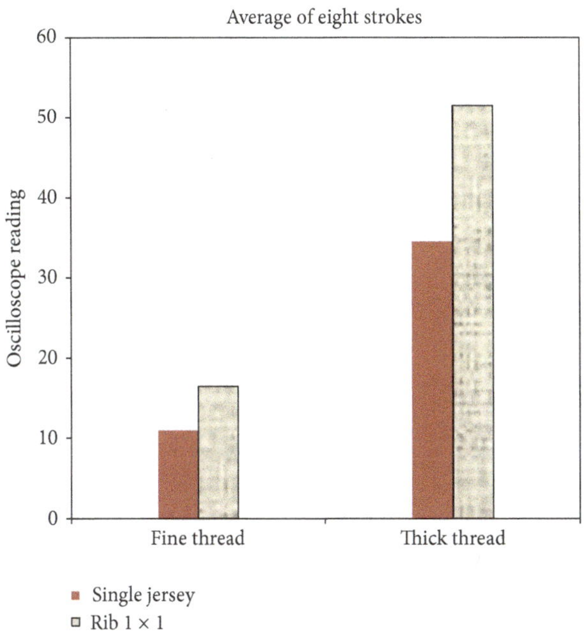

FIGURE 5: Sensitivity of the measuring system.

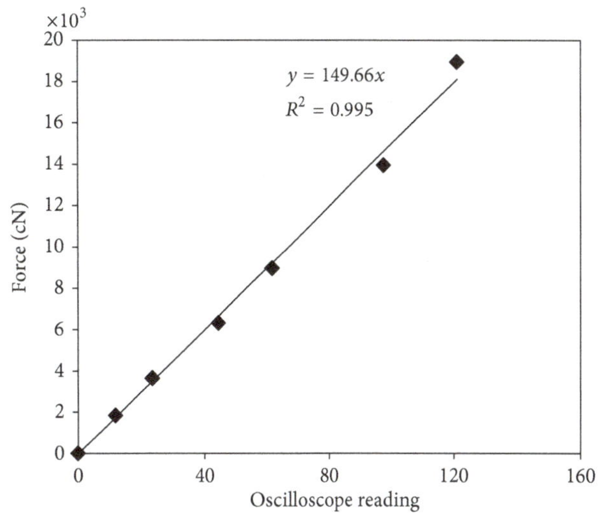

FIGURE 6: Calibration of the measuring system.

The overall measuring circuit is shown in Figure 4 and the output voltage of this circuit is recorded by PCSU1000 digital storage oscilloscope.

2.2. Sensitivity and Calibration of the Measuring System. To insure the sensitivity of the measuring system at different parameters, the signals are recorded for two knitted fabric structures and two different yarn counts. The tested fabric structures are single jersey and Rib 1 × 1.

Figure 5 displays the absolute oscilloscope readings for the average of eight knitting strokes. It is clear that the oscilloscope reading for Rib 1 × 1 structure is higher than the single jersey structure. Also, for each structure the thick yarns give higher reading than the fine yarns. So, it can be insured that the measuring system is sensitive for any change in knitting parameters.

In order to convert the absolute display values of the digital oscilloscope to force unite (cN), a static calibration is executed. A connecting piece is fixed on a straight bar and weights are suspended on the end of a cantilever beam. The weights are increased gradually and then decreased gradually, and the signals are recorded for each weight step.

Figure 6 shows the curve represents the relation between the absolute display values of the digital oscilloscope and the knitting force (cN). With the help of curve fitting, the calibration equation is determined and given by

$$Y = 149.66\,X. \tag{1}$$

2.3. Signal Analysis. In this section, the analysis of the measured signals is illustrated.

Figure 7 shows the recorded signal during the movement of knitting machine carriage from right to left and vice versa. It is clear that there is symmetry between both halves of the recorded signal around the zero line. Therefore, the analysis of the signal is applied for one upper half of the signal which represents the movement of the carriage from right to left as shown in Figure 8.

Figure 8 shows the recorded signal for four strokes for the movement of the carriage from right to left. It should be noted that each stroke has two peaks A and B. Peak A represents the stresses in the knitting zone while peak B represents the stresses applied on the sensing element during the reverse of the carriage direction.

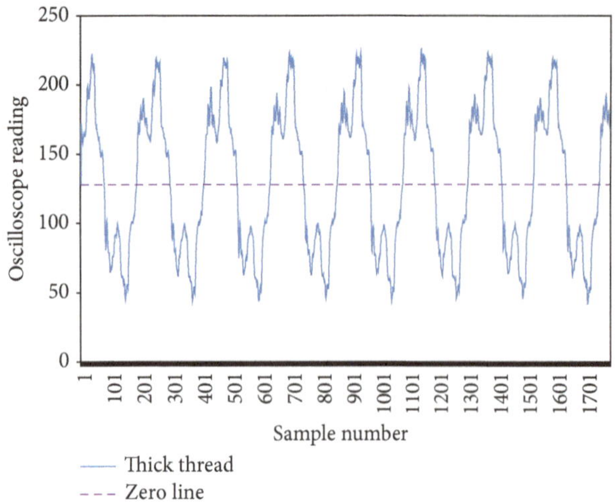

FIGURE 7: Recorded signal in two directions.

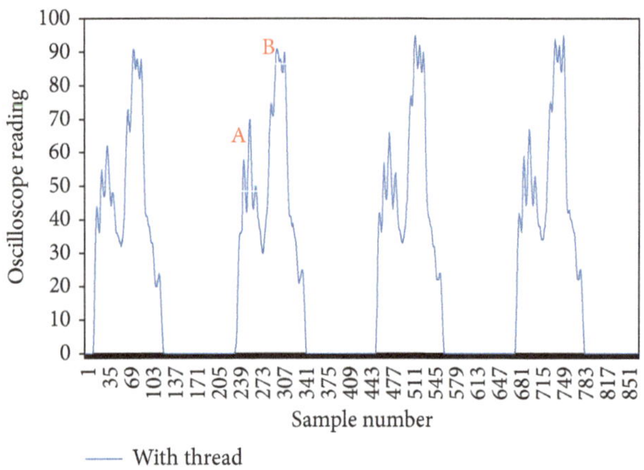

FIGURE 8: Recorded signal during carriage movement from right to left.

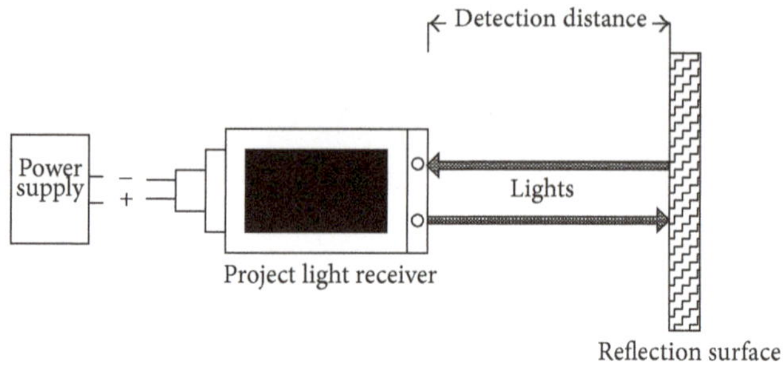

FIGURE 9: Arrangement of the position sensor.

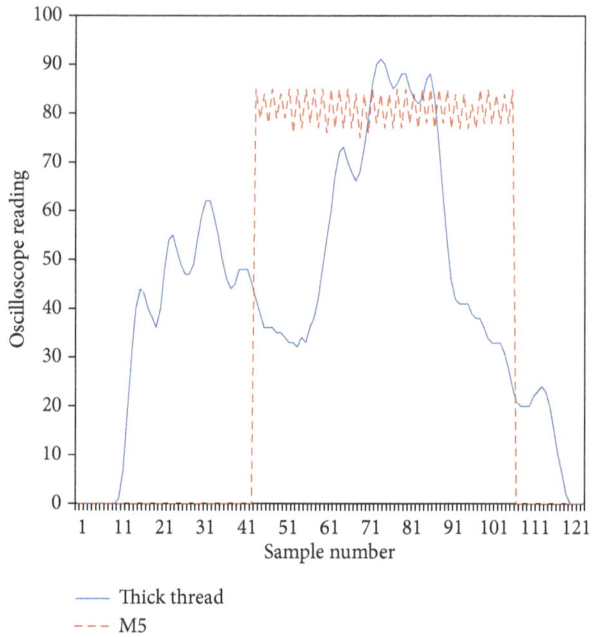

FIGURE 10: Determination of the knitting zone.

For the purpose of study the stresses are during the knitting zone, and two position sensors (project light receiver) are used to determine the knitting zone. The arrangement of the position sensor is shown in Figure 9.

Figure 10 shows the recorded signal from the measuring system for one stroke and the position sensor output during the same stroke.

During the knitting zone the knitting force results from the yarns (S_T) that are estimated by the following steps.

Step 1. The machine is operated without yarns, and the signals are recorded for fifty strokes. The highest value of each stroke (N) is used to estimate the average value (N_a) that is shown in

$$N_a = \frac{\sum_{i=1}^{50} N_i}{50}. \tag{2}$$

Step 2. The machine is operated with yarns, and the signals are recorded for fifty strokes also. With the help of the calibration equation (1), the values of the recorded signals for the operation with and without yarns cases are converted from oscilloscope readings to knitting force (cN). The knitting force results from the yarns (S_T) that are calculated by subtracting the average value (N_a) from the highest value of the signal during operation with yarns (T) for the fifty strokes as shown in

$$S_{Ti} = T_i - N_a. \tag{3}$$

Step 3. The maximum (Max), minimum (Min), average, and the coefficient of variation (C.V) values of (S_{Ti}) are calculated.

The average value of (S_{Ti}) is given in

$$\text{average} = \frac{\sum_{i=1}^{50} S_{Ti}}{50}, \tag{4}$$

where i is the number of stroke from 1 to 50.

In order to analyze the recorded signals directly, a subroutine program is written in Matlab environment as shown by steps in Figure 11.

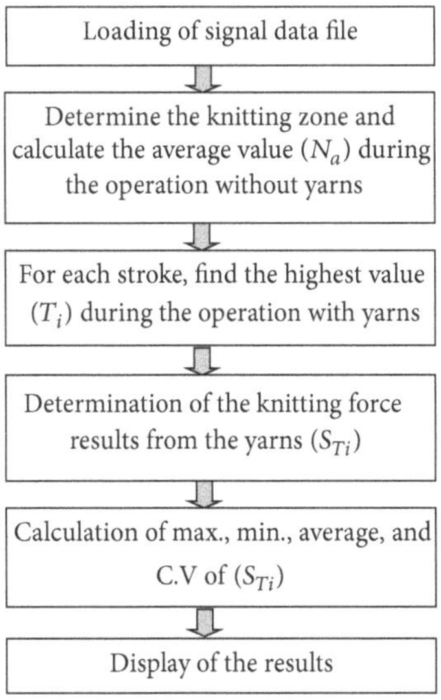

FIGURE 11: Flowchart of the subroutine program steps.

Figure 12 shows the output of the program for a test. As shown in this figure, graph (a) represents the oscilloscope recorded signal of the knitting machine carriage movement in two directions. Graph (b) represents two signals of one stroke for knitting machine carriage with and without yarns. Graph (c) shows the signal of the dynamic knitting forces during the operation with yarns and the average value of the knitting force (N_a) during the operation without yarns. The knitting forces of the fifty successive strokes (S_{Ti}) are shown in graph (d). Finally, table (e) shows the maximum, minimum, average, and the coefficient of variation of the knitting forces.

3. Experimental Work

In this section, the proposed measuring system is used to measure the knitting force for the main knitted fabric structures on PASSAP 600 flat knitting machine.

Table 1 demonstrates the experiment plan for the purpose of testing. Three factors are taken into account during the experimental work. Firstly, three knitted fabric structures (single jersey, Rib 1 × 1, and Full cardigan) were used. Secondly, three different loop lengths are used by changing the stitch cam setting. Finally, five different twists of ply yarn are implemented by varying the twist inserted per meter {0, 100, 150, 200, and 250 (T/m)}. The material used is acrylic and the used yarn count is 8.4/5 Ne.

Figure 13 shows the relationship between the stitch cam setting and loop length for the three knitted fabric structures. It is clear that the loop length is increased linearly with the stitch cam setting for all structures. It should be noted that each structure has a specific range of the stitch cam setting and consequently a range of the loop length {see Figure 13 and Table 1}.

4. Results and Discussions

The effect of the yarn ply twist factor on the knitting force is studied in this section. The test results for three weft knitted fabric structures at three different loop lengths are presented and discussed.

TABLE 1: Experimental plan.

(a)

Structure	Yarn ply twist (T/m)	Stitch Cam Setting		
		7	7.5	8
Single jersey	No yarn	√	√	√
	0	√	√	√
	100	√	√	√
	150	√	√	√
	200	√	√	√
	250	√	√	√

(b)

Structure	Yarn ply twist (T/m)	Stitch Cam Setting		
		5.5	6	7
Rib 1 × 1	No yarn	√	√	√
	0	√	√	√
	100	√	√	√
	150	√	√	√
	200	√	√	√
	250	√	√	√

(c)

Structure	Yarn ply twist (T/m)	Stitch Cam Setting		
		3.5	4.5	5
Full cardigan	No yarn	√	√	√
	0	√	√	√
	100	√	√	√
	150	√	√	√
	200	√	√	√
	250	√	√	√

4.1. Single Jersey Fabrics. One sample of a single jersey fabric structure is produced and the signal is recorded by the proposed measuring circuit and processed by the written program as discussed in the previous sections. Figure 14 illustrates a printout of the written program output for tested single jersey fabric structure at yarn ply twist factor 4 (α_e) and loop length 16 mm.

The test is repeated for the same fabric structure and different yarn ply twist factor and loop length as discussed in Section 3. To study the knitting force and to determine the optimum setting, the average knitting force is recorded during each test. This test plan is implemented for all produced fabric structures in this research.

Figure 15 shows the effect of yarn ply twist factor on the knitting force in single jersey knitted fabrics at different levels of loop length.

It should be noted that the knitting force decreases significantly as the yarn ply twist factor increases for the range from 0 to 1.98 (α_e) of the ply twist factor for the three levels

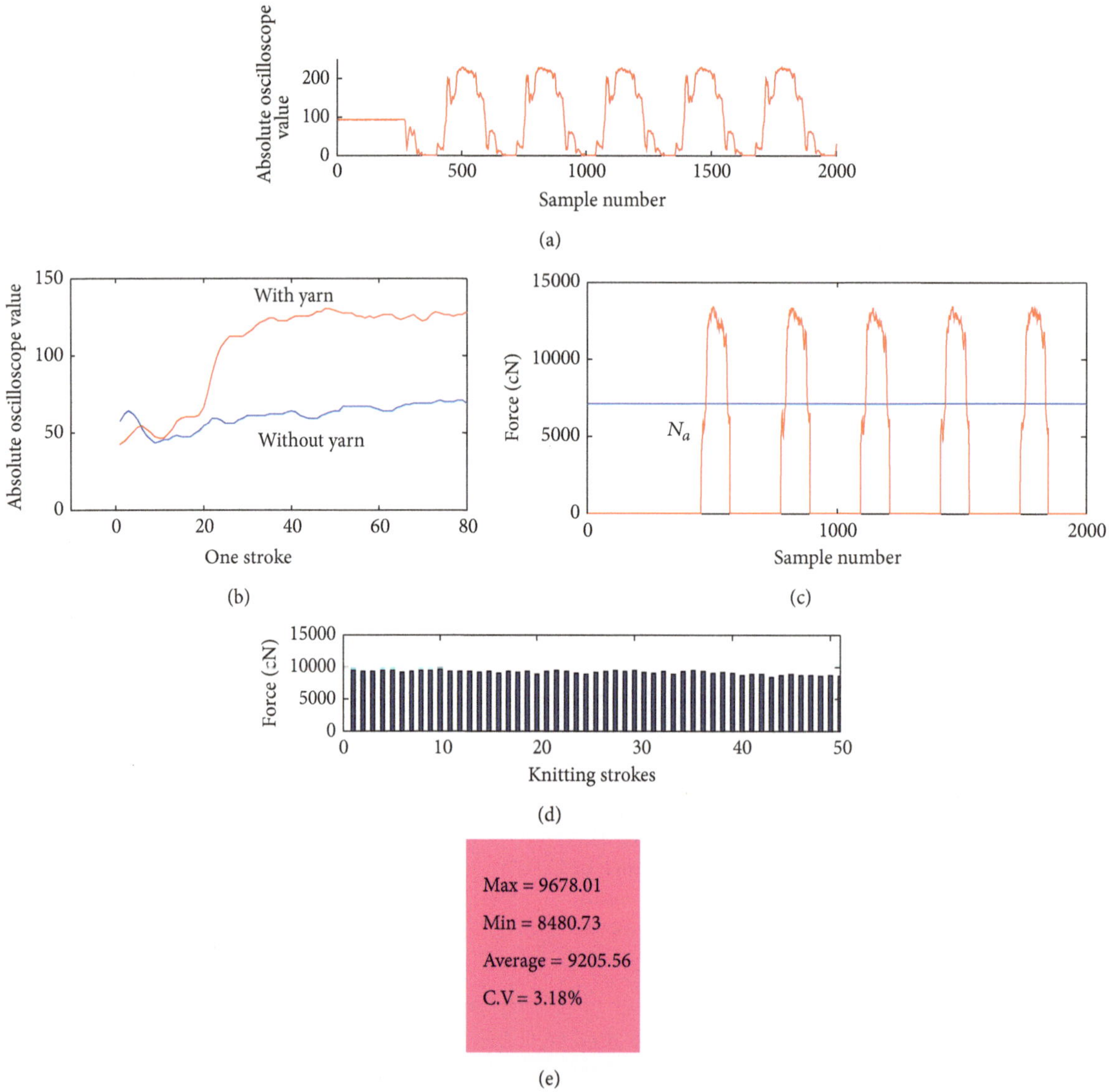

FIGURE 12: The output of the subroutine Matlab program.

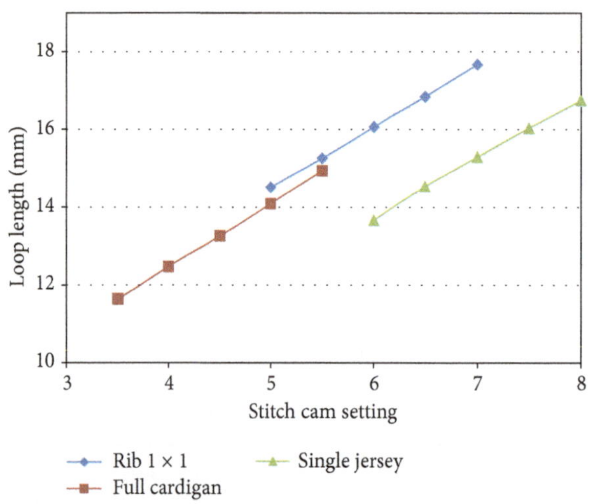

FIGURE 13: Effect of cam setting on loop length.

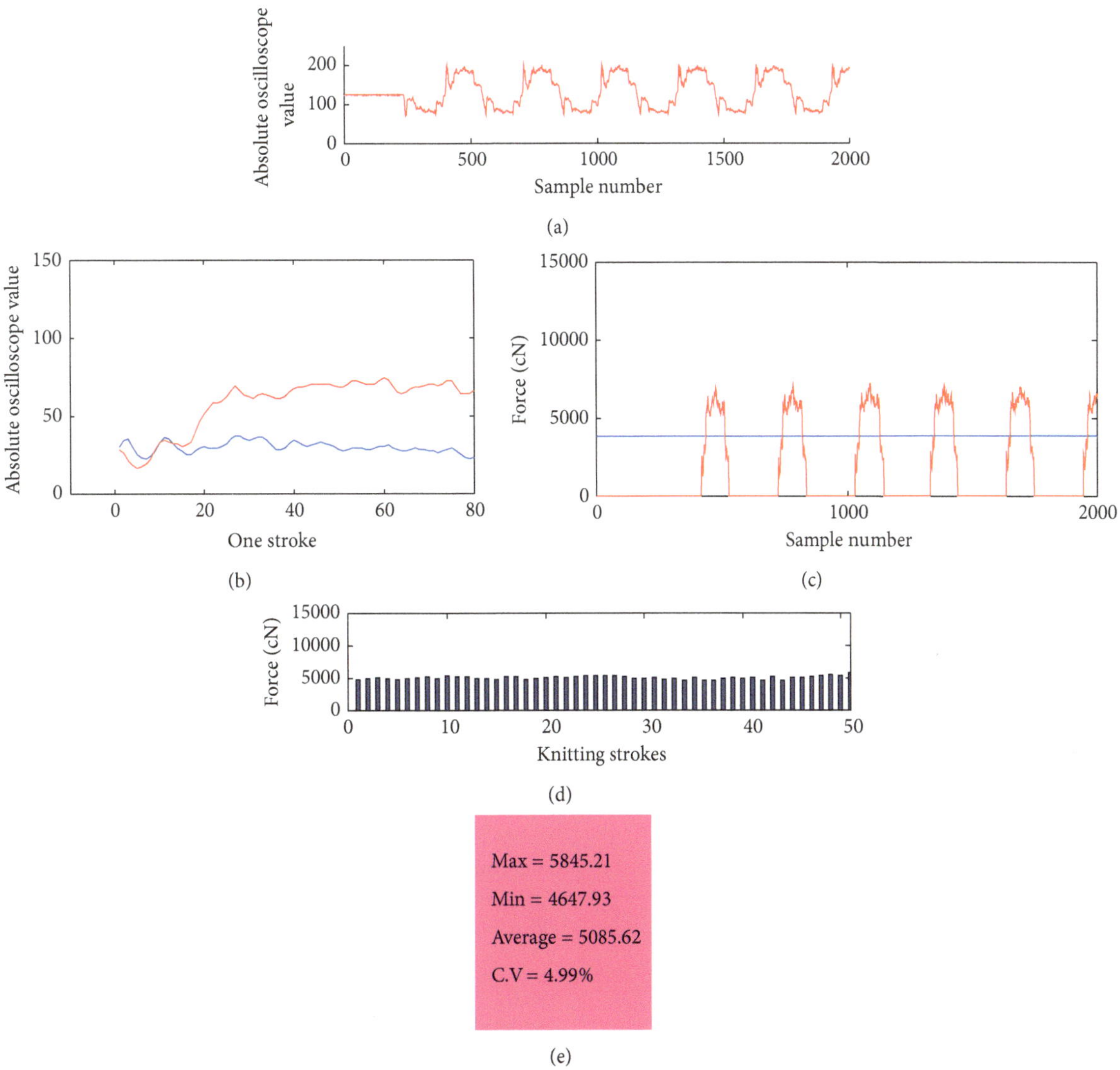

FIGURE 14: Output program for one single jersey samples.

of loop length. It is may be justified as the increase of the yarn ply twist factor for a specific level decreases the friction area between the yarn and the needle hook and consequently decreases the knitting force.

It also can be found that if the yarn play twist factor is increased than 1.98 (α_e), the knitting force ascends in a fluctuated trend. It is because increasing of the yarn ply twist factor for higher levels increases the pending rigidity of the yarn.

It is evident from Figure 15 that the behavior of the knitting forces is nearly the same for the three levels of the loop length. However, as the loop length is increased, the knitting force increased, due to increased moved distance of old loop on needle stem.

It should be noted that the yarn ply twist factor of 1.98 (α_e) has less knitting force for all levels of loop lengths especially at 15.3 mm. It is obvious that at the loop length of 15.3 mm the knitting force is decreased by 36% compared with its value for the loop length of 16.7 mm and it is decreased by 32% approximately compared with its value at zero yarn ply twist factor. This point can be considered as the optimum setting for single jersey fabric structure which gives minimum knitting force.

4.2. Rib 1 × 1 Fabrics. It is well known that Rib 1 × 1 knitted fabrics has a different method of the yarn track during the formation compared to single jersey knitted fabrics. In Rib 1 × 1 knitted fabrics the yarns pass from the front needles to the back needles and vice versa, and therefore, the knitting force increases.

Figure 16 illustrates the effect of yarn ply twist factor on the knitting force in Rib 1 × 1 knitted fabrics at different levels of loop length.

It can be seen that the knitting force decreases dramatically as the yarn ply twist factor increases for the range from 0 to 1.98 (α_e) of the ply twist factor and it increases significantly for yarn ply twist factor over 1.98 (α_e) regardless of the loop length. Also, it can be observed that the loop length has no significant effect on the knitting force.

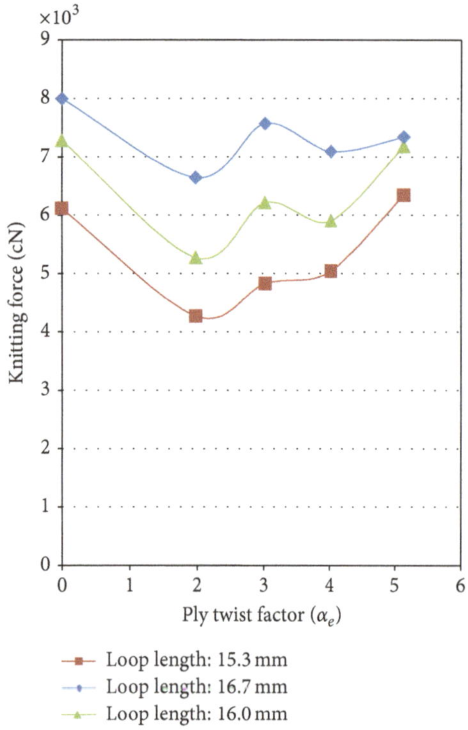

Figure 15: Effect of yarn ply twist factor on knitting force in single jersey fabrics.

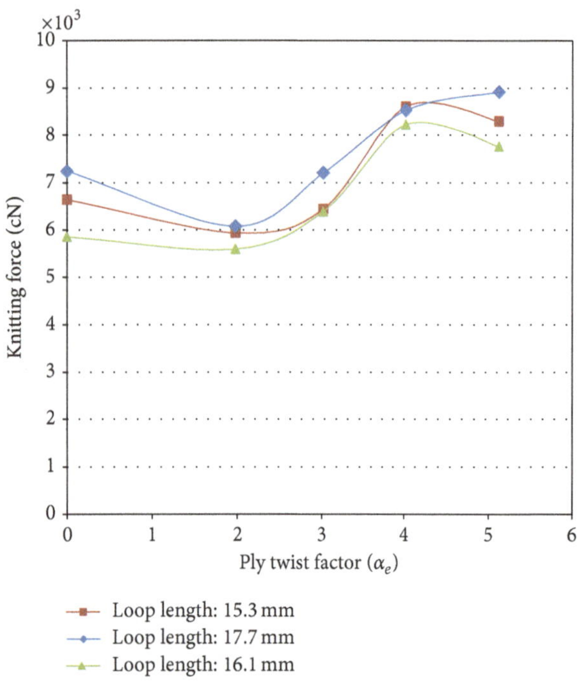

Figure 16: Effect of yarn ply twist factor on knitting force in Rib 1 × 1 fabrics.

It should be noted that the knitting force in Rib 1 × 1 is higher than single jersey fabric structure {see Figures 15 and 16}.

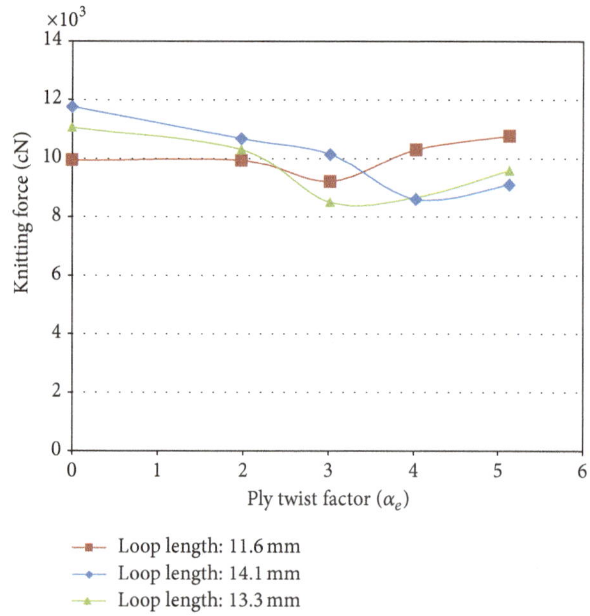

Figure 17: Effect of yarn ply twist factor on knitting force in full cardigan fabrics.

It is clear that, from Figure 16, at yarn ply twist factor of 1.98 (α_e) the knitting forces decreased by 8% compared with its value at zero yarn ply twist factor. This point can be considered as the optimum setting for Rib 1 × 1 fabric structure which gives minimum knitting force regardless of the loop length.

4.3. Full Cardigan Fabrics. The knitting force for full cardigan knitted fabric is higher than the tested previous knitted fabrics (single jersey and Rib 1 × 1). That is because there is an accumulation of yarns inside the needles hooks, as a result the friction area between the yarn and the needle hook during the casting-off and loop formation is increasing.

Figure 17 demonstrates the effect of yarn ply twist factor on the knitting force in full cardigan knitted fabrics at different levels of loop length.

It is clear from Figure 17 that there is a behavior similarity of the knitting force at different yarn ply twist factors and different loop lengths. It can be seen that the loop length level has no significant effect on the knitting force.

It should be noted that the minimum knitting force can be achieved over the range from 3 to 4 (α_e) of yarn ply twist factors because the knitting force is decreased by 27% approximately compared with its value at zero yarn ply twist factor. This range of the ply twist factor can be considered as the optimum setting for full cardigan regardless of the loop length.

It obvious that the optimal setting is achieved at a higher value of ply twist factor because there is an accumulation of yarns and it is likely to increase the friction area as discussed later, but yarn twisting decreases this area and consequently decreases the knitting force.

5. Conclusion

In this research, a novel measuring system for the knitting force in a flat weft knitting machine has been presented and implemented practically. The recorded data from the measuring system has been calibrated to match the knitting force. The analyzed results demonstrated that the proposed system worked effectively and efficiently. In addition, the optimum machine setting has been determined for three well-known fabric structures taking into account the yarn ply twist factor and loop length level.

Conflict of Interests

The authors declare that there is no conflict of interests regarding the publication of this paper.

References

[1] J. J. F. Knapton and D. L. Munden, "A study of the mechanism of loop formation on Weft-knitting machinery—part I: the effect of input tension and cam setting on loop formation," *Textile Research Journal*, vol. 36, no. 12, pp. 1072–1080, 1966.

[2] Wray and Burns, "Analysis of the dry-relaxed knitted loop configuration, part:1 two-dimensional analysis," *Journal of the Textile Institute*, vol. 58, no. 8, pp. 329–351, 1967.

[3] S. Dönmez and A. Marmarali, "A model for predicting a yarn's knittability," *Textile Research Journal*, vol. 74, no. 12, pp. 1049–1054, 2004.

[4] H. Hu and M. Zhu, "A study of the degree of breakage of glass filament yarns during the weft knitting process," *Autex Research Journal*, vol. 5, no. 3, pp. 141–148, 2005.

[5] X.-M. Liu, N.-L. Chen, and X.-W. Feng, "Effect of yarn parameters on the knittability of glass ply yarn," *Fibres and Textiles in Eastern Europe*, vol. 16, no. 5, pp. 90–93, 2008.

[6] K. Kowalski, B. Włodarczyk, and T. M. Kowalski, "Probabilistic model of dynamic forces in thread in the knitting zone of weft knitting machines, allowing for the heterogeneity of viscoelasticity yarn properties," *Fibres and Textiles in Eastern Europe*, vol. 81, no. 4, pp. 61–67, 2010.

25

Defect Detection Techniques for Airbag Production Sewing Stages

Raluca Brad,[1] Lavinia Barac,[2] and Remus Brad[2]

[1] *Industrial Machinery and Equipment Department, Lucian Blaga University of Sibiu, B-dul Victoriei 10, 550024 Sibiu, Romania*
[2] *Computer Science and Electrical Engineering Department, Lucian Blaga University of Sibiu, B-dul Victoriei 10, 550024 Sibiu, Romania*

Correspondence should be addressed to Raluca Brad; raluca.brad@ulbsibiu.ro

Academic Editor: Nabil Ibrahim

Airbags are subjected to strict quality control in order to ensure passengers safety. The quality of fabric and sewing thread influences the final product and therefore, sewing defects must be early and accurately detected, in order to remove the item from production. Airbag seams assembly can take various forms, using linear and circle primitives, with threads of different colors and length densities, creating lockstitch or double threads chainstitch. The paper presents a framework for the automatic detection of defects occurring during the airbag sewing stage. Types of defects as skipped stitch, missed stitch, or superimposed seam for lockstitch and two threads chainstitch are detected and marked. Using image processing methods, the proposed framework follows the seams path and determines if a color pattern of the considered stitches is valid.

1. Introduction

The term of product quality refers primarily to the product ability to fulfill the purpose for which it was designed, while performing a quality control assumes a regular check of quality products variables. Therefore, the quality features which will be subsequently tested need to be defined. The textiles products are assessed according to several criteria, depending on their destination. For instance, technical textiles have to meet close specifications of weight, flexibility, and strength, while in the case of apparel, quality characteristics as design, color, and comfort are important. The airbag is a safety device designed for modern vehicles, required to inflate quickly in the case of a collision, preventing the passengers from crashing onto the cars' interior [1]. The concept of airbag was introduced in the late 1960s, but it was increasingly employed after 1990, due to market requirements and imposed legislation. Generally, the air cushion is woven from nylon filament yarns suitable due to the high strength-to-weight proportion, favorable elongation, adequate thermal properties, and relatively low cost of production. In average, to produce a driver's airbag, about 1.42 meters of material is needed.

Other mandatory features for airbag are high tensile strength, seam stability, air permeability, and ability to be folded and stored in dark for years, without suffering any damage [2]. To fulfill the designed functionality, it is necessary that all these properties are valid. Therefore, the quality control of airbag production is extremely important, because a minor defect can influence the cushions performance. Quality control is very strict for airbags, especially in the areas of propulsion, static, and dynamic inflation and of fabrics and seams quality.

In the case of fabric quality, critical parameters refer to: tensile strength and elongation of weaving yarns, tensile strength of material, fabric density and weight, air permeability (both for static and dynamic air) and slip resistance of the seam [3]. Thus, a rigorous testing and quality control starting from the yarns used in weaving to the final sewing operation is absolutely necessary.

In the past decades, seams quality control was visually performed by human experts pursuing two goals: a continuous control to prevent the sale of poor quality products and setting the optimal parameters for sewing machines. One of the first automated techniques for seam control was reported in [4], employing a laser to scan the

Table 1: Stitch types in airbag sewing process [14, 23, 24].

Stitch type	Seam class	Seam line	Sewing operation
301 lockstitch	SSa-1/1.01	Straight	Two reinforcements sewing on the lower panel
	LSbj-1/5.30.01	Circular	Vent reinforcements sewing on the lower panel
	SSv-1/5.01	Circular	Generator reinforcements and cover sewing on the lower panel subassembly
401 two threads chainstitch	SSa-1/1.01	Straight	The bonding sewing of the ends of the upper and lower panels
	SSa-1/1.01	Straight	Perimeter sewing
	SSa-1/1.01	Straight	Reinforcement sewing on the bottom panel
	SSa-2/1.01	Straight	The closure sewing

stitches. Also, piezoelectric or early vision methods were presented in the referenced paper. Among the electromechanical detection systems for irregular formed stitches, the counting of thread consumption and tension is presented in [5, 6].

In the area of image processing, an automatic detection of stitch defects was presented in [7], where two classes have been identified: seam and cloth defects. The texture features have been used to classify the defects. Another contribution is presented in [8], where the seam line is detected using the Radon transform, offering a robust recognition: invariant to illumination, motion, and noise. While trying to automate the control process, Bahlmann et al. [9] proposed a method using neural networks for the detection and seams classification. The solution consists of a features' extraction using Discrete Fourier Transform and a neural network training/recognition stage. The Fourier spectrum will reflect the fabric puckering due to a tight stitch. The Hough transform for line detection is used in the paper of Yuen et al. [10] for stitch inspection. The difficulty of the method lies in the binarization of acquired images before the Hough detection. The authors of [11] offer a solution for seams assessment employing a feature extraction and a self-organizing map. In this case, the seam pucker defect is emphasized, as another major nonconformity caused by uneven thread tension or material feed. The applications of neural networks in the field of textile are subjected to a comprehensive review paper in [12].

To prevent the spread of sewing defects in the subsequent stages, a visual inspection point is inserted before achieving the final closure seam. In this checkpoint, the fixing stitches of the 3 reinforcement and the upper and lower panels' assembly are inspected. Our application can successfully perform sewing defect detection, considering the seams paths and detecting defects such as threads breakage, stitch skipping, double stitching, dimensional nonconforming, uneven stitch, or missing stitch.

The paper is organized as follows: Section 2 presents the materials implied in the airbag sewing stage, together with seams types and possible defects. In Section 3, the image processing stages for nonconformities detection are described, while the results of linear and circular seam inspection are shown in Section 4, followed by conclusions.

2. Seam Nonconformities in Airbag Assembly Stage

The material that fulfills the constructive function of the product consists of a plain fabric of 6.6 polyamide coated with thin film of silicon. During the weaving process, the following specifications have to be satisfied: a density of 46×46 ends and picks/inch, 470 dtex yarn count, and a total of 68 filaments S twisted per yarn. As soon as a roll was removed from the loom, the fabric is coated in order to acquire certain properties: an improved abrasion resistance, a better processability in cut and sewn technology, and controlled air porosity [13]. The material follows the technological flow and gets to the cutting stage, after then parts of the product are assembled using the sewing threads.

Assembling the components of a passenger airbag is accomplished with seams that have different paths depending on the part shape and can be straight or circular (Table 1). In order to be accurate, sewing patterns are used during the process; however 0 defects are not achievable.

Two types of stitches are being used: the 301-type lockstitch [14], a resistant stitch which stiffens more layers and the 401-type double-locked chainstitch [14], imposing a seam with high elasticity and strength, which is usually placed where shock is applied. However, compared with the lockstitch, the chainstitch is easier frayed. The lockstitch is formed by a red needle thread passing through the fabric and interlocking with the green bobbin thread, used for reinforcement grip seam on the bottom, vent, and generator reinforcements seam. The chainstitch is obtained with the needle red thread and the orange looper thread, being used for perimeter seam and bonding of panels seam.

The failure of a component, such as the seams, can result in failure of the whole product, regardless the quality of other components or fabrics used in the product. Factors like wear or break of sewing threads before the fabric degradation, punching of fabric during the sewing process, and a low slip resistance of seam can lead to a faulty seam. The potential causes can be controlled by the sewing variables: type of sewing machine, size of sewing needle, sewing speed, sewing thread type, and stitch length [15]. The problems caused by faulty seams could be avoided by a novel technology applying a laser welding process as an alternative joining technique of PA or PES fabrics [16].

Whatever the stitch type is, a sewing defect is an unaesthetic appearance of a joint as a result of incorrect stitching and will produce one of the following nonconformities.

(i) Seam puckering—a local defect appearing as material wrinkling near the stitch.

(ii) Open seam—undesirable gap in a seam caused by thread breakage or fabric yarn breakage.

(iii) Seam slippage—fabric yarns loosened from the seam by pulling or pressing.

(iv) Stitch skipping—stitches are incompletely formed or skipped during the assembly.

Seam appearance and performance are influenced by thread features like type, structure, twist, ply, color matching, finishes, and size [17–19]. Due to a specific requirement of durability and abrasion resistance, nylon twisted multifilament thread is used in assembling the passenger airbag parts (top panel, lower panel, reinforcements, and cover). For a reliable detection and strength seam monitoring, sewing threads of different colors and linear density are used: green thread 1401 dtex, orange thread 932 dtex, and red thread 1401 dtex. In the most important points, which directly affect the functionality of the airbag, less fine threads with high tensile strength are employed.

3. A Method for Airbag Seam Inspection

In our practical approach, we have considered two types of seam paths: linear and circular or partial circular (arcs). Three sewing nonconformities have been considered for detection: missing stitches, skipped stitches, and superimposed seam. The corresponding image processing steps for each class of defects are presented in Figure 1.

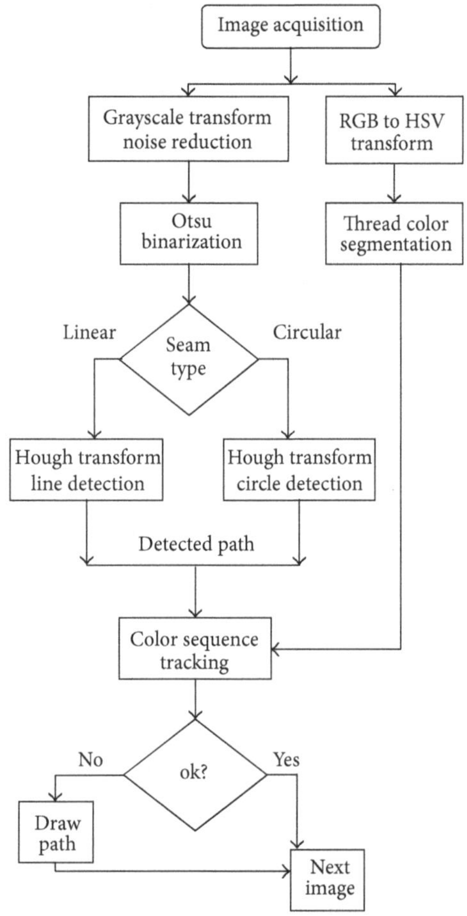

Figure 1: Flowchart of image processing.

In view of linear or circular path recognition, the image is converted from color to a grayscale format and smoothed for noise attenuation, using a Gaussian filter. The image binarization was performed using a version of the Otsu algorithm, which assumes that each of the two classes (object and background) is characterized by a range of intensities. Thus, the image histogram is bimodal and the optimal threshold will separate the two classes of pixels [20]. This process involves iteration for all possible threshold values, creating classes of objects (foreground) and background. The optimal threshold minimizes the interclass variance, defined as the weighted sum of the variances of each class (1). Consider

$$\sigma_w^2(t) = w_1(t)\sigma_1^2(t) + w_2(t)\sigma_2^2(t), \quad (1)$$

Where t is the threshold and $w_i(t)$, $i = \overline{1,2}$ are the class probabilities estimated as

$$w_1(t) = \sum_{i=1}^{t} p(i), \qquad w_2(t) = \sum_{i=t+1}^{I} p(i) \quad (2)$$

and the average of each class are given by

$$\mu_1(t) = \sum_{i=1}^{t} \frac{ip(i)}{w_1(t)}, \qquad \mu_2(t) = \sum_{i=t+1}^{I} \frac{ip(i)}{w_2(t)}, \quad (3)$$

where the individual class variances are

$$\sigma_1^2(t) = \sum_{i=1}^{t} [i - \mu_1(t)]^2 \frac{p(i)}{w_1(t)},$$
$$\sigma_2^2(t) = \sum_{i=t+1}^{I} [i - \mu_2(t)]^2 \frac{p(i)}{w_2(t)}. \quad (4)$$

The detection of linear seams is made using the Hough transform [21]. The basic idea behind is to detect straight lines in an image assessing the pixels position and their collinearity. All the pixels belonging to the object class, from the previous stage of binarization, will be considered based on their coordinates. It allows passing from the Cartesian coordinate system to the polar system characterized by the distance ρ between the line and the origin and the angle θ of the vector from the origin to the nearest point of the line using

$$\rho = x\cos\theta + y\sin\theta. \quad (5)$$

Each line in the Cartesian space will have a unique pair of coordinates (ρ, θ) in the Hough space. Therefore, a point of coordinates (x_0, y_0) and all the lines passing through it will determine an infinity of pairs (ρ, θ) arranged on a sinusoidal

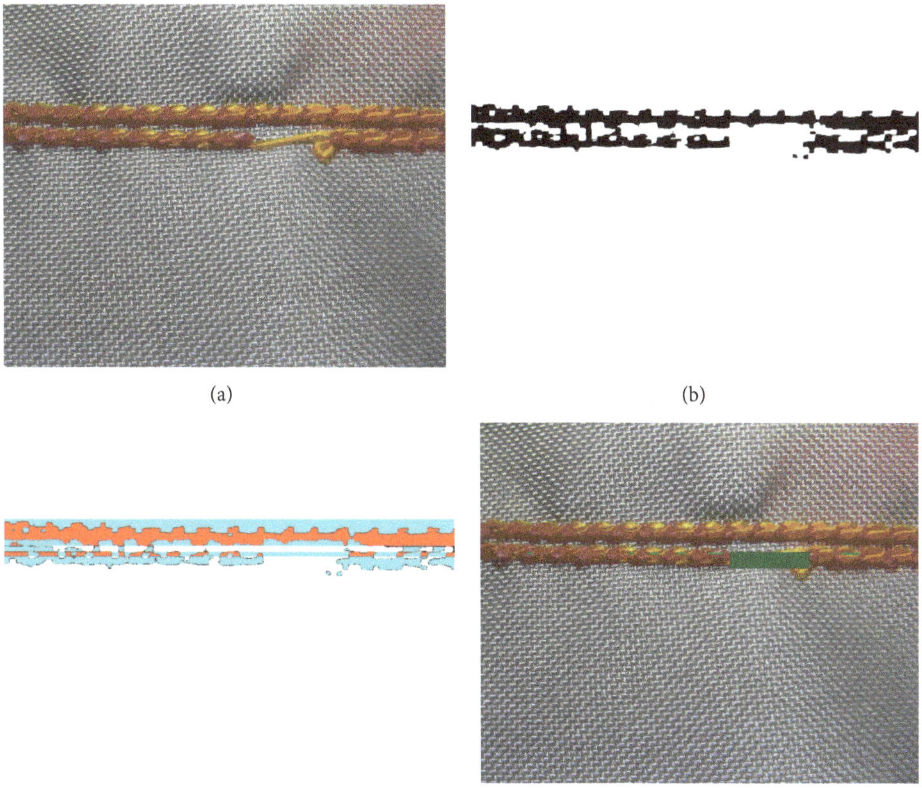

(a) (b) (c) (d)

FIGURE 2: Original image of an airbag double seam with (a) missing stitches; (b) binarized image; (c) color segmentation; and (d) defect detection result.

curve in the Hough space. A set of points that forms a line in the Cartesian space will produce a set of sinusoids intersecting in the point establishing the line parameters.

Using pairs of binarized pixels, the transform is applied for the area of interest and the intersections of sinusoids in Hough space are discovered. To retain the information about intersections, a data structure called accumulator will be used. It records the intersections coordinates and the number of sinusoids superimposed. Finally, the accumulator will be inspected for values greater than a predetermined threshold. Their corresponding coordinates will be converted back into the Cartesian space, to be displayed as lines.

The detection of circles is useful in determining the circular contour seams; therefore, the Hough transform for circles uses the same accumulator structure to retain information regarding the detected circle center [22]. Iteratively, for each pixel belonging to the structure binarized in the previous step, a circle of radius R is drawn in the accumulator (a two-dimensional array of image size). The local maxima represent the cells with highest drawn circle intersections, consequently the center of the detected circle of radius R. A priori knowledge of the circular seams radius will require less iteration for R. As a result, detecting the center and the radius of the circle will describe the circular seam.

As presented in the previous section, colored threads are used for the airbags seams. In view of a sewing tension control and monitoring and stitch line detection, color segmentation has to be used. As the image is acquired in RGB mode, precise color segmentation was performed in HSV color space. The corresponding red and green values for the threads are characterized by precise hue and saturation parameters and were employed in the classification of pixels. Colors are scanned on a local neighborhood sized 5×1, perpendicular to and along the previously detected shape, following three types of patterns, and corresponding to the nonconformities to be processed. If the rule is not respected, the pixels corresponding to the linear path are displayed in green, while red boxes mark the missing pattern for the circular seam.

4. Results

The algorithm for linear or circular seams control is analyzing the colors being present in the acquired image along the recognized lines or arcs by the Hough transform. A sequence of color appearances will be pursued and if an incorrect one is detected, a defective seam region will be revealed.

The image processing stages of straight seams detection are presented in Figure 2. First, two operations will take place: an image conversion to gray levels and noise removal by smoothing, in order to diminish the sensor noise. The next step will be the image binarization using the Otsu threshold. In this case, the object of interest is the seam and will be displayed with black pixels, while the fabric will be labeled as background (Figure 2(b), with a computed threshold value

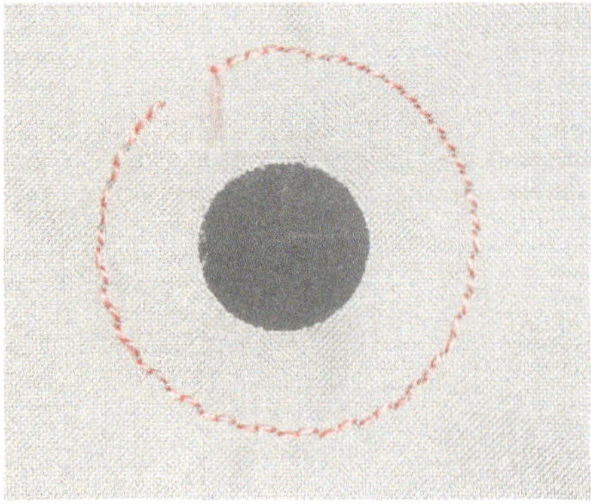

(a)

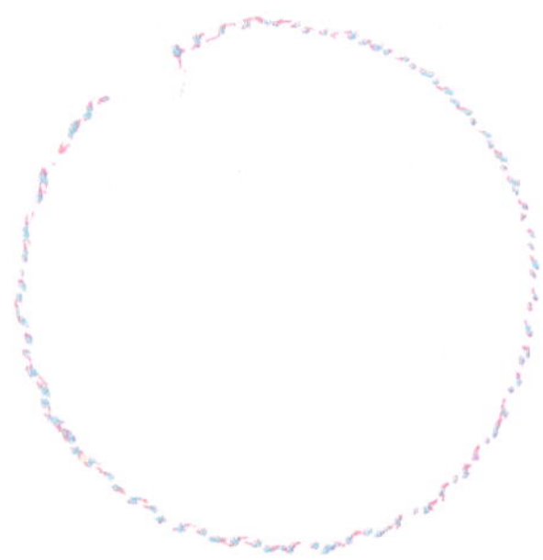

(b)

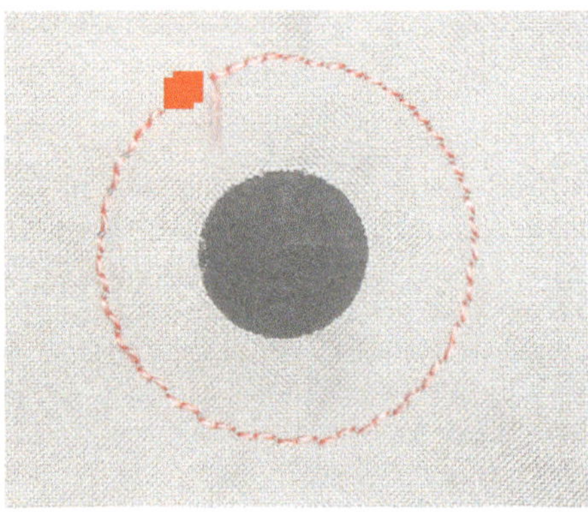

(c)

Figure 3: Circular seam with thread missing stitches: (a) vent reinforcement seam image, (b) color segmentation, and (c) defect detection.

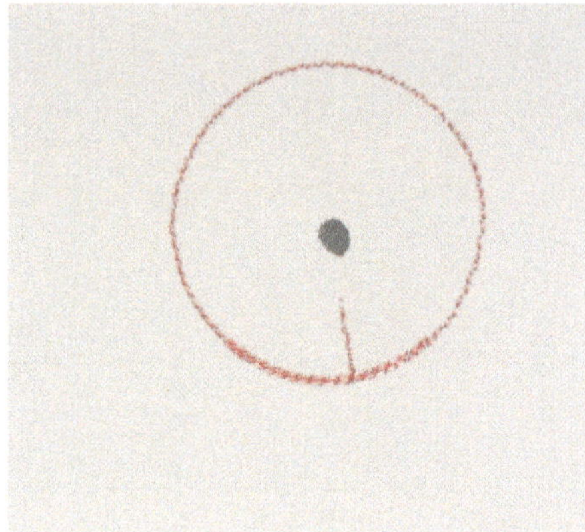

(a)

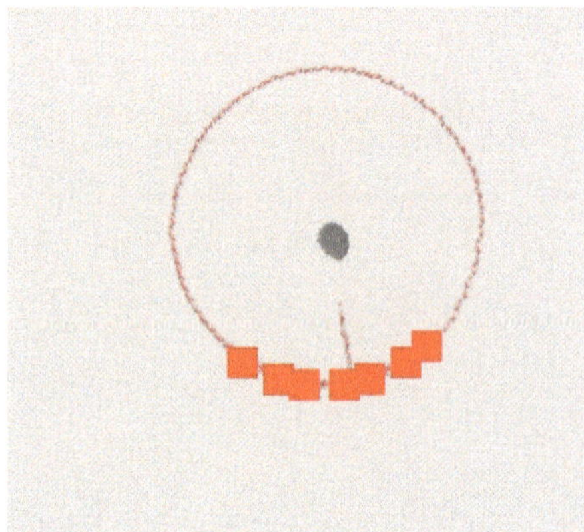

(b)

Figure 4: A long superimposed circular seam: (a) original image, (b) detected nonconformity.

of 81). After thresholding binarization, the Hough transform will detect lines in the image as the seams should be straight and complete (Figure 2(d), using a specified threshold of 55 for the accumulator).

The color image segmentation has the purpose of detecting correct formed stitches. In this respect, a transformation in the HSV color space will be performed, followed by a segmentation using descriptors for the tracked colors. The evaluation will consist of a color sequence inspection. If the sequence is not correct, a negative diagnosis result will be displayed as the seam is either incomplete or broken, either the direction is deviated.

The assessing of circular or curved seams is based on the arcs/circle recognition using the corresponding Hough transform (Figure 3). The detection of the circular outline of the stitching is performed in two steps: first, the estimated calculations of the radius of the circle determined by seam and then the calculating of the center using the circle equation. The stitches are represented by binarized pixels located in predefined shapes. Furthermore, a neighboring block scans along the detected form and searches for a color sequence in the HSV space (Figure 3(b)). The presence of two colors per stitch is checked and points out an incorrect seam (Figure 3(c)).

Using this type of fabric, two tests were made, both being diagnosed as faulty seams. For the case presented in Figure 3, the defect consists of a seam thread break, which means incomplete stitching. The lack of seam is marked correspondingly. In Figure 4, the circular seam is superimposed; therefore, the defect is symbolized with red marks. Another example of linear skipped stitch detection using the color segmentation is presented in Figure 5.

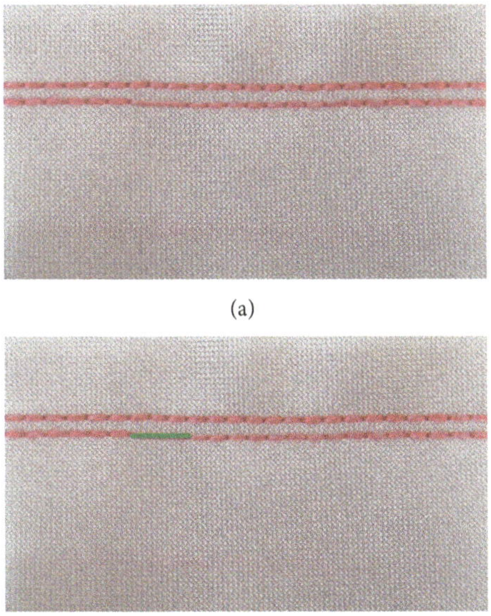

(a)

(b)

FIGURE 5: Skipped stitch detection: (a) input image, (b) detected seam.

5. Conclusions

The airbag, jointly with the seat belts, must assure passengers safety in the case of a car accident. Besides the properties of fabric, the assembly seams play an important role in airbag functionality. Due to the fact that the seams cannot be repaired during the production process, any stitching defect will cause a nonconforming product. Intermediate inspection of stitches before the final closure of the airbag is an important stage, since it prevents noncompliant subassembly to pass on the next production step.

In the airbag sewing stage, fabric and reinforcements parts are transformed into a final product using threads of different linear densities and colors. Seams paths can be linear or circular (including arcs and combined forms).

The process of automatic defect detection depends on the seams paths; therefore, two different methods were employed in the cases of straight and circular seams and the most common defects of sewing were considered: missed stitch, skipped, and superimposed stitches. Using a grayscale combined with a color image processing, the correct rule of seam formation is checked along the detected paths and the corresponding nonconformities are highlighted.

The processing framework succeeded to inspect correctly all the collected defects. Automatic defects detection of seams keeps the sewing process in control and makes the final product comply with the client requirements.

Conflict of Interests

The authors declare that there is no conflict of interests regarding the publication of this paper.

Acknowledgment

The Authors would like to acknowledge TAKATA PETRI Sibiu, for providing the samples used in this study.

References

[1] K. Singha, "Strategies for in automobile: strategies for using automotive textiles-manufacturing techniques and applications," *Journal of Safety Engineering*, vol. 1, no. 1, pp. 7–16, 2012.

[2] S. K. Mukhopadhyay and J. F. Partridge, "Automotive textiles," *Automotive Textiles*, vol. 29, no. 1-2, pp. 1–125, 1999.

[3] A. R. Horrocks and S. C. Anand, *Handbook of Technical Textiles, Textiles in Passenger Cars*, Woodhead Publishing, 2000.

[4] T. Clapp, M. Gunner, J. Dorrity, and L. Olson, "The online inspection of sewn seams," Project S94-41, National Textile Center, Annual Report, 1994.

[5] W. D. Samuel and V. Poojitha, "A simple system for the online detection of SKIP/LOOP stitches in single needle lockstitch sewing machines," *AUTEX Research Journal*, vol. 10, no. 3, pp. 69–72, 2010.

[6] H. Carvalho, A. Rocha, and J. L. Monteiro, "An innovative device for bobbin thread consumption measurement on industrial lockstitch sewing machines," in *Proceedings of the IEEE International Conference on Industrial Technology (ICIT '04)*, vol. 2, pp. 1034–1039, December 2004.

[7] K. Yin and W. Yu, "Image processing for the use of garment production detection system," in *Proceedings of the 1st International Congress on Image and Signal Processing (CISP '08)*, vol. 3, pp. 349–352, May 2008.

[8] I. G. Mariolis and E. S. Dermatas, "Illumination invariant seam line detection in real garments," *International Journal of Clothing Science and Technology*, vol. 21, no. 5, pp. 286–299, 2009.

[9] C. Bahlmann, G. Heidemann, and H. Ritter, "Artificial neural networks for automated quality control of textile seams," *Pattern Recognition*, vol. 32, no. 6, pp. 1049–1060, 1999.

[10] C. W. M. Yuen, W. K. Wong, S. Q. Qian, D. D. Fan, L. K. Chan, and E. H. K. Fung, "Fabric stitching inspection using segmented window technique and BP neural network," *Textile Research Journal*, vol. 79, no. 1, pp. 24–35, 2009.

[11] K. L. Mak and W. Li, "Objective evaluation of seam pucker on textiles by using self-organizing map," *IAENG International Journal of Computer Science*, vol. 35, no. 1, pp. 47–54, 2008.

[12] C. L. P. Hui, N. S. Fun, and C. Ip, "Review of application of artificial neural networks in textiles and clothing Industries over last decades," in *Artificial Neural Networks—Industrial and Control Engineering Applications*, K. Suzuki, Ed., INTECH, 2011.

[13] W. Fung and M. Hardcastle, *Textiles in Automotive Engineering*, Woodhead Publishing Limited, Cambridge, UK, 1st edition, 2001.

[14] ISO, 4915:1991, Textiles-Stitch Types-Classification and Terminology.

[15] B. P. Saville, *Physical Testing of Textiles*, Woodhead Publishing Limited, Cambridge, UK, 2000.

[16] M. Hustedt, J. Stein, D. Herzog, and O. Meier, "Laser-based joining of technical textiles for airbag production," in *Proceedings of the 3rd World Automotive Congress Plastics-in-Motion*, pp. 1–11, Prague, Czech Republic, May 2008.

[17] R. S. Rengasamy, V. K. Kothari, R. Alagirusamy, and S. Modi, "Studies on air-jet textured sewing threads," *Indian Journal of Fibre and Textile Research*, vol. 28, no. 3, pp. 281–287, 2003.

[18] R. Mohanta, "A study on the influence of various factors on seam performance," *Asian Textile Journal*, vol. 15, no. 10, pp. 57–62, 2006.

[19] S. Gribaa, S. B. Amar, and A. Dogui, "Influence of sewing parameters upon the tensile behavior of textile assembly," *International Journal of Clothing Science and Technology*, vol. 18, no. 4, pp. 235–246, 2006.

[20] J. Zhang and J. Hu, "Image segmentation based on 2D Otsu method with histogram analysis," in *Proceedings of the International Conference on Computer Science and Software Engineering (CSSE '08)*, pp. 105–108, December 2008.

[21] J. Illingworth and J. Kittler, "A survey of the hough transform," *Computer Vision, Graphics and Image Processing*, vol. 44, no. 1, pp. 87–116, 1988.

[22] S.-C. Pei and J.-H. Horng, "Circular arc detection based on Hough transform," *Pattern Recognition Letters*, vol. 16, no. 6, pp. 615–625, 1995.

[23] ISO, 4916:1991, Textiles-Seam types-Classification and Terminology.

[24] ASTM, D6193-11, Standard Practice for Stitches and Seams.

Environmental Sustainability Evaluation of Apparel Product: A Case Study on Knitted T-Shirt

Md. Mazedul Islam[1] and Md. Mashiur Rahman Khan[2]

[1] *Department of Textile Engineering, Dhaka University of Engineering & Technology, Gazipur 1700, Bangladesh*
[2] *Department of Apparel Manufacturing Engineering, Bangladesh University of Textiles, Tejgaon, Dhaka 1208, Bangladesh*

Correspondence should be addressed to Md. Mazedul Islam; mazed.butex77@gmail.com
and Md. Mashiur Rahman Khan; mrktex@yahoo.com

Academic Editor: Anindya Ghosh

The work reported in this paper presents the assessment of environmental performances of selected branded apparel T-Shirt products made by Bangladesh. The study is based on a standard evaluation tool named Higg Index which is basically used widely to measure the environmental sustainability of various apparel products. Higg Index is an internal self-assessment tool created by the outdoor apparel industry and Nike's apparel environmental design tool which aims to aggregate information on the environmental performance of products. The Index considers performance across the full life-cycle of a product, including impacts from "input materials, manufacturing, packaging, transportation, use, and end-of-life." Selected apparel branded T-Shirt products from S. Oliver, BUTex-Innovation, PUMA, Esprit, Aarong, and Yellow were taken into consideration. The results indicate that newly developed ecofriendly T-shirt and foreign branded products named S. Oliver, PUMA, and Esprit gained higher score but local branded product like Aarong and Yellow gained lower score in terms of environmental sustainability based on Higg Index assessment tool. Moreover, many weaknesses and opportunities for improvement of both local and foreign branded T-Shirt products have been identified and suggested which would eventually lead the fashion designer, apparel manufacturer, stakeholder, and consumer towards greener apparel products.

1. Introduction

Sustainability is the major concern in the age of modern world. For textile and apparel sector, this has been a burning issue for many related concerned bodies. Over the past few years, increasing awareness of the environmental and social concerns surrounding the fashion industry have led to a rise in the implementation of sustainability initiatives. There has been a growing concern over apparel brands in improving their environmental impact and the social responsibility throughout their supply chains. Environmental sustainability refers to the ability of something to continue without upsetting earth's ecological balance. Sustainable apparel products can be defined as a part of the growing design philosophy and trend of sustainability, the goal of which is to create a system which can be supported indefinitely in terms of environmentalism and social responsibility [1]. Environmental sustainability in business refers to longevity, but in terms of which natural resources the production process might draw upon, how resources are used and replenished, the overall impact of the final production on the environment, and where the product ends up following its disposal. The textile and apparel products impact the environment at every point along the product's lifecycle [1]. However, rapid growth in the industrial sector is playing a vital role in the economy of Bangladesh. Despite the growing necessity of environmental practices in the apparel industry, little research has explored methods and standard tools to guide apparel designers, manufacturers, merchandisers, and consumers in adapting environmental issues in their work. Many designers do not recognize how their designs impact the environment. The role of the designer must expand dramatically in light of current environmental concerns with the contamination and destruction of the ecosystem [2]. Today's designer, manufacturer, merchandiser, and consumer must be well informed on a vast array of topics, from methods of production to governmental regulations to life cycle analysis. They must account for the environmental impacts of the materials uses, the resulting waste from the forms they chooses, how products are produced and packaged, where they will be made and then sold causing energy use for transportation, and how consumers may use and dispose of the product [3]. Hence, there is a growing need to measure the environmental sustainability of apparel products by standard methods, tools, and technique to minimize the environmental impact. The Higg Index by sustainable apparel coalition is such a standard tool by which the environmental sustainability of apparel products could be measured. This study attempts to identify and analyse a number of these issues related to apparel product sustainability taking several local and foreign branded T-shirts into consideration as case study.

2. Literature Review

2.1. Sustainable Apparel Product. Sustainability is defined as the design of human and industrial systems to ensure that humankind's use of natural resources and cycles do not lead to diminished quality of life due either to losses in future economic opportunities or to adverse impacts on social conditions, human health, and the environment [3]. To achieve sustainable development, designers need to be aware of environmental impacts and incorporate environmental awareness into the design [2]. Product sustainability is the easiest aspect to alter for an apparel brand, as this is where a company has the most and direct control through design and product development [2]. Transforming product sustainability may be achieved via various aspects such as fiber/textile selection, processing methods, use behaviors, and reuse/recycle strategies. Fiber/textile selection is often the first step that designers and product developers will take in reducing the environmental impact of a garment. Environmentally preferred fibres/textiles can significantly reduce the environmental impact and increase the resourcefulness of an apparel product throughout the garments life cycle without change to design practice or product development processes [3].

2.2. Environmental Impact of Textiles Apparel Products. The apparel industry has many negative environmental and social impacts that are complex and occur at different stages of the apparel life cycle. The development of fast fashion has amplified the impacts due to the increased volume of apparel produced and sold at low prices [4]. Apparel has a long and complicated production chain consisting of many phases including resource production and extraction, fibre and yarn manufacturing, textile manufacturing, apparel assembly, packaging, transportation and distribution, consumer use, recycling, and ultimate disposal [5]. The environmental impacts of apparel are varied across the phases, difficult to assess for individual garments, and subject to type of raw material used, dyeing, and laundering. It was not until the 1990s that there was a greater awareness as to the severity of the negative environmental impacts of apparel production [3]. The major environmental impacts associated with the production and use of apparel throughout its life cycle includes wastewater emissions from dyeing, finishing, and washing processes, increase in pollution, solid waste production, and significant depletion of resources from consumption of water, fossil fuels, and raw materials [2, 6]. Energy is used for laundering, transportation, operations of machines for various processes, production of primary materials especially man-made fibres such as polyester (a petroleum based product) and yarn manufacturing of natural fibres such as cotton. Conventional cotton production has high water consumption and employs the use of toxic chemicals that may harm human health and the environment [1, 7]. Chemicals are also released in wastewater from processes such as pretreatments, dyeing, finishing, and laundry. These chemicals are disruptive to both the environment and aquatic-based life [8]. Solid waste is produced during natural fibre yarn, textile, and apparel manufacturing and disposal of apparel products at the end of their life. There are significant issues with clothing waste as the majority of clothing and textile waste ends up in landfills as opposed to being recycled or reused [2].

2.3. Sustainable Apparel Coalition and Higg Index Tool. The SAC is a multistakeholder engagement, formed in 2011, by a group of global apparel and footwear companies and nonprofit organizations representing nearly one third of the global market share for apparel and footwear. The SAC seeks to build a common approach for measuring and evaluating apparel and footwear product sustainability performance [9]. It aims to develop common measurements and a common environmental understanding of products' impacts across the industry by building on the Outdoor Industry Association (OIA)'s Eco Index and Nike's Environmental Design tool [10]. The Eco Index is a standardized tool for measuring the environmental impacts of outdoor products such as boots, clothing, and tents and evaluates the impacts in six key areas of a product's life cycle: materials, packaging, product manufacturing, and assembly, transport and distribution, use of service, and end of life [9]. Nike's Environmental Design tool measures the environmental impacts of apparel. Measuring performance of apparel products will spotlight priorities for action and opportunities for technological innovation [9]. The Higg index 1.0 is an "indicator based tool for apparel that enables companies to evaluate material types, products, facilities, and processes based on a range of environmental and product design choices" [9]. The scope for the desired outcomes of the SAC Higg Index 1.0 tool includes improvements to reduce water use and improve quality, reduce energy and emissions, minimizing waste, reduce chemicals and toxicity, and increase transparency for social and labour issues [6]. The first version of the Higg Index 1.0 tool which was released in June 2012 was based on life cycle thinking and is publicly available for any organization. The Index tool was developed to measure the environmental impacts of apparel products. The tool evaluates material type, products, and facilities and processes. By utilizing practice based, qualitative binary yes/no questions, assessments can be made as to the sustainability performance of product and drive behavior for improvement. The Higg Index RDM-Beta is a prototype aims at educating and providing quick directional guidance to apparel designers during the product creation process about the potential environmental impacts of their design solutions [11]. It is to engage designers in the Higg Index's produce life cycle thinking and how we directionally assess materials sustainability through the MSI. By helping organizations standardize measurement and evaluation of environmental performance, the tool creates a starting point for engagement, education, and collaboration among stakeholders [12]. Through self-assessment, organizations can better understand the environmental impacts that occur throughout the life cycle and the effect of design choices such as material type. However, there are many apparel brands belonging to the SAC such as H&M, Gap Inc., Nike, adidas, Puma, Patagonia, Mountain Equipment Co-op (MEC), Levi's, Hanesbrand, Marks & Spencer, Esprit, Columbia, Timberland and Loomstate, and S. Oliver. These brands are widely using this Higg Index tool to measure environmental performance of their apparel products towards better sustainability [11].

3. Materials and Methodology

To carry out this study, selected knitted men's T-shirt products of 100% cotton single jersey fabric were taken into consideration as case study. Both foreign and local branded knitted men's "M" sized short sleeve T-shirt products, namely, S. Oliver, BUTex Innovation, PUMA, Levi's, Aarong, and Yellow, were selected on random basis to measure their environmental performance. The PUMA, S. Oliver, and Esprit branded T-shirts were collected from various organizations, namely, Viyellatex group Ltd. And Yellow and Aarong T-shirts were collected from Beximco fashion wear and Aarong and BUTex Innovation T-shirts were developed by the researcher as ecofriendly. Higg Index provides a value that represents how environmentally responsible an apparel item is in terms of sustainability. For this study, the tool was used to obtain product scores for the above mentioned T-Shirt. This score was obtained by answering questions relating to the materials, packaging, manufacturing, product care, and repair service, and product end of life. The data used during this process includes information readily available to the general consumers, manufacturers and merchandisers including hangtags, care labels, and online product descriptions and throughout the LCA phase of the developed T-Shirt. The product Tec-pack details information was the important sources for various Higg Index required information towards sustainability measurement. Higg Index 1.0 requires that all materials present be entered into a materials input table. This includes a description of all materials used in addition to information obtained from follow-up questions. By utilizing practice based, qualitative binary yes/no questions, assessments can be made as to the sustainability performance of product and drive behavior for improvement. Higg Index 1.0 is a Microsoft Office Excel and wed based tool that has predetermined standard values based on global survey for specific materials that are used in determining a score. The higher the Higg Index score, the better the environmental performance of an apparel products.

4. Results and Discussions

In this study, all the five selected foreign and local branded knitted T-shirts alone with new developed one were assessed using the Higg Index 1.0 to obtain a value that represents a product's environmental impact. A higher score implies a more sustainable product. Upon completion of the excel and web based Higg Index tool, the BUTex-Innovation, S. Oliver, PUMA, Esprit, Yellow, and Aarong T-shirts received total score of 75, 57, 79, 53, 47, and 36, respectively, on a scale of 1–100. The superior score of the BUTex Innovation and foreign banded product especially PUMA T-shirt was due to the organic material content, ecofriendly raw materials and process used product end of life options and confirmed traceability. On the other hand the local branded T-shirt gained comparatively lower score in most of the areas due not using ecofriendly raw materials, manufacturing processes, high impact finishes, improper or insufficient care and repair instructions, traceability issues, and so forth. This also confirms that by increasing the transparency of a product and offering more information about a product's development, the more this product will represent ecofriendliness and will be superior in terms of sustainability.

4.1. Materials Selection. The materials selected for the knitted T-shirt for both foreign and local branded product were entered into the material inputs table in the Higg Index 1.0. tool. The values obtained by various products are shown in Table 1. The developed men's T-shirt and foreign brands PUMA, S. Oliver, and Esprit obtained higher score. This was due to the ecofriendly raw materials used, that is, BUTex Innovation and PUMA t-shirt included 100% organic cotton fibre. During processing of fabric, BUTex Innovation T-Shirt, PUMA, and other foreign branded T-shirts used ecofriendly and certified and environmentally low impacted dyes chemicals. But in case of local branded Yellow and Aarong T-shirts Higg Index score was lower due to environmentally hazardous conventional raw materials, processing, dyes used, no third party verification, and so forth. For better environmental sustainability these issues must be addressed for local branded apparel products.

4.2. Manufacturing Stages. The manufacturing stages are obviously an important area for better product sustainability. Some major issues have great impact on product environmental sustainability as per Higg Index tool. The developed men's T-shirt, PUMA, and foreign branded T-shirt used low impact garment finish, that is, process combination enzyme wash or basic rinse that scored higher. But local branded Aarong and Yellow T-shirt used high impact garment finish, namely, traditional bleach, acid chemical wash, and so forth and scored lower. Besides, marker efficiency has great impact on product sustainability. The higher the marker efficiency, the better the product sustainability and lower marker efficiency indicate its direct impact on supply chain including more wastages of material like fabric, dyes chemicals, process loss, land fill, and so forth. As developed T-shirt, PUMA, S. Oliver, and Esprit brand had more than 92% marker

TABLE 1: Higg Index evaluation score for T-shirts products.

Scoring summary stages	Section score each out of 100						Section weight %	Score obtained					
	BUTex-Innovation	S. Oliver	PUMA	Esprit	Yellow	Aarong		BUTex-Innovation	S. Oliver	PUMA	Esprit	Yellow	Aarong
Materials selection	100	71	100	71	66	57	30%	30	21	30	21	20	17
Manufacturing (i) Marker efficiency (ii) Garment finish (iii) Applied graphics and logo (iv) Screen printing size and color	53	55	53	53	40	24	25%	13	14	13	13	10	6
Packaging	90	90	90	50	40	0	10%	9	9	9	5	4	0
Product use and care service	90	20	90	20	20	20	20%	14	4	18	4	4	4
End of life	60	60	60	60	60	60	15%	9	9	9	9	9	9
Total							100%	75	57	79	53	47	36

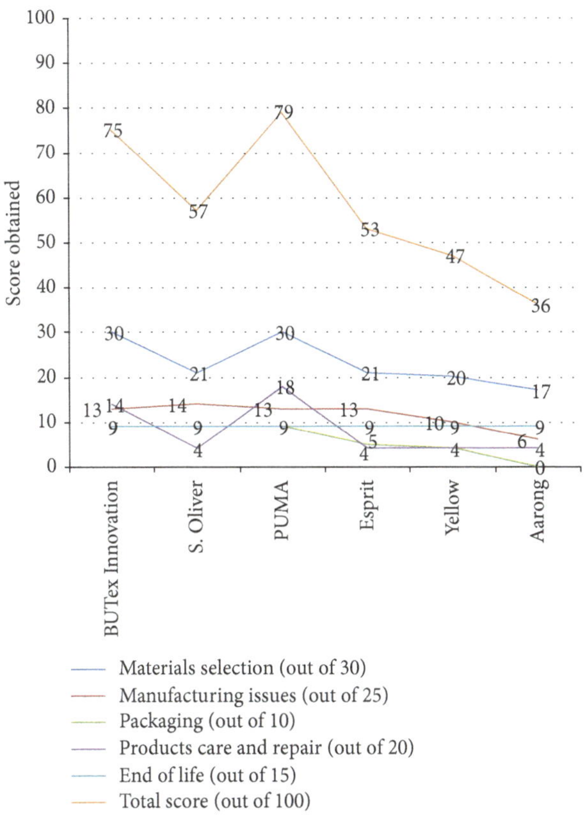

FIGURE 1: Higg Index assessment score for various branded T-Shirt.

efficiency, they scored better. But Aarong and Yelllow brands had marker efficiency of 78% and 84% and scored lower in the Higg Index tool. Similarly, a single colored, screen printed logo with small area received a higher score of for foreign branded T-shirt and these issues scored lower for local branded T-shirt. The total points for manufacturing stages are shown in Figure 1.

4.3. Packaging. Product packaging was included in this assessment. Conventional and environmentally harmful materials were used for local branded T-shirt especially in case of Aarong, which did not provide any score points rather than zero. But PUMA and other T-shirt used only sustainable, third party verified and reduced packaging materials and obtained better score. The packaging scores are shown in Table 1.

4.4. Product Care and Repair Services. For PUMA, new developed men's T-Shirt, S. Oliver, and Esprit T-shirts, best class to low impact details care instructions were provided due to the required care of the chosen materials that scored higher, that is, 18 for PUMA. But for Yellow and Aarong T-shirts, there were no in details care instructions that leaded to lower score in Higg Index, that is, only 4. Proper care instructions should be provided to the product label as most of the water and energy is used by the customer during the care and repair service stages.

4.5. End of Life. All T-shirt products received 60 points and a base score 9 for product end of life. The PUMA and BUTex Innovation product is made from 100% organic cotton and other branded T-shirt made from conventional cotton. These individual material types can be accurately identified by end of use facilities or processes. The material types are recycled via existing infrastructure and processes or materials are used in closed-loop recycling processes were not known. In response to the questionnaires regarding end of life of products as per assessment Higg Index tools, the all products responded in the same way to the questionnaires. So the EOL stages obtained a base score, that is, 60%, and no additional points were given as shown in Table 1.

5. Technical Issues for Apparel Products Sustainability Improvement

The researcher has identified many technical requirements issues for apparel products sustainability improvement based on Higg Index assessment tool.

Sourcing Fair Production and Practice Transparency. Sourcing fair production can be performed by selecting manufacturers that are known to operate in a socially responsible manner. For the technical requirement of practicing transparency, the researcher suggests a hangtag including maximum in details information to the products that includes the name and contact information of a hypothetical manufacturer. This would allow a customer to verify socially responsible methods were practiced when the product was produced.

Sourcing Local Production. Local production is a valuable asset to improve product sustainability. This will also improve the social and economic sustainability of a product.

Consider Waste Reducing Design and Clean Production Adjustments. To reduce waste, the researcher suggested using ecofriendly dyeing method like reduced water dyeing or waterless dyeing method for the fabric processing. An on-product label with instructions to recycle, reuse of the product, thus reducing waste, should be added to the label. Clean production can be achieved by selecting manufacturers that make a commitment to the use of alternative or renewable energies.

Consider Organic and Natural Fibers. Natural fibers are often viewed as the most environmentally responsible as they come from renewable sources and use less energy and chemicals to produce than manufactured fibers. Ecofriendly fibers, production methods that demand fewer natural resources and use nature to their benefit should be used.

Source of CMT Production. In order to replicate the possible effects of sourcing CMT production, the researcher decided to identify only one manufacturing facility instead of two or more in the Higg Index 1.0 to determine if fewer facilities results in a better score.

Consider Natural and Low Impact Dyes and Chemicals. The designers, manufacturers, and merchandiser should introduce natural and environmentally low impact dyes in materials processing to better product sustainability.

6. Conclusion

The textile and apparel industries are notorious for their excessive contribution of waste and pollution to our environment. The unsustainable practices trends of poorly designing, raw materials selection, processing, manufacturing, and disposing of apparel product is a major contributing factor to this problem. The purpose of this study was to evaluate the environmental sustainability of various branded T-shirts and find out various sustainable issues for greener apparel products. Based on Higg Index environmental assessment standard tool, the findings of this study could be helpful for many organizations and customers who attempt to produce and consume sustainable textiles and clothing products. Both the local and foreign branded products should be produced by strictly maintaining sustainability issues. Hence, this study is a little approach towards better sustainable practices and eventually would drive the textile and clothing designer, manufacturers, merchandisers, and consumers to introduce various sustainable issues in their works. Hence, growing awareness and leading them towards sustainable practices in the textile and clothing business for greener world.

Conflict of Interests

The authors declare that there is no conflict of interests regarding the publication of this paper.

References

[1] J. M. Allwood, S. E. Laursen, C. M. de Rodriguez, and N. M. P. Bocken, *Well Dressed? The Present and Future Sustainability of Clothing and Textiles in the United Kingdom*, Institute of Manufacturing, University of Cambridge, Cambridge, UK, 2006.

[2] K. Fletcher, *Sustainable Fashion & Textiles*, Earthscan, London, UK, 2008.

[3] C. A. Rusinko, "Green manufacturing: An evaluation of environmentally sustainable manufacturing practices and their impact on competitive outcomes," *IEEE Transactions on Engineering Management*, vol. 54, no. 3, pp. 445–454, 2007.

[4] J. K. Steinberger, D. Friot, O. Jolliet, and S. Erkman, "A spatially explicit life cycle inventory of the global textile chain," *The International Journal of Life Cycle Assessment*, vol. 14, no. 5, pp. 443–455, 2009.

[5] V. D. Lewis and C. Chen, "The life of a piece of cloth: developing garments into a sustainable service system," *International Journal of Environmental, Cultural, Economic and Social Sustainability*, vol. 2, no. 1, pp. 197–207, 2006.

[6] H. Goworek, "Social and environmental sustainability in the clothing industry: a case study of a fair trade retailer," *Social Responsibility Journal*, vol. 7, no. 1, pp. 74–86, 2011.

[7] J. Speer, "Organic cotton: where, why and how. Appare," *Journal of Sustainable Textiles Materials*, vol. 46, no. 9, pp. 29–34, 2005.

[8] S. E. Laursen, J. Hansen, H. H. Knudsen, H. Wenzel, H. F. Larsen, and F. M. Kristensen, *EDIPTEX: Environmental Assessment of Textiles*, Danish Environmental Protection Agency, Copenhagen, Denmark, 2007.

[9] The higg index, "Sustainable Apparel Coalition," 2012, http://www.apparelcoalition.org/higgindex.

[10] K. Niinimäki and L. Hassi, "Emerging design strategies in sustainable production and consumption of textiles and clothing," *Journal of Cleaner Production*, vol. 19, no. 16, pp. 1876–1883, 2011.

[11] A. Reuben, *The Higg Index for Sustainable Apparel Environmental Performance Index*, 2013, http://epi.yale.edu/indicators/indicator%E2%80%90case%E2%80%90studies/reports/higg%E2%80%90index.

[12] C. A. Adams and G. R. Frost, "Integrating sustainability reporting into management practices," *Accounting Forum*, vol. 32, no. 4, pp. 288–302, 2008.

Synthesis of Malononitrile-Condensed Disperse Dyes and Application on Polyester and Nylon Fabrics

Yusuf Y. Lams,[1] P. O. Nkeonye,[2] K. A. Bello,[2] M. K. Yakubu,[2] and A. O. Lawal[3]

[1]*Department of Textile Technology and Fashion Design, College of Science and Technology, Kaduna Polytechnic, PMB 2021, Kaduna, Nigeria*
[2]*Department of Textile Science and Technology, Ahmadu Bello University, Zaria, Nigeria*
[3]*Department of Applied Science, College of Science and Technology, Kaduna Polytechnic, PMB 2021, Kaduna, Nigeria*

Correspondence should be addressed to Yusuf Y. Lams; yyakubulams@yahoo.com

Academic Editor: Gajanan S. Bhat

An active methylene compound, malononitrile, was introduced into the structures of a series of disperse dyes previously prepared by coupling phloroglucinol, barbituric acid, and α and β-naphtho to 4-amino-3-nitrobenzaldehyde. The dyes were purified by recrystallization from ethanol. The purity of the dyes was examined by thin-layer chromatography (TLC) and the dyes were characterized by visible absorption and Fourier transform infrared spectroscopy (FTIR). The malononitrile-condensed dyes produced deeper colours and shades with better fastness to wash, light, and perspiration on application to polyester and nylon fabrics when compared to their uncondensed analogue.

1. Introduction

Disperse dyes are substantially water-insoluble nonionic dyes which are applied to hydrophobic fibres (cellulose acetate, polyester, and nylon) from aqueous dispersions [1–5].

Disperse dyes are characterized by the absence of solubilizing groups and low molecular weight. From a chemical point of view, more than 50% of disperse dyes are simple azo compounds. About 25% are anthraquinones and the rest are methine, nitro, and naphthoquinone dyes [6]. The most dominant group of these dyes are the azo disperse dyes which account for approximately 60–70% of all disperse dyes manufactured [7–16].

The use of malononitrile (propanedinitrile, dicyanomethane; malonic (acid) dinitrile; propiodinitrile; methylene cyanide; cyanoacetic acid nitrile; malonitrile; malodinitrile) either as a coupling or diazo component or condensing agent with different chromophoric compounds in the synthesis of disperse dyes have been reported [1, 5, 11–16]. However, there seem to be no literatures on the condensation of disperse dyes derived from 4-amino-3-nitrobenzaldehyde coupled to phloroglucinol, barbituric acid, and α and β-naphthols with malononitrile. The present study reports the synthesis of new malononitrile-condensed disperse dyes and their application onto polyester and nylon fabrics.

Malononitrile was successfully introduced into the structures of four dyes (**D1–4**) previously prepared by the conventional diazotization and coupling methods to give dyes (**D5–8**) (see Figure 2). This was achieved by the Knoevenagel condensation in which a molecule of water was eliminated by way of nucleophilic addition of an active hydrogen compound (malononitrile) to a carbonyl group (C=O) contained in 4-amino-3-nitrobenzaldehyde followed by a dehydration reaction. The base in the reaction was piperidine. The products have larger molecular weight with usually higher melting points when compared with their parent compounds.

2. Experimental

2.1. Synthesis of 4-Amino-3-nitrobenzaldehyde (Diazo Component). 4-Acetylaminobenzaldehyde (60 g) was added to 150 mL acetic anhydride and heated to 105°C with vigorous stirring until it dissolved. The solution was cooled rapidly in ice-water to 30°C to produce a fine precipitate. 24 mL nitric acid and 60 mL acetic anhydride were carefully mixed, and

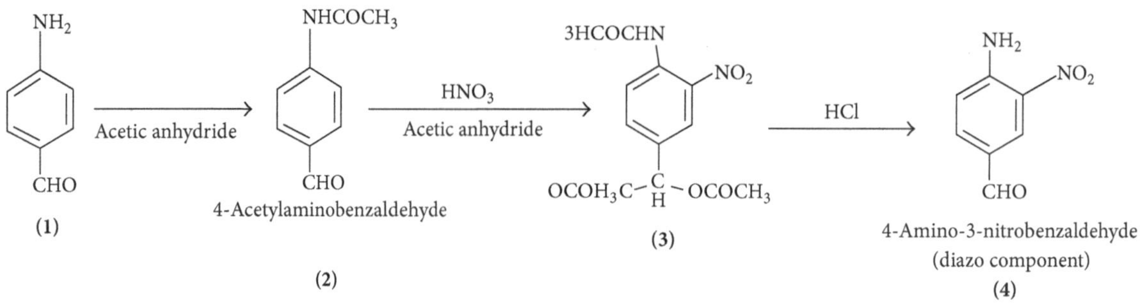

Scheme 1: Scheme for the synthesis of 4-amino-3-nitrobenzaldehyde.

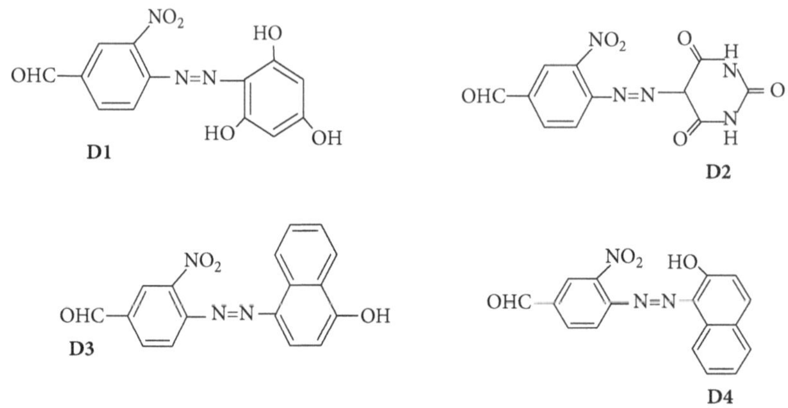

Figure 1: Structures of dyes **1–4**.

the mixture was added dropwise to the stirred suspension of 4-Acetylaminobenzaldehyde **(2)** at a rate that the temperature did not rise above 35°C. The addition of about one-third of the acid produced colourless crystal deposits which have been shown by Hodgson&Beard to be 4-acetylbenzalidene diacetate **(3)** [10]. Further drops of the acid dissolved the crystals with a gradual temperature rise up to 50°C. The mixture was stirred at this temperature for a further 15 minutes and poured into 800 mL ice-water. The precipitated 4-acetylamino-3-nitrobenzalidene diacetate was filtered off, washed thoroughly with ethanol, and then with water before it was dried. Recrystallisation from ethanol produced 93% of the diacetate as pale yellow leaflets (64 g) with a m.p. of 112-113°C.

The diacetate (60 g) was added to concentrated hydrochloric acid (170 mL) and heated on a water bath for 15 min. After cooling, 257 mL water was added and the precipitated 4-amino-3-nitrobenzaldehyde **(4)** was filtered off, washed with water and dried. It was purified by recrystallisation from water, to produce orange needle like substance with m.p. of 188–190°C (see Scheme 1).

2.2. Diazotisation of 4-Amino-3-nitrobenzaldehyde. 4-amino-3-nitrobenzaldehyde (3.2 g, 0.02 mol) was added to a mixture of acetic acid (24 mL) and water (6 mL) and cooled with stirring to 12°C. Sodium nitrite (1.5 g, 0.02 mol), dissolved in 10 mL of water, was added in one portion and 10 mL of concentrated hydrochloric acid was added immediately. The mixture was maintained at 15–20°C for 15 min, and then excess nitrous acid was destroyed by addition of small amount of urea. The solution was diluted with 200 mL ice-water to achieve 0–5°C and produced a clear solution of the diazonium salt **(5)**. See Figure 3.

2.3. Coupling of Diazotised 4-Amino-3-nitrobenzaldehyde to Various Coupling Components. The solution of the various coupling components was cooled to 0–5°C and was treated with a cold (0–5°C) of the diazonium salt (Scheme 2) as prepared above, to produce dyes **D1–4** (Figure 1). Same procedure was followed to prepare coupling solutions with the substituted coupling components.

2.4. Condensation of Aldehyde Dyes Derived from 4-Amino-3-nitrobenzaldehyde with Malononitrile. 0.006 mole of the azo dye and 0.006 mole of malononitrile were dissolved with warming in absolute ethanol (150 cm^3). One drop of piperidine was added and the mixture was refluxed for 15 minutes. On cooling, the product generally crystallized out and was filtered off and recrystallized from ethanol (Scheme 3).

2.5. Purification of Synthesized Dyes. The dyes were purified by recrystallization from ethanol and their purity examined by thin-layer chromatography (TLC).

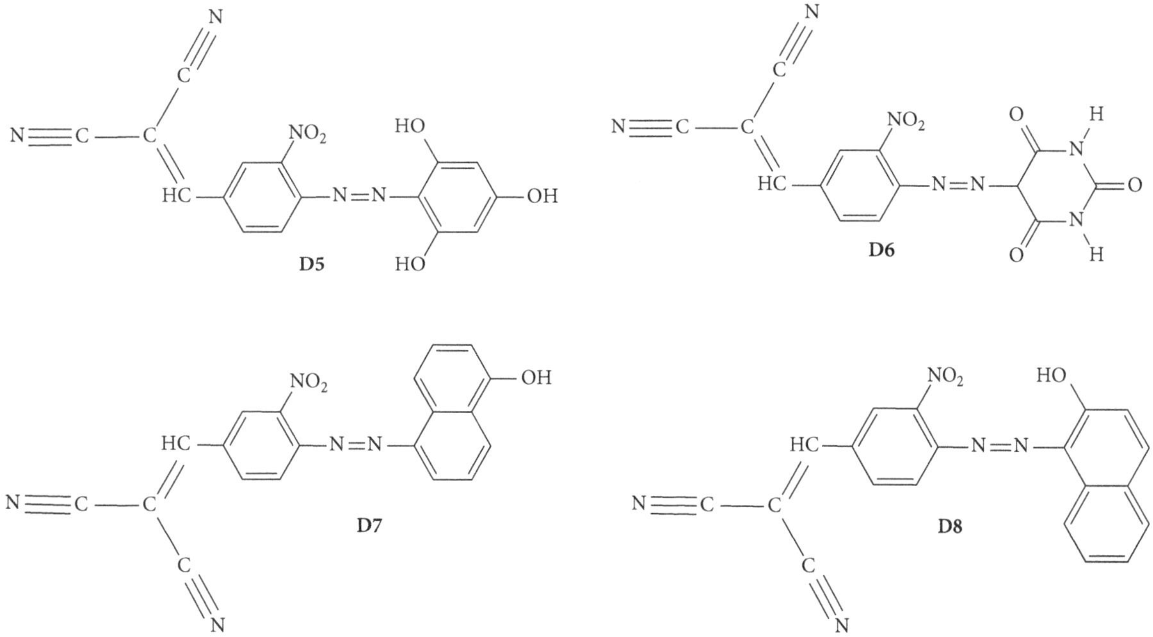

FIGURE 2: Structures of malononitrile-condensed disperse dyes (**D5–8**) derived from 4-amino-3-nitrobenzaldehyde.

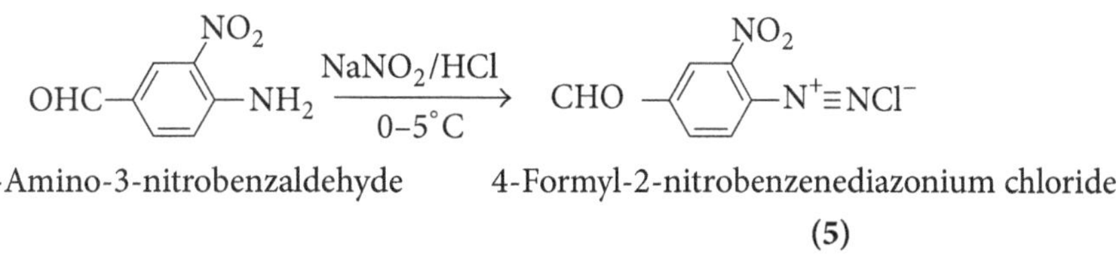

FIGURE 3

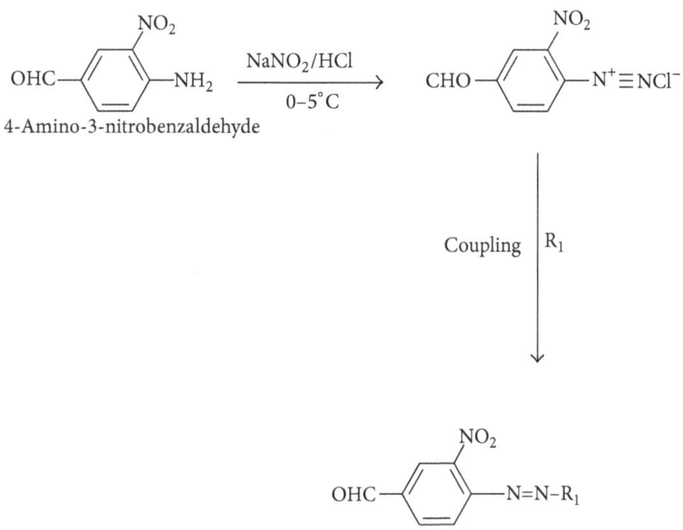

Where R_1 is the coupling component

SCHEME 2: Scheme for the synthesis of aldehyde dyes (**D1–4**) derived from 4-amino-3-nitrobenzaldehyde.

TABLE 1: Yield, colour, molecular weight and melting points of dyes derived from Diazo component.

Dye	Coupling component R_1	Weight of Dye	% yield	Colour of crystals	Mol. wt (g)	Melting point (°C)
D1	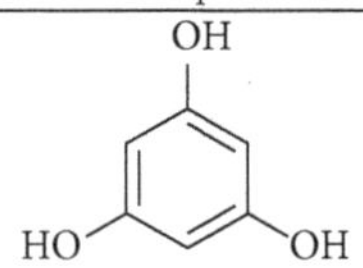	1.969	65	Coffee-brown	303	205–207
D2	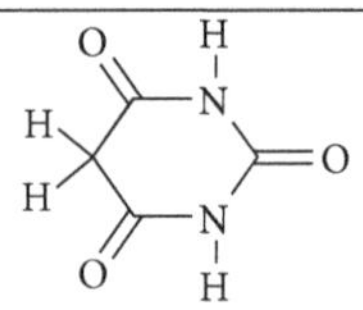	2.288	75	Yellow	309	222
D3	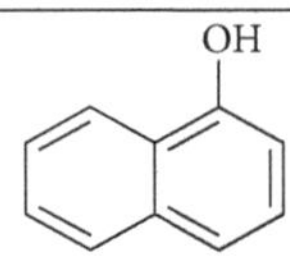	2.022	63	Light-orange	321	128–130
D4	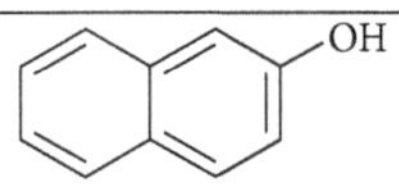	2.183	68	Dark-red	321	110–112
Diazo Component	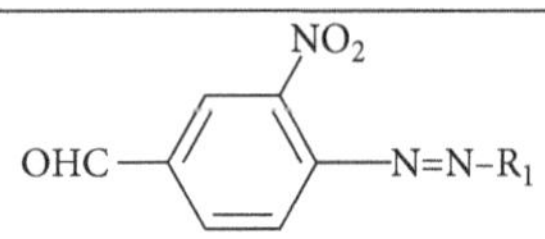					

R_1: the coupling component.

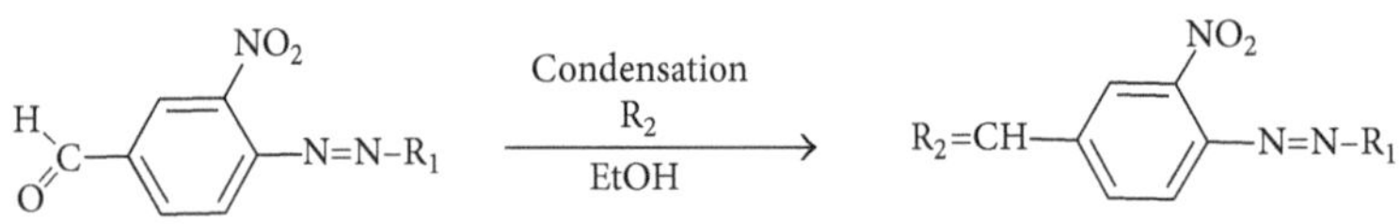

Where R_1 is the coupling component and
R_2 is the active methylene compound (malononitrile)

SCHEME 3: Scheme for the synthesis of malononitrile-condensed disperse dyes derived from 4-amino-3-nitrobenzaldehyde.

2.6. Characterization of the Dyes. The dyes were characterized by visible absorption spectroscopy using Unicam SP 800 spectrophotometer, IR spectral bands obtained from Pekin Elmer 100 FT-IR spectrophotometer.

The molar extinction coefficient presented in Table 3 was calculated using Beer Lambert's law.

2.7. Application of the Synthesized Dyes onto the Substrates (Dyeing)

2.7.1. Application to Polyester. The dye bath was made up of 2% o.w.f. (dye), 1.5 mL chlorobenzene (carrier), and 50 : 1 (liquor ratio) at 100°C. The fabric (10 cm by 10 cm), after being wetted and thoroughly squeezed to remove excess water, was immersed into the bath at 50°C and allowed to reach the boil within 15 minutes. Dyeing was continued at the boil for 1 hour with constant agitation. At the end of dyeing, the substrate was removed, squeezed, and rinsed thoroughly under a running tap and allowed to dry at room temperature.

2.7.2. Application to Nylon. The dye bath was made up of 2% o.w.f. (dye) and 50 : 1 (liquor ratio) at 100°C. The fabric (10 cm by 10 cm), after being wetted and thoroughly squeezed to remove excess water, was immersed into the bath at 50°C and allowed to reach the boil within 15 minutes. Dyeing was continued at the boil for 1 hour with constant agitation. At the end of dyeing, the substrate was removed, squeezed, and rinsed thoroughly under a running tap and allowed to dry at room temperature.

2.7.3. Reduction Clearing. The dyed material was treated in a bath containing 1.5 g/L dispersing agent, 2 g/L caustic soda, and 2 g/L sodium dithionite at 60°C for 30 minutes. This was aimed at removing unfixed dye and carrier residues that may be left on the fabric after dyeing.

2.8. Results and Discussion. The yield, colour, molecular weight, and melting points of synthesized dyes 1–4 and condensed dyes 5–8 are presented in Tables 1 and 2, respectively.

TABLE 2: Yield, colour, molecular weight and melting points of condensed dyes derived from the Diazo component.

Dye	Condensed Dye	Coupling component R_1	Condensing component R_2	Weight of Dye	% yield	Colour of crystals	Mol. wt (g)	Melting point (°C)
D5	D1	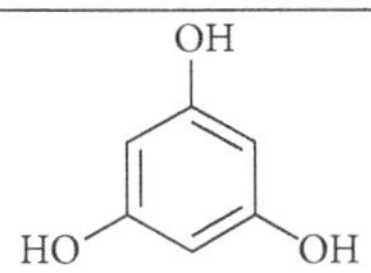	$NC-CH_2-CN$	1.279	62.9	Coffee-brown	339	315-316
D6	D2	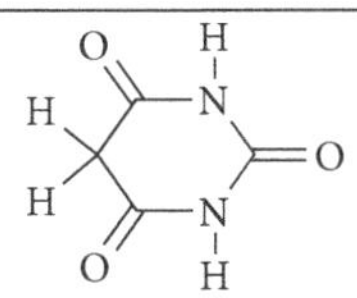	$NC-CH_2-CN$	1.390	70	Orange-yellow	331	298-299
D7	D3	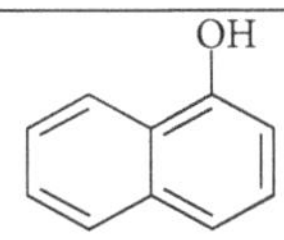	$NC-CH_2-CN$	1.573	73.48	Reddish-brown	357	154–156
D8	D4	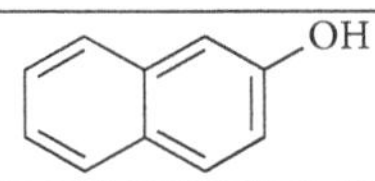	$NC-CH_2-CN$	1.665	74.0	Intense-red	357	160-161
Diazo Component	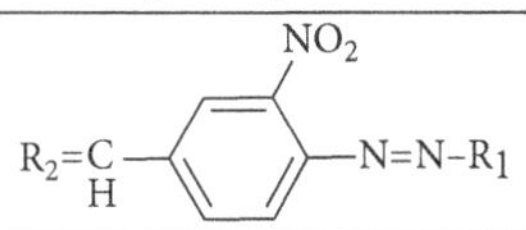							

R_1: the coupling component, R_2: the condensing component.

TABLE 3: Absorbance and λ_{max} of dyes derived from 4-amino-3-nitrobenzaldehyde in different solvents (UV spectral properties of dyes).

Dye	Ethanol A	Acetone B	Dimethyl Formamide C	Ethanol + HCL D	Change in λ_{max} A – D	Molar extinction Coefficient in DMF LIT MOL^{-1} CM^{-1} $\times 10^4$
D1	680	440	549	436	−244	4.361
D2	450	451	459	447	−3.0	2.640
D3	489	532	470	532	+43	3.478
D4	532	500	548	486	−46	3.969
D5	602.0	551.5	651	780.5	+178.5	1.520
D6	455.5	553.0	484	456.5	+1.0	2.540
D7	456	461	667	453	−3.0	5.016
D8	561.0	550.0	559	557	−4.0	4.395

The UV/visible and IR spectral characteristics of the dyes indicate that they absorbed at very high wavelengths up to 680 nm for dye 1 and near infra-red region up to 780.5 nm for dye 5. The presence of the main functional groups such as carbonyl (C=O), azo (N=N), nitrile C≡N, nitro (NO_2), and hydroxyl (OH), in the structures of the dyes as presented in Table 3 and Figure 1 was established.

The solvatochromic study of the dyes was measured in acetone, dimethylformamide (DMF), and ethanol [9]. The results showed that the dyes absorbed at much higher wavelengths in DMF, an indication that DMF has greater polarity towards the dyes than the other solvents. The halochromism (effect of few drops of acid on the visible absorption band) of the dyes was studied in ethanol. The dyes exhibited both positive and negative halochromism on addition of one drop of hydrochloric acid to ethanol as indicated in Table 3.

The condensation of the aldehyde dyes with the active methylene compound (malononitrile) imparted significantly on light absorption of the corresponding dyes. It is believed that malononitrile caused increase in resonance which in turn increased the intensity of absorption of light and consequently shifting absorption bands to longer wavelengths.

2.8.1. Infrared Spectral Characteristics of the Synthesized Dyes. Infrared spectra of the synthesized dyes were measured in KBr pellets on a Perkin-Elmer 100 FTIR spectrophotometer. The results are shown in Table 4.

The spectral results of dyes 5, 6, 7, and 8 as presented in Table 4 showed characteristic bands around 2900–3088.67 cm^{-1} indicating the presence of aromatic C–H stretching vibration. The aliphatic C–H bending vibration

TABLE 4: Infra-red spectral data for the dyes derived from 4-amino-3-nitrobenzaldehyde.

DYE	Aromatic C–H	Aliphatic C–H	Carbonyl C=O, Lactum HC(=O)–NH–CH(=O)	Azo N = N	Aldel-hyde CHO,	Cyano C≡N	Hydroxyl OH	Aromatic C–H	Aromatic Nitro NO_2	α-sub Naphthol	β-sub Naphthol
	Stretch	Bend	Stretch	Stretch	Stretch	Stretch	Stretch	Bending	Bending		
D1	2925.00	2852.00		1406.78	1600.07	—	3410.53	833.81	1509.10	—	—
D2	3088.67	2849.95	1615.31	1425.08	1689.18	—		821.06	1573.38	—	—
D3	2924.40	—	—	1404.27	1697.93	—	3439.49	—	1573.76	835.85 872.72	—
D4	2960.00	2815.00	—	1407.95	1693.76	—	3390.47	814.22	1557.56	—	844.00 878.14
D5	2923.49	2880.00	—	1428.00	—	2357.50	3620.08	672.36	1513.80	—	—
D6	2900.00	—	1620.00^{+} 1678.93^{-}	1427.00	—	2357.61	—	672.98	1365.00	—	—
D7	2923.50	2850.00	—	1420.00	—	2359.89	3600.00	672.65	1567.64	820.00	—
D8	2923.94	—	—	1420.00	—	2359.81	3623.46	672.16	1550.62	—	814.00

TABLE 5: Wash, light and perspiration fastness of dyes derived from 4-amino-3-nitrobenzaldehyde on polyester and nylon fabrics.

Dye	Fastness to Wash				Fastness to Light		Fastness to Perspiration							
	Polyester		Nylon				Polyester				Nylon			
					Polyester	Nylon	ACD		ALK		ACD		ALK	
	CSH	STN	CSH	STN			CSH	STN	CSH	STN	CSH	STN	CSH	STN
D1	4-5	3-4	4-5	4-5	3	7	5	4-5	5	4-5	4	1-2	5	4-5
D2	4-5	4-5	4-5	5	7	7	4	4	4	4-5	5	4-5	5	5
D3	4-5	4-5	4-5	3-4	7	5	4	3-4	5	4-5	4	4	5	5
D4	5	5	4-5	3-4	7	7	3	2-3	4	3-4	4	4-3	5	4-5
D5	5	5	5	5	4	7	4	3-4	5	4-5	4	4-5	5	5
D6	5	5	4-5	5	7	7	5	4	5	4	5	5	5	4
D7	5	5	4-5	4-5	5	5	3	2	3	3	2	3-4	2	3
D8	5	5	5	4-5	7	7	5	3	5	4-5	5	5	5	5

occurred at 2815–2880 cm^{-1}. Absorption peaks due to lactum group in dyes 2 and 6 appeared at 1620 cm^{-1} and 1678.93 cm^{-1}, respectively. The strong bands which appeared around 1615–1695 cm^{-1} are due to carbonyl (C=O) stretching vibration. The bands 1404–1428 cm^{-1} arise from the presence of azo group (N=N) while the broad peak around 3390–3623 cm^{-1} is due to aromatic (ArOH) stretching vibration. Other important absorption bands that appeared in the aldehyde dyes include 2357.50–2359.89 cm^{-1} (C≡C) stretching vibration for Dyes 5, 6, 7, and 8; 1509.10–1573.38 cm^{-1} nitro group (NO_2) bending vibration for Dyes 1 and 3; 1600.07–1697.93 cm^{-1} aromatic (CHO) stretching vibration for Dyes 1 and 3; 820–872.72 cm^{-1} (α-substituted naphthol) for Dyes 3 and 7 and 814–878 cm^{-1} (β-substituted naphthol) for Dyes 4 and 8 respectively. Absorption peaks due to N-H bending vibration were observed at 1509 $cm^{-1}$and 1504 cm^{-1} for Dyes 5 and 6, respectively. Further studies with NMR will help or likely confirm the observations made based on the FTIR.

2.8.2. Fastness Properties of Synthesized Dyes. The dyes 1–8 were applied at 2% depth on the polyester and nylon fabrics. Their wash, light, and perspiration fastness are given in Table 5.

A remarkable degree of levelness and brightness was observed after washing. This is a clear indication of good penetration and excellent affinity of these dyes to the fabrics.

The dyeing showed very good (4-5) to excellent (5) washing fastness generally on both polyester and nylon fabrics. The results from Table 5 show good (5) to excellent (7) light fastness properties for most of the dyes, except for **dye 1** that was relatively poor (3). The observed higher fastness can be explained in terms of the reduction of the electron density at the azo group due to the inductive effect exerted by the CN group.

Fastness to perspiration averaged very good (4) to excellent (5) for most of the dyes. However, dye 7 gave poor fastness on both polyester and nylon. This is perhaps due to the weak hydrogen bond on α-position on the naphthalene structure. The results indicate that majority of the dyes showed better perspiration properties to alkali than acid.

Generally, the malononitrile-condensed dyes provided better fastness to wash, light, and perspiration when compared with their uncondensed counterparts. This is thought to be due to increased molecular size and conjugation.

3. Conclusion

A series of new monoazo disperse dyes were produced by coupling diazotized 4-amino-3-nitrobenzaldehyde and their derivatives to different coupling components. The dyes provide a wide range of attractive colours varying from peach, yellow, and orange, to amber, brown, purple, violet, bluish, and greenish shades on fibre and offer excellent affinity and intensity of colour. The intrinsic conjugation in the nitrobenzaldehyde structure results in high bathochromic shifts and lead to brightness of shades. The nature of the substituent in the coupling components has a great influence on the visible absorption and shade of dying. Most dyes showed very good exhaustion and leveled and uniform dyeing. The remarkable degree of levelness after washing is a clear evidence of the good penetration and affinity of the dyes to the fabrics on which they were applied. The barbituric and naphthalene substituted dyes generally showed better fastness properties than their hydroylated analogues. The condensed dyes provided better fastness to wash, light, and perspiration when compared with their uncondensed counterparts. This is thought to be due to increased molecular size and conjugation. The compactness of the polyester fibre structure accounts for these dyes' superior fastness properties on it than nylon.

Conflict of Interests

The authors declare that there is no conflict of interests regarding the publication of this paper.

Acknowledgment

The authors are grateful to the Department of Chemistry, University of Kwazulu-Natal, South Africa, for granting them permission to carry out the FT-IR analysis of the dyes in the instrumentation laboratory.

References

[1] A. M. Al-Etaibi, M. A. El-Apasery, M. R. Ibrahim, and N. A. Al-Awadi, "A facile synthesis of new monoazo disperse dyes derived from 4-hydroxyphenylazopyrazole-5-Amines: evaluation of microwave assisted dyeing behavior," *Molecules*, vol. 17, no. 12, pp. 13891–13909, 2012.

[2] Y. Y. Lams, P. O. Nkeonye, K. A. Bello, M. K. Yakubu, and A. O. Lawal, "Synthesis of disperse dyes from pyridone and resorcinol coupled to diazotized 2-amino-4-chloro-5-formylthiazole and application to polyester," *Advances in Chemistry*, vol. 2014, Article ID 864286, 7 pages, 2014.

[3] A. A. Fadda and K. M. Fadda, "Synthesis of novel azo disperse dyes derived from 4-aminoantipyrine and their applications to polyester fabrics," *The American Journal of Organic Chemistry*, vol. 2, no. 3, pp. 52–57, 2012.

[4] Y. S. Choi, K. S. Lee, H. J. Kim et al., "Synthesis, spectral property and dyeing assessment of azo disperse dyes containing carbonyl and dicyanovinyl groups," *Bulletin of the Korean Chemical Society*, vol. 34, no. 3, pp. 863–867, 2013.

[5] J. Koh, M.-J. Oh, H. Kim, and S. D. Kim, "Alkaline dissolution monitoring of radial-type polyester microfiber fabrics by a cationic dye-staining method," *Journal of Applied Polymer Science*, vol. 99, no. 1, pp. 279–285, 2006.

[6] P. A. Eigenmann and C. A. Haegelliand, "Food colouring and preservative—allergy and hyperactivity," *Archives of Disease in Childhood*, vol. 90, p. 875, 2004.

[7] A. S. Abd-El-Aziz and T. H. Afifi, "Novel azo disperse dyes derived from aminothiophenes: synthesis and UV—visible studies," *Dyes and Pigments*, vol. 70, no. 1, pp. 8–17, 2006.

[8] Internet International Union of Pure and Applied Chemistry, "Azo compounds," Compendium of Chemical Terminology, 2009.

[9] A. Marini, M. Aurora, B. Alessendro, and M. Benedetha, "What is solvatochromism?" *Chemistry Journal of Physical Chemistry B*, vol. 114, no. 51, pp. 117128–117135, 2010.

[10] K. A. Bello, "Long wavelength absorbing azo dyes derived from 2-amino-4-chloro-5-formylthiazole as diazo component," *Dyes and Pigments*, vol. 27, no. 1, pp. 45–54, 1995.

[11] H. R. Maradiya and V. S. Patel, "Synthesis and application of disperse dyes based on 2-aminothiazole derivatives," *Chemistry of Heterocyclic Compounds*, vol. 39, no. 3, pp. 357–363, 2003.

[12] M. M. El-Molla, Z. H. Ismaeil, F. M. A. Soliman, H. Shaimaa, and A.-E. Monem, "Synthesis of several newly disperse dyes and their application in textile printing," *Journal of the Textile Association*, vol. 74, no. 1, pp. 18–25, 2013.

[13] S. K. Zadafiya, J. H. Tailor, and G. M. Malik, "Disperse dyes based on thiazole, their dyeing application on polyester fiber and their antimicrobial activity," *Journal of Chemistry*, vol. 2013, Article ID 851418, 5 pages, 2013.

[14] P. O. Nkeonye, *Fundamental Principles of Textile Dyeing, Printing and Finishing*, Ahmadu Bello University Press, 1987.

[15] A. Marini, A. Muñoz-Losa, A. Biancardi, and B. Mennucci, "What is solvatochromism?" *Journal of Physical Chemistry B*, vol. 114, no. 51, pp. 17128–17135, 2010.

[16] P. Parvizi, A. Khosravi, S. Moradian, and K. Gharanjig, "Synthesis and application of some alkali-clearable azo disperse dyes based on naphthalimide derivatives," *Journal of the Chinese Chemical Society*, vol. 56, no. 5, pp. 1035–1042, 2009.

28

Evaluation of Combined Effect of Mercerized and Dyed Yarns on Physical Properties of Plain Single Jersey Knitted Fabrics

S. M. Elrys, A. El-Hossini, and A. M. EL-Hadidy

Mansoura University, Faculty of Engineering, Textile Department, Mansoura, Egypt

Correspondence should be addressed to S. M. Elrys; samah_elrys@yahoo.com

Academic Editor: Hu Hong

A combined effect of mercerized and dyed yarns on physical properties of plain single jersey knitted fabrics has been studied. Plied cotton yarn was produced and mercerized at different NaOH concentrations and temperatures. Then, this yarn was bleached and dyed with two types of dyes, vat dyeing and reactive dyeing. Each of mechanical and color properties were measured to these treated yarns. Plain single jersey knitted fabrics were produced from these treated yarns. The total evaluation of the properties for plain single jersey fabrics was measured by using radar chart method. It was found that the highest area of radar chart, which represents the total physical properties, is achieved at 32°Be′, in case of vat dyeing. Besides, analysis of variance (AVOVA) and MANOVA were applied to illustrate which of the properties would be affected by NaOH con. and types of dyeing for yarns and fabrics.

1. Introduction

The quality of end products like fabrics depends on the quality of their components like yarns. Textile finishing processes are usually used to improve the quality of materials. Among these, mercerization is one of the most important wet finishing processes of cotton materials, as it improves handling and appearance of cotton fibers to simulate the superior properties of synthetic fibers. In addition, Mercerization treatment improves dye ability and luster required for appearance, smoothness and strength required for ability of processing such as sewing threads.

These properties are the most important ones needed to materials for competition with other products.

There are many methods of mercerization where it can be applied to yarns or fabrics. Depending on tension, it can be done in the slack state or under tension state. Depending on temperature, it can be done under high temperatures which is called hot mercerization or under normal temperatures which is called normal mercerization. Also, it can be applied in hank, cheese, and single-end mercerization.

However, all of these methods of mercerization are applied by treating the materials with caustic soda solution at a certain concentration and temperature depending on the type of applied method. Each of these methods has its merits and demerits. The extent of changes that occur depends on the processing time, caustic concentration, temperature, and degree of polymerization, source of cellulose, the physical state of the cellulose, slack or tension treatment, and the degree of tension [1].

Dyeing is another wet process that influences yarn and fabric properties. Cotton and other cellulosic fibers are dyed with direct, sulphur, vat, reactive, and more types than for any other fiber. Each of these classes of dye has its own application methods, dyeing characteristics, cost, fastness properties, colour range, advantages, and disadvantages. Reactive dyes, particularly those used for dyeing cotton, have become one of the major classes of dyes because of their good washing fastness, their bright shades, and their versatile batch and continuous dyeing methods. Furthermore, vat dyes are one of the oldest types of dye. Vat dyes in particular give dyeing on cellulosic fibers with the best overall fastness properties [2].

Single jersey knitted fabrics are used to make underwear and outerwear garments such as T-shirts. The knitted fabrics undergo a series of different processing treatment like bleaching, scouring, dyeing, softener padding, and relax drying. These processes are carried out to impart a particular property related to that process like scouring for absorbency, bleaching for whiteness, dyeing for imparting colour to fabric, and finishing for improving softness and handling the fabric [3].

It can be clear that each wet process has its special influence on the properties of the treated yarns or fabrics. However, studying the combined effect of these wet processes on the physical properties of yarns and plain singles jersey fabrics produced from these treated yarns is rare.

2. Review of Literature

Mercerization is originally discovered by John Mercer in 1844. From this time, mercerization has a major interest to researchers because of its advantages.

Akhbari et al. investigated the parameters influencing mercerization using RSM method, in order to increase the tensile strength of mercerized yarn [4]. Samei et al. compared the effects of hot mercerization on open-end and ring spun yarns in slack and under tension conditions. Mercerized yarns were bleached and dyed with reactive dye [5]. Gemci applied mercerization process on cotton fabrics with three different weaving patterns using three different time periods and three different processes under factory conditions to examine the dimension stability [6]. Moghassem and Valipour investigated the effects of four mercerization treatment parameters, namely, alkali concentration, time of treatment, bath temperature, and mercerization tension on dimensional properties of plain knitted fabric produced from cotton yarn [7]. Hasani studied the effect of different finishing processes like bleaching, dyeing and softening on mechanical and surface properties of cotton knitted fabrics [8].

Sameii et al. studied the effect of mercerization parameters on cotton fabrics made up of open-end yarns. Reactive dyes were used for dyeing. The results showed that tensile strength of fabrics could be improved in mercerization process [9]. The effect of different wet processing stages and sequences on the physical, dimensional, and dyeing properties of the single jersey knitted fabrics are produced by Murugesh and Selvadass [3].

From the previous studies, there are many researches about the effect of mercerization conditions only on the mercerized yarns. Some of the researchers studied the effect of mercerization and dyeing on mercerized yarns. Others studied the effect of mercerization parameters only on properties of fabrics. Also, little researchers studied the relation between properties of mercerized yarns and properties of plain single jersey knitted fabrics produced from them.

However, the combined effect of mercerization variables and variables of the following finishing operations is scanty. Consequently, this research attempts to expand the study of the effect of mercerization besides some types of dyeing on the produced yarns and properties of plain single jersey knitted fabrics produced from these yarns.

3. Experimental

3.1. Materials. Plied ring spun cotton yarn with a count of 20/2 Ne was produced. It was spun from Egyptian cotton G86 with single twist factor $\alpha e = 3.6$ and plied twist factor $\alpha e = 3.5$.

3.2. Methods

3.2.1. Mercerization. Mercerization was done on Jaeggli mercerizing yarn machine. Yarn hanks were treated with caustic soda concentration ranging from 26°Be′ to 34°Be′ with 2°Be′ intervals, using wetting agent [(6–8) g/L of FLORANIT 4028 of Pulcra Chemical]. Temperature of treatment was firstly changed by using 19°C and 24°C. Both applied tension and time of treatment were constant. To remove the excess caustic soda after the treatment, the yarn hanks were washed with hot and cold water. The hanks were then neutralized with acetic acid solution (0.5 cm/L) to remove any residual alkali. Yarns were finally rinsed with cold water, squeezed with centrifugal forces, dried at 110°C for 2 hrs.

3.2.2. Bleaching. Mercerized yarns were bleached in an exhaustion procedure, the bleaching bath containing 0.5% hydrogen peroxide, 2% sodium hydroxide, 2% asbicon (detergent used as assistant factor), and 1% Egypttool (wetting agent). The bleaching was carried out at 100°C for 1 hr. Acetic acid was used for neutralization.

3.2.3. Dyeing. Dyeing with vat and reactive dyes were carried out by exhaustion procedure.

(a) Reactive Dyeing. EcoFix.Blue.R (1.5%), Na_2CO_3 (5 g/L), and NaCl (20 g/L) were adjusted at 50°C and then raised to 60°C and maintained at this temperature for 90 min to dye as reactive dyeing. Dyed samples were then washed in a soap solution and then boiled and softened by using fatty acid and, finally, squeezed and dried at 100°C.

(b) Vat Dyeing. Yarns were dyed in a solution containing India Blue CLF (vat dye) (1.5%), NaOH (0.1%), and sodium hydrosulphite (5 g/L). As known, vat dyeing is based on the principle of converting water-insoluble vat dye by alkaline reduction to a water-soluble leuco compound having affinity to cotton. Sodium hydrosulphite is used in the reduction process at 60°C and then raised to 80°C for 20 min and then washed. Finally, it is oxidized by using H_2O_2 at 50°C for 30 min, washed, and softened by using fatty acid.

3.2.4. Knitting Process. Plain single jersey knitted fabrics were produced on Stoll flat knitting machine, G 12, and loop length 0.6 cm. The specification of the pretreated produced samples can be illustrated in Table 1.

In addition, two samples of Plain single jersey knitted fabrics were produced from without mercerized, bleached, vat, and reactive dyed yarns. These two samples were used for comparison.

TABLE 1: Fabrics production plan.

Types of dyeing	NaOH conc. (Be′)				
	26	28	30	32	34
Vat dyeing	√	√	√	√	√
Reactive dyeing	√	√	√	√	√

3.3. Testing Methods. Testing methods were carried on both of yarns and finished plain single jersey knitted fabrics as follows.

3.3.1. Testing Methods of Yarns. Breaking load and extension of yarns were measured by using USTER Tensorapid 4. Color strength of yarns was observed by using Data color International SF 600 at D65. Yarn abrasion resistance was measured according to ASTM D6611.

3.3.2. Testing Methods of Fabrics. A sensitive digital balance device with accuracy of two digits was used to measure the weight of the samples and then fabric weight in g/m^2 was estimated. Thickness of fabric samples was measured by using K094 SDL Atlas digital thickness gauge tester according to ASTM D1777, Standard Test Method for Thickness of Textile Material.

The fabric bursting strength and elongation due to burst were tested using H5KT Tinius Olsen universal testing machine using ASTM D 3787 bursting strength in knitted goods standard test method. Fabric abrasion resistance of fabrics was measured by using Taber Abrasion (Rotary Platform) tester according to ASTM D3884. The air permeability was tested using M021A SDL Atlas air permeability tester using ASTM D 737 standard test method. Fabric stiffness was measured by using the circular bend procedure according to ASTM D 4032. K/S was measured using D412G1 color i5 spectrophotometer using daylight condition (D65). K/S is used to evaluate the depth of dyed color after mercerization.

The Shrinkage value was defined by (1), for American AATCC number 135-1987

$$\mathrm{SH}\,(\%) = \left[\frac{(L - L_0)}{L_0}\right] \times 100, \tag{1}$$

where L_0 is the length of the sample before laundering and drying and L is the length of the sample after laundering and drying.

4. Experimental Results and Discussion

4.1. Results of Yarns. At the end of each treatment process, the yarn samples were collected and tested for some of the mechanical and color properties. The experimental results of yarns are compared to grey yarn as illustrated in the following equation:

$$\text{Change}\,(\%) = \left[\frac{(T - G)}{G}\right] \times 100, \tag{2}$$

where T is the treated samples of yarns and G is the grey sample.

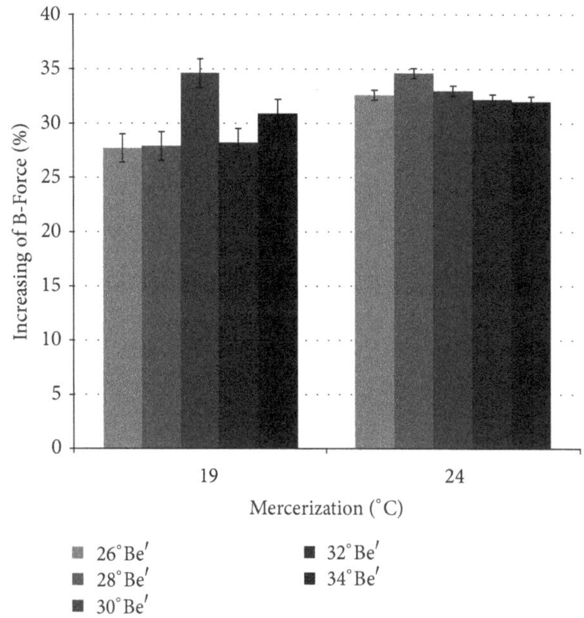

FIGURE 1: Effect of mercerization temperature and NaOH conc. on yarn breaking load.

4.1.1. Effect of Mercerization Temperature and NaOH Concentration on Yarn Breaking Load. Figure 1 shows the effect of temperature of mercerizing and NaOH concentration on the breaking load of yarns. It can be clearly seen that breaking load rose from 27.7% to 34.6% with an average value of 31.37%. In general, there is an increase in the strength in case of temperature 24°C more than temperature 19°C. In case of temperature 19°C, the highest value of breaking load occurs at 30°Be′, which is nearly equal to the highest value at 24°C which occurs at 28°Be′. This means that there is a relationship between mercerization temperature and NaOH concentration. In AS, with increasing temperature and with lower NaOH concentration, the same value can be obtained with increasing NaOH concentration at lower temperature.

Of course, lower concentration with higher temperature is economically better. Consequently, the temperature 24°C is selected to be studied where the average increase value reaches 32.88%. The main reason of increasing the breaking load after mercerization process because cotton fibers are modified during mercerizing process and transformed from cellulose I to cellulose II. Fibers become more amorphous and less crystalline.

4.1.2. Effect of Mercerization and Dyeing on Breaking Load of Mercerized Yarns. Figure 2 shows the effect of both vat and reactive dyeing on the breaking load for the mercerized yarn at 24°C and at different NaOH concentrations. As shown, there is an increase generally in breaking load after vat dyeing more than reactive dyeing compared to bleaching.

Results showed that the breaking load rose from 8.6% to 27.9% with the average value 16.24% in case of vat dyeing.

Also, results showed that the breaking load increased to 24.1% in case of reactive dyeing because of the effect of

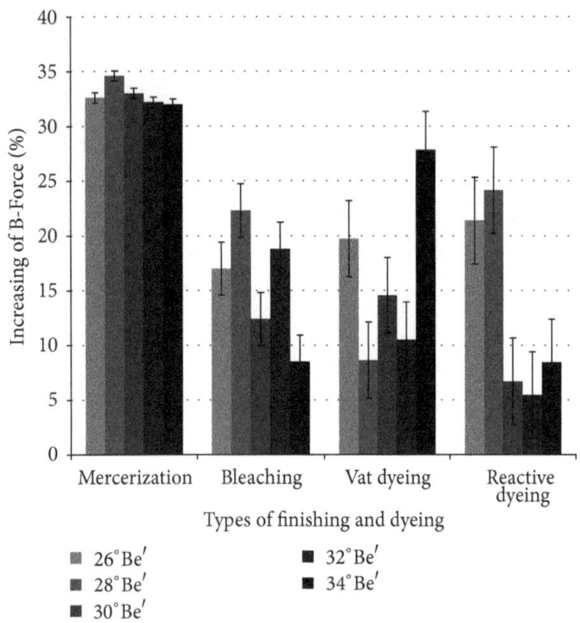

FIGURE 2: Effect of vat and reactive dyeing on breaking load of mercerized yarns.

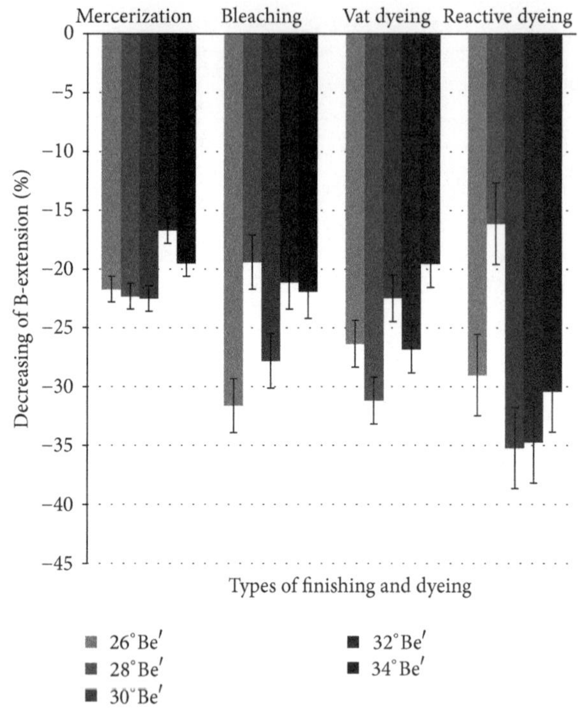

FIGURE 3: Effect of types of finishing on yarn breaking extension.

chemical bonds between the dye and the fibers. Moreover, reactive dye can also be bonded with more than one fiber consequently increasing the tensile strength. The increase in breaking load in case of reactive ranges from 5.5% to 24.1% with an average value of 13.22%.

The highest value in the breaking load for vat dyeing occurs at 34°Be′ (27.9%) but for reactive dyeing it occurs at 28°Be′ (24.1%).

On the other hand, the lowest value in breaking load for vat dyeing occurs at 28°Be′ (8.6%) but for reactive dyeing it occurs at 32°Be′ (5.5%).

As mercerizing process leads to the increase of dyeing absorption because of breaking the hydrogen bonds between the fibers in the crystalline regions and converting them to amorphous regions. This increase in vat dyeing is arranged during washing with alkali soap after dyeing process. In addition, the increase of these spaces between fibers with the presence of vat dyeing may be the reason for the rise of the breaking load.

4.1.3. Effect of Mercerization and Dyeing on Yarn Breaking Extension. Figure 3 shows the effect of mercerization and dyeing on the mercerized yarn at different NaOH concentrations and at 24°C. As shown mercerizing process affects the breaking extension negatively.

From the results after mercerization, breaking extension decreases from −16.7% to −22.5% with average value −20.5%. However, after bleaching breaking extension decreases from −19.4% to −31.4% with average value −24.4%.

Also there is a decrease after vat dyeing from −19.5% to −31.2% with average value −25.3%. Moreover, breaking extension decreases sharply after reactive dyeing from −16.1% to −35.2% with average value −29.1%. The decrease percentage of breaking extension can be illustrated due to the axis parallel alignment of fibers in the yarn centre [10].

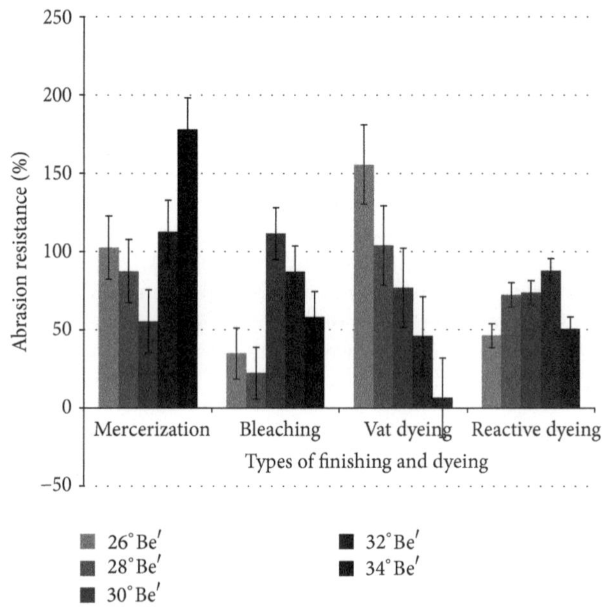

FIGURE 4: Effect of mercerization and dyeing on yarn abrasion resistance.

4.1.4. Effect of Mercerization and Dyeing on Yarn Abrasion Resistance. Figure 4 shows the effect of both mercerization and dyeing on abrasion resistance of yarns at different NaOH concentrations. It can be clearly seen that there is an increase in abrasion resistance of yarns after mercerization from 55%

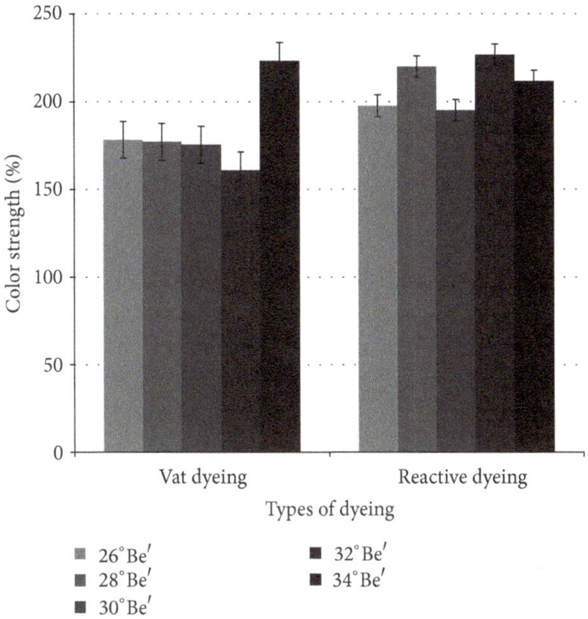

FIGURE 5: Effect of vat and reactive dyeing on color strength of mercerized yarns.

to 178%. There is an increase after vat dyeing from 7% to 155%, and after reactive dyeing from 46% to 87%.

Abrasion of the yarn is followed by the gradual removal of fibers from the yarns when they are subjected to repeated distortion. Factors affecting the cohesion of the fibers in the yarn and yarn to yarn friction will have influence on the abrasion resistance. They mainly depend on the fiber surface, supramolecular structure of fibers, yarns surfaces, and yarn compressibility in addition to the magnitude of the normal pressure acting between them [11].

4.1.5. Effect of Vat and Reactive Dyeing on Color Strength of Mercerized Yarns. Figure 5 shows the effect of vat and reactive dyeing after mercerization at different NaOH concentrations and at 24°C on the color strength of yarns. In general, there is an increase in color strength in reactive dyeing more than vat dyeing. It can be clearly seen that color strength increased from 160.7% to 223.2% with average value 183% in case of vat dyeing because mercerizing process converts fibers to cylindrical shape leading to the increase of the spaces between the fibers, so the dye molecules can easily penetrate into the fibers and increase the color strength of this vat dye. Also, results showed an increase in color strength of reactive dyeing from 195.1% to 226.6% with average value 210.3% because molecules of this reactive dye are bonded with covalent bonds with the fibers, so the color strength of the dye increases.

The highest value in vat dyeing occurs at 34°Be′ (223.2%) and in reactive dyeing occurs at 32°Be′ (226.6%).

4.2. Results of Knitted Fabrics. Experimental results of fabric physical and colour properties are compared to the two unmercerized samples as illustrated in the following equation:

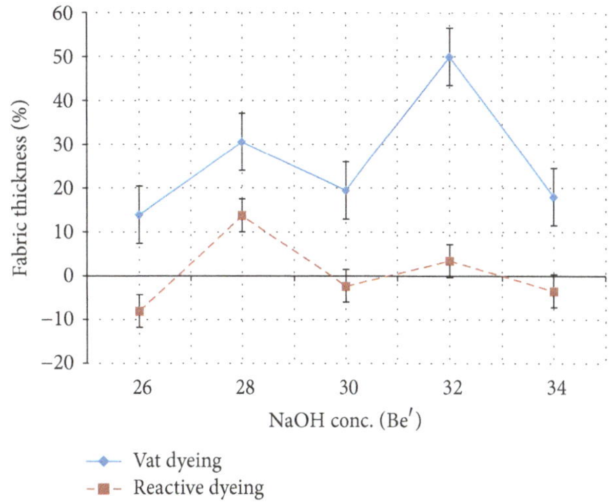

FIGURE 6: Effect of mercerization and dyeing on thickness of plain single jersey knitted fabrics.

$$\text{Change}\,(\%) = \left[\frac{(\text{MD} - \text{BD})}{\text{BD}}\right] \times 100, \qquad (3)$$

where MD is the mercerized, bleached, and dyed samples and BD is the unmercerized, bleached, and dyed samples, with the knowledge that each dyed sample is compared to its same type of dyed sample.

4.2.1. Fabric Thickness. Figure 6 shows the effect of both of mercerization and dyeing processes on the thickness of plain single jersey knitted fabrics. As shown, there is generally an increase in thickness in vat dyeing more than reactive dyeing with 26%. The same effect of the thickness is obtained in both vat dyeing and reactive dyeing depending on NaOH concentration.

There is an increase in case of vat dyeing from 13.9% to 50% compared to vat standard sample, with average increase value 26.4%, while, in case of reactive dyeing, the average increase value of thickness was 0.7% compared to reactive standard sample. The highest value of thickness at vat dyeing occurs at 32°Be′ and at reactive dyeing occurs at 28°Be′.

4.2.2. Fabric Weight. Figure 7 shows the effect of mercerization and dyeing on the weight of plain single jersey knitted fabrics. As shown, the same effect on the weight of fabrics in both dyeing with increasing NaOH concentration. As in the case of the thickness, the weight of fabrics at vat dyeing is more than that at reactive dyeing. As in the case of vat dyeing, increasing temperature of mercerization leads to increase of the swelling, converting crystalline regions to amorphous regions. Consequently, there are greater opportunities to soak up water moisture because of the increase of the free amounts of hydroxyl groups which forms the hydrogen bonds. On the other hand, there is a decrease in the weight in the case of reactive dyeing because increasing the temperature leads

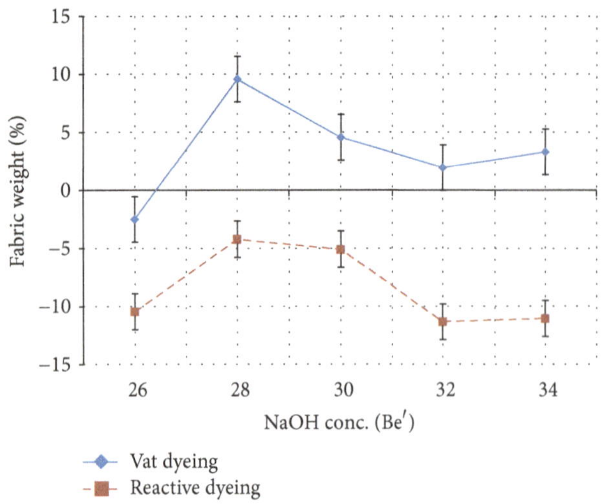

FIGURE 7: Effect of mercerization and dyeing on weight of plain single jersey knitted fabrics.

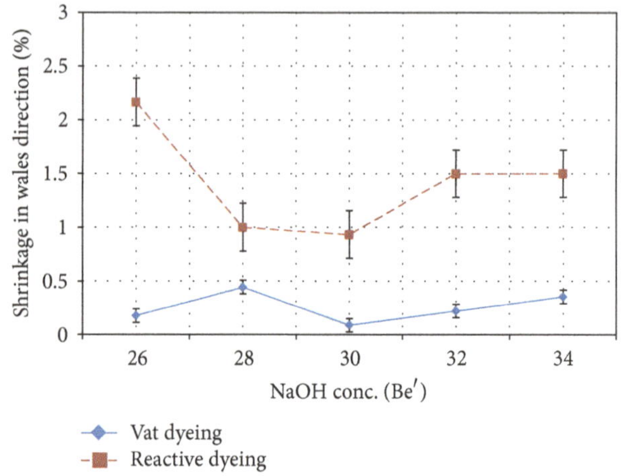

FIGURE 9: Effect of mercerization and dyeing on shrinkage in wales direction of plain single jersey knitted fabrics.

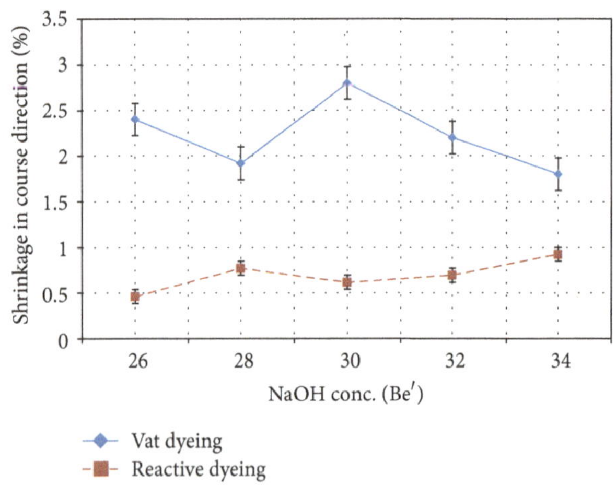

FIGURE 8: Effect of mercerization and dyeing on shrinkage in courses direction of plain single jersey knitted fabrics.

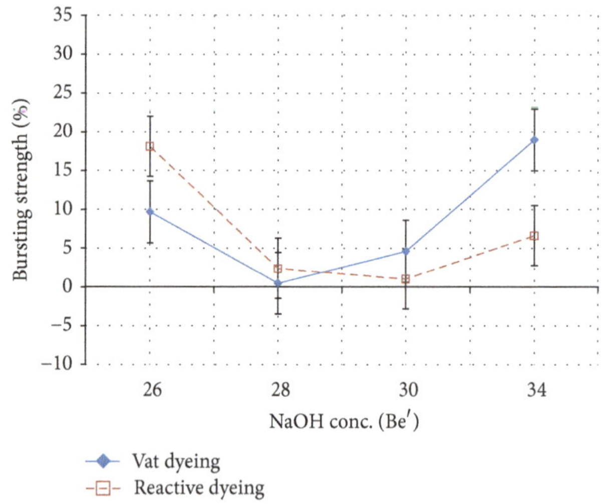

FIGURE 10: Effect of mercerization and dyeing on bursting strength of plain single Jersey knitted fabrics.

to the removal of starching materials and some of natural impurities materials.

The average increase value in case of vat dyeing is 3.4%, but the average decrease value in case of reactive dyeing is 8.4%.

4.2.3. Fabric Shrinkage. Figures 8 and 9 show the effect of mercerization and dyeing on the shrinkage in both wales and courses directions of plain single jersey knitted fabrics. It can be clearly seen that there is a rise in the shrinkage in length direction in case of vat more than reactive dyeing. This means that reactive dyeing is better in shrinkage in length direction. The increase in shrinkage in length in case of vat dyeing ranges from 1.8% to 2.8% with average value 2%, but the increase in shrinkage in case of reactive ranges from 0.5% to 0.9% with average value 0.7%.

Of course the opposite happened in courses direction for both vat dyeing and reactive dyeing. As shown in Figure 9, the increase in shrinkage in courses direction in case of reactive dyeing is more than vat dyeing. This means that vat dyeing is better in the shrinkage in width more than reactive dyeing. The best values for shrinkage in wales direction occur at 26°Be′ for reactive dyeing and at 34°Be′ for vat dyeing. Besides, the best values for shrinkage in courses direction occur at 26°Be′ for vat dyeing and at 30°Be′ for reactive dyeing.

4.2.4. Bursting Strength of Plain Single Jersey Fabrics. Figure 10 shows the effect of mercerization and dyeing on the bursting strength of plain single jersey knitted fabrics. The results showed that the same effect on the tensile strength happens for the two dyeing until a certain point, and then the opposite happens. In general, there is an increase in tensile strength of knitted fabrics in case of vat dyeing more than reactive dyeing. The increase of tensile strength in case of vat dyeing rises from 0.5% to 19% with average value 8.4%.

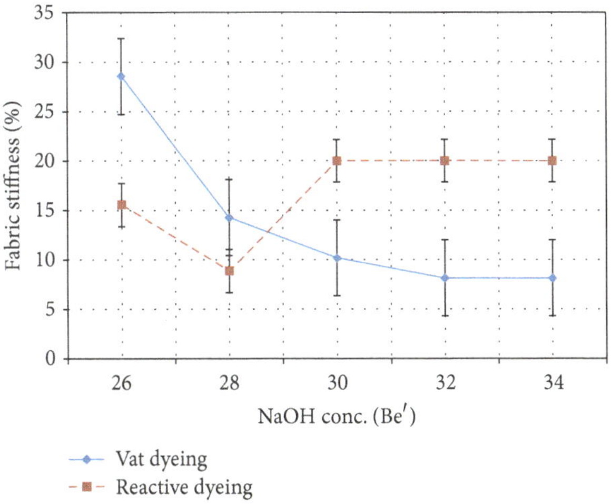

FIGURE 11: Effect of mercerization and dyeing on stiffness of plain single jersey knitted fabrics.

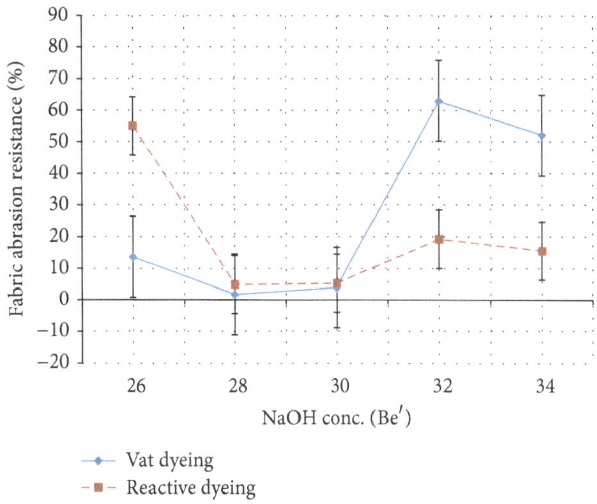

FIGURE 12: Effect of mercerization and dyeing on abrasion resistance of plain single jersey knitted fabrics.

Also, increase of tensile strength in case of reactive dyeing ranges from 1.1% to 18.1% with average value 7.1%. In fact, the increase in tensile strength is because of strengthening of the weak points along the fiber and also the modifications in orientation and consolidation of weak points.

The highest value of breaking load in case of vat dyeing occurs at 34°Be′, and the highest value in case of reactive dyeing occurs at 26°Be′.

4.2.5. Fabric Stiffness of Plain Single Jersey Fabrics. Figure 11 shows the effect of mercerization and dyeing on the stiffness of plain single jersey fabrics.

For vat dyeing, stiffness of plain jersey fabrics decreases with the increasing NaOH concentration. However, for reactive dyeing, the stiffness is decreased then increased and finally remained constant. The lowest value of stiffness at vat dyeing occurs at 34°Be′ and at reactive dyeing occurs at 28°Be′.

4.2.6. Abrasion Resistance of Plain Single Jersey Fabrics. Figure 12 shows the effect of mercerization and dyeing on the abrasion resistance of plain single jersey fabrics. As before in the bursting strength, the same effect on fabric abrasion resistance for the vat and reactive dyeing happens till 30°Be′; after that the opposite happens. Abrasion resistance fabric of vat dyeing increases from 1.7% to 62.9% with average value 26.8%. In addition, abrasion resistance fabric of reactive dyeing increases from 4.8% to 55.1% with average value 20%. The highest value of fabric abrasion resistance in case of vat dyeing occurs at 32°Be′, and the highest value of fabric abrasion resistance in case of reactive dyeing occurs at 26°Be′.

4.2.7. Air Permeability of Plain Single Jersey Fabrics. Figure 13 shows the effect of mercerization and dyeing on air permeability of plain single jersey fabrics. As shown, the increase in air permeability in case of vat dyeing is more than reactive dyeing. The results show that there is significant increase in air permeability in case of vat dyeing from 35% to 47% with average value 42.3%. Also, air permeability in case of reactive dyeing rises from 18.7% to 40% with average value 31.8%. The highest value for air permeability in case of vat dyeing occurs at 28°Be′, and in case of reactive dyeing it occurs at 34°Be′. The increase of air permeability in both vat dyeing and reactive dyeing can be explained due to the fiber arrangement in the cross-section of the yarn.

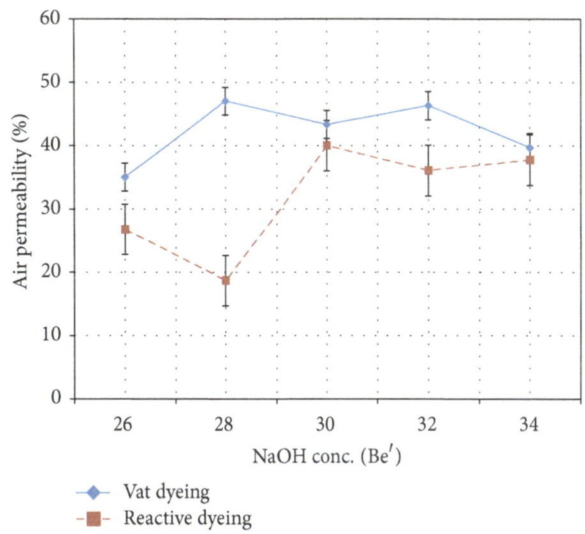

FIGURE 13: Effect of mercerization and dyeing on air permeability of plain single jersey knitted fabrics.

4.2.8. K/S of Plain Single Jersey Fabrics. Figure 14 shows the effect of mercerization and dyeing on K/S of plain single jersey fabrics. As shown, vat dyeing is higher than reactive dyeing regarding K/S values. The increase in K/S values in case of vat dyeing ranges from 53% to 90% with average value 72.5%.

Besides, the K/S values in case of reactive dyeing rise from 4.8% to 43.7% with average value 27.5%. The highest value

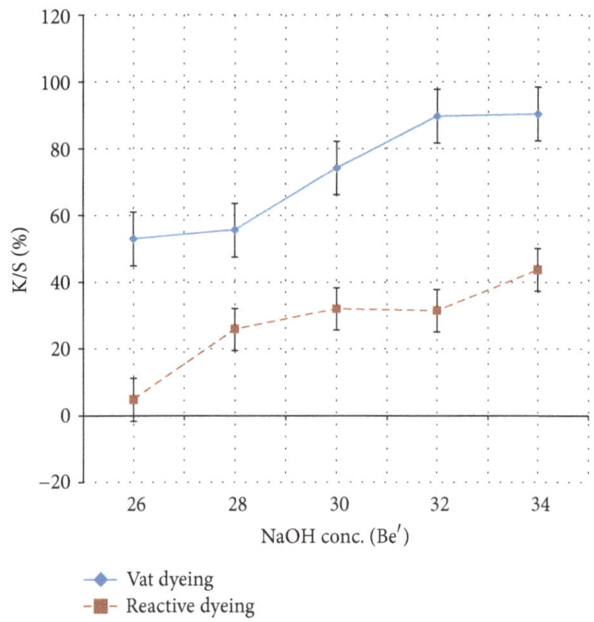

FIGURE 14: Effect of mercerization and dyeing on K/S values of plain single jersey knitted fabrics.

of K/S in case of vat dyeing occurs at 34°Be′, and in case of reactive dyeing also occurs at 34°Be′.

The increase in color properties can be explained generally because of the destruction of crystalline regions during swelling and changes in microstructure and morphology.

5. Analysis of Variance

The influence of NaOH concentration on the properties of yarns after mercerization, bleaching, and vat and reactive dyeing was analyzed by the main effects analysis of variance (ANOVA). The results of P values (variables have significant effect on the measured properties $P < 0.05$) are given in Table 2.

It was found that NaOH conc. had a significant effect on B-Force (after bleaching and dyeing), elongation (after each treatment), and abrasion resistance (after mercerization, bleaching, and only vat dyeing) of yarns.

Besides, the effect of both NaOH concentration and type of dyeing on the properties of plain knitted fabrics was analyzed by the main effects analysis of variance (MANOVA). The results of P values of fabrics are shown in Table 3.

The results showed that NaOH concentration had a significant effect on the fabric thickness, bursting strength, fabric weight, abrasion resistance, and air permeability of plain knitted fabrics. On the other hand, the types of dyeing had a significant effect on fabric weight, shrinkage in course direction, bursting strength, and abrasion resistance of plain knitted fabrics.

TABLE 2: Effect of NaOH conc. on yarn properties.

Type of test	P value of NaOH conc.
B-Force after mercerization	0.358
B-Force after bleaching	0.007
B-Force after vat dyeing	0.000
B-Force after reactive dyeing	0.013
Elongation after mercerization	0.000
Elongation after bleaching	0.000
Elongation after vat dyeing	0.000
Elongation after reactive dyeing	0.000
Abrasion resistance after mercerization	0.000
Abrasion resistance after bleaching	0.000
Abrasion resistance after vat dyeing	0.004
Abrasion resistance after reactive dyeing	0.147

TABLE 3: Effect of NaOH conc. and types of dyeing on the properties of plain knitted fabrics.

Types of test	P value of NaOH conc.	P value of types of dyeing
Fabric thickness	0.000	0.084
Fabric weight	0.002	0.027
Shrinkage in wales direction	0.228	0.070
Shrinkage in courses direction	0.719	0.032
Bursting strength	0.000	0.001
Fabric stiffness	0.684	0.166
Abrasion resistance	0.000	0.000
Air permeability	0.031	0.257

6. Overall Evaluation of Yarn and Fabric Properties

Hypothesis testing was used to know if the best properties of yarns occur at the same variables for the produced knitted fabrics or not. This was considered the null hypothesis, while the alternative hypothesis occurred when the yarns highest area of radar chart does not occur at the same variables which represent the fabric highest area of radar chart.

Figures 15 and 16 show the total evaluation of the properties for yarns by using Performance Diagram method based on the absolute values. It was found that the highest area of radar chart occurred at 26°Be′ in case of vat dyeing as shown in Table 4. The lowest area of Performance Diagram in case of vat dyeing occurred at 32°Be′. The highest area of Performance Diagram in case of reactive dyeing occurred at 32°Be′, but the lowest area is achieved at 34°Be′.

Figures 17 and 18 show the total evaluation of the properties for plain single jersey fabrics by using radar chart method based on absolute values. It was found that the highest area of Performance Diagram, which represents the total physical properties, is achieved at 32°Be′, in case of vat dyeing as shown in Table 5. On the contrary, the lowest area

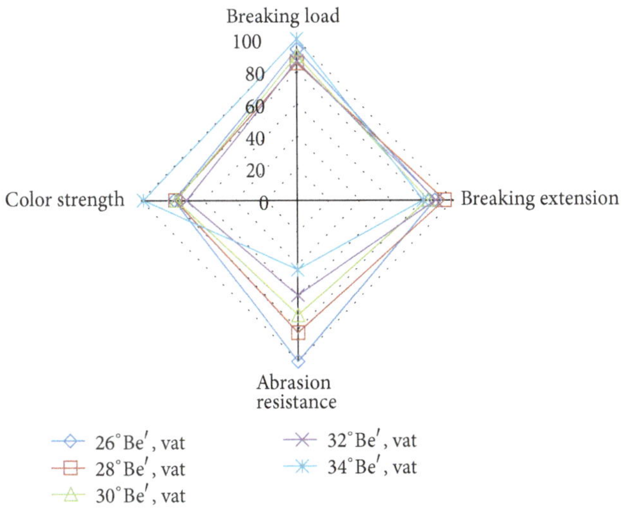

Figure 15: Overall evaluation of yarn properties at vat dyeing.

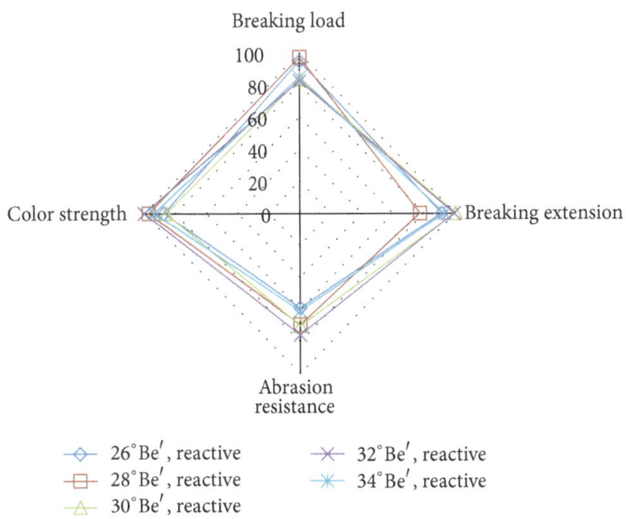

Figure 16: Overall evaluation of yarn properties at reactive dyeing.

Table 4: Total area of overall yarn properties.

Types of dyeing	NaOH conc. (Be′)				
	26	28	30	32	34
Vat dyeing	81%*	72%	65%	58%	64%
Reactive dyeing	69%	73%	71%	79%	68%

*Actual area/ideal area.

of Performance Diagram in case of vat dyeing occurred at 26°Be′.

The average of improvement in the overall physical properties in case of vat dyeing is 65%.

The highest area of Performance Diagram in case of reactive dyeing is achieved at 26°Be′; the lowest area occurred at 34°Be′. The average improvement in the overall physical properties in case of reactive dyeing is 57%.

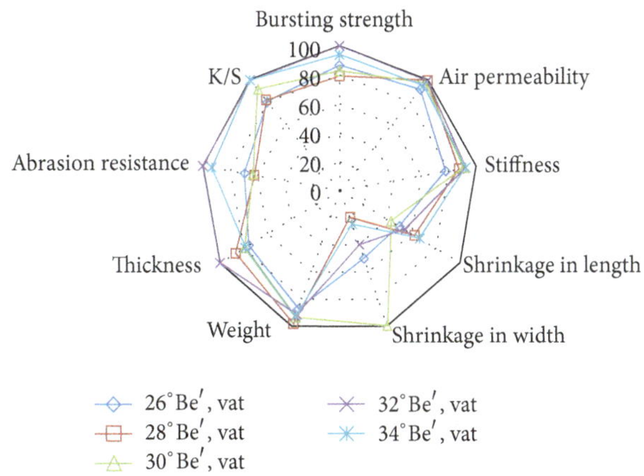

Figure 17: Overall evaluation of fabric properties at vat dyeing.

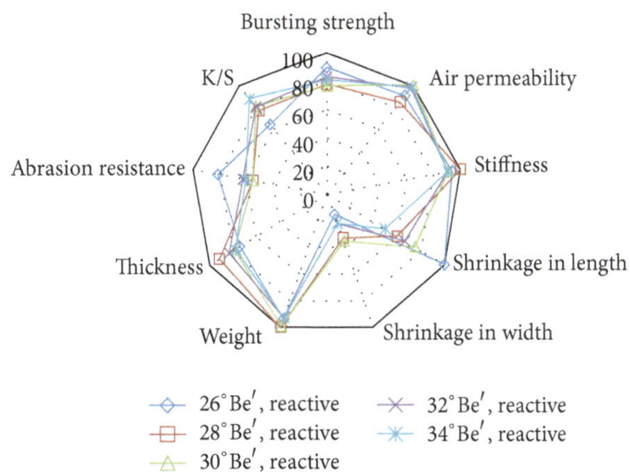

Figure 18: Overall evaluation of fabric properties at reactive dyeing.

Table 5: Total area of overall fabric properties.

Types of dyeing	NaOH (Be′)				
	26	28	30	32	34
Vat dye	55%	56%	67%	77%	68%
Reactive dye	59%	56%	58%	57%	54%

This means that the null hypothesis was not obtained as the yarns properties are not necessarily equal to the properties of fabrics produced from them.

It was noticed that the vat dyeing maintains the physical properties generally more than reactive dyeing.

7. Conclusion

(i) Mercerizing process of yarns at different NaOH concentrations, at 24°C, followed by bleaching and dyeing with vat and reactive dyeing, has an influence on the final properties of plain jersey fabrics.

(ii) There is an increase of thickness by about 3.4% to 50%, but mercerized dyed fabrics with vat dyeing are thicker than the reactive ones compared to unmercerized samples.

(iii) The average increase value of weight is 3.4% in case of vat dyeing. On the contrary, the average decrease value of weight in case of reactive dyeing is 8.4% compared to unmercerized samples.

(iv) For the shrinkage in wales direction, the increase in shrinkage for reactive dyeing is less than that for vat dyeing. On the contrary in courses direction, the increase in shrinkage for vat is less than reactive dyeing compared to unmercerized samples.

(v) There is an increase of bursting strength in case of vat dyeing from 0.5% to 19% and in case of reactive dyeing from 1.1% to 18.1% compared to unmercerized samples.

(vi) Consequently, mercerized plain jersey fabrics dyed with vat dyeing are stiffer than those of the reactive ones.

(vii) There is an increase of the fabric abrasion resistance for vat dyeing from 1.7% to 62.95% and for reactive dyeing from 4.8% to 55.1% compared to unmercerized samples.

(viii) Mercerized dyed fabrics with vat dyeing have higher air permeability values than reactive dyeing.

(ix) In general, there is an increase of the color properties for both vat dyeing and reactive dyeing compared to unmercerized samples because of the mercerization effect.

(x) From the overall evaluation of physical properties by Performance Diagram, the highest area occurred at $32^{\circ}\mathrm{Be}'$ in case of vat dyeing.

(xi) Vat dyeing maintains the physical properties of the fabric generally more than reactive dyeing.

(xii) Analysis of variance showed the properties of yarns and fabrics affected by NaOH conc. and types of dyeing.

Conflict of Interests

The authors declare that there is no conflict of interests regarding the publication of this paper.

References

[1] T. Wakida, M. Lee, S. J. Park, and A. Hayashi, "Hot mercerization of cottons," *Fiber*, vol. 58, pp. 304-307, 2002.

[2] A. D. Broadbent, *Basic Principles of Textile Coloration*, Society of Dyers and Colourists, Bradford, UK, 2001.

[3] B. K. Murugesh and M. Selvadass, "Influence of wet processing on properties of single jersey knitted fabrics," *International Journal of Fiber and Textile Research*, vol. 3, no. 1, pp. 18–30, 2013.

[4] M. Akhbari, A. Zahiri, and S. J. E. Bassam, "Optimization of parameters influencing mercerization using the RSM method in order to increase the tensile strength of mercerized yarn," *Fibres & Textiles in Eastern Europe*, vol. 94, no. 5, pp. 30–35, 2012.

[5] N. Samei, S. M. Mortazavi, A. Rashidi, and S. S. Najjar, "Changes in physical properties of hot mercerized ring and open-end spun cotton yarns," *Iranian Polymer Journal*, vol. 17, no. 12, pp. 937–945, 2008.

[6] R. Gemci, "Examining the effects of mercerization process applied under different conditions to dimensional stability," *Scientific Research and Essays*, vol. 5, no. 6, pp. 560–571, 2010.

[7] A. R. Moghassem and P. Valipour, "An extensive look in to the effect of mercerization treatment on dimensional properties of cotton plain knitted fabric," *Fibers and Polymers*, vol. 14, no. 2, pp. 330–337, 2013.

[8] H. Hasani, "Effect of different processing stages on mechanical and surface properties of cotton knitted fabrics," *Indian Journal of Fibre and Textile Research*, vol. 35, no. 2, pp. 139–144, 2010.

[9] N. Sameii, S. M. Mortazavi, A. S. Rashidi, and S. Sheikhzadah-Najar, "An investigation on the effect of hot mercerization on cotton fabrics made up of open-end yarns," *Journal of Applied Sciences*, vol. 8, no. 22, pp. 4204–4209, 2008.

[10] Y. Huh, Y. R. Kim, and W. Oxenham, "Analyzing structural and physical properties of ring, rotor, and friction spun yarns," *Textile Research Journal*, vol. 72, no. 2, pp. 156–163, 2002.

[11] I. Jordanov, B. Mangovska, and P. F. Tavčer, "Mechanical and structural properties of mercerized cotton yarns, bio-scoured with pectinases," *Tekstil*, vol. 59, no. 10, pp. 439–446, 2010.

29

Antimicrobial Activity of Perspiration Pads and Cotton Cloth Fabricated with the Ethyl Acetate Extract of *Eichhornia crassipes* (Mart.) Solms

P. Lalitha,[1] P. Jayanthi,[1,2] and R. Sujitha[1]

[1] *Department of Chemistry, Avinashilingam Institute for Home Science and Higher Education for Women, Coimbatore, Tamil Nadu 641043, India*
[2] *Department of Science and Humanities, Dr. Mahalingam College of Engineering and Technology, Pollachi, Coimbatore, Tamil Nadu 642 003, India*

Correspondence should be addressed to P. Lalitha; goldenlalitha@gmail.com

Academic Editor: Ajoy K. Sarkar

Waterhyacinth is one of the most dangerous aquatic weeds causing various ecological and environmental issues. The plant has several pharmaceutical properties in spite of its harmful nature. In the present study, perspiration pads and cotton cloth fabricated with the ethyl acetate extract of waterhyacinth were evaluated for its activity against *Corynebacterium*. Perspiration pads and samples of cotton cloth were fabricated with ethyl acetate extract of *Eichhornia crassipes* by three methods, namely, dipping, sonication, and homogenization. The antimicrobial activity of the fabricated textile materials was tested against *Corynebacterium*, a microorganism commonly seen in human sweat. Disc diffusion method (AATCC 147) was used for evaluating the antimicrobial nature of the fabricated samples. Sonication was found to be efficient for coating of the extract on the cotton cloth, whereas dipping method was found to be efficient in case of perspiration pads. No bacterial growth was observed under and on the fabricated cloth and perspiration pads indicating that the fabrics possess bacteriostatic property and not bactericidal property and the absence of leaching of the extract. The results showed significant antimicrobial activity of the ethyl acetate extract of *Eichhornia crassipes* coated onto fabrics against *Corynebacterium* with no growth under and on the test sample.

1. Introduction

Human body represents a multitude of microenvironments, each with its own normal flora. In general, intact skin serves as a natural protective barrier against invasion by most infectious disease growth. Species found repeatedly on the skin of groups of individuals are referred to as resident flora or microbiota of the skin. The conditions of the skin favour the growth of *Streptococcus*, *Micrococcus*, *Staphylococcus*, fungi, diphtheroid bacilli, and yeast [1]. Human axillary odour is formed by the action of *Corynebacteria* on odourless axilla secretions. The skin in human armpits contains a dense arrangement of sweat glands. Volatile substances evaporating from these areas make a key contribution to human body odour. However, sweat secreted from apocrine glands in these skin areas is initially odourless, and skin bacteria release the odoriferous principles from nonsmelling substrates present in these secretions. The axilla is a skin region colonized by an unusually dense bacterial population, with a species composition dominated by the two genera *Staphylococcus* and *Corynebacterium*. Most individuals carry a flora that is dominated by either one of these two genera, and there is a strong correlation between a high population of *Corynebacteria* and strong axillary odour formation. Subjects whose axillary skin is mainly colonized by *Staphylococci* emit only low levels of odour [2].

The three major components contributing to odour production are sweat secretion, bacterial population, and a moist environment. There are many ways to inhibit sweat odour and one among them is the use of antimicrobial fabrics. Many products are available in the market for the consumers but most of them are synthetic based and may not be environmentally benign. Their compliance to the international bodies such as EPU is essential [3]. Plant extracts have long been used as antimicrobial textile finishes owing to its ecocompatible nature. Natural antimicrobial agents are nontoxic and nonallergenic and do not cause the problems of microbial resistance [4] Perspiration pads are widely used by men and women to keep clothes clean and dry. Several commercial pads are in the market. Utilization of waterhyacinth, a throw away weed in the fabrication of sweat pads, will help in the switch over of a waste to wealth.

Eichhornia crassipes, popularly called waterhyacinth [5], is considered as a menace to the environment due to its rapid growth. In spite of the many problems that it causes, the solvent extracts of the plant possesses good antimicrobial activity, antioxidant activity, wound healing activity, and larvicidal activity [5] and is not toxic up to a tested dosage of 2000 mg/kg body weight [6].

Hence in the current study, the ethyl acetate extract of fresh *Eichhornia crassipes* was coated on to the cotton cloth and perspiration pad by three different methods, namely, sonication, dipping, and homogenization and was tested for its antimicrobial activity against *Corynebacterium* adopting standard procedures.

2. Materials and Methods

2.1. Collection of Plant. Eichhornia crassipes was collected from the lakes near Kurichi at Coimbatore between January 2012 and March 2012. The plant sample was identified by Dr. G. V. S. Murthy, Scientist F & Head of Office, Botanical Survey of India, Southern Regional Centre, Coimbatore, with the number BSI/SRC/5/23/2011-12/Tech.

2.2. Extraction of Plant Material. Eichhornia crassipes (80 g) was extracted with ethyl acetate (100 mL) for 30 minutes and desolvetised yielding ethyl acetate extract.

2.3. Phytochemical Tests. Phytochemical screening was carried out for the chloroform, ethanol and methanol extracts of waterhyacinth [7].

2.4. Preparation of Fabrics

2.4.1. Fabrication of Cotton Cloth with Ethyl Acetate Extract of Eichhornia crassipes. The white cotton cloth material was purchased with respect to the expected quality requirements. Cotton cloth (8 × 8 cm size) and perspiration pads (commercially available ones) were washed with Millipore water. The gel absorbent inside the perspiration pads was removed. The cotton cloth and the perspiration pad were air-dried. The samples were cut into small pieces; the diameter and size of the samples were measured. The samples were fabricated with the ethyl acetate extract of *Eichhornia crassipes* by sonication, dipping, and ultrasonic homogenizer methods. Similar procedure was adopted for coating standard sample ciprofloxacin.

2.4.2. Fabrication of Perspiration Pads with Ethyl Acetate Extract of Eichhornia crassipes. Ethyl acetate extract of *Eichhornia crassipes* was taken in a 500 mL beaker and then dissolved in Millipore water. The samples were subjected to sonication, dipping, and homogenization for 1 hour separately. After 1 hour, the samples were air-dried and used for the antibacterial assessment against the *Corynebacterium* strain.

2.5. Determination of Antimicrobial Effect of Fabrics

2.5.1. Test Organism. The inoculums for the experiment were prepared in fresh nutrient broth from preserved slant culture. The inoculums were standardized by adjusting the turbidity of the culture to that of McFarland standards by the addition of sterile saline.

2.5.2. Disc Diffusion Method (AATTC 147). A sterile swab was dipped in the standardized inoculums and inoculated onto the plates prepared earlier (aseptically). The excess of inoculum was removed by pressing and rotating the swab firmly against the side of the culture tube above the level of the liquid and finally streaking the swab all over the surface of the medium 3 times rotating the plate through an angle of 60°C after each application. Finally the swab was passed round the edge of the agar surface. The inoculums were left to dry at room temperature with the lid closed. Each Petri dish was divided into 2 quadrants. In one quadrant, discs coated with the extract were placed and in the other quadrant standard Ciprofloxacin 10 μg were placed with the help of sterile forceps. Then Petri dishes were placed in the refrigerator at 4°C for 1 hour for diffusion and incubated at 37°C for 24 hours. The zone of inhibition produced by leached substance was measured. Unleached samples do not give zone of inhibition but no growth should be observed above and below the sample cloth and perspiration pads.

3. Results and Discussion

3.1. Phytochemical Screening. The phytochemical screening tests carried out for the ethyl acetate extract of fresh waterhyacinth show that the plant possesses phytochemicals of commercial importance. The results of the phytochemical screening tests for the ethyl acetate extract of fresh waterhyacinth are given in Table 1. The extract tested positive for alkaloids, flavonoids, sterols, terpenoids, anthraquinones, anthocyanins, phenolics, quinones, and carbohydrates.

3.2. Fabrication of Cotton Cloth and Perspiration Pads with Ethyl Acetate Extract of Eichhornia crassipes. The amount of ethyl acetate extract coated onto the cotton cloth and perspiration pad by different methods is given in Table 2. Sonication method resulted in better coating of the extract (35 mg) onto the cotton cloth, whereas dipping was found to be better method of coating for perspiration pads (13.3 mg).

Growth of moulds, yeasts, and bacteria causes discolouration, quality deterioration, and formation of odour. This necessitates protection of textile fabrics from microbial attack

TABLE 1: Phytochemical screening of the ethyl acetate extract of fresh waterhyacinth.

S. number	Phytochemicals	Ethyl acetate extract
1	Alkaloids	
a	Meyers test	+
b	Wagner's test	+
c	Hager's test	+
2	Flavonoids	
a	Lead acetate test	−
b	H_2SO_4 test	−
3	Sterols	
a	H_2SO_4 test	−
b	Liebermann-Burchard test	+
4	Terpenoids	
a	Chloroform test	+
b	Liebermann-Burchard test	−
5	Anthraquinone	
a	Borntrager's test	+
6	Anthocyanins	
a	NaOH test	−
7	Proteins	
a	Ninhydrin test	−
8	Phenolics	
a	Ferric chloride test	−
b	Liebermann's test	+
9	Quinones	
a	HCl test	+
10	Carbohydrates	
a	Molisch's test	+
b	Fehling's test	+
11	Saponin	
a	Foam test	−
12	Tannin	
a	Braymer's test	−

and this warrants antimicrobial treatment for textile materials to avoid cross infection by pathogenic microorganisms. The antimicrobial agents can be applied to the textile substrates by techniques like exhaust, pad-dry-cure, coating, spray, and foam techniques. Various test procedures have been used to test the effectiveness of the antimicrobial activity, to evaluate antimicrobial fabrics, and to protect users and textile against action mechanisms of antimicrobial compounds [8].

The cotton cloth was fabricated with the ethyl acetate extract of fresh plant of *Eichhornia crassipes* by three methods, namely, sonication, dipping, and using ultrasonic homogenizer. It is reasonably apparent from the results that sonication method yields good coating of the extract onto the fabric. Nearly three times the extract was coated onto the cloth material even after a wash, compared to dipping method and use of ultrasonic homogenizer. The coated cloth was washed by dipping in water after drying it for about 30 minutes.

Sonication in a sonic bath disperses the extract making it fine for efficient coating onto the sample cloth. Dipping involves slack adherence of the extract onto samples that it is removed in washing. The results attest the efficacy of sonication method in coating of extracts onto cotton cloth.

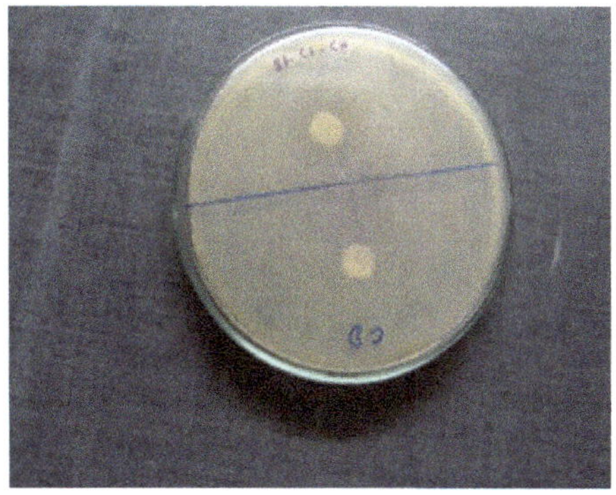

FIGURE 1: Antibacterial activity of cotton cloth fabricated by dipping method against *Corynebacterium*.

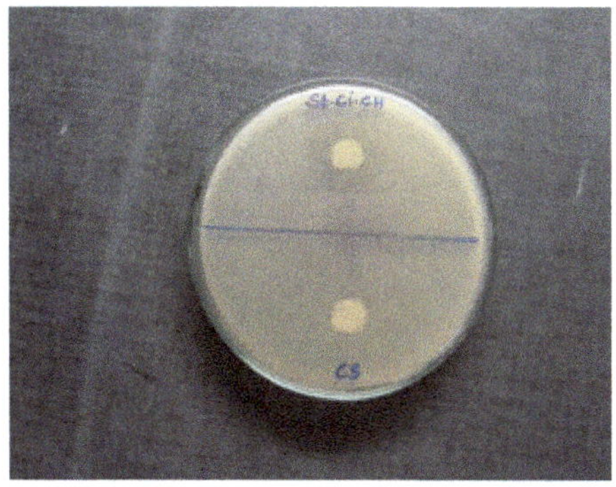

FIGURE 2: Antibacterial activity of cotton cloth fabricated by sonication method against *Corynebacterium*.

3.3. Determination of Antimicrobial Effect of Fabrics Coated with Ethyl Acetate Extract of Eichhornia crassipes. The antimicrobial activity of the cotton cloth (8 × 8 cm) and perspiration pads coated with ethyl acetate extract of fresh leaves and shoot portion of *Eichhornia crassipes* against *Corynebacterium* adopting AATCC 147 method are shown in Figures 1, 2, 3, 4, 5, and 6. No bacterial growth was observed under and on the fabricated cloth and perspiration pads indicating that the fabrics possess bacteriostatic property and not bactericidal property and the absence of leaching of the extract.

The perspiration pads were fabricated with the ethyl acetate extract of fresh plant of *Eichhornia crassipes* by sonication, dipping, ultrasonic homogenizer methods as in the case of cotton cloth. The commercial samples of perspiration pads were purchased and used as blank samples. The perspiration pads were washed with water to remove the existing gel coating in the commercial samples and were coated with the extract to test the efficacy of the extract. Similar procedure was adopted to quantify the amount of sample coated onto the pads.

Table 2: Various methods of finishing cotton cloth and perspiration pads.

Sample	Methods of coating of extract onto the sample	Weight of the cloth/perspiration pads (g)		Difference of weight (mg) coating
Cotton cloth [8 × 8 cm]	Sonication	137	172	35
	Dipping	199	207	8
	Ultrasonic homogenizing	262	276	14
Perspiration pads [8 × 8 cm]	Sonication	180	191	11
	Dipping	206	219	13
	Ultrasonic homogenizing	154	162	8

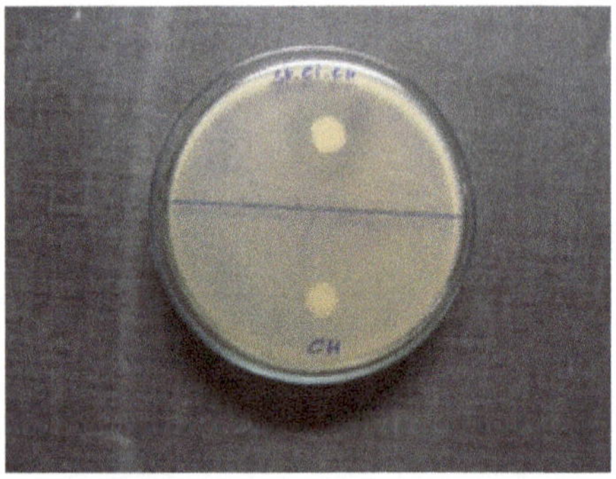

Figure 3: Antibacterial activity of cotton cloth fabricated by homogenizer method against *Corynebacterium*.

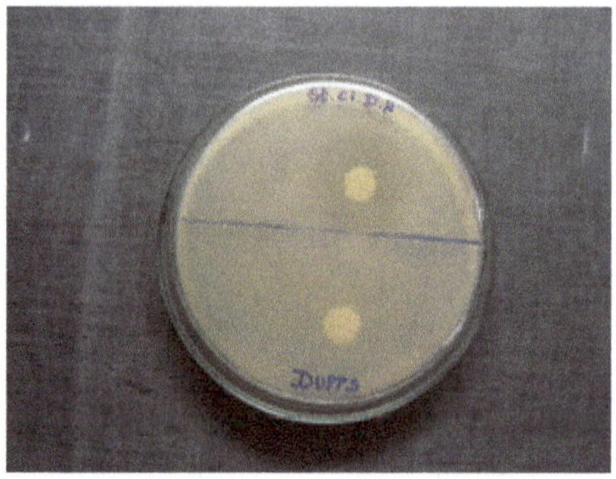

Figure 5: Antibacterial activity of perspiration pads fabricated by sonication method against *Corynebacterium*.

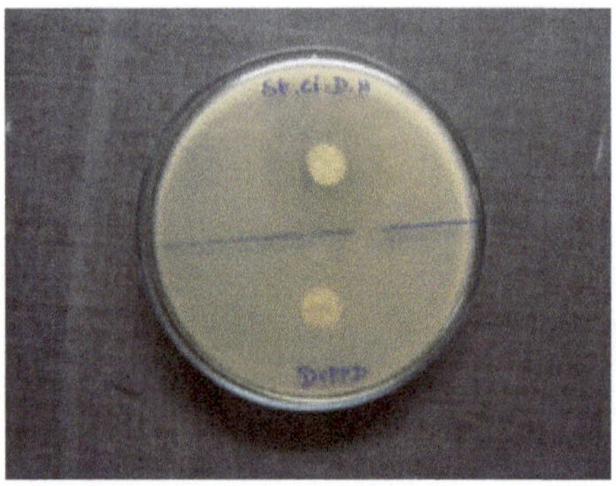

Figure 4: Antibacterial activity of perspiration pads fabricated by dipping method against *Corynebacterium*.

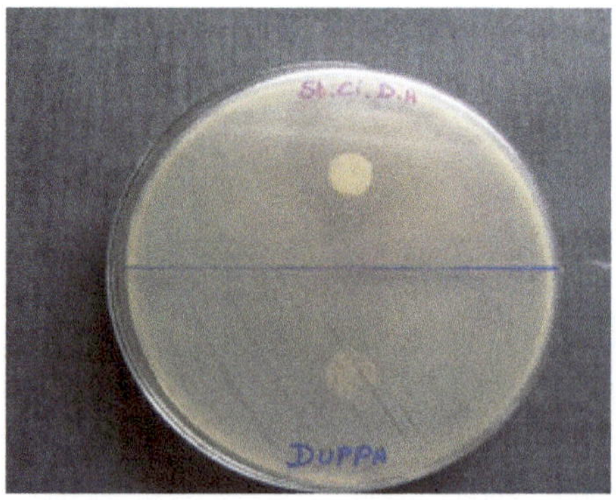

Figure 6: Antibacterial activity of perspiration pads fabricated by homogenizer method against *Corynebacterium*.

Ciprofloxacin was used as a standard and was also coated onto the cotton cloth and perspiration pads for one hour using homogenizer method. After washing it was found that 11.4 mg of the standard was coated onto the cloth and 11.5 mg of the standard was coated onto the perspiration pads.

In the case of coating of perspiration pads, dipping proved to be more efficient than the other three methods. This may be attributed to the nature of the pad materials which, being more of absorbing tendency, may have taken up the extracts immediately after dipping and retained it. In the case of sonication, desorption would have taken place since sonication would have dispersed off the extracts from the fabric in the one-hour period. This can be minimized by reducing the time of coating using sonic waves.

All the fabricated cotton cloth and perspiration pads were tested for antimicrobial activity against *Corynebacterium*. The fabricated cotton cloth and perspiration pads were placed on the surface of the Petri dish that has been inoculated with *Corynebacterium* and incubated for 24 hours. No bacterial growth was observed under and on the fabricated cloth and perspiration pads. According to qualitative AATCC 147 method for a successful result, there should be no bacterial growth observed under and on the test sample. This test is very easy to assess with antimicrobials that leach, as a zone of inhibition. With antimicrobials that are fixed to the fabric, there is no zone of inhibition and no bacterial growth under the test sample indicating that the fabrics possess bacteriostatic property and not bactericidal property.

In the present study, there was no leaching of the extracts as was seen from the absence of zone of inhibition. Good antimicrobial activity was noted since there was no growth in all the fabric samples, suggesting the success of the study.

The antimicrobial activity of the cotton cloth and perspiration pads may be attributed to the presence of phytochemicals like alkaloids, flavonoids, sterols, terpenoids, anthraquinone and anthocyanins, phenolics, quinones, and carbohydrates in the ethyl acetate extract of fresh waterhyacinth. Duangsri et al. [9] have reported the antimicrobial activity of the fabric coated with tobacco leaf containing higher polyphenol content. Steroids and phenols have been attributed to the antimicrobial activity of the cotton fabrics coated with peony extract and clove extract, respectively [10]. Flavonoids have also been demonstrated to possess bacteriocidal property [11].

Thymus vulgaris and *Citrus paradise* have shown significant antimicrobial activity when tested against the sweat bacteria including *Corynebacterium xerosis* [12]. In case of standard ciprofloxacin, the zone of inhibition around the cloth samples indicate the leaching of ciprofloxacin.

4. Conclusion

The results of the present study indicate that the ethyl acetate extract possesses activity against *Corynebacterium*. Further studies may be carried out to prove the effectiveness of the treatment. Hence, the plant may be successfully employed in the pharmaceutical industry as antibacterial finishes.

Conflict of Interests

The authors declare that there is no conflict of interests regarding the publication of this paper.

Acknowledgment

The authors thank the authorities of the Avinashilingam Institute for Home Science and Higher Education for Women, Coimbatore, for providing the facilities to carry out this research work.

References

[1] http://www.hccfl.edu/media/568216/19-exercise%20xiv.pdf.

[2] R. Emter and A. Natsch, "The sequential action of a dipeptidase and a β-lyase is required for the release of the human body odorant 3-methyl-3-sulfanylhexan-1-ol from a secreted Cys-Gly-(S) conjugate by *Corynebacteria*," *The Journal of Biological Chemistry*, vol. 283, no. 30, pp. 20645–20652, 2008.

[3] M. P. Sathianarayanan, N. V. Bhat, S. S. Kokate, and V. E. Walunj, "Antibacterial finish for cotton fabric from herbal products," *Indian Journal of Fibre and Textile Research*, vol. 35, no. 1, pp. 50–58, 2010.

[4] D. Gupta and A. Laha, "Antimicrobial activity of cotton fabric treated with *Quercus infectoria* extract," *Indian Journal of Fibre and Textile Research*, vol. 32, no. 1, pp. 88–92, 2007.

[5] P. Lalitha, S. K. Sripathi, and P. Jayanthi, "Secondary metabolites of *Eichhornia crassipes* (Waterhyacinth): a review (1949 to 2011)," *Natural Product Communications*, vol. 7, no. 9, pp. 1249–1256, 2012.

[6] P. Lalitha, S. K. Sripathi, and P. Jayanthi, "Acute toxicity study of extracts of *Eichhornia crassipes* (Mart.) Solms," *Asian Journal of Pharmaceutical and Clinical Research*, vol. 5, no. 4, pp. 59–61, 2012.

[7] S. Firdhouse and P. Alam, "Phytochemical investigation of extract of amorphophallus campanulatus tubers," *International Journal of Phytomedicine*, vol. 3, no. 1, pp. 32–35, 2011.

[8] K. A. Erdem and S. N. O. Yurudu, "The evaluation of antibacterial activity of fabrics impregnated with dimethyltetradecyl (3-(trimethoxysilyl) propyl) ammonium chloride," *IUFS Journal of Biology*, vol. 67, pp. 115–122, 2008.

[9] P. Duangsri, K. Juntarapun, and C. Satirapipathkul, "The Tobacco leaf extract and antibacterial activity in textile," in *Proceedings of the RMUTP International Conference: Textiles & Fashion*, July 2012.

[10] Y.-H. Lee, E.-K. Hwang, and H.-D. Kim, "Colorimetric assay and antibacterial activity of cotton, silk, and wool fabrics dyed with peony, pomegranate, clove, *coptis chinenis* and gallnut extracts," *Materials*, vol. 2, no. 1, pp. 10–21, 2009.

[11] S. Mohanraj, P. Vanathi, N. Sowbarniga, and D. Saravanan, "Antimicrobial effectiveness of vitex negundo leaf extracts," *Indian Journal of Fibre and Textile Research*, vol. 37, no. 4, pp. 389–392, 2012.

[12] R. Kozlowski and J. Walentowska, "Role of biocides from plant origin in protection of natural fibers against biodeterioration," in *International Conference on Flax and Other Bast Plants*, pp. 326–331, 2008.

Application of Purified Lawsone as Natural Dye on Cotton and Silk Fabric

Md. Mahabub Hasan, Khandakar Abu Nayem, Abu Yousuf Mohammad Anwarul Azim, and Nayon Chandra Ghosh

Department of Textile Engineering, Primeasia University, Dhaka 1213, Bangladesh

Correspondence should be addressed to Md. Mahabub Hasan; merajtex2008@yahoo.com

Academic Editor: Sambandam Anandan

The color which is obtained from the leaves of Henna, that is, *Lawsonia inermis* L., is used frequently in hair coloring. It is the chemical lawsone that is responsible for the reddish brown color. Its content makes it a substantive dye for dyeing the textile materials. This work concerns with the extraction and purification of natural dyestuff from a plant *Lawsonia inermis* L. and dyeing of cotton and silk fabric in exhaust dyeing method. The dye portion is isolated from the total extract by column chromatography and is evaluated by dyeing cotton and silk under different dyeing conditions. The color strength and fastness properties of the dye are undertaken by changing mordant and techniques of mordanting. The changes of colors have been noticed by using different types of mordant. The dye exhaustion percentage, wash, rubbing, and light fastness results reveal that the extract of henna can be used for coloration of cotton and silk fabric.

1. Introduction

As compared to natural dyes, synthetic dyes have been widely used due to their lower prices and wide range of bright shades with considerable improved color fastness properties. The use of synthetic dyes has gained higher acceptability in the field of food [1], cosmetic [2], and more importantly textile industries [3] after the invention of synthetic dyes by William Henry Perkin in 1856 [4]. However, recently the concern for the environment has created an increasing demand for natural dyes. The ancient people used these dyes in cosmetics, food, leather, and also in medicine [5–7]. The use of natural dyes is found severely, when the weaving technique for dyeing textile materials is developed [8]. In the situation of environmental concerns, it is time to reconsider the use of renewable, nontoxic, and gentle colors from nature. Eco-friendly natural dyes are familiar substances that spark creative ideas. They could serve as a viable option for our dye needs. Indigo, the blue dye still used in jeans, is chemically the same as the dye extracted from the indigo plant. Natural dyes are very old, perhaps when people found that something changed the color of white fibers.

The worldwide demand for natural dyes is nowadays of great interest due to the increased awareness on therapeutic properties of the dyes. Natural dyes are derived from naturally occurring sources such as plants, insects, animals, and minerals. Several synthetic colorants have been banned because they cause allergy-like symptoms or are carcinogens [9–11]. Although known for a long time for dyeing as well as medicinal properties of the dye, the structures and protective properties of natural dyes have been recognized only in the recent past. Many of the plants are used for dye extraction.

The present study focuses on the development of the optimum extraction conditions of coloring component from the natural material Henna, that is, *Lawsonia inermis* L. and then the major component of the dye, that is, lawsone, is purified. The chemical structure of lawsone is shown in Figure 1. Finally the purified dye is applied on cotton and silk fabric by exhaust dyeing method. The color strength value, dye exhaustion %, and color fastness properties of the dyed fabrics are evaluated and compared with each other.

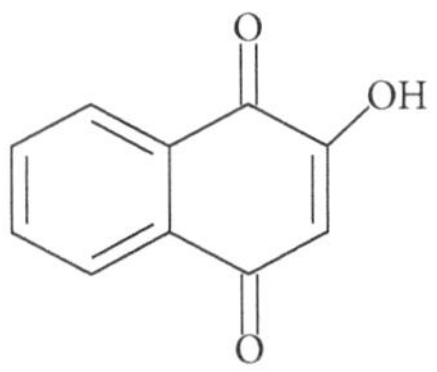

FIGURE 1: 2-Hydroxy-1,4-naphthoquinone.

2. Materials and Experimental Methods

2.1. Dyestuff Used. Henna, that is, *Lawsonia inermis* L., is used to obtain the dyestuff of lawsone, which is reddish brown in color. Henna is collected from the local market of Bangladesh.

2.2. Fabrics Used. Cotton and silk fabrics selected for the present study are collected from Sinha Textile Group, Bangladesh. Cotton fabric is scoured-bleached and degumming is done for silk fabric before dyeing.

2.3. Chemicals Used. Ethanol (C_2H_5OH), chloroform ($CHCl_3$), acetone (C_3H_5OH), and n-hexane (C_6H_{14}) are used as solvent to extract the dye from *Lawsonia inermis* L. The chemicals are collected from Sigma- Aldrich GmbH and Carl Roth GmbH. They were pure grade chemicals used for laboratory purposes.

2.4. Mordants Used. The following mordants are used: potassium aluminium sulphate ($AlK(SO_4)_2{\cdot}12H_2O$), copper sulphate ($CuSO_4$), and tartaric acid ($C_4H_6O_6$). They were pure grade chemicals used for laboratory and are collected from Carl Roth GmbH.

2.5. Solvent Extraction of Dye from Henna Powder. Plant tissues are difficult to extract because of active lipases which hydrolyze rapidly phospholipids and glycolipids and increase the amount of free fatty acids in the extract. Thus a solvent extraction is preferred to inhibit these enzymes.

Fresh leaves of henna are washed and cleaned with distilled water. They are then dried in the sunlight for one day and again dried at 80°C for one hour in a hot air oven. Dried leaves are made in powder form for getting proper extraction result. 10 gm of henna powder is taken into a thimble and placed in a soxhlet apparatus. 200 mL of solvent is added in soxhlet for 10 gm of sample and the extraction is carried out according to the boiling point (B.P.) of solvent for 3 hours. After extraction the brown extract is cooled, filtered, and concentrated by evaporation. Finally, a crude dried extract is obtained, which is brown in color. The dye is extracted from henna using different solvent according to their boiling point by the same method and yield is calculated. Finally color strength % of extracted dye is compared for getting the best extraction result.

2.6. Purification of Lawsone Dye from Extract by Column Chromatography. The column chromatography consists of a glass tube with a diameter of 30 mm and a height of 70 cm with a tap at the bottom. The column is clamped vertically. The stopcock of the column is closed and it is fitted with fritted disk. The silica gel is weighed out in a flask and enough solvent (eluent) is added with stirring to form slurry. Then, slurry is slowly transferred into the column using a glass funnel until the silica gel level is about 30–40 cm. When the solvent reached the silica gel surface, we slowly added the mixture solution of henna extract into the column using a dropper. After the sample is loaded, a small layer of sand is added to the top of the column. This helped to keep the top of the column level when added solvent eluent. Once the mixture is added to the column and the protective layer of sand is placed, the solvent (eluent) is added continuously. Further the small fraction of the eluent is collected from the bottom of the column. The components are retained by the stationery phase and the lawsone part is separated through the column with the eluent.

2.7. Optimization of Dyeing Conditions for Cotton and Silk Fabrics. Cotton and silk fabrics are dyed with the lawsone dye at a liquor ratio of 1:40. For optimizing the dyeing conditions, at first, experiments are carried out to optimize the dyeing pH. Dyeing processes are carried out with 2% (o.w.f) concentration of purified dye at pH 3, 5, 7, and 9. To understand the effect of dyeing temperature and dyeing time on the color strength, another set of experiments is carried out in optimized dyeing pH at 60, 75, 90, and 100°C for different time periods, that is, 30, 45, 60, and 90 minutes. Another set of experiments is also carried out at different dye concentrations such as 2%, 4%, 6%, and 8% in optimum dyeing conditions. Based on the color strength (K/S) values of the dyed samples, optimum dyeing pH, temperature, and time are selected and taken for further study. To study the effect of different mordants and their varying concentrations on color strength, mordanting is carried out with three metallic salts such as aluminium potassium sulphate, copper sulphate, and tartaric acid. Mordanting is carried out through two ways: (1) premordanting and (2) postmordanting. The sample is taken as 1 gm for every experiment.

After dyeing, the dyed samples are rinsed with cold water and washed in a bath with a liquor ratio of 1:50 using 1 gm/liter of the soaping agent at 60°C for 10 minutes and then they are rinsed and finally dried in a dryer.

2.8. Evaluation of Color Strength. Estimation of color strength of the dyed fabrics is carried out by determining the K/S values using a computer color matching system (CS-5, Applied color system, USA). The reflectance value (R) in the visible wavelength region is measured by means of the ACS spectrophotometer. The value of reflectance (R) of the dyed fabric is measured at the wavelength of 420 nm and also the K/S value of the sample is found directly from the instrument. Every dyed sample is measured in the same way and the K/S values are obtained directly from the instrument, which followed the Kubelka-Munk theory as

$$\frac{K}{S} = \frac{(1-R)^2}{2R}, \quad (1)$$

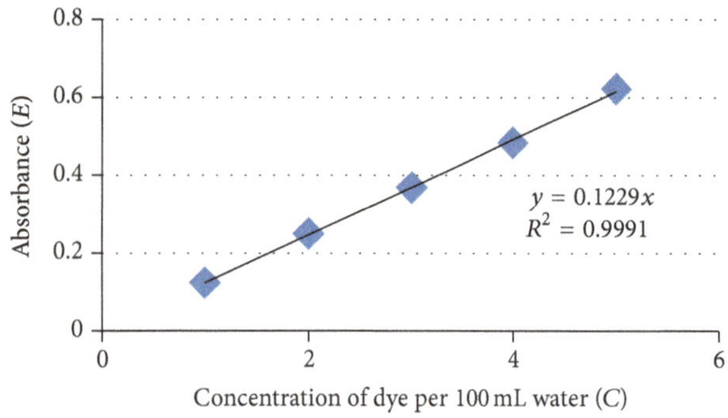

FIGURE 2: Calibration curve of lawsone dye.

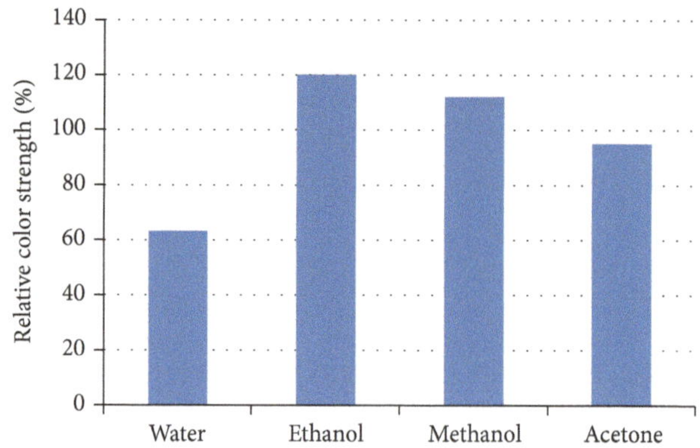

FIGURE 3: The relative color strength of henna extract with different solvent.

where K is absorption coefficient of the sample, S is scattering coefficient of the sample, and R is the degree of reflection. The value of K/S is directly proportional to the concentration of dye in the dyed fabric.

2.9. Measurement of Exhaustion of Dyes by UV/VIS Spectroscopy. Degree of exhaustion is the amount of dyestuff, which is diffused in the fiber from the dye bath at the time of dyeing. UV/Vis spectrophotometer (DT A 01, Perkin Elmer, Singapore) is used to measure the exhaustion and fixation of the dyestuff. By measuring the concentration of dye bath before and after the dyeing process, the percentage of exhaustion can be estimated with (2) considering the color concentration of mordants:

$$E\% = \frac{C_1 - C_2}{C_1} \times 10, \tag{2}$$

where C_1 and C_2 are the concentrations of the dye bath before and after dyeing process, respectively.

The concentrations of the dye solution before and after dyeing are measured using UV/VIS spectrometer at the wavelength of 420 nm. Before measuring the absorbency, the wavelength of maximum absorbency is determined for the dye by using the calibration standard solutions. For the calibration standard solutions, a dye stock solution of 20 mg in 20 mL water is prepared. By pipetting 1 mL to 5 mL of this stock solution is taken and diluting to 100 mL, 5 different calibration standard solutions are prepared. The concentrations of the calibration standard solutions ranged from 0.01 g/L to 0.05 g/L. The dye solution is taken into a cuvette with a width of 1 cm and subsequently the absorption spectrum is recorded, which are presented in Figure 2.

The value of extinction coefficient is obtained from the linear regression line ($y = mx$), where the value of "m" corresponds to the extinction coefficient of the dyes. The extinction coefficient of the dye is determined as per Beer-Lambert Law, based on (3):

$$\log \frac{I_0}{I} = E = \varepsilon c d, \tag{3}$$

where I_0 is intensity of the initial light, I is intensity of the light after passing through the dye liquor, E is extinction at a specific wavelength of λ, ε is extinction coefficient, c is concentration of the dye liquor, and d is cuvette width.

2.10. Evaluation of Color Fastness Properties. Wash fastness of the samples dyed under the optimized conditions is tested according to ISO 105-CO3 method. ISO 105-X 12 test methods are followed to measure the rubbing fastness. Color fastness of textile material to day light is of considerable importance to the consumer. The specimen should be tested according to ISO test method. The light fastness of the dyed test samples (1 × 3 cm) is exposed to UV light of a Xeno tester for 40 hours. The temperature of the machine was 30°C, R.H. 65%, and irradiation dose 980 KJ/m^2. The fastness is assessed by comparing the fading of the specimen with that of blue wool patterns.

3. Results and Discussion

3.1. Optimization of Extraction Conditions and Purification of Lawsone Dye. Soxhlet extracts are weighted after drying and color strength percentage of lawsone dye is calculated; those are shown in Figure 3. The relative color strength of dye extracts obtained in distilled water is not appreciable, which is also shown in [12]. Maximum amount of lawsone dye is obtained by using ethanol. So the ethanol extract is used to purify the dye and then dyeing of cotton and silk fabric is performed with the purified dye.

The ethanol extracted dye is subjected to column chromatography by using the mobile phase of chloroform and n-hexane and the fractions are collected and tested by thin layer chromatography (TLC). The absorbance of the purified lawsone dye is analyzed by UV-visible spectroscopy at 420 nm. The test showed that the lawsone dye is separated perfectly by the mixture of chloroform and n-hexane in the ratio of 4 : 1.

3.2. Effect of Dye Concentration on Color Strength of Cotton and Silk Fabrics. The color strength of the dyed cotton and silk fabrics is dependent on the concentration of the extracted lawsone dye as shown in Figure 4. The color of the dyed cotton and silk fabrics is reddish, when the percentage of dye (o.w.f) is less than 2. When the percentage of dye is more than 2, the fabrics showed reddish brown color. From the results, it is seen that the color strength, expressed as percentage,

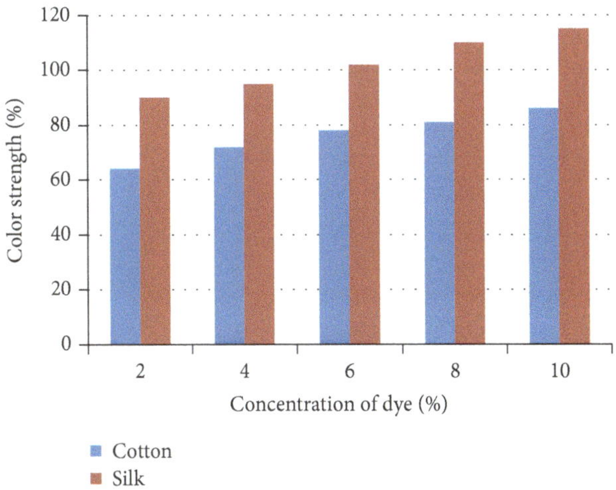

FIGURE 4: Effect of concentration of lawsone dye on color strength of cotton and silk fabrics.

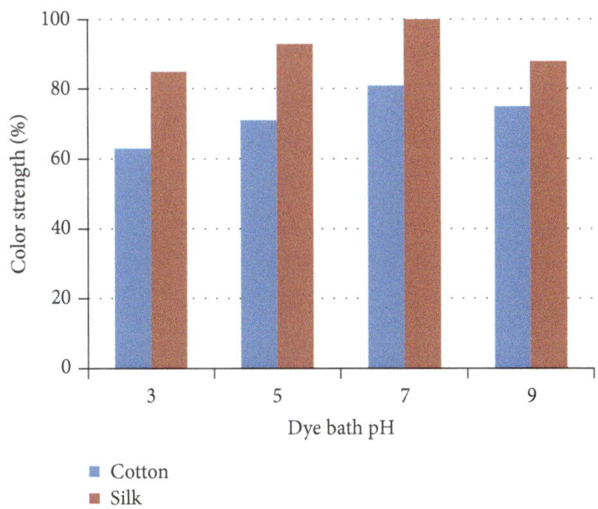

FIGURE 5: Effect of pH on color strength of cotton and silk fabrics dyed with lawsone (4% shade).

increased with increasing concentration of lawsone dye up to a certain limit. It is also observed from the result that there is a significant difference in the depth of shade between cotton and silk fabrics, though the dyeing condition and concentration of dye are the same. It is found that the color strength of the silk fabric is higher than the cotton fabric at the same concentration. The color strength of silk is more because of its functional groups in its structure.

3.3. Effect of pH on Color Strength of Fabric. Figure 5 showed that the color strength is increased with the increase of pH value in the dyeing bath from pH 3 to 7 for both the fabrics. The maximum dye uptake is obtained at pH 7 for both. The maximum dye uptake of cotton fabric is also found at neutral pH in [12], which is almost similar to the obtained result, though the dye extraction condition is not the same.

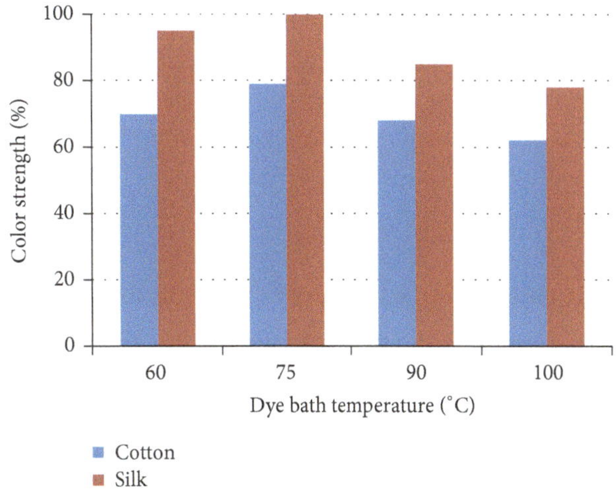

FIGURE 6: Effect of temperature on color strength of fabrics dyed with lawsone dye (4% shade).

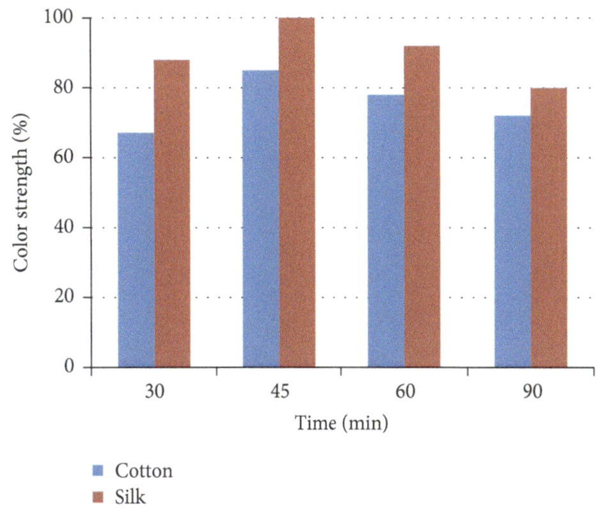

FIGURE 7: Effect of time on color strength of fabrics dyed with lawsone dye (4% shade).

3.4. Effect of Temperature on Color Strength of Fabric. The effect of dyeing temperature on color strength of cotton and silk fabric is demonstrated in Figure 6. As evident, the maximum color strength is obtained at 75°C for both cotton and silk fabrics. The shade is also very uniform. The maximum color strength of cotton dyed with henna dye is found at the temperature of 70°C in [12], which is very much similar to the obtained result. If the temperature is increased to more than 75°C, then the depth of shade is decreased. The result is also not good, if the fabrics are dyed less than 75°C. The same result is found in [12]. The depth of shade is more in case of silk fabric than cotton fabric at the same concentration and temperature because of the structure of the fiber.

3.5. Effect of Time on Color Strength of Fabric. The effect of dyeing time on color strength is shown in Figure 7. The best results with respect to time for dyeing cotton and silk fabrics

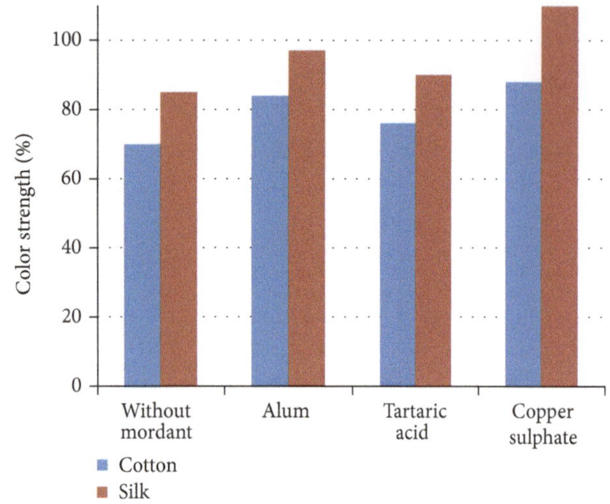

FIGURE 8: Effect of premordanting (2 gm/L) on color strength of fabrics (4% shade).

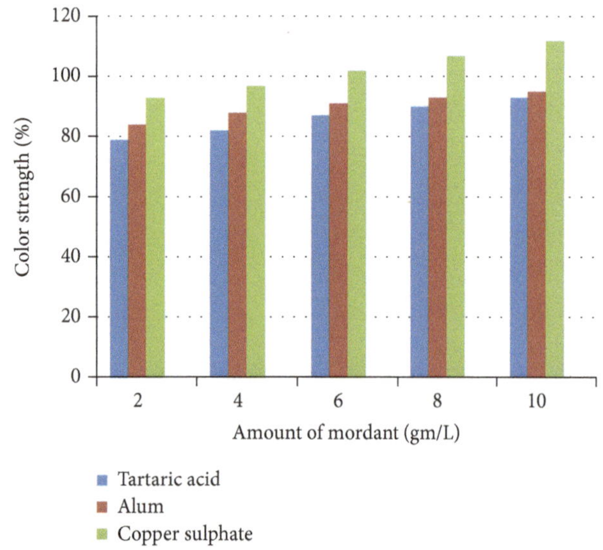

FIGURE 9: Effect of premordanting on color strength of cotton fabric (4% shade).

are obtained at 45 minutes. The color strength is decreased, when the fabrics are dyed one hour or more than one hour. The decrease in color strength may be attributed to the fact that the desorption of the dye molecules could happen as a consequence of over dyeing, if the process is carried out for more than 45 minutes. So, it can be said that the time for maximum exhaustion of dyes depends on some factors such as shade % and dyeing conditions.

3.6. Effect of Mordanting on Color Strength of Cotton and Silk Fabrics. The effects of mordant on relative color strength for the premordanting method are shown in Figure 8. It is observed that relative color strength values are higher for samples dyed with premordanted fabric at 100°C in comparison with samples dyed without mordanting. The premordanted fabric dyed with henna dye showed better results than unmordanted fabric, which is also demonstrated in [12]. Since the mordant has affinity for lawsone, it attracts more coloring component from dye bath to fabric. It is also observed in Figures 9 and 10 that the relative color strength is increased with increasing the mordant concentration and reached a maximum at 10 gram per liter. It is also seen that the effect of different mordants is not the same. The results of alum and copper sulphate are very good in terms of color strength value. But the deviations in shade tone and brilliancy among the abovementioned dyed sample cannot be measured in terms of color strength value. It is seen that different brilliant shade can be produced on cotton and silk fabrics with lawsone dye by using same amount of different mordants, which is also discussed in [12].

It is clearly seen in Figure 11 that the relative color strength is decreased by the application of mordant. The similar result is observed in [12], when the postmordanting technique is applied to dye the cotton fabric with henna dye. Thus, the premordanting method showed a higher depth of shade in comparison with the postmordanting method.

Copper sulphate, alum, and tartaric acid are well known for their ability to form coordination complexes and to readily chelate with the dye. As a result, these mordants helped to increase the color strength of the dyed fabric. It is also seen that tartaric acid ions form weak coordination complexes with dye than copper sulphate and alum; so the color strength of the dyed fabrics mordanted with tartaric acid is less than the dyed fabrics mordanted with copper sulphate and alum.

TABLE 1: Maximum exhaustion of lawsone dye by cotton and silk fabrics.

Fabric dyed with lawsone	Dye exhaustion (%)
Cotton without mordant	52
Cotton premordanted with tartaric acid	61
Cotton premordanted with $CuSO_4$	72
Cotton premordanted with alum	68
Silk without mordant	71
Silk premordanted with tartaric acid	77
Silk premordanted with $CuSO_4$	82
Silk premordanted with alum	80

3.7. Result of Dye Exhaustion. The highest degree of exhaustion of lawsone by cotton and silk fabrics, which are dyed in exhaust method, is determined by UV-Visible spectroscopy and the results are shown in Table 1.

The degree of exhaustion of silk fabric dyed with lawsone dye is found higher than that of cotton fabric as shown in Table 1. This is because of the molecular structure of the fiber. Silk fiber has more functional groups in its structure than cotton fiber. As a result, the dye exhaustion is more in case of silk fiber than the cotton. The results also showed that higher dye exhaustion occurred in mordanted cotton and silk fabrics than unmordanted samples.

TABLE 2: Fastness properties of cotton and silk fabrics dyed with lawsone dye.

Dyeing technique	Shade% (o.w.f)	Fastness rating									
		Wash fastness				Rubbing fastness				Light fastness (40 hours)	
		Color change		Staining (cotton)		Dry rubbing		Wet rubbing		Cotton	Silk
		Cotton	Silk	Cotton	Silk	Cotton	Silk	Cotton	Silk		
Without mordant	2	3	3-4	3	3	3-4	3-4	3-4	3-4	3-4	3-4
	6	3-4	3-4	3	3-4	3	3-4	3-4	3	3	3
With tartaric acid	2	4-5	4-5	4	4-5	4	4-5	3-4	3-4	4	4
	6	4	4	3-4	3-4	3-4	4	3-4	3-4	3-4	3-4
With alum	2	4-5	4-5	4-5	4	4-5	4-5	3-4	3-4	4	4-5
	6	4	4	4-5	4	4	3-4	3	3	3-4	3-4
With copper sulphate	2	4-5	4-5	4-5	4-5	4-5	4-5	3-4	3-4	4-5	4-5
	6	4-5	4-5	4-5	4-5	4-5	4-5	3	4	4-5	4

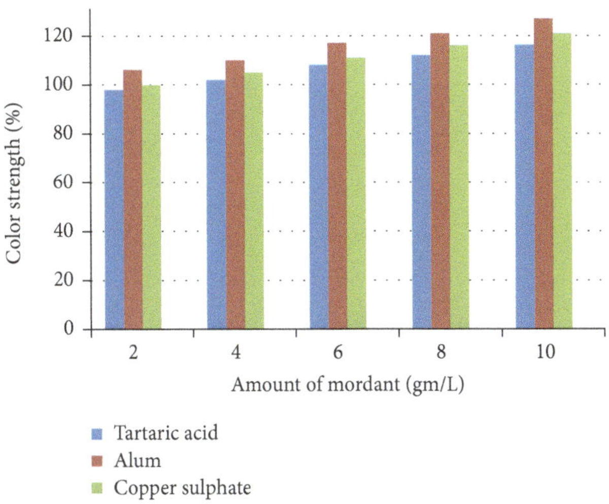

FIGURE 10: Effect of premordanting on color strength of silk fabric (4% shade).

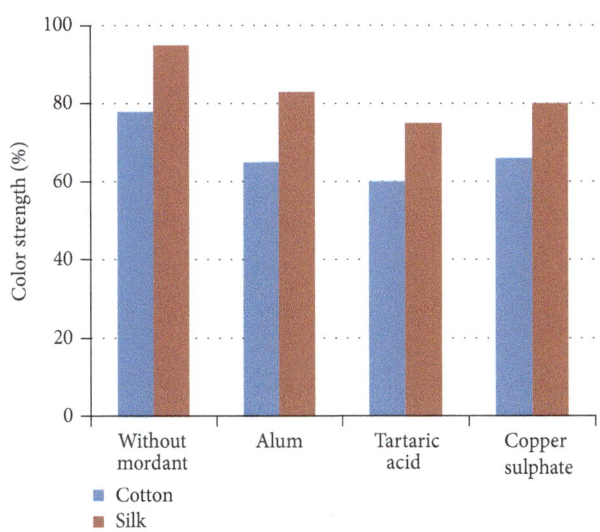

FIGURE 11: Effect of postmordanting (2 gm/L) on color strength of cotton and silk fabrics (4% shade).

3.8. Result of Fastness Properties. The wash fastness rating of cotton and silk fabrics dyed with or without mordants at the dye concentrations of 2% and 6% (o.w.f) is presented in Table 2. The result showed that the color fastness ratings of the cotton and silk dyed fabrics without mordanting are not good. The fastness properties have improved significantly, when different types of mordants are used before dyeing. The mordanting with copper sulphate demonstrated better result than the others. It is due to the formation of strong bond between the dye and fiber. It is observed that the light fastness of lawsone dyed fabrics is lower than other fastness properties. It is because of the chromophore group structure of the lawsone dye.

3.9. Economical Aspects for Dyeing with the Dyes of Natural Purified Lawsone. The cost of natural dye, that is, lawsone dye, is more than synthetic dye. But the auxiliaries cost and dyeing process cost of fabric by natural lawsone dye are less than synthetic dye. Moreover the natural dye is not harmful for environment, so effluent treatment plant is not necessary for such type of technology. So the technique can be applied commercially for dyeing protein and cellulosic fabric like silk and cotton.

4. Conclusion

The main goal of the work is to extract and purify the lawsone dye from the plant *Lawsonia inermis* L., which is used to dye the cotton and silk fabric to compare the properties of the dyed fabric. The result shows that the color strength values of silk dyed fabrics are higher than the cotton dyed fabrics, though they are dyed in same method. The dye exhaustion percentage of the silk fabric is also higher than cotton fabric. It is observed that mordant has a great effect to increase the color strength of the fabric and the effects of different mordants are not the same. Copper sulphate shows the better result than the other two. Depending on the dyeing procedure and the types of mordant used, a variety of colors also can be produced using the same percentage of lawsone dye. At the same time, the dye processing cost and the cost for effluent treatment plant (ETP) are also lower than that of other synthetic dyes. So, extraction and purification of natural dyes and their application can be of great significance in the future of the commercial and domestic dyeing industries.

Conflict of Interests

The authors declare that there is no conflict of interests regarding the publication of this paper.

References

[1] T. Fossen, L. Cabrita, and O. M. Andersen, "Colour and stability of pure anthocyanins influenced by pH including the alkaline region," *Food Chemistry*, vol. 63, no. 4, pp. 435–440, 1998.

[2] C. D. Calnan, "Quinazoline yellow SS in cosmetics," *Contact Dermatitis*, vol. 2, no. 3, pp. 160–166, 1976.

[3] P. Sacvarino, G. Viscardi, P. Quarliotto, E. Montoneri, and E. Barni, "Reactivity and effects of cyclodextrins in textile dyeing," *Dyes and Pigments*, vol. 42, no. 2, pp. 143–147, 1999.

[4] S. C. Druding, *Dye History from 2600 BC to the 20th Century, Washington at Convergence*, 1982.

[5] D. Taylor and L. Hofer, *Hand Book of Paint & Body*, 1994.

[6] K. Othmer, *Kirk Othmer Encyclopedia of Chemical Technology, Canada*, vol. 8, Wiley, 1998.

[7] N. Rungruangkitkrai and R. Mongkholrattanasit, *The 4th RMUTP International Conference, Textiles & Fashion, Bangkok, Thailand, 3-4 July 2012*, 2012.

[8] K. Othmer, "Dyes, natural," in *Kirk-Othmer Encyclopedia of Chemical Technology*, Wiley, 2009.

[9] Ratna and B. S. Padhi, "Pollution due to synthetic dyes toxicity & carcinogenicity studies and remediation," *International Journal of Environmental Sciences*, vol. 3, no. 3, pp. 940–955, 2012.

[10] R. M. Melgoza, A. Cruz, and G. Buitrón, "Anaerobic/aerobic treatment of colorants present in textile effluents," *Water Science & Technology*, vol. 50, no. 2, pp. 149–155, 2004.

[11] S. S. Kulkarni, A. V. Gokhale, U. M. Bodake, and G. R. Pathade, "Cotton dyeing with natural dye extracted from Pomgranate (*Punica granatum*) peel," *Universal Journal of Environmental Research and Technology*, vol. 1, no. 2, pp. 135–139, 2011.

[12] S. Ali, T. Hussain, and R. Nawaz, "Optimization of alkaline extraction of natural dye from Henna leaves and its dyeing on cotton by exhaust method," *Journal of Cleaner Production*, vol. 17, no. 1, pp. 61–66, 2009.

31

Prediction of Color Properties of Cellulase-Treated 100% Cotton Denim Fabric

C. W. Kan, W. Y. Wong, L. J. Song, and M. C. Law

Institute of Textiles and Clothing, The Hong Kong Polytechnic University, Hung Hom, Kowloon, Hong Kong

Correspondence should be addressed to C. W. Kan; tccwk@polyu.edu.hk

Academic Editor: Jiri Militky

Artificial neural network (ANN) model was used for predicting colour properties of 100% cotton denim fabrics, including colour yield (in terms of K/S value) and CIE L^*, a^*, b^*, $\mathbf{C}^*$, and $\mathbf{h}^\circ$ values, under the influence of cellulase treatment with various combinations of cellulase processing parameters. Variables examined in the ANN model included treatment temperature, treatment time, pH, mechanical agitation, and fabric yarn twist level. The ANN model was compared with a linear regression model where the ANN model produced superior results in the prediction of colour properties of cellulase-treated 100% cotton denim fabrics. The relative importance of the examined factors influencing colour properties was also investigated. The analysis revealed that cellulase treatment processing parameters played an important role in affecting the colour properties of the treated 100% denim cotton fabrics.

1. Introduction

Denim jeans provide durability and vintage look for fashionable appearance which is the reason why they are welcomed by most people over the world. Conventional technologies involve creating designs by fading the colour of fabric by making the use of enzymatic treatment and bleach washing. Among these technologies, enzymatic treatment using cellulase is a commonly used method to achieve good colour fading effect with good fabric softness for 100% cotton denim fabrics. Although this technology could produce the desirable colour effects, there are a number side effects such as (i) difficulty in application, (ii) time consuming in processing, (iii) difficulty in creating standard and reproducible designs, (iv) effect cannot be applied to all textile surfaces, and (v) loss of quality if the process is not carefully controlled [1, 2]. The literature review shows that artificial neural network (ANN) model has been used in many engineering fields [3–5]. In the case of the textile industry, ANN model is mainly used in yarn and fabric technologies [6–10]; no comprehensive ANN model has been used for predicting colour properties of 100% cotton denim material after cellulase treatment. Therefore, in this paper, we use artificial neural network (ANN) model for prediction of colour properties, including colour yield (in terms of K/S value) and CIE L^*, a^*, b^*, $\mathbf{C}^*$, and $\mathbf{h}^\circ$ values of 100% cotton denim fabrics under the influence of cellulase treatment with the consideration of variables such as treatment temperature, treatment time, pH, mechanical agitation, and fabric yarn twist level.

2. Experimental

2.1. Material. In this study, 100% desized 3/1 right hand twill cotton denim fabrics were used, and their specifications were shown in Table 1.

2.2. Cellulase Treatment. A neutral cellulase enzyme (Lava Cell NNM, supplier: DyStar) in powder form was used for treating the denim fabric. Denim fabric samples (size of 22 cm × 22 cm) were treated with 0.1% cellulase with a liquor ratio of 50 : 1. Different cellulase treatments were conducted according to the different combination of treatment temperature (50°C, 55°C, and 60°C), treatment time (30 minutes, 45 minutes, 60 minutes, 75 minutes, and 90 minutes), mechanical agitation (simulated by 0, 50, and 100 steel balls for no mechanical agitation, mild mechanical agitation and severe mechanical agitation, resp.), and pH value (6, 7, and 8). After cellulase treatment, the denim fabrics were rinsed with deionised water to lower the temperature quickly and

Table 1: Specifications of denim fabrics.

Fabric	1	2	3	4	5	6
Fabric weight (g/m^2)	378 ± 2.0	380 ± 1.8	407 ± 2.1	400 ± 2.0	170 ± 1.8	178 ± 1.7
Density						
Warp (ends/cm)	25	25	26	26	27	27
Weft (picks/cm)	20	20	18	19	20	20
Yarn count						
Warp (Tex)	88	88	88	88	88	88
Weft (Tex)	57	57	57	57	57	57
Number of warp twists (per cm)	4.3	3.5	4.7	3.9	4.2	3.7
Number of weft twists (per cm)	4.7	4.7	4.7	4.7	4.7	4.7

to wash away abraded substances attached on fabrics. The denim fabrics were then treated with deionised water at 80°C to deactivate the enzyme action. After that, the denim fabrics were squeezed to remove excessive water and were dried in an oven at a temperature of 70°C for two hours. Then, the denim fabrics were placed for conditioning under standard atmospheric pressure at 65 ± 2% relative humidity and temperature of 20±2°C for at least 24 hours before further tests.

2.3. Colour Measurement. Colour properties were measured by a spectrophotometer of GretagMacbeth Color-Eye 7000A with D65 daylight and 10° standard observer. Four measurements were taken for each fabric sample with 10 mm aperture and specular-excluded mode. The fabric samples were conditioned at 20 ± 2°C and relative humidity of 65 ± 2% before taking the measurements. K/S sum in terms of summation of individual K/S value over the wavelength from 400 nm to 700 nm was calculated, and K/S sum value can help to determine the colour yield of the fabric samples. In addition, the CIE L*, a*, b*, $\mathbf{C}^*$, and $\mathbf{h}^\circ$ values were also obtained which represents lightness/darkness, redness/greenness, yellowness/blueness, chroma, and hue, respectively.

2.4. Artificial Neural Network (ANN) Model. After cellulase treatment, six color-related data, that is, K/S sum value (color-ks), CIE L* value (color-l), CIE a* value (color-a), CIE b* value (color-b), CIE $\mathbf{C}^*$ value (color-c), and CIE $\mathbf{h}^\circ$ (color-h) value, were obtained. In order to find the optimal neural network, ANN models having different topologies were formulated. The basic topology of the network is shown in Figure 1. In the ANN model, cellulase treatment processing factors, including treatment temperature (temperature), treatment time (time), pH (pH), and mechanical agitation (MA) were put in the input layer. Besides, fabric yarn twist level (twist) was also an important input node in case of cellulase-treated samples. For the output layer, the six nodes were corresponding to color-ks, color-l, color-a, color-b, color-c and color-h.

In order to train the ANN model, a typical three-layer network with Bayesian regulation backpropagation was used. It is considered as the most popular learning algorithm for the learning of a multilayered feed forward neural network

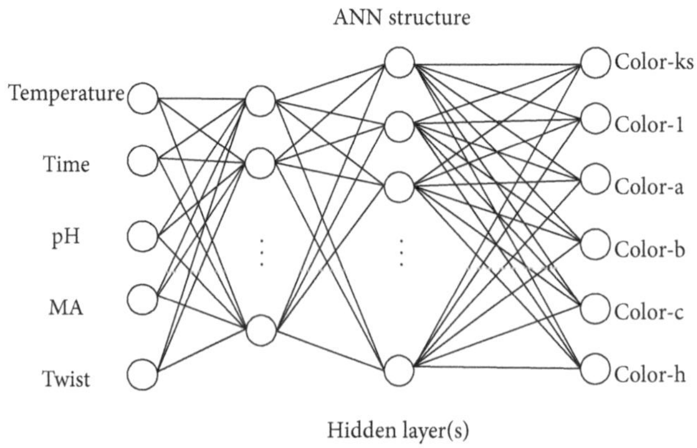

Figure 1: Basic structure of ANN model.

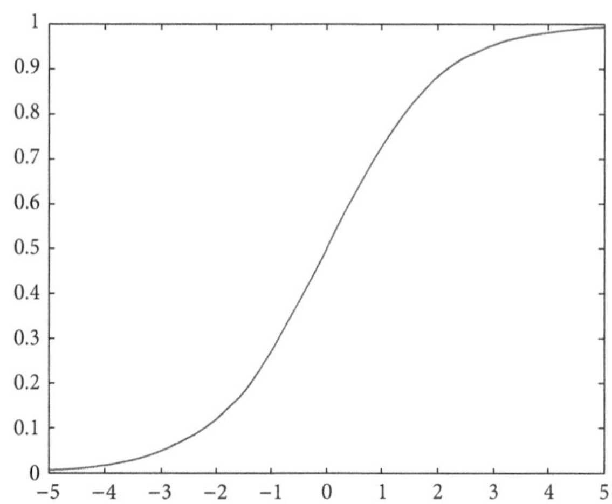

Figure 2: Sigmoid function.

[11]. For activation function, the sigmoid function is typically used [12], and in this case, was given by

$$\text{sigmoid}(a) = \frac{1}{1+e^{-a}}. \quad (1)$$

This function can range between 0 and 1 and is differentiable. Figure 2 illustrates the function. Other parameters of the network were set as shown in Table 2.

TABLE 2: Artificial neural network parameters.

Parameter name	Value	Description
net.trainParam.epochs	1000	Maximum number of epochs to train
net.trainParam.goal	1e − 6	Performance goal
net.trainParam.mu	0.005	Marquardt adjustment parameter
net.trainParam.mu_dec	0.1	Decrease factor for mu
net.trainParam.mu_inc	10	Increase factor for mu
net.trainParam.mu_max	1e − 10	Maximum value for mu
net.trainParam.max_fail	10	Maximum validation failures
net.trainParam.min_grad	1e − 10	Minimum performance gradient

TABLE 3: Artificial neural network topology.

	Topology								
Number	N1	N2	N3	N4	N5	N6	N7	N8	N9
Hidden layer	1	1	1	1	1	2	2	2	2
Nodes	10	15	20	25	30	10-5	10-10	15-10	20-10

TABLE 4: ANN statistical data for color-ks prediction.

Network number	N1	N2	N3	N4	N5	N6	**N7**	N8	N9
MSE for different sets									
Train error	0.2926	0.0174	0.2721	0.9069	0.0351	0.0001	0.0001	0.0001	0.0001
Validate error	21.394	18.295	8.945	12.862	4.991	0.011	0.003	0.011	0.005
Test error	14.059	7.058	11.220	16.309	5.785	0.007	0.004	0.019	0.016
For total data									
MAE	1.1736	1.1202	1.1968	1.5670	0.7956	0.0309	0.0235	0.0383	0.0332
MSE	7.1800	5.0185	4.1479	6.3110	2.1497	0.0037	**0.0015**	0.0060	0.0043
RMSE	2.6796	2.2402	2.0366	2.5122	1.4662	0.0610	0.0381	0.0776	0.0653
Minimum error	0.0017	0.0010	0.0043	0.0110	0.0009	0.0001	0.0002	$8.7E-05$	0.0001
Maximum error	22.295	10.227	10.001	12.494	7.508	0.322	0.162	0.503	0.431
Relative error distribution									
0-1%	155	153	157	152	160	162	162	162	162
1%–5%	7	9	5	10	2	0	0	0	0
5%–10%	0	0	0	0	0	0	0	0	0
10%–50%	0	0	0	0	0	0	0	0	0
>50%	0	0	0	0	0	0	0	0	0

Network number 7 gives the best performance in the ANN models related to color-ks prediction with the smallest MSE of 0.0015.

In addition, determining the number of hidden layers and the number of nodes in each layer is not straightforward and still has not been solved perfectly. There are a number of theoretical results concerning the number of hidden layers in a network. Specifically, Hetcht-Nielsen has shown that a network with two hidden layers can approximate any arbitrary nonlinear function and generate any complex decision region for classification problems [12]. Later, it was shown by Cybenko that a single layer is enough to form an arbitrarily close approximation to any nonlinear decision boundary [13]. Hornik and Stinchombe have come up with a more general theoretical result. They have shown that a single hidden layer feed forward network with arbitrary sigmoid hidden layer activation functions can well approximate an arbitrary mapping from one finite dimensional space to another [11].

Generally speaking, with more nodes in hidden layer, the network's ability of approximation is increased while the generalization ability is decreased. Therefore, the structure of the ANN is usually decided by experience together with trials [14]. On top of that, the size of the training set and the number of input/output nodes also affect the topology of the optimal neural network [15].

In this study, nine different topologies of the neural network were constructed and tested. According to the experience and practice, considering the fact that the data size and the number of input/output nodes are small, ANN with more than 30 nodes in the hidden layer is not necessary in this case. This is because more nodes cannot get better performance and a lot of training time will be spent, which may be computationally wasteful. Smaller networks require

Table 5: ANN statistical data for color-l prediction.

Network number	N1	N2	N3	N4	N5	**N6**	N7	N8	N9
MSE for different sets									
Train error	$1.9E-04$	$2.3E-05$	$1.1E-04$	$9.8E-06$	$1.8E-04$	$3.8E-08$	$3.9E-08$	$2.6E-08$	$3.5E-08$
Validate error	0.006	0.004	0.002	0.002	0.004	$3.4E-05$	$9.7E-05$	$6.2E-05$	$6.8E-05$
Test error	0.003	0.002	0.005	0.005	0.004	$1.6E-05$	$4.9E-05$	$5.6E-05$	$6.3E-05$
For total data									
MAE	0.025	0.018	0.022	0.018	0.025	0.002	0.002	0.002	0.003
MSE	0.002	0.001	0.001	0.001	0.002	$\mathbf{9.9E-06}$	$2.9E-05$	$2.3E-05$	$2.6E-05$
RMSE	0.0422	0.0346	0.0383	0.0368	0.0402	0.0031	0.0054	0.0048	0.0051
Minimum error	$2.0E-05$	$4.8E-05$	$4.4E-05$	$1.0E-05$	$4.2E-05$	$3.0E-07$	$2.0E-06$	$3.4E-07$	$1.6E-06$
Maximum error	0.231	0.154	0.218	0.205	0.177	0.014	0.041	0.021	0.024
Relative error distribution									
0-1%	160	161	160	161	161	162	162	162	162
1%–5%	2	1	2	1	1	0	0	0	0
5%–10%	0	0	0	0	0	0	0	0	0
10%–50%	0	0	0	0	0	0	0	0	0
>50%	0	0	0	0	0	0	0	0	0

Network number 6 gives the best performance in the ANN models related to color-l prediction with the smallest MSE of $9.9E-06$.

Table 6: ANN statistical data for color-a prediction.

Network number	N1	N2	N3	N4	N5	N6	N7	**N8**	N9
MSE for different sets									
Train error	$3.0E-09$	$2.7E-10$	$1.1E-09$	$1.1E-10$	$3.8E-10$	$1.4E-10$	$3.6E-10$	$3.4E-10$	$1.2E-10$
Validate error	$2.5E-08$	$1.4E-08$	$1.5E-08$	$1.4E-08$	$1.6E-08$	$5.3E-09$	$1.1E-08$	$6.4E-09$	$7.3E-09$
Test error	$2.4E-08$	$1.9E-08$	$1.3E-08$	$3.4E-08$	$1.5E-08$	$1.2E-08$	$4.6E-09$	$6.9E-09$	$7.3E-09$
For total data									
MAE	$7.5E-05$	$4.1E-05$	$5.2E-05$	$4.8E-05$	$4.7E-05$	$3.5E-05$	$3.6E-05$	$3.5E-05$	$3.1E-05$
MSE	$1.2E-08$	$6.5E-09$	$6.2E-09$	$9.5E-09$	$6.4E-09$	$3.6E-09$	$3.3E-09$	$\mathbf{2.8E-09}$	$3.0E-09$
RMSE	0.0001	$8.1E-05$	$7.9E-05$	$9.8E-05$	$8.0E-05$	$6.0E-05$	$5.7E-05$	$5.3E-05$	$5.4E-05$
Minimum error	$3.1E-07$	$1.4E-07$	$4.3E-07$	$4.8E-08$	$3.2E-07$	$1.5E-07$	$4.8E-07$	$9.7E-08$	$3.4E-08$
Maximum error	0.0004	0.0006	0.0004	0.0005	0.0004	0.0002	0.0002	0.0002	0.0002
Relative error distribution									
0-1%	162	162	162	162	162	162	162	162	162
1%–5%	0	0	0	0	0	0	0	0	0
5%–10%	0	0	0	0	0	0	0	0	0
10%–50%	0	0	0	0	0	0	0	0	0
>50%	0	0	0	0	0	0	0	0	0

Network number 8 gives the best performance in the ANN models related to color-a prediction with the smallest MSE of $2.8E-09$.

less memory to store the connection weights and can be implemented in hardware easier and more economically. Training a smaller network usually requires less computation because each iteration is computationally less expensive. Smaller networks have also very short propagation delays from their inputs to their outputs. This is very important during the testing phase of the network, where fast responses are usually required. The details of the ANN topology are shown in Table 3.

In the experiment, there were totally 810 datasets for each group. These data were randomly divided into three sets: training (60%), cross-validation (20%), and test (20%) in order to meet the requirements of both accuracy and generalization.

3. Results and Discussion

3.1. ANN Model. In order to evaluate the effect of ANN model, the MAE (mean absolute error), MSE (mean square

TABLE 7: ANN statistical data for color-b prediction.

Network number	N1	N2	N3	N4	N5	N6	**N7**	N8	N9
MSE for different sets									
Train error	$3.3E-05$	$3.5E-05$	$1.4E-05$	$2.6E-05$	$2.4E-05$	$6.9E-09$	$7.0E-09$	$7.0E-09$	$6.8E-09$
Validate error	0.0005	0.0004	0.0005	0.0003	0.0005	$3.0E-06$	$1.5E-06$	$2.0E-06$	$2.8E-06$
Test error	0.0008	0.0006	0.0007	0.0006	0.0005	$4.4E-06$	$1.7E-06$	$4.7E-06$	$3.8E-06$
For total data									
MAE	0.0111	0.0098	0.0088	0.0090	0.0089	0.0006	0.0004	0.0006	0.0005
MSE	0.0003	0.0002	0.0002	0.0002	0.0002	$1.5E-06$	$\mathbf{6.2E-07}$	$1.3E-06$	$1.3E-06$
RMSE	0.0168	0.0151	0.0153	0.0142	0.0142	0.0012	0.0008	0.0012	0.0011
Minimum error	$8.5E-06$	$2.1E-04$	$2.6E-05$	$5.9E-05$	0.0001	$9.5E-07$	$1.7E-07$	$1.5E-06$	$6.5E-08$
Maximum error	0.0607	0.0677	0.0696	0.0701	0.0586	0.0059	0.0045	0.0060	0.0080
Relative error distribution									
0-1%	152	157	152	157	155	162	162	162	162
1%–5%	10	5	10	5	7	0	0	0	0
5%–10%	0	0	0	0	0	0	0	0	0
10%–50%	0	0	0	0	0	0	0	0	0
>50%	0	0	0	0	0	0	0	0	0

Network number 7 gives the best performance in the ANN models related to color-b prediction with the smallest MSE of $6.2E-07$.

TABLE 8: ANN statistical data for color-c prediction.

Network number	N1	N2	N3	N4	N5	N6	**N7**	N8	N9
MSE for different sets									
Train error	$2.0E-05$	$3.6E-06$	$2.5E-06$	$3.6E-06$	$3.6E-06$	$4.1E-09$	$3.3E-09$	$4.8E-09$	$4.5E-09$
Validate error	$1.6E-04$	$1.6E-04$	0.0002	0.0001	$1.5E-04$	$1.5E-07$	$1.1E-07$	$2.2E-07$	$1.8E-07$
Test error	$2.1E-04$	0.0001	0.0001	0.0001	0.0001	$2.0E-07$	$9.8E-08$	$1.1E-07$	$2.6E-07$
For total data									
MAE	0.0062	0.0043	0.0048	0.0041	0.0041	0.0001	0.0001	0.0002	0.0002
MSE	$8.6E-05$	$6.2E-05$	$7.6E-05$	$5.0E-05$	$5.4E-05$	$7.2E-08$	$\mathbf{4.3E-08}$	$6.8E-08$	$9.3E-08$
RMSE	0.0093	0.0079	0.0087	0.0070	0.0073	0.0003	0.0002	0.0003	0.0003
Minimum error	$6.8E-06$	$2.5E-06$	$1.5E-05$	$3.5E-05$	$1.1E-05$	$2.8E-06$	$2.8E-09$	$5.5E-07$	$3.0E-07$
Maximum error	0.0360	0.0382	0.0409	0.0320	0.0446	0.0016	0.0010	0.0012	0.0022
Relative error distribution									
0-1%	162	162	161	162	161	162	162	162	162
1%–5%	0	0	1	0	1	0	0	0	0
5%–10%	0	0	0	0	0	0	0	0	0
10%–50%	0	0	0	0	0	0	0	0	0
>50%	0	0	0	0	0	0	0	0	0

Network number 7 gives the best performance in the ANN models related to color-c prediction with the smallest MSE of $4.3E-08$.

error), and RMSE (rooted mean square error) are calculated. These three functions are widely used in evaluating the effect of fitting. Their definitions are as follows:

$$\mathrm{MAE} = \frac{1}{n}\sum_{i=1}^{n}\left(abs\left(t_i - o_i\right)\right),$$

$$\mathrm{MSE} = \frac{1}{n}\sum_{i=1}^{n}\left(t_i - o_i\right)^2, \qquad (2)$$

$$\mathrm{RMSE} = \sqrt{\mathrm{MSE}}.$$

In the function, t_i and o_i are the target output and the ANN predicted output, respectively.

As mentioned before, the whole dataset were divided into training, validation, and test set. After training the networks, the models are expected to be used in large scale prediction works. They were validated by an unseen testing data set to predict the color values. Therefore, besides MAE and MSE, the test error, which is usually a symbol for the robustness and generalization of the model, is also an important factor in judging the quality of neural network.

Besides, some other statistical information was also collected. Detailed results for colour properties, that is, K/S Sum value (color-ks), CIE L^* (color-l), CIE a^* (color-a), CIE b^* (color-b), CIE $\mathbf{C}^*$ (color-c) and CIE $\mathbf{h}^\circ$ (color-h) values were shown in the Tables 4, 5, 6, 7, 8, and 9, respectively. Here,

Table 9: ANN statistical data for color-h prediction.

Network number	N1	N2	N3	N4	N5	**N6**	N7	N8	N9
MSE for different sets									
Train error	$3.2E-04$	$7.7E-05$	$2.1E-05$	$7.1E-05$	$1.9E-04$	$4.3E-06$	$2.1E-06$	$1.5E-06$	$1.9E-06$
Validate error	0.0037	0.0030	0.0032	0.0013	0.0030	$2.4E-05$	0.0001	0.0001	$9.1E-05$
Test error	0.0036	0.0018	0.0013	0.0042	0.0023	$5.8E-05$	$4.8E-05$	$4.0E-05$	$7.4E-05$
For total data									
MAE	0.0263	0.0185	0.0151	0.0181	0.0224	0.0028	0.0033	0.0033	0.0033
MSE	0.0016	0.0010	0.0009	0.0011	0.0012	**$1.9E-05$**	$3.3E-05$	$3.4E-05$	$3.4E-05$
RMSE	0.0402	0.0315	0.0300	0.0337	0.0341	0.0044	0.0058	0.0058	0.0058
Minimum error	$1.4E-04$	0.0002	$1.1E-05$	$8.1E-06$	0.0002	$1.1E-06$	$3.1E-05$	$6.3E-06$	$4.5E-05$
Maximum error	0.1859	0.1570	0.1982	0.2245	0.1437	0.0265	0.0279	0.0302	0.0316
Relative error distribution									
0-1%	162	162	162	162	162	162	162	162	162
1%–5%	0	0	0	0	0	0	0	0	0
5%–10%	0	0	0	0	0	0	0	0	0
10%–50%	0	0	0	0	0	0	0	0	0
>50%	0	0	0	0	0	0	0	0	0

Network number 6 gives the best performance in the ANN models related to color-h prediction with the smallest MSE of $1.9E-05$.

Table 10: Relative importance of the input variables.

Excluded variable	Temperature	Time	pH	MA	Twist
Color-ks (optimized MSE is 0.0015)					
MSE	1443.3	549.3	117.0	222.9	2630.4
% Increase	96219900	36619900	7799900	14859900	175359900
Color-l (optimized MSE is $9.9E-06$)					
MSE	0.5835	0.2833	0.0629	0.0961	0.9708
% Increase	5893839.4	2861516.2	635253.5	970607.1	9805960.6
Color-a (optimized MSE is $2.8E-09$)					
MSE	0.0006	0.0003	$1.2E-05$	0.0001	0.0010
% Increase	21428471.4	10714185.7	428471.4	3571328.6	35714185.7
Color-b (optimized MSE is $6.2E-07$)					
MSE	0.0972	0.0625	0.0071	0.0099	0.1470
% Increase	14507362.7	9328258.2	1059601.5	1477511.9	21940198.5
Color-c (optimized MSE is $4.3E-08$)					
MSE	0.0557	0.0347	0.0024	0.0034	0.1007
% Increase	129534783.7	80697574.1	5581295.3	7906876.7	234185946.5
Color-h (optimized MSE is $1.9E-05$)					
MSE	8.3680	4.9060	0.2621	0.4304	9.7639
% Increase	44042005.3	25820952.6	1379373.7	2265163.2	51388847.4

for simplicity, only the network with the smallest MSE is highlighted as best network model.

It can be seen from Tables 4 to 9 that for two hidden layer neural networks, the train errors are normally smaller than one hidden layer network. However, it does not matter for the test error. On the other hand, the one layer network can also perform well in test error, which means it is more robust in most cases. From the Tables 4 to 9, it can also note that in ANN model, the relative error distributes mostly in 0–5%, which is an acceptable range.

3.2. The Relative Importance of the Input Variables. In order to find out the relative importance of these five input variables, an additional experiment was performed. In this experiment, all the input variables were remained except one designated. Then, the MAE, MSE, and other statistical information of the

TABLE 11: LR model statistical data.

	Color-ks	Color-l	Color-a	Color-b	Color-c	Color-h
MAE	29.344	0.564	0.004	0.204	0.159	2.146
MSE	1313.6	0.496	$3.09E-05$	0.079	0.051	8.195
RMSE	36.244	0.704	0.006	0.281	0.227	2.863
Minimum error	0.0211	0.0019	$5.40E-05$	0.0003	0.0001	$3.40E-04$
Maximum error	86.687	1.946	0.0189	0.928	0.825	8.757
Relative error distribution						
0-1%	21	29	159	26	32	120
1%–5%	67	101	3	76	91	42
5%–10%	60	27	0	33	28	0
10%–50%	14	5	0	27	11	0
>50%	0	0	0	0	0	0

new ANN model were calculated and further compared with the optimized one. Here, for simplicity, the increase in the MSE value was treated as the indicator of the importance of the excluded input. The results are shown in Table 10.

It can be seen that the five input variables are all of great importance. Among them, fabric yarn twist level is the most significant one. This is because if all the process parameters are kept constant, the twist level in the yarn would affect the penetration of cellulase into the fabric to function the colour fading effect. A high level of twist in a yarn holds the fibres together and hence restricts the access of water to the yarn interior. Therefore, such a high yarn twist level would affect the cellulase absorption during the treatment. For other four variables in the process, treatment temperature plays a key role followed by treatment time, mechanical agitation, and pH. Treatment temperature is the most important factor on cellulase (enzyme) treatment because enzyme reaction increases with temperature but it is only activated within a temperature range in which the enzyme structure maintains stable and unchanged. Beyond this optimum range, the enzyme activity will decrease sharply as the protein structure of enzyme is tangled through thermal agitation. Longer enzyme treatment time will prolong enzymatic degradation of cellulose and the time for further abrasion. With a longer treatment time, cellulase effectively hydrolyses fragments of cotton fibrils [16], and thus the fabric would be less fuzzy than the original [17]. Therefore, shorter treatment time with enzyme can preserve desired colour depth in denim fabrics. The level of mechanical action is altered by a varying number of steel balls applied in enzyme treatment. By increasing the number of steel balls applied, the abrasion of the fabric surface will increase which creates more accessible sites for enzyme attack. Denim fabric treated under no agitation has lighter shade than that treated with mild and vigorous agitation. It is because the cellulose fibres are weakened after treating with enzyme alone and are not well removed by mechanical agitation, [17]. The surface of fabric becomes more hairy and forming a layer of fuzz on surface [18]. For specimen treated under vigorous agitation, a biopolishing effect is obtained. The weaken fibres are well removed by strong mechanical agitation, and thus a cleaner and smoother surface is resulted which produce a lighter shade of fabric as dye particles are also removed with cellulose fibres. For specimen treated with mild agitation (50 steel balls), the situation is between zero agitation and vigorous agitation. Mild agitation is not strong enough to remove as many weaken surface fibres as vigorous agitation can, thus, rather, an uneven surface is produced under mild agitation. The uneven surface causes higher light diffusion on surface, and thus darker shade of specimen is resulted as they achieve lower reflectance on surface. Enzyme performs its maximal activity at a particular range of pH values. Deviation of the optimum pH range will alter the internal structure of enzyme due to the electrostatic interactions within the enzyme. Thus, the enzyme will no longer perform its normal function. As observed from Table 10, there is an increase in MSE when compared with the values in the optimized ANN model as shown in Tables 4 to 9. This means, if comparing them with normal models, removing some input variables may enhance the quality of ANN model. This is possible and reasonable because our experiment data is not sufficient enough. More variables in the input layer may cause the ANN to be overfitting.

3.3. Linear Regression (LR) Model. In this study, a traditional linear regression (LR) model was also established for the sake of comparison with ANN model. In general, response variable y may be related to k regressor variables. The function is shown in as follows:

$$y = \beta_0 + \beta_1 x_1 + \beta_2 x_2 + \cdots + \beta_k x_k + \varepsilon. \quad (3)$$

The parameters β_i are called the regression coefficients. This model describes a hyperplane in k-dimensional space of the regressor variables. The method of least square is typically used to estimate the regression coefficients in a multiple linear regression model. Here, the linear regression model is applied in aforementioned experiment data [19, 20]. The results are shown in Table 11.

When comparing the ANN model and the LR model, it is evident that the predictive power of the ANN model is better than the LR model. From Table 11, all the evaluation numbers are obviously much greater than those of Tables 4 to 9, which mean that applying LR model for prediction in the recent case

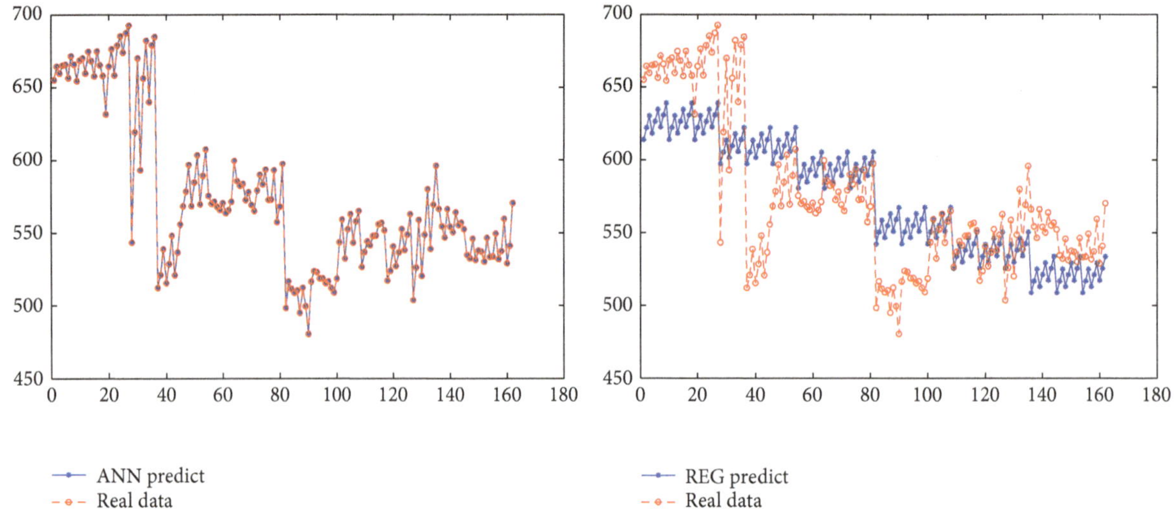

(a) (b)

Figure 3: (a) ANN model fitting for color-ks. (b) LR model fitting for color-ks.

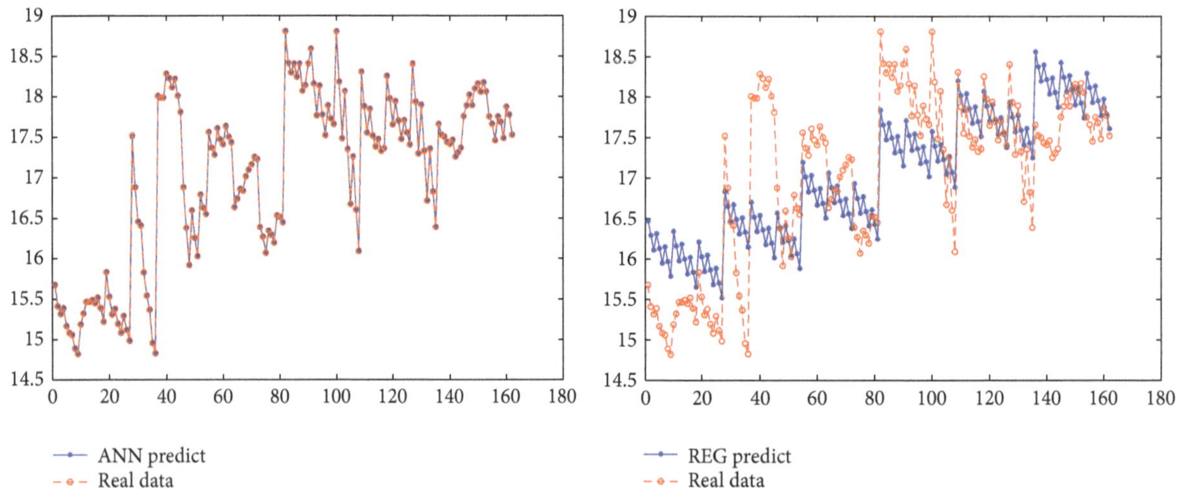

(a) (b)

Figure 4: (a) ANN model fitting for color-l. (b) LR model fitting for color-l.

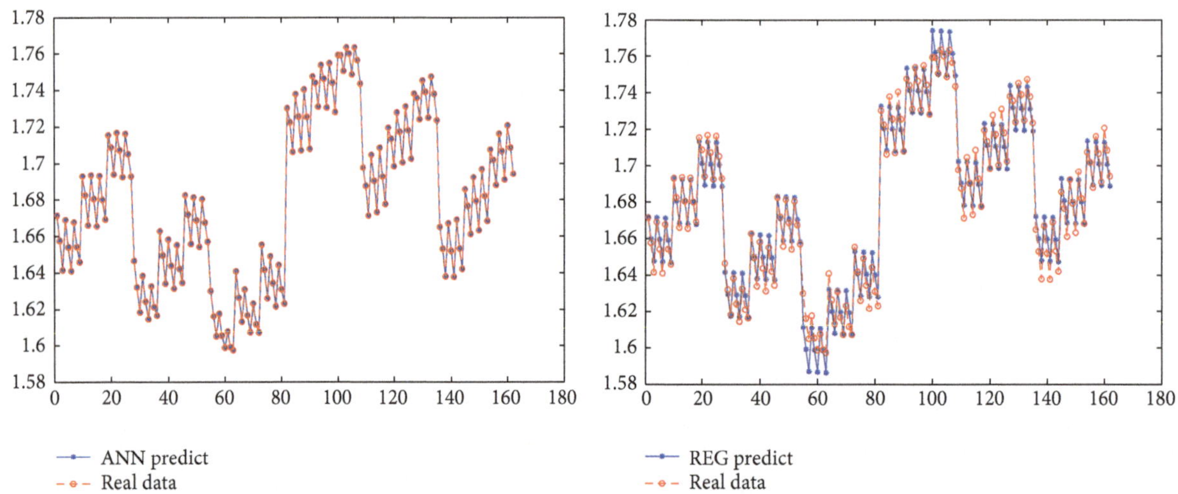

(a) (b)

Figure 5: (a) ANN model fitting for color-a. (b) LR model fitting for color-a.

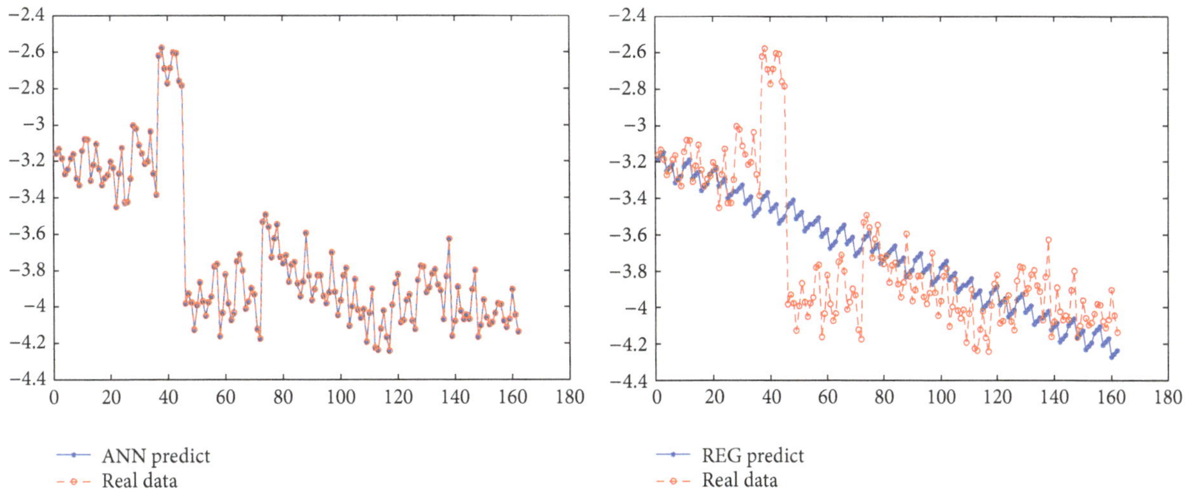

(a) (b)

FIGURE 6: (a) ANN model fitting for color-b. (b) LR model fitting for color-b.

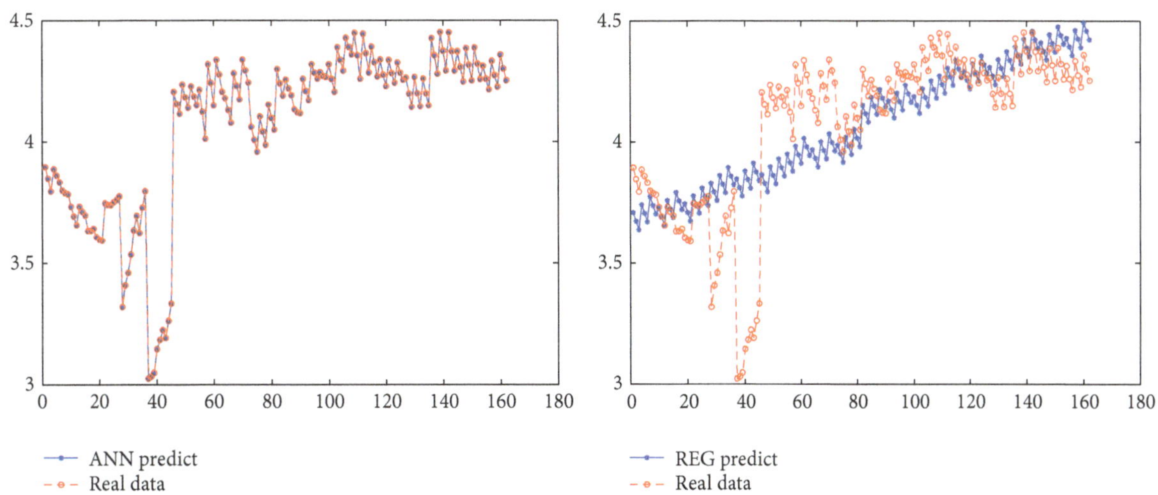

(a) (b)

FIGURE 7: (a) ANN model fitting for color-c. (b) LR model fitting for color-c.

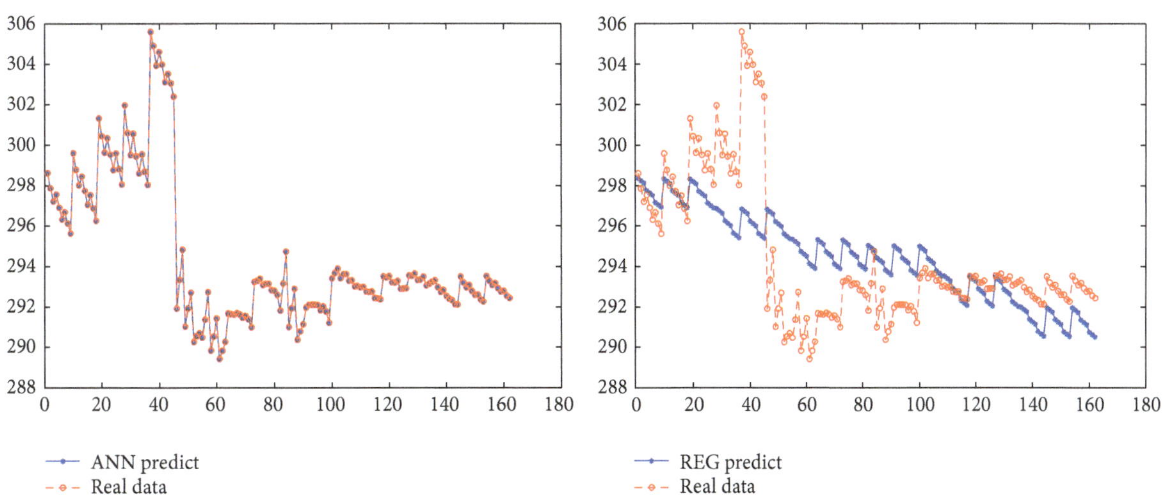

(a) (b)

FIGURE 8: (a) ANN model fitting for color-h. (b) LR model fitting for color-h.

is worse than ANN model. Figures 3, 4, 5, 6, 7 and 8 depict the comparison graphically.

4. Conclusion

In the paper, colour properties of 100% cotton denim fabrics, including colour yield (in terms of K/S value), CIE L^*, a^*, b^*, $\mathbf{C}^*$, and $\mathbf{h}^\circ$ values, after cellulase enzyme treatment were predicted by using both ANN and LR models. Several statistical tests were conducted to examine the performance of these experiments and models. Experimental results revealed that changes in the number of nodes of the neural network model affected the performance of the model. Results also reflected that the colour properties could be predicted accurately with the help of the ANN model. These prediction results demonstrated the usefulness of cellulase treatment before coloring and might find good applications for future use by the textile industry. When a comparison was made between ANN and LR models, the LR model did not perform well in the experiment due to limitations of the model itself.

Acknowledgment

The Authors would like to acknowledge the financial support from The Hong Kong Polytechnic University for this work.

References

[1] Z. Gao, L. Zhang, and J. Zhao, "Application of laser technology in textile industry," *Journal of Textile Research*, vol. 27, no. 8, pp. 117–120, 2006.

[2] Z. Ondogan, O. Pamuk, E. N. Ondogan, and A. Ozguney, "Improving the appearance of all textile products from clothing to home textile using laser technology," *Optics and Laser Technology*, vol. 37, no. 8, pp. 631–637, 2005.

[3] L. C. Chang, F. J. Chang, and H. C. Hsu, "Real-time reservoir operation for flood control using artificial intelligent techniques," *International Journal of Nonlinear Sciences and Numerical Simulation*, vol. 11, no. 11, pp. 887–902, 2010.

[4] T. W. Lau, P. C. L. Hui, F. S. F. Ng, and K. C. C. Chan, "A new fuzzy approach to improve fashion product development," *Computers in Industry*, vol. 57, no. 1, pp. 82–92, 2006.

[5] L. Wu, K. L. Yick, S. P. Ng, J. Yip, and K. H. Kong, "Parametric design and process parameter optimization for bra cup molding via response surface methodology," *Expert Systems With Applications*, vol. 39, pp. 162–171, 2012.

[6] R. Beltran, L. Wang, and X. Wang, "Measuring the influence of fibre-to-fabric properties on the pilling of wool fabrics," *Journal of the Textile Institute*, vol. 97, no. 3, pp. 197–204, 2006.

[7] J. Fan and L. Hunter, "A worsted fabric expert system. II. An artificial neural network model for predicting the properties of worsted fabrics," *Textile Research Journal*, vol. 68, no. 10, pp. 763–771, 1998.

[8] P. K. Majumdar and A. Majumdar, "Predicting the breaking elongation of ring spun cotton yarns using mathematical, statistical, and artificial neural network models," *Textile Research Journal*, vol. 74, no. 7, pp. 652–655, 2004.

[9] C. M. Murrells, X. M. Tao, B. G. Xu, and K. P. S. Cheng, "An artificial neural network model for the prediction of spirality of fully relaxed single jersey fabrics," *Textile Research Journal*, vol. 79, no. 3, pp. 227–234, 2009.

[10] F. Pynckels, P. Kiekens, S. Sette, L. van Langenhove, and K. Impe, "Use of neural nets for determining the spinnability of fibres," *Journal of the Textile Institute*, vol. 86, no. 3, pp. 425–437, 1995.

[11] K. Hornik and M. Stinchombe, *Multilayer Feed-Forward Networks Are Universal Approximators in Artificial Neural Networks: Approximation and Learning Theory*, Blackwell Press, Oxford, UK, 1992.

[12] R. Hetcht-Nielsen, "Theory of the backpropagation neural networks," in *Proceedings of International Joint Conference on Neural Networks (IJCNN '89)*, vol. 1, pp. 593–611, 1989.

[13] G. Cybenko, "Approximation by superpositions of a sigmoidal function," *Mathematics of Control, Signals, and Systems*, vol. 2, no. 4, pp. 303–314, 1989.

[14] S. C. Huang and Y. F. Huang, "Bounds on the number of hidden neurons in multilayer perceptrons," *IEEE Transactions on Neural Networks*, vol. 2, no. 1, pp. 47–55, 1991.

[15] M. A. Sartori and P. J. Antsaklis, "A simple method to derive bounds on the size and to train multilayer neural networks," *IEEE Transactions on Neural Networks*, vol. 2, no. 4, pp. 467–471, 1991.

[16] A. S. Aly, A. B. Moustafa, and A. Hebeish, "Bio-technological treatment of cellulosic textiles," *Journal of Cleaner Production*, vol. 12, no. 7, pp. 697–705, 2004.

[17] C. W. Kan, C. W. M. Yuen, and S. Q. Jiang, "The effect of cellulase treatment on hydrolysis of linen," *Fibers and Polymers*, vol. 7, no. 3, pp. 241–244, 2006.

[18] A. Cavaco-Paulo, "Mechanism of cellulase action in textile processes," *Carbohydrate Polymers*, vol. 37, no. 3, pp. 273–277, 1998.

[19] O. N. Hung, L. J. Song, C. K. Chan, C. W. Kan, and C. W. M. Yuen, "Using artificial neural network to predict color properties of laser-treated 100% cotton fabric," *Fibers and Polymers*, vol. 12, pp. 1069–1076, 2011.

[20] O. N. Hung, L. J. Song, C. K. Chan, C. W. Kan, and C. W. M. Yuen, "Predicting the laser-engraved color properties on cotton-spandex fabric by artificial neural network," *AATCC Review*, vol. 12, no. 3, pp. 57–63, 2012.

32

Effect of Some Navels on Properties of Cotton/Nylon66 Blend (1 : 1) Rotor Spun Yarn and Wrapper Formation: A Comparison between Rotor and Ring Spun Yarn

Babak Yadollah Roudbari and Safdar Eskandarnejad

Textile Department, Science and Research Branch, Islamic Azad University, Tehran 1477893855, Iran

Correspondence should be addressed to Babak Yadollah Roudbari; babak.roudbari@yahoo.com

Academic Editor: Phillip W. Gibson

Use of nylon/cotton blend yarn in military uniform is common and due to advantages in its fabric in comparison to 100% cotton fabrics, capabilities of military uniforms have been improved. In this study the effects of navel type on properties of (50%-50%) nylon/cotton blend yarn and wrapper formation were investigated and compared with similar ring spun yarn. Rotor spun yarn was produced on a single head laboratory rotor spinning machine with four navels (smooth, spiral, 3 grooved, and 4 grooved) and ring yarn was produced on a zinser 319 ring spinning machine. Test result showed that navel type has a significant effect on yarn strength and strength of smooth navel yarn was maximum. Elongation of a 100% cotton rotor spun yarn is more than similar ring yarn, but it was not observed in cotton/nylon blend. Yarn irregularity and imperfections varied significantly with navel type and for rotor yarns were more than the ring yarn. Navel type had significant effect on yarn hairiness but it didnot have an effect on yarn abrasion significantly.

1. Introduction

The use of blended yarn has increased significantly these days. Blending is used as a means of substituting a less expensive fiber for a highly priced one and to produce yarns with qualities that cannot be obtained by using one type of material alone.

El-sheikh [1] reported that the properties of blended yarns depend on their constituent fibers and also showed that blended yarn's tenacity decreases as the percentage of high tenacity fiber is increased. Kemp and Owen [2] investigated the stress/strain characteristics and cotton fiber breakage for a series of nylon/cotton blended yarns at strain above the breaking strain of all cotton and also studied the strength and mechanical behavior of nylon/cotton blended yarn and they found that a dependence exists between the behavior of the two fiber types. Gibson et al. [3] worked on military uniform and reported that nylon/cotton blend is used as an outer shell in different version of military chemical protective uniforms. And some researchers [4–6] worked on battle dress uniforms (BDU) and found that nylon/cotton blend is more suitable than those of any other blends. Kong et al. [7] reported that by adding 2–4% elastan yarn (lycra) through the nylon/cotton fabrics, the ability of extension of these kinds of fabrics can be increased up to 50% and it is appropriate for military uniforms. Javadiyan et al. [8] worked on seven different nylon/cotton blend ratios and reported that by increasing the nylon fiber blend ratio, the yarn elongation, abrasion resistance, tensile work of rupture, and hairiness significantly increase, whereas yarn tensile modulus decrease.

Navels differ in their material, forms, number of grooves, existence of fluted inserts, length of ceramic parts, and grooves [9]. Smooth navels often give better yarn characteristics; the yarn is more resistant to rubbing and has good heat conductivity and generally less false twist; grooved navels can operate at lower twist levels; the running performance is better because of the greater false twist effect but hairiness does increase; ceramic navels with 4–6 grooves have proved advantageous in the spinning of blended yarns and fibers that are not strongly sensitive to heat [10]. The configuration of

TABLE 1: Properties of fibers.

No	Fiber	Blend ratio	Length (mm)	Linear density (fineness)
1	Cotton	50%	Effective 28	0.15 tex (3.8 micronaire)
2	Nylon66	50%	Cutting 38	0.17 tex (1.5 den)

the navel has quite a considerable effect on both spinning stability and yarn appearance [11]. Nawaz et al. [12] reported that the effect of draw-off navel type and yarn count was highly significant while for yarn evenness, the effect of yarn count and rotor diameter was highly significant, whereas the effect of draw-off navel was at par. Erbil et al. [13] worked on the effect of navel type on hairiness of rotor-spun blend yarn and reported that the form and structure of navel have an important effect on hairiness, except the number of notches.

The work reported here is concerned about the evaluation of the quality and mechanical properties of five different nylon/cotton blends of yarn that are particularly used in military uniforms, apparel and textile industry.

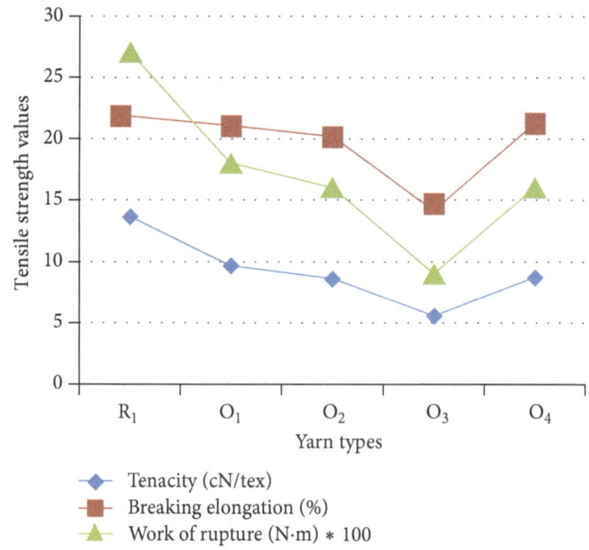

FIGURE 1: Effect of navel type on tensile properties of nylon/cotton rotor spun yarn.

2. Materials and Method

Nylon66 and Cotton have been used in the present investigation (Table 1). Cotton and nylon66 fibers in carded sliver were prepared. The carded sliver in the same proportion passed through two stages of drawing processes and created 4.5 Ktex nylon/cotton blend sliver and then the blended sliver feed to flyer frame for making 528 tex roving. Finally, draw sliver and roving were produced on a RU04 rotor spinning (Shubert & Salzer) with four different navels (smooth, spiral, 3 grooved, and 4 grooved) and Zinser 319 Ring spun machine respectively, to produce yarn of 20 Ne (29.5 tex).

Spindle speed, nominal count, and twist were 6700 rpm, 20 Ne, and 637 T.P.M, respectively. Also the speed of rotor, opening roller, take-up roller, and feed roller was 49300 rpm, 9200 rpm, 71.3 m/min, and 0.48 m/min, respectively.

The codings of the four different rotor spun yarns and ring spun yarn (witness yarn) are listed in Table 2.

All the samples were tested for the following: tensile strength, evenness, hairiness, and abrasion resistance.

2.1. Tensile Strength. Yarn tensile properties (tenacity, breaking elongation, and work of rupture) were determined on a SDL tensile tester with a jaw speed of 1 m/min and a test length of 50 mm, and 30 tests were conducted for each yarn.

2.2. Evenness. The irregularity and yarn imperfection (thick, thin, nep, and CV%) were measured by using a Keisokki uster over a length of 1000 m of yarn speed at 25 m/min and 5 tests were conducted for each yarn. The degree of sensitivity of the thinness, thickness, and neps was −50%, +50% and +280% respectively.

2.3. Hairiness and Abrasion Resistance. A SDL hair tester was used to determine the hairiness with a testing speed of 20 m/min and a test length of 30 m for 5 samples (totally 150 m); only hairs with a length greater than or equal to 3 mm were measured. The yarn abrasion resistance was measured by using a SDL yarn abrasion tester. A abradant (P1200) was used and 30 tests were conducted for each yarn sample. All tests were conducted under the standard laboratory conditions (22 ± 2°C and 65±% r·h).

2.4. Scanning Electron Microscopy (SEM). Wrapper formation of the samples was observed by scanning electron microscopy (VEGA TESCAN, Czech Republic). Samples were fixed in SEM holders and coated with a thin layer of gold prior to SEM investigation.

2.5. Analysis of Data. The data was analyzed statistically by using CRD (complete random design) method for the interpretation of data. Analysis of variance (ANOVA) and Duncan test were provided at α value of 0.05%. All statistical analysis steps were performed on SPSS software.

Table 3 shows test results for the properties of yarn quality. A summary of ANOVA statistical results is tabulated in Tables 4, 5, and 6. As shown in these tables, the effect of navel type on yarn physical properties is almost statistically significant.

3. Results and Discussion

3.1. Tensile Properties. The effect of navel type on nylon/cotton blended rotor spun yarn tensile properties is illustrated in Figure 1.

From the test result in Tables 3 and 4 and Figure 1, we can observe that navel types have statistically significant difference according to tensile properties. Duncan test revealed that the highest and the lowest yarn tenacities were recorded for R_1 (13.6 cN/tex) and O_3 (5.7 cN/tex), respectively. These

Table 2: Navel types.

Navel type	Yarn code	Properties	Processed material	Spinning system	Navel design
—	R_1	—	—	Ring	—
KN	O_1	Ceramic, Smooth	Cotton and blend	Rotor	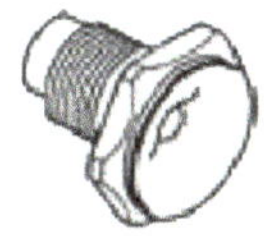
KS	O_2	Ceramic, Spiral	Cotton and blend	Rotor	
K3KK	O_3	Ceramic, 3 grooved, short	Cotton and blend	Rotor	
K4KK	O_4	Ceramic, 4 grooved, short	Cotton and blend	Rotor	

Table 3: Test result for yarn quality properties.

Yarn type	Tensile strength			Evenness				Hairiness	Abrasion resistance
	Tenacity (cN/tex)	Breaking elongation (%)	Work of rupture (N·m)	CV%	Thin (−50%) per Km	Thick (+50) per Km	Nep (+280) per Km		
R_1	13/65	21/80	0/27	13/94	0	18	8	6	6/87
O_1	9/68	21/05	0/18	16/13	8	34	2	1/9	8/53
O_2	8/62	20/16	0/16	14/60	6	48	4	3	8/50
O_3	5/58	14/13	0/09	17/83	38	70	33	7/6	5/77
O_4	8/74	21/27	0/16	17/05	16	74	2	1/7	8/20

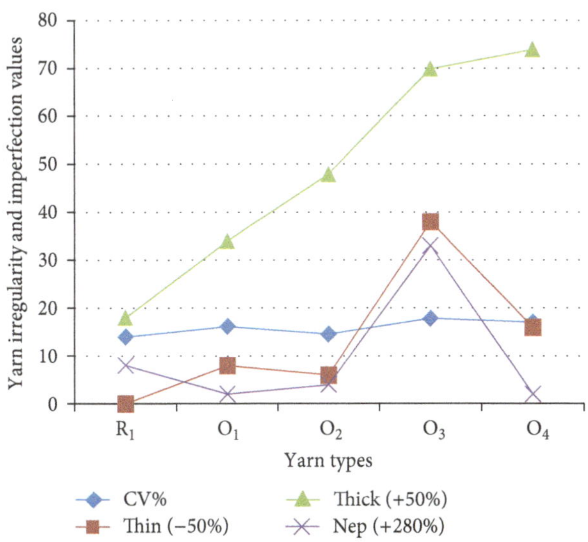

Figure 2: Effect of navel type on yarn irregularity and imperfection of nylon/cotton rotor spun yarn.

values were nonsignificant for O_2 and O_4. Also between the navel types, smooth navel (O_1) has the highest value.

Figure 1 and Duncan test indicated that the highest and the lowest yarn breaking elongations were recorded for R_1 (21.8%) and O_3 (14.13%), respectively; also there is a significant difference between R_1, O_2 and O_3, and these values were non-significant for the others (Table 4). In contrast with breaking elongation for 100% cotton in written text book [14], the breaking elongation in nylon/cotton rotor spun yarn with all navel types (except 3 grooved) is lower than similar ring spun yarn.

Figure 1 and Duncan test showed that the highest and the lowest yarn works of rupture were recorded for R_1 (0.27 N·m) and O_3 (0.087 N·m), respectively. There is no significant difference between O_2 and O_4.

3.2. Evenness. The effects of navel type on nylon/cotton blended rotor spun yarn irregularity (CV%) and imperfection are shown in Figure 2. From the test results in Table 3 and Figure 2, it is clear that navel type has statistically significant

TABLE 4: A summary of ANOVA statistical result (Duncan test) for tensile strength*.

Yarn type	Tensile Strength														
	Tenacity					Breaking elongation					Work of rupture				
	R_1	O_1	O_2	O_3	O_4	R_1	O_1	O_2	O_3	O_4	R_1	O_1	O_2	O_3	O_4
R_1	—	+	+	+	+	—	–	+	+	–	—	+	+	+	+
O_1	+	—	+	+	+	–	—	–	+	–	+	—	+	+	+
O_2	+	+	—	+	–	+	–	—	+	–	+	+	—	+	–
O_3	+	+	+	—	+	+	+	+	—	+	+	+	+	—	+
O_4	+	+	–	+	—	–	–	–	+	—	+	+	–	+	—

*At 5% confidence limit, "+" means statistically significant and "–" means statistically in-significant.

TABLE 5: A summary of ANOVA statistical result (Duncan test) for evenness*.

Yarn type	Evenness																			
	CV%					Thin					Thick					Nep				
	R_1	O_1	O_2	O_3	O_4	R_1	O_1	O_2	O_3	O_4	R_1	O_1	O_2	O_3	O_4	R_1	O_1	O_2	O_3	O_4
R_1	—	+	–	+	+	—	–	–	+	–	—	–	–	+	+	—	–	–	+	–
O_1	+	—	+	+	+	–	—	–	+	–	–	—	–	+	+	–	—	–	+	–
O_2	–	+	—	+	+	–	–	—	+	–	–	–	—	+	+	–	–	—	+	–
O_3	+	+	+	—	+	+	+	+	—	[illegible]	[illegible]	+	+	—	–	+	+	+	—	+
O_4	+	+	+	+	—	–	–	–	+	—	+	+	+	–	—	–	–	–	+	—

*At 5% confidence limit, "+" means statistically significant and "–" means statistically insignificant.

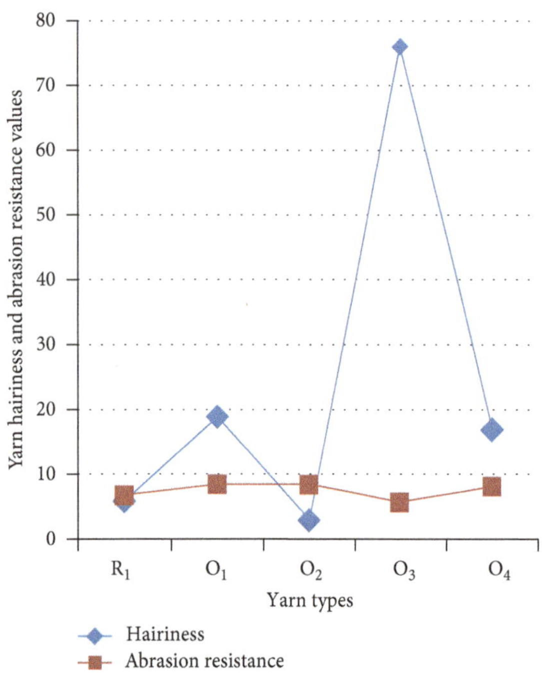

FIGURE 3: Effect of navel type on hairiness and abrasion resistance of nylon/cotton rotor spun yarn.

difference at the level of 0.05 according to yarn irregularity and yarn imperfection.

Figure 2 and Duncan test revealed that the highest and the lowest yarn CV% were O_3 (17.82) and R_1 (13.93), respectively. It is found that in 3 grooved navels, the highest yarn irregularity was obtained. Also between the navels, spiral navel has the lowest CV%. In contrast with result for 100% cotton yarn, rotor spun yarns have more CV% than similar ring spun yarn.

As shown in Figure 2, in thin point, there is no significant difference among spun yarn except 3 grooved yarns; also among navels, spiral navel has the lowest thin point but in thick point; it is found that 4 grooved navel has the highest and ring yarn has the lowest one. Fibers in surface layer of rotor spun yarn partially evaded from twisting operation in spinning process that resulted in yarn taking lower twist; thus, few fibers participated in yarn strength and lead to creating the thick points.

As shown in Figure 2 and Duncan table revealed that the highest and lowest yarn neps were O_1 and O_3, respectively, and there is no significant difference between R_1, O_2, O_4, and O_1. The high amount of irregularity of 100% cotton rotor spun yarn is attributed to the high amount of short fiber content as well as to the relative short effective length of cotton fiber compared to nylon66; it is deduced that the wrapper fibers existing around the yarn body are an undesired factor which in turn increases the yarn irregularity.

3.3. Hairiness. Figure 3 shows the effect of navel type on nylon/cotton blended rotor spun yarn hairiness. Test result shows that 4 grooved navel has the lowest and 3 grooved navel has the highest hairiness. Also Duncan test revealed that there in no significant difference between O_1 and O_4 but other navels and ring spun yarns differ significantly from each other.

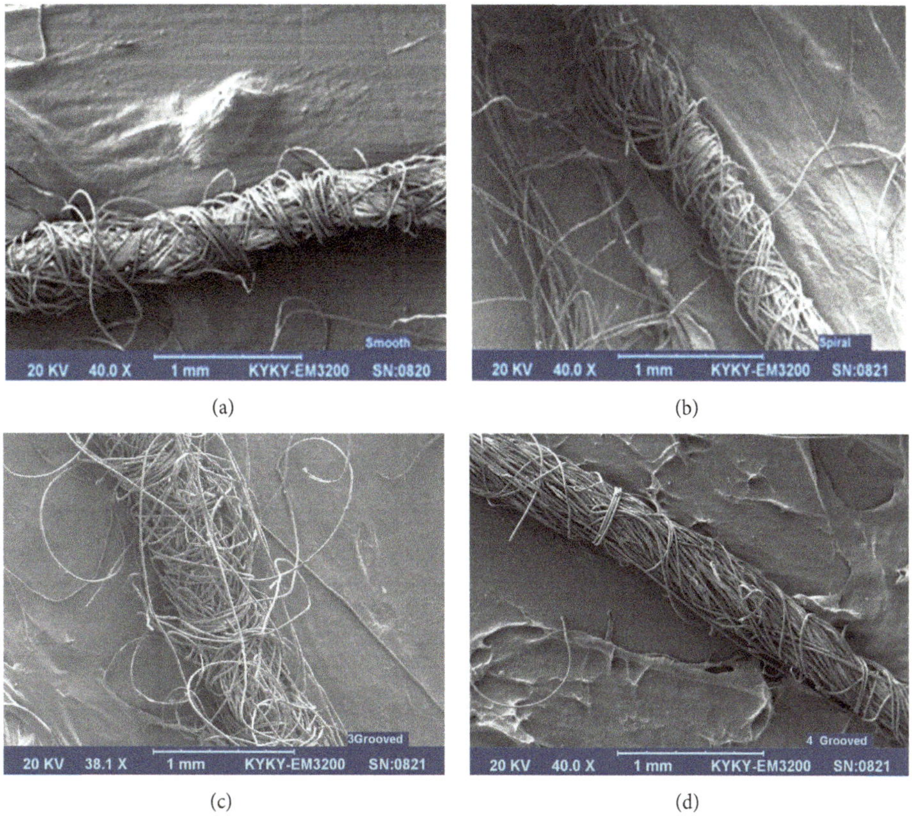

FIGURE 4: Effect of navel types on wrapper formation (a) smooth navel, (b) spiral navel, (c) 3 grooved navel, (d) 4 grooved navel.

TABLE 6: A summary of ANOVA statistical result (Duncan test) for hairiness and abrasion resistance*.

Yarn type	Hairiness					Abrasion resistance				
	R_1	O_1	O_2	O_3	O_4	R_1	O_1	O_2	O_3	O_4
R_1	—	+	+	+	+	—	+	+	–	+
O_1	+	—	+	+	–	+	—	–	+	–
O_2	+	+	—	+	+	+	–	—	+	–
O_3	+	+	+	—	+	–	+	+	—	+
O_4	+	–	+	+	—	+	–	–	+	—

*At 5% confidence limit, "+" means statistically significant and "–" means statistically in-significant.

3.4. Abrasion Resistance. Effect of navel type of nylon/cotton blended rotor spun yarn is shown in Figure 3; it is found that all navel types (except O_3) have higher abrasion resistance than ring spun yarn. According to Figure 3 and Duncan table, it is clear that the highest and the lowest abrasion resistances were recorded for smooth and 3 grooved navels, respectively. And there is no significant difference between O_1, O_2 and O_4.

3.5. Wrapper Formation. Figure 4 shows the effect of navel types on wrapper formation. Hajilari et al. [15] worked on wrapper types and classified them into four group as follows.

First group is loose and concentrated wrapper fiber; second is loose and distributed wrapper fiber; third, tough and concentrated wrapper fiber; fourth, tough and distributed wrapper fiber.

As shown in Figure 4, smooth navel has considerable effect on wrapper formation and this type of navel has the most wrapper density compared to the others (i.e., the third group, tough and concentrated). Spiral and 4 grooved navels have tough and distributed wrapper fiber but as it is clear from Figure 4, 4 grooved navel yarn is more oriented than smooth and spiral navel yarns. It is clear that the fiber orientation and parallelization of 3 grooved navel yarn is too low due to existing loose and distributed wrapper fibers and it is a potential reason for deteriorating yarn quality. As we observed in all physical tests, a physical property of this yarn was the worst.

4. Conclusions

Test result shows that navel types statistically have significant effect on tenacity and smooth navel has the highest value. According to the result of breaking elongation of 100% cotton rotor spun yarn, cotton rotor yarn has more elongation than similar ring yarn but in nylon/cotton blend it was not observed (except 3 grooved navel). Also navel type has considerable effect on yarn irregularity and imperfection; in contrast with written result for 100% cotton yarn, the higher CV% of nylon/cotton rotor spun yarn was obtained and generally, yarn imperfection of rotor spun yarn is more than similar ring spun yarn. In contrast with this subject that says that 3 grooved navel has lower yarn bulk and hairiness than 4 grooved navel [16], in nylon/cotton blend these properties were not obtained. Navel types havenot significant effect on yarn abrasion resistance and all types of navels (except 3 grooved) have more abrasion resistance than ring yarn. Totally it is found that in yarn irregularity and breaking elongation tests, the effect of initial material is more than the effect of yarn structure and also 3 grooved navel is not appropriate for nylon/cotton blend. The SEM investigation also shows the effect of navel types on nylon/cotton rotor yarn and among these yarn, 4 grooved navel was more oriented than the others.

References

[1] A. El-Sheikh, "The dynamic modulus and some other properties of viscos-ployester blends," *Textile Research Journal*, vol. 44, no. 5, pp. 343–351, 1974.

[2] A. Kemp and J. D. Owen, "The strength and behavior of Nylon/Cotton blended yarns undergoing strain," *Journal of Textile Insititute*, vol. 46, pp. 684–698, 1960.

[3] P. Gibson, D. Rivin, C. Kendrick, and H. Schreuder-Gibson, "Humidity-dependent air permeability of textile materials," *Textile Research Journal*, vol. 69, no. 5, pp. 311–317, 1999.

[4] P. Gibson, "Water-repellent treatment on military uniform fabrics: physiological and comfort implications," *Journal of Industrial Textiles*, vol. 38, no. 1, pp. 43–54, 2008.

[5] G. Alon and L. I. Weiner, "Prediction of fabric wear," *Textile Research Journal*, vol. 41, no. 5, pp. 462–463, 1971.

[6] H. G. Schutz, A. V. Cardello, and C. Winterhalter, "Perceptions of fiber and fabric uses and the factors contributing to military clothing comfort and satisfaction," *Textile Research Journal*, vol. 75, no. 3, pp. 223–232, 2005.

[7] X. Kong, S. W. Birk, and K. A. Frankel, "Stretchable Fabrics Comprising Elastic Incorporated into Nyco in Combat Uniforms," U.S. Patent, 7405170, 2008.

[8] F. Javadiyan, S. M. Etrati, and S. Shaikhzadeh Najar, "An invesitigation on Cotton/Nylon66 blended rotor spun yarn properties," in *Proceedings of the 7th International Conference of Textile Science (TEXSCI '10)*, 2010.

[9] S. Kaplan, C. Araz, and Ö. Göktepe, "A multicriteria decision aid approach on navel selection problem for rotor spinning," *Textile Research Journal*, vol. 76, no. 12, pp. 896–904, 2006.

[10] W. Klein, "New spinning system," in *Manual of Textile Technology*, vol. 5, of *Short Stable Spinning Series*, pp. 42–44, The Textile Institute, England, UK, 1993.

[11] Y. Erbill, *Effect of take-off nozzles type on yarn quality during spinning open end rotor blend yarn production [M.S. thesis]*, Institute of Basic and Applied Science, Cukurova University, 2005.

[12] S. H. Muhammad Nawaz, N. Ahmad Jamil, M. Iftikhar, and B. Farooqi, "Effect of multiple open-end processing variables upon yarn quality," *International Journal of Agriculture & Biology*, pp. 256–258, 2002.

[13] Y. Erbil, O. Babaarslan, and P. D. Baykal, "Influence of navel type on the hairiness properties of rotor-spun blend yarns," *Fibres and Textiles in Eastern Europe*, vol. 16, no. 2, pp. 31–34, 2008.

[14] A. A. Alamdar Yazdi, *The Modern Spinning of Short Staple*, Yazd University Publication, 2003.

[15] M. Hajilari, S. Eskandarnejad, H. Behzadan, and M. B. Moghadam, "Effect of two separate fibre feed systems in rotor spinning on yarn properties," *Fibers and Polymers*, vol. 8, no. 5, pp. 543–549, 2007.

[16] H. Deussen, *Rotor Spinning Technology*, Schlafhorst, North Carolina, NC, USA, 1993.

33

Dyeing of Polyester and Polyamide Synthetic Fabrics with Natural Dyes Using Ecofriendly Technique

Khaled Elnagar,[1] Tarek Abou Elmaaty,[2] and Sally Raouf[2]

[1] *Chemical Metrology, National Institute of Standards, Cairo 12211, Egypt*
[2] *Department of Textile Printing, Dyeing and Finishing, Faculty of Applied Arts, Damietta University, Damietta 34512, Egypt*

Correspondence should be addressed to Tarek Abou Elmaaty; tasaid@du.edu.eg

Academic Editor: Ricardo Molina

This work presents an ecofriendly method for dyeing synthetic fabrics with natural dyes using UV/ozone pretreatment to activate fiber and improve dyeability of polyester and nylon. Fabrics are pretreated with UV/ozone for different periods of time ranged from 5 min to 120 min. Effect of pretreatment on surface morphology was studied by scanning electron microscope (SEM). Mechanical behavior was studied by testing tensile strength and elongation percentage. Chemical modification of the surface was studied using attenuated total reflection Fourier transform infrared spectrometer (ATR-FTIR). Dyeability of the treated samples was investigated in terms of their colour strength expressed as *K/s* in addition to fastness to washing and light. This research showed the increment of the affinity of the studied synthetic fabrics towards curcumin and saffron natural dyes using ecofriendly technique.

1. Introduction

With the continued increase in the use of synthetic fibers in the last fifteen years, many investigations have been carried out with the objective of obtaining a better understanding of improving chemical and physical properties of synthetic fibers.

These fibers surpass the production of natural fibers with a market share of 54.4%. The advantages of these fibers are their high modulus and strength, stiffness, stretch or elasticity, wrinkle and abrasion resistances, relatively low cost, convenient processing, tailor-able performance, and easy recycling. The downside to synthetic fibers use are reduced wearing comfort, build-up of electrostatic charge, the tendency to pill, difficulties in finishing, poor soil release properties, and low dyeability. These disadvantages are largely associated with their hydrophobic nature. To render their surfaces hydrophilic, various physical, chemical, and bulk modification methods are employed to mimic the advantageous properties of their natural counterparts [1].

The classical chemical modification of synthetic polymers, using strong alkaline or acid agents, requires high amounts of energy and chemicals (binders, coupling agents, etc.), which are partially discharged to the environment. Furthermore, some of the substances used, due to their weak bonding, are released from the end-products, presenting potential health risks and reducing the technical lifetime of the products. Although alkaline products render synthetic fibres more hydrophilic, they also lead to the deterioration of other properties causing irreversible yellowing and loss of resistance. In order to reduce the use of such chemicals and their environmental impact, new technologies have been developed to replace classical methodologies [2, 3].

One of the potentially more effective and economical methods for synthetic fabrics surface modification is ultraviolet ozone (UV/O_3) treatment with significant depth. In UV/O_3 treatment, two different UV radiations are present. One has a wavelength of 184.9 nm ($\lambda 1$), and the other has a wavelength of 253.7 nm ($\lambda 2$). When the molecules on the surface of synthetic polymers are exposed to the UV lights, excitation and dissociation of the polymeric molecules by the short-wavelength UV radiation take place. This is known as a photosensitized oxidation process. This treatment, also, changes the surface morphology and topography of the polymer. The SEM (scanning electron microscopy) images of the samples showed a marked increase in the grain size and the surface roughness of the irradiated polymer in comparison to the surface of the untreated samples. This method is suitable for synthetic fabric pretreatment and does not cause damage by heat to the fabrics because the modification is carried out under mild thermal conditions (or low temperatures). Thus, beneficial improvements to synthetic fabrics dyeability, printability, and shrink resistance properties can be achieved [4, 5].

In the recent years, there has been a trend to revive the art of natural dyeing. This is a result of the stringent environmental standards imposed by many countries in response to the toxic and allergic reactions associated with synthetic dyes. Natural dyes are more ecofriendly than synthetic dyes and can exhibit better biodegradability and generally have a higher compatibility with the environment [6, 7].

This study is to improve the hydrophilicity of synthetic fabrics via UV/O_3 irradiation and to impart deep coloring effect of the irradiated synthetic fabrics when dyed with natural dyes.

2. Experimental

2.1. Materials

2.1.1. Fabrics. In this study, 100% scoured and bleached plain weave polyester (109 g/m^2), nylon (119 g/m^2) fabrics were used. Fabrics were purchased from Misr for Spinning and Weaving Company, Mahalla El-Kobra, Egypt.

2.1.2. Natural Dyes. Curcumin and saffron were purchased from Riedel de Haen Ag Seelze Hannover Company, Germany, and were used as natural dyes.

2.1.3. Chemicals. Ferrous sulphate ($FeSO_4$) was used as a mordant, and sodium sulphate and sodium carbonate were purchased from Algmohria Company, Egypt, and were used as chemicals reagent.

2.2. Methods

2.2.1. UV/Ozone Treatment. UV/O_3 irradiation was carried out using high intensity; low-pressure mercury lamp without outer envelope (LRF 02971, 200 watts, 220 volts, made in Poland) was placed in a cubic box of side length 60 cm. Samples' strips were hanged around the source at a distance (~20 cm), for different periods 5, 20, 50, 110, and 120 minutes, respectively [8, 9].

2.2.2. Dyeing Procedure

(A) Mordanting. The polyester and nylon fabrics were premordanted by treating in a bath containing ferrous sulphate 6% (owf) keeping material to liquor ratio 1:15 at 60°C for 60 min followed by rinsing with cold water and well drying [10].

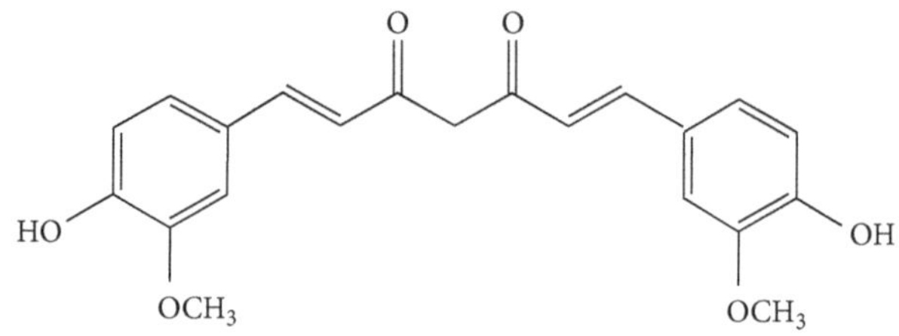

FIGURE 1: Chemical structure of curcumin.

(B) Dyeing with Natural Dyes. All the dyeing experiments were carried out by exhaustion method in a water bath containing the dye extract.

(1) Dyeing with Curcumin Dye. The premordanted polyester and nylon fabrics were introduced, respectively, into the dye bath containing 2% curcumin dye (Figure 1) at a material-to-liquor ratio 1:15 at 70°C in the presence of 50 gm/L sodium sulphate to promote exhaustion. Dyeing was maintained for 100 min. The dyed samples were washed with cold water and dried well at room temperature [10, 11].

(2) Dyeing with Saffron Dye. The premordanted polyester and nylon fabrics were introduced, respectively, into the dye bath containing 5% saffron dye (Figure 2) at room temperature which then increased to 85°C with gentle stirring, keeping the liquor ratio 1:30. The dyeing was then continued for one hour at pH 7-8 using 2 gm/L of sodium carbonate. After dyeing the samples were taken out of the bath and thoroughly washed with water [12].

2.3. Measurements and Analysis Testing

2.3.1. SEM Analysis. The scanning electron microscope (SEM) photographs were carried out for samples under our study using SEM Model Philips XL 30, with accelerating voltage 30 kV, magnification 10x up to 400.000x, and resolution for wavelength (3.5 nm). Samples are coated with gold. A scanning electron microscope (SEM) is used to examine the surface morphology of conventionally and UV/ozone treated polyester and nylon fabrics.

2.3.2. Mechanical Testing. The tensile strength and percentage elongation at break of untreated and treated samples for different periods were carried out according to the Standard Method (ASTM 2000) using a Shimadzu Universal Tester of (C.R.T)-type S-500 Japan.

2.3.3. Fourier Transform Infrared Spectroscopy with Attenuation Total Reflection (ATR-FTIR). ATR-FTIR spectra of the above mentioned samples were recorded by means of Nicolet 380 Spectrometer using a zinc selenide crystal, in the wavelength range 650–4000 cm^{-1}. To ensure reproducible contact between the crystal face and the fabric, a pressure of about 18 Kpa is applied to the crystal holder. The ATR-FTIR absorbance frequencies for the treated samples are recorded with an average of 128 scans using a resolution of 4 cm^{-1}.

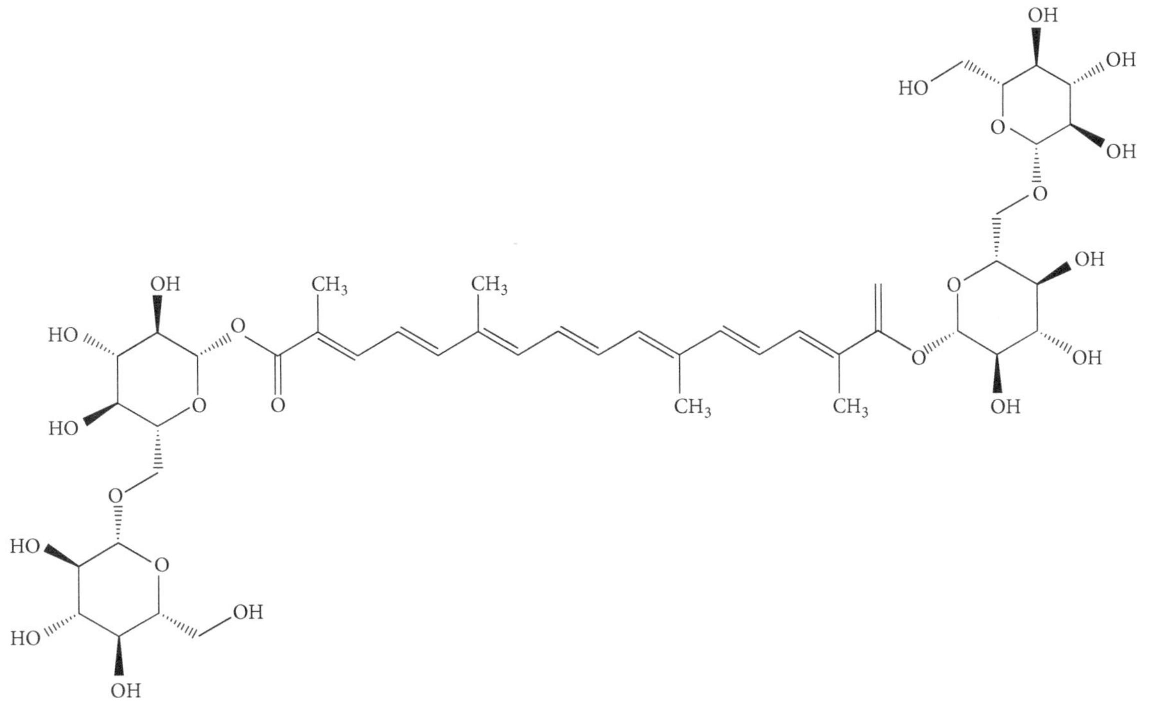

FIGURE 2: Chemical structures of saffron.

2.3.4. Colour Strength (K/s) and Colour Parameters Determination. These measurements were carried out using Colour Eye 3100 Spectrophotometer SDL, England. The colour strength (K/s) values of the different examined samples were determined from the reflection spectra values at the proper wavelength according to Kubelka Munk equation (AATCC, 1991) as follows:

$$\frac{K}{s} = \frac{(1-R)^2}{2R}, \tag{1}$$

where "R" is the decimal fraction of the reflectance of dyed fabric.

These (K/s) values represent the dyeability of the different treated and untreated samples.

2.3.5. Assessment of Color Fastness

(1) Fastness to Washing. The different (treated and untreated with UV/Ozone) dyed samples were stitched between two pieces of undyed cotton fabric (all fabric pieces had the same diameter) and then washed at 50°C for 30 min. The staining of the undyed adjacent fabric was assessed according to the following grey scale: 1, poor; 2, fair; 3, moderate; 4, good; 5, excellent (ISO 105-C01:1998).

(2) Fastness to Light. The different (treated and untreated with UV/O_3) dyed samples are exposed to artificial daylight for 100 hours using Tera Light Fastness Tester [9] at temperature 25 ± 2°C and relative humidity 65 + 5% alongside with a Standard Grey Scale. The light fastness values are assessed visually according to the Standard Method (ISO 105:1997).

3. Results and Discussion

3.1. SEM Analysis. The SEM images of the polyester and nylon fabrics in Figures 3 and 4 show that the untreated fabric has fairly smooth surface, and the existence of few small particles on the surface of the fibers are probably due to the insufficient removal of the chemicals during the washing stage. However, the surface morphology of the treated fabrics has been changed dramatically and there are different features on the surface of the fibers, such as grooves and voids and nanoscale surface roughness [13]. This modification may be owed to the etching of the fiber surface happening after the bombardment of the surface of the fibers with ions and free radicals. This surface roughening effect leads to either a light trapping phenomenon due to destructive interference or increase in the dyeing rate of the fiber, which is a consequence of increasing the surface area of the fiber. Both of these effects may be the reason for the increase in the K/s of the fabrics after the treatment. UV/Ozone irradiation long period may also cause degree of oxidation with significant depth.

3.2. The Effect of the UV/O_3 Treatment on the Mechanical Testing of the Synthetic Fabrics. The changes in the mechanical properties (tensile strength and percentage elongation at break) of blank unexposed polyester and nylon fabrics and those exposed to UV/O_3 treatments are evaluated and the obtained results are represented in Table 1. These results showed that both the tensile strength and percentage elongation decreased. The changes in the mechanical properties of the examined fabric during exposure reflect not only the changes in the chemical structure of its polymer system but its supper molecular organization, that is, its morphology. Thus,

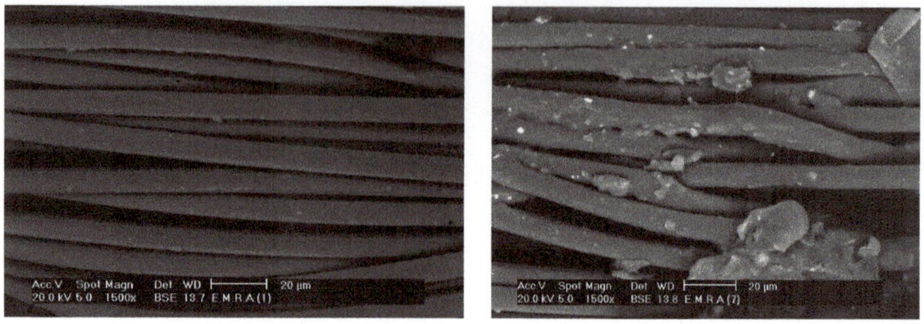

(a) (b)

FIGURE 3: The SEM image of the polyester fabrics, (a) untreated sample, (b) treated sample, ×1500.

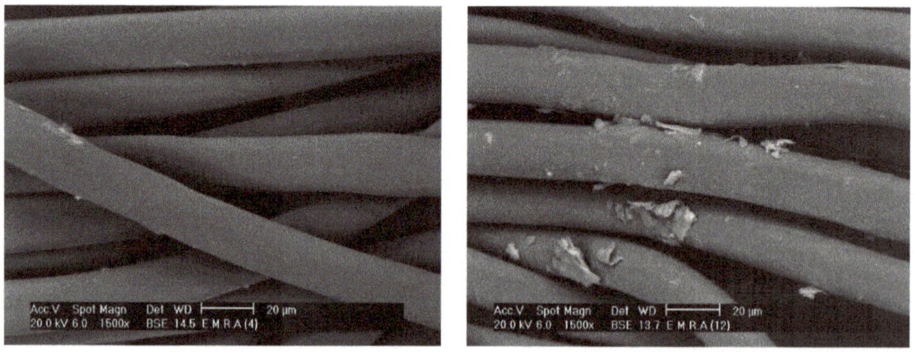

(a) (b)

FIGURE 4: The SEM image of the nylon fabrics, (a) untreated sample, (b) treated sample, ×1500.

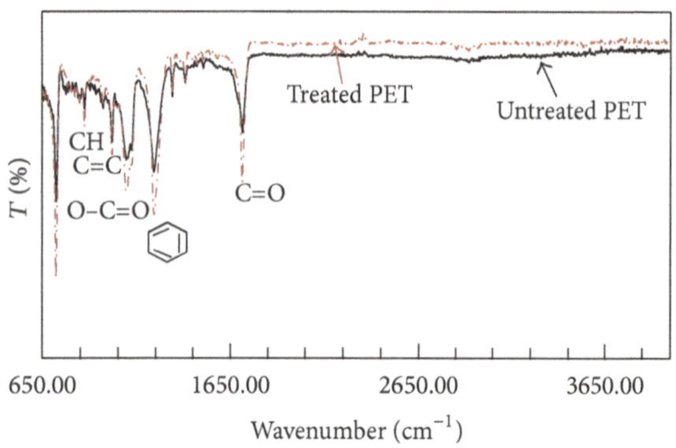

FIGURE 5: ATR-FTIR spectra of untreated and treated PET fabric.

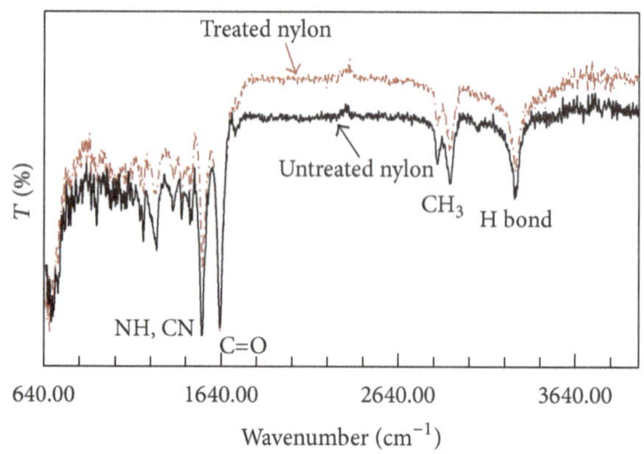

FIGURE 6: ATR-FTIR spectra of untreated and treated nylon fabric.

these results assumed that the examined polyester and nylon samples had lost part of its strength by UV/Ozone treatment especially for the nylon fibers which may be due to the fact that such treatment had caused a removal of the surface layer of the synthetic fabrics. Due to the irradiation to UV/Ozone a degree of oxidation of the bulk polymer may also occur.

3.3. The Effect of UV/Ozone Treatment on the Fourier Transform Infrared Spectroscopy with Attenuation Total Reflection (ATR-FTIR) of Synthetic Fabrics. The ATR-FTIR spectroscopic technique was employed to study the effect of UV/Ozone treatment on the % reflection values as a function of the wave number in the range from 4000 to 500 cm^{-1} of the major function time. Figures 5 and 6 show the spectra of untreated and UV/Ozone treated polyester and nylon fabrics. Figure 5 shows IR spectra of the polyester fabric. The waves were assigned as follows: 1717 cm^{-1} (C=O), 1407 cm^{-1} (aromatic ring), 1339 cm^{-1} and 1014 cm^{-1} (carboxylic ester or anhydride), 1092 cm^{-1} and 1014 cm^{-1} (O=C–O–C or secondary alcohol), 971 cm^{-1} (C=C), 872 cm^{-1} (five substituted H in benzene), 847 cm^{-1} (two neighboring H in benzene), and 724 cm^{-1} (heterocyclic aromatic ring). The

TABLE 1: The changes in the mechanical properties of the tested fabrics with UV/Ozone irradiation.

Fabric	Polyester		Nylon	
	Tensile (Kg·f)	Elongation (%)	Tensile (Kg·f)	Elongation (%)
Untreated	425.5	18.30	678.0	22.80
Treated	382.8	12.40	240.3	12.40

TABLE 2: Effect of exposure time on washing fastness of polyester dyed samples.

Exposure time (min)	Curcumin dye		Saffron dye	
	Staining	Alteration	Staining	Alteration
0 (untreated)	5	2/3	5	4
5	4/5	2/3	5	4
20	4/5	3	5	4/5
50	4/5	3	5	4/5
110	4/5	3	5	4/5
120	4/5	3	5	4/5

main structure of the polyester sample had ester, alcohol, anhydride, aromatic ring, and heterocyclic aromatic rings. Alcohol was able to react with anhydride and produce ester groups. That was the reason there was still alcohol and anhydride as residual reactants left in the polyester. The carboxyl, ester, anhydride, and alcohol groups showed the polyester fabric was not pure PET. The peak at 1407 cm^{-1} corresponded to the aromatic ring which was a stable group. It was the characteristic absorption peak of PET. The peaks at 1717 and 1092 cm^{-1} were assigned to the ester group which preferred to break under certain conditions. From Figure 5 the increment of the intensity of the most characteristic groups due to oxidation effect caused by photochemical interactions and polymer degradation [14–19] is clear.

Figure 6 showed that the characteristic bands of the nylon fibers were observed at 3300 cm^{-1} with an overtone at 3080 cm^{-1} due to the stretching of hydrogen bonded to the N–H and the N–H bending. The CH_2 symmetric and asymmetric stretching were found at 2930 and 2854 cm^{-1}, associated at 720 cm^{-1} with the vibration of 4 or more $-CH_2$ groups in the polymer chain. The band at 1634 cm^{-1} is associated with the C=O stretching and the 1535 cm^{-1} band is due to a combination of the bending N–H band and the stretching of the C–N [20–25]. It is clear that the intensity of C=O and CO increased due to the UV/Ozone irradiation while the other characteristic groups were slightly degreased.

3.4. The Effect of Different UV/Ozone Treatment Times on the K/s Values of the Dyed Synthetic Fabrics. Molecular oxygen was subjected to irradiation at 184.9 nm and 253.7 nm leading to absorption of hydrocarbons and the formation of ozone [26] and form active chemical sites. The products underwent UV/ozone excitation with atomic oxygen to form simpler, volatile molecules, which desorbed from the surfaces that has roughness at the nanoscales [13]. Therefore, when both wavelengths were present atomic oxygen was continuously generated and ozone was continually formed and then destroyed [26–29].

Figures 7 and 8 show the effect of different UV/Ozone treatment times on the K/s values of the two examined synthetic fabrics (polyester and nylon) dyed with saffron and curcumin dyes. As is evident, increasing the UV/Ozone treatment time (5, 20, 50, 110, and 120 min) brings about a noticeable increase in the K/s value; this probably due to the diffusion of larger amount of dye into fabrics under study.

Increasing K/s of the UV/ozone treated fabrics than untreated fabrics may be attributed to the fact that interaction of UV in presence of oxygen on the polymeric materials (polyester and nylon fabrics) produces many free radicals ($P^\bullet$, $POO^\bullet$, $HOO^\bullet$, $H^\bullet$, $HO^\bullet$) as explained in Figure 9. Increase of K/s values can be also be attributed to the affinity increase of dye molecules with the new functional groups generated during the treatment (C–O, C=O, COOH) but not in a radical state. These free radicals are responsible for making active sites for the attachment of treatment solutions (saffron or curcumin dye molecules) to the polyester and nylon fabrics.

3.5. The Effect of the UV/O_3 Treatment on the Fastness Properties of the Synthetic Fabrics

3.5.1. Washing Fastness. The colour fastness to washing of the different examined dyed samples was determined according to standard methods (AATCC, 61-1975, 8-1977).

The results of washing fastness for all of the examined samples are summarized in Tables 2 and 3. These data showed that there was a slight effect of UV/O_3 treatment in improving washing fastness. Washing fastness assessment of obtained colors on polyester and nylon fibers have shown good levels.

3.5.2. Light Fastness. The light fastness results are good and acceptable showing little fading of light shades in both the treated and untreated dyed nylon and polyester fabrics. The

Table 3: Effect of exposure time on washing fastness of nylon dyed samples.

Exposure time (min)	Curcumin dye		Saffron dye	
	Staining	Alteration	Staining	Alteration
0 (untreated)	3	4	5	4
5	3	4	5	4
20	3/4	4/5	5	4/5
50	3/4	4/5	5	4/5
110	3/4	4/5	5	4/5
120	3/4	4/5	5	4/5

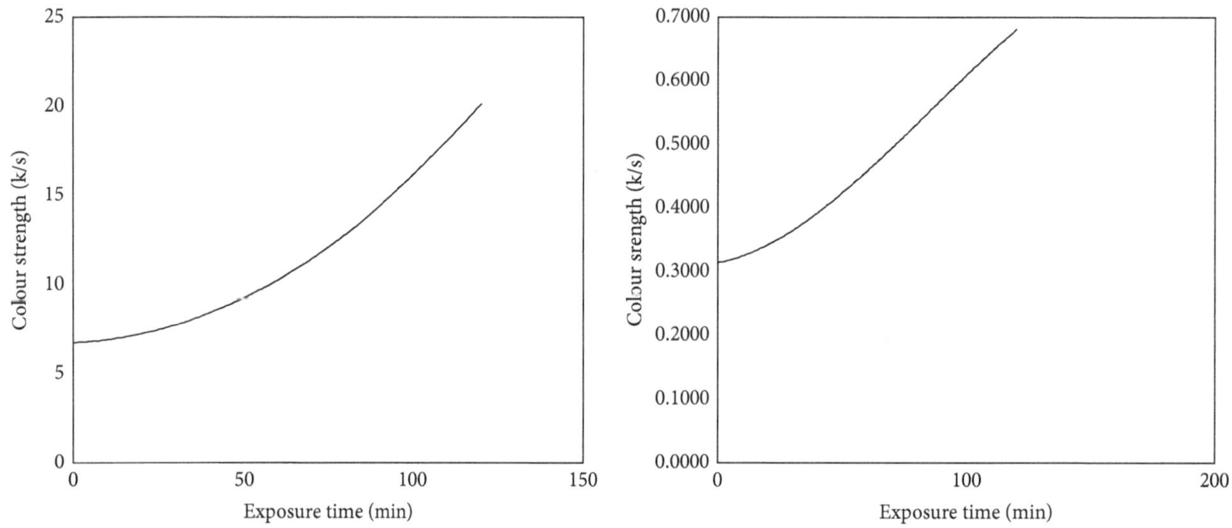

(a) (b)

Figure 7: The K/s values of polyester fabrics, (a) curcumin dye, (b) saffron dye.

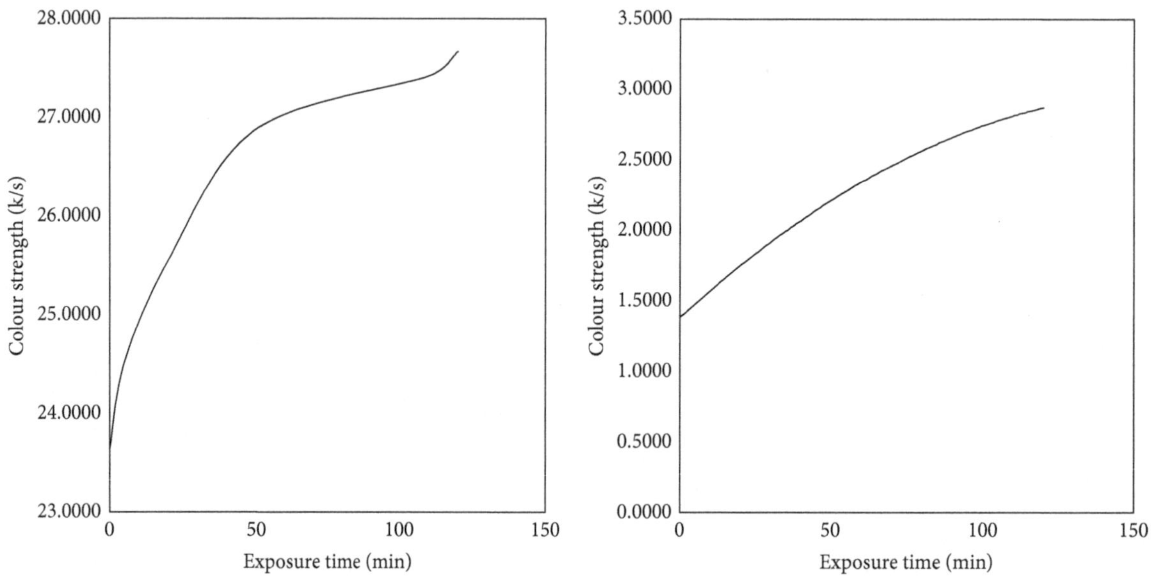

(a) (b)

Figure 8: The K/s values of nylon fabrics, (a) curcumin dye, (b) saffron dye.

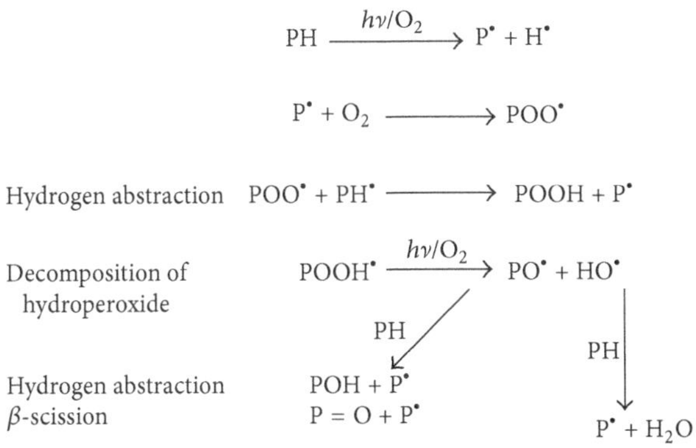

FIGURE 9: Effect of UV/O_3 radiation on polymeric materials.

TABLE 4: Effect of exposure time on light fastness of polyester dyed samples.

Exposure time (min)	Curcumin dye	Saffron dye
0 (untreated)	2	2
5	2/3	2/3
20	2/3	2/3
50	3	2/3
110	3	3
120	3	3

TABLE 5: Effect of exposure time on light fastness of nylon dyed samples.

Exposure time (min)	Curcumin dye	Saffron dye
0 (untreated)	2	2
5	2/3	2
20	2/3	2/3
50	3	2/3
110	3	3
120	3	3

results of light fastness for all of the examined samples are summarized in Tables 4 and 5.

4. Conclusion

This work can conclude that treating nylon and polyester under study by UV/Ozone for different time ranged from 0 to 120 min increased the dyeability of both fabrics with curcumin and saffron natural dyes. Dyeability of both fabrics was measured according to colour strength (K/s) and fastness to washing and light fastness. Treatment with UV/ozone showed marked change of the surface chemical structure as shown by ATR-ATR-FTIR measurements in addition to the surface morphology as shown by SEM images.

Conflict of Interests

The authors declare that there is no conflict of interests regarding the publication of this paper.

Acknowledgments

The authors thank Faculty of Applied Arts, Damietta University, and National Institute for Standards, Ministry of Scientific Research, for offering the facilities to conduct most of the experimental part of this study.

References

[1] M. P. Gashti, J. Willoughby, and P. Agrawal, "Surface and bulk modification of synthetic textiles to improve dyeability," in *Textile Dyeing*, P. J. Hauser, Ed., chapter 13, InTech, Rijeka, Croatia, 2011.

[2] C. Silva and A. Cavaco-Paulo, "Biotransformations in synthetic fibres," *Biocatalysis and Biotransformation*, vol. 26, no. 5, pp. 350–356, 2008.

[3] T. A. Elmaaty, K. Elnagar, S. Hassan, and H. Gamal, "Antibacterial activity and dyeing characteristics of some azo-pyazole disperse dyes using eco-friendly ultrasound energy for PET fabric," *International Journal of Scientific & Engineering Research*, vol. 5, no. 5, pp. 1156–1161, 2014.

[4] K. J. Lee, C. Kim, K. Ho, and K. Constant, "Improved surface wettability of polyurethane films by ultraviolet ozone treatment," *Journal of Applied Polymer Science*, vol. 118, no. 5, pp. 3024–3303, 2010.

[5] N. A. Elzaher and M. N. Micheal, "Time optimization of ultraviolet-ozone pretreatment for improving wool fabrics properties," *Journal of Applied Polymer Science*, vol. 85, no. 7, pp. 1469–1476, 2002.

[6] A. K. Samanta and A. Konar, "Dyeing of textiles with natural dyes," in *Natural Dyes*, E. P. A. Kumbasar, Ed., chapter 3, p. 132, InTech, Rijeka, Croatia, 2011.

[7] F. A. Nagia and R. S. EL-Mohamedy, "Dyeing of wool with natural anthraquinone dyes from *Fusarium oxysporum*," *Dyes and Pigments*, vol. 75, no. 3, pp. 550–555, 2007.

[8] S. F. Ibrahim, K. El-Nagar, and F. M. Tera, "Improving wool printability at reduced energy using UV/ozone treatment," *Research Journal of Textile and Apparel*. In press.

[9] K. Elnagar, S. M. Reda, H. E. Ahmed, and S. Kamal, "Studying irradiation homogeneity in light aging for historical textile conservation," *Fibers and Polymers*, vol. 14, no. 9, pp. 1581–1585, 2013.

[10] S. Urbeen, S. Ali, T. Hussain, and R. Nawaz, "Dyeing properties of natural dyes extracted from turmeric and their comparison with reactive dyeing," *Research Journal of Textile and Apparel*, vol. 12, no. 4, 2008.

[11] K. Sachan and V. P. Kapoor, "Optimization of extraction and dyeing conditions for traditional turmeric dye," *Indian Journal of traditional Knowledge*, vol. 6, no. 2, pp. 270–278, 2007.

[12] A. S. Raja, P. K. Pareek, D. B. Shakyawar, S. A. Wani, F. A. Nehvi, and A. H. Sofi, "Extraction of natural dye from Saffron flower waste and its application on pashmina fabric, Pelagia Research Library," *Advances in Applied Science Research*, vol. 3, no. 1, pp. 156–161, 2012.

[13] J. Jang and Y. Jeong, "Nano roughening of PET and PTT fabrics via continuous UV/O_3 irradiation," *Dyes and Pigments*, vol. 69, no. 3, pp. 137–143, 2006.

[14] X. Song, H. Liu, L. Cheng, and Y. Qu, "Surface modification of coconut-based activated carbon by liquid-phase oxidation and its effects on lead ion adsorption," *Desalination*, vol. 255, no. 1–3, pp. 78–83, 2010.

[15] D. L. Pavia, G. M. Lampman, and G. S. Kriz, *Introduction to Spectroscopy*, Thomson Business Information India Pvt., Bangalore, India, 3rd edition, 2006.

[16] V. Bajpai, A. Dey, S. Ghosh, S. Bajpai, and M. K. Jha, "Quantification of bacterial adherence on different textile fabrics," *International Biodeterioration and Biodegradation*, vol. 65, no. 8, pp. 1169–1174, 2011.

[17] L. Li, M. Frey, and K. J. Browning, "Biodegradability study on cotton and polyester fabrics," *Journal of Engineered Fibers and Fabrics*, vol. 5, no. 4, pp. 42–53, 2010.

[18] Y. Zhang, Y. Su, J. Peng et al., "Composite nanofiltration membranes prepared by interfacial polymerization with natural material tannic acid and trimesoyl chloride," *Journal of Membrane Science*, vol. 429, pp. 235–242, 2013.

[19] M. Minhalma and M. N. de Pinho, "Tannic-membrane interactions on ultrafiltration of cork processing wastewaters," *Separation and Purification Technology*, vol. 22-23, pp. 479–488, 2001.

[20] Z. Wu, C. Zhou, R. Qi, and H. Zhang, "Synthesis and characterization of nylon 1012/clay nanocomposite," *Journal of Applied Polymer Science*, vol. 83, no. 11, pp. 2403–2410, 2002.

[21] S. G. Gholap, D. A. Musale, S. S. Kulkarni, S. K. Karode, and U. K. Kharul, "Protein and buffer transport through anionically grafted nylon membranes," *Journal of Membrane Science*, vol. 183, no. 1, pp. 89–99, 2001.

[22] E. M. Enlow, J. L. Kennedy, A. A. Nieuwland, J. E. Hendrix, and S. L. Morgan, "Discrimination of nylon polymers using attenuated total reflection mid-infrared spectra and multivariate statistical techniques," *Applied Spectroscopy*, vol. 59, no. 8, pp. 986–992, 2005.

[23] R. Iwamoto and H. Murase, "Infrared spectroscopic study of the interactions of nylon-6 with water," *Journal of Polymer Science. Part B: Polymer Physics*, vol. 41, no. 14, pp. 1722–1729, 2003.

[24] B. Stuart, *Infrared Spectroscopy: Fundamentals and Applications*, John Wiley & Sons, 2004.

[25] A. A. Younis, M. A. Nour, and K. El-Nagar, "Studying the effect of phosphorylated sol-gel on polyester fabric as flame-retardant coating," *Applied Chemistry-Elixir International Journal*, vol. 70, pp. 24230–24235, 2014.

[26] M. N. Michael and N. A. El-Zaher, "Invistigation into the effect of UV/ozone treatment on dyeing properties of natural fabrics," *Egyptian Journal of Chemistry*, vol. 47, no. 1, pp. 93–100, 2004.

[27] F. Fattahi, H. Izadan, and A. Khoddami, "Investigation into the effect of UV/Ozone irradiation on dyeing behaviour of poly(lactic acid) and poly(ethylene terephthalate) substrates," *Progress in Color, Colorants and Coatings*, vol. 5, pp. 15–22, 2012.

[28] S. F. Ibrahim, *Investigation into the effect of UV/ozone treatments on physical properties and dyeing kinetics of some polymeric fabrics [Ph.D. thesis]*, Faculty of Science, Cairo University, Cairo, Egypt, 2003.

[29] E. M. Osman, M. N. Michael, and H. Gohar, "the effect of both UV/Ozone and chitosan on natural fabrics," *International Journal of Chemistry*, vol. 2, no. 2, pp. 28–39, 2010.

Permissions

All chapters in this book were first published in JTEX, by Hindawi Publishing Corporation; hereby published with permission under the Creative Commons Attribution License or equivalent. Every chapter published in this book has been scrutinized by our experts. Their significance has been extensively debated. The topics covered herein carry significant findings which will fuel the growth of the discipline. They may even be implemented as practical applications or may be referred to as a beginning point for another development.

The contributors of this book come from diverse backgrounds, making this book a truly international effort. This book will bring forth new frontiers with its revolutionizing research information and detailed analysis of the nascent developments around the world.

We would like to thank all the contributing authors for lending their expertise to make the book truly unique. They have played a crucial role in the development of this book. Without their invaluable contributions this book wouldn't have been possible. They have made vital efforts to compile up to date information on the varied aspects of this subject to make this book a valuable addition to the collection of many professionals and students.

This book was conceptualized with the vision of imparting up-to-date information and advanced data in this field. To ensure the same, a matchless editorial board was set up. Every individual on the board went through rigorous rounds of assessment to prove their worth. After which they invested a large part of their time researching and compiling the most relevant data for our readers.

The editorial board has been involved in producing this book since its inception. They have spent rigorous hours researching and exploring the diverse topics which have resulted in the successful publishing of this book. They have passed on their knowledge of decades through this book. To expedite this challenging task, the publisher supported the team at every step. A small team of assistant editors was also appointed to further simplify the editing procedure and attain best results for the readers.

Apart from the editorial board, the designing team has also invested a significant amount of their time in understanding the subject and creating the most relevant covers. They scrutinized every image to scout for the most suitable representation of the subject and create an appropriate cover for the book.

The publishing team has been an ardent support to the editorial, designing and production team. Their endless efforts to recruit the best for this project, has resulted in the accomplishment of this book. They are a veteran in the field of academics and their pool of knowledge is as vast as their experience in printing. Their expertise and guidance has proved useful at every step. Their uncompromising quality standards have made this book an exceptional effort. Their encouragement from time to time has been an inspiration for everyone.

The publisher and the editorial board hope that this book will prove to be a valuable piece of knowledge for researchers, students, practitioners and scholars across the globe.

List of Contributors

A. K. Choudhary
Department of Textile Technology, National Institute of Technology, Jalandhar, Punjab 144011, India

Amit Goel
Department of Textile Technology, National Institute of Technology, Jalandhar, Punjab 144011, India
Punjab Institute of Textile Technology, Amritsar, Punjab 144011, India

Phillip Gibson, Heidi Schreuder-Gibson, Pearl Yip and Brendan Denker
U.S. Army Natick Soldier Research, Development and Engineering Center, Natick, MA 01760, USA

Hamid Benaddi
Stedfast, Inc., Granby, QC, Canada J2G 3Y3

Sa Wang, Lev Bromberg and T. Alan Hatton
Department of Chemical Engineering, Massachusetts Institute of Technology, Cambridge, MA 01239, USA

Arobindo Chatterjee and Pratibha Singh
Department of Textile Technology, National Institute of Technology, Jalandhar 144 011, India

H. F. Cheung, C. W. Kan, C. W. M. Yuen, J. Yip and M. C. Law
Institute of Textiles and Clothing, The Hong Kong Polytechnic University, Hung Hom, Kowloon 852, Hong Kong

Mukesh Kumar Singh
Uttar Pradesh Textile Technology Institute (Formerly Known as GCTI), Kanpur 208001, India

Annika Singh
Institute of Biosciences and Biotechnology, CSJMU, Kanpur, India

Ghada Ali Abou-Nassif
Fashion Design Department, Design and Art Faculty, King Abdul Aziz University, Jeddah, Saudi Arabia

Rattanaphol Mongkholrattanasit
Department of Textile Chemistry Technology, Faculty of Industrial Textiles and Fashion Design, Rajamangala University of Technology Phra Nakhon, 517 Nakhonsawan Road, Kwang Suan Chitladda, Dusit District, Bangkok 10300, Thailand

Charoon Klaichoi
Department of Textile Product Design, Faculty of Industrial Textiles and Fashion Design, Rajamangala University of Technology Phra Nakhon, 517, Nakhonsawan Road, Kwang Suan Chitlada, Dusit District, Bangkok 10300,Thailand

Nattadon Rungruangkitkrai
Department of Textile Science, Faculty of Agro-Industry, Kasetsart University, 50 Phaholyotin Road, Chatuchak District, Bangkok 10900,Thailand

Kamolkan Sriharuksa and Nattaya Punrattanasin
Department of Textile Science and Technology, Faculty of Science and Technology,Thammasat University, Rangsit Center, Pathumthani 12121,Thailand

Monthon Nakpathom
National Metal and Materials Technology Center, National Science and Technology Development Agency, Pathumthani 12120, Thailand

Mohammad Gias Uddin
Department of Textile Engineering, Faculty of Engineering, Ahsanullah University of Science and Technology, Dhaka 1208, Bangladesh

Wenlong Zhou and Huayin Wang
Key Laboratory of Advanced Textile Materials and Manufacturing Technology (Ministry of Education), Zhejiang Sci-Tech University, Hangzhou, Zhejiang 310018, China

Ping Li
Yancheng Institute of Industry Technology, Yancheng, Jiangsu 224005, China

A. M. Rehan Abbasi
Technical University of Liberec, Studentska 2, 46117 Liberec, Czech Republic
Department of Textile Engineering, BUITEMS, 87100Quetta, Pakistan

J. Militky and J. Gregr
Technical University of Liberec, Studentska 2, 46117 Liberec, Czech Republic

Kiruthika Deivasigamani and Kavitha Krishnamoorthi
Kumaraguru College of Technology, Coimbatore 641049, India

Siva Kumar Kolandaivel
Angel College of Engineering and Technology, Tirupur 641665, India

Phillip Gibson and Michael Sieber
U.S. Army Natick Soldier Research, Development, and Engineering Center, Natick, MA 01760, USA

Jerry Bieszczad, John Gagne and David Fogg
Creare, Inc., Hanover, NH 03755, USA

Jintu Fan
Cornell University, Ithaca, NY 14850, USA

Anupama Sargur Ranganath
Department of Design and Merchandising, Colorado State University, Fort Collins, CO 80523, USA

Ajoy K. Sarkar
Textile Development and Marketing, Fashion Institute of Technology, State University of New York, NY 10001-5992, USA

R. Abd El-khalek, R. El-Bealy and A. El-Deeb
Textile Engineering Department, Faculty of Engineering, Mansoura University, Mansoura 35516, Egypt

A. Fouda, A. El-Hadidy and A. El-Deeb
Textile Engineering Department, Faculty of Engineering, Mansoura University, Mansoura 35516, Egypt

Suvalee T. Tangboonritruthai, William Oxenham and Nancy L. Cassill
North Carolina State University, 1000 Main Campus Drive, Raleigh, NC 27695, USA

Erin D. Parrish
East Carolina University, Rivers Building, East Fifth Street, Greenville, NC 27858, USA

Amandeep Kaur and J. N. Chakraborty
National Institute of Technology, Jalandhar 144 011, India

M. D. Teli, Javed Sheikh and Pragati Shastrakar
Department of Fibers and Textile Processing Technology, Institute of Chemical Technology, Mumbai 400019, India

A.A. Salama, A. S. El-Deeb and I.M. El-shahat
Textile Engineering Department, Mansoura University, Mansoura 35516, Egypt

Mine Akgun, Behcet Becerir and Halil Rifat Alpay
Textile Engineering Department, Faculty of Engineering, Uludag University, Gorukle, 16059 Bursa, Turkey

Jayant Udakhe, Neeraj Shrivastava and Smita Honade
Textile Chemical and Colour Department, Wool Research Association, Thane 400607, India

Dhanashree Banait and Namita Sonawane
Department of Fibres and Textile Processing Technology, I.C.T., Mumbai 400019, India

Lopamudra Nayak
Orissa University of Agriculture & Technology, Bhubaneswar 751003, India

M. Forouharshad and B. Yadollah Roudbari
Young Researchers Club, Islamic Azad University, South Tehran Branch, Tehran 1981916791, Iran

M. Montazer
Textile Department, Center of Excellence, Amirkabir University of Technology, Tehran, Iran

A. Fouda, A. El-Hadidy and A. El-Deeb
Textile Engineering Department, Faculty of Engineering, Mansoura University, Mansoura 35516, Egypt

Raluca Brad
Industrial Machinery and Equipment Department, Lucian Blaga University of Sibiu, B-dul Victoriei 10, 550024 Sibiu, Romania

Lavinia Barac and Remus Brad
Computer Science and Electrical Engineering Department, Lucian Blaga University of Sibiu, B-dul Victoriei 10, 550024 Sibiu, Romania

Md. Mazedul Islam
Department of Textile Engineering, Dhaka University of Engineering & Technology, Gazipur 1700, Bangladesh

Md. Mashiur Rahman Khan
Department of Apparel Manufacturing Engineering, Bangladesh University of Textiles, Tejgaon, Dhaka 1208, Bangladesh

Yusuf Y. Lams
1Department of Textile Technology and Fashion Design, College of Science and Technology, Kaduna Polytechnic, PMB 2021, Kaduna, Nigeria

P. O. Nkeonye, K. A. Bello and M. K. Yakubu
Department of Textile Science and Technology, Ahmadu Bello University, Zaria, Nigeria

A. O. Lawal
Department of Applied Science, College of Science and Technology, Kaduna Polytechnic, PMB 2021, Kaduna, Nigeria

S. M. Elrys, A. El-Hossini and A. M. EL-Hadidy
Mansoura University, Faculty of Engineering, Textile Department, Mansoura, Egypt

P. Lalitha and R. Sujitha
Department of Chemistry, Avinashilingam Institute for Home Science and Higher Education for Women, Coimbatore, Tamil Nadu 641043, India

P. Jayanthi
Department of Chemistry, Avinashilingam Institute for Home Science and Higher Education for Women, Coimbatore, Tamil Nadu 641043, India
Department of Science and Humanities, Dr. Mahalingam College of Engineering and Technology, Pollachi, Coimbatore, Tamil Nadu 642 003, India

Md. Mahabub Hasan, Khandakar Abu Nayem, Abu Yousuf Mohammad Anwarul Azim and Nayon Chandra Ghosh
Department of Textile Engineering, Primeasia University, Dhaka 1213, Bangladesh

C. W. Kan, W. Y. Wong, L. J. Song and M. C. Law
Institute of Textiles and Clothing, The Hong Kong Polytechnic University, Hung Hom, Kowloon, Hong Kong

Babak Yadollah Roudbari and Safdar Eskandarnejad
Textile Department, Science and Research Branch, Islamic Azad University, Tehran 1477893855, Iran

Khaled Elnagar
Chemical Metrology, National Institute of Standards, Cairo 12211, Egypt

Tarek Abou Elmaaty and Sally Raouf
Department of Textile Printing, Dyeing and Finishing, Faculty of Applied Arts, Damietta University, Damietta 34512, Egypt